# Genetic and Morphological Variation in Tropical and Temperate Plant Species

# Genetic and Morphological Variation in Tropical and Temperate Plant Species

Editors

**W. John Kress**
**Nancai Pei**

MDPI • Basel • Beijing • Wuhan • Barcelona • Belgrade • Manchester • Tokyo • Cluj • Tianjin

*Editors*
W. John Kress
Department of Botany, National Museum of Natural History,
Smithsonian Institution
USA

Nancai Pei
Research Institute of Tropical Forestry,
Chinese Academy of Forestry
China

*Editorial Office*
MDPI
St. Alban-Anlage 66
4052 Basel, Switzerland

This is a reprint of articles from the Special Issue published online in the open access journal *Forests* (ISSN 1999-4907) (available at: https://www.mdpi.com/journal/forests/special_issues/Plant_Genetic_Morphological).

For citation purposes, cite each article independently as indicated on the article page online and as indicated below:

| LastName, A.A.; LastName, B.B.; LastName, C.C. Article Title. *Journal Name* **Year**, *Article Number*, Page Range. |
|---|

**ISBN 978-3-03936-756-6 (Hbk)**
**ISBN 978-3-03936-757-3 (PDF)**

# Contents

# About the Editors

**W. John Kress**, Ph.D., Visiting Scholar, Dartmouth College and The Arnold Arboretum of Harvard University Distinguished Scientist and Curator Emeritus National Museum of Natural History Smithsonian Institution, P.O. Box 37012 Washington, DC 20013-7012; Tel: 202-633-0939 (DC Office); 202-372-7745 (Mobile); Email: kressj@si.edu; kressjohn@gmail.com; Research Interests: systematic biology; evolutionary biology; conservation biology.

**Nancai Pei**, Dr. and Associate Research Professor in Research Institute of Tropical Forestry, Chinese Academy of Forestry; Add: No. 682, Guangshan Road 1, Tianhe District, Guangzhou 510520, P. R. China; E-Mail: nancai.pei@gmail.com; Research Interests: forest biology; plant DNA barcoding; urban forestry.

*forests*

MDPI

*Editorial*

# Research in Forest Biology in the Era of Climate Change and Rapid Urbanization

**Nancai Pei [1,*] and W. John Kress [2,*]**

1 Research Institute of Tropical Forestry, Chinese Academy of Forestry, Guangzhou 510520, China
2 Department of Botany, MRC-166, National Museum of Natural History, Smithsonian Institution, P.O. Box 37012, Washington, DC 20013-7012, USA
* Correspondence: nancai.pei@gmail.com (N.P.); kressj@si.edu (W.J.K.)

Received: 16 December 2019; Accepted: 21 December 2019; Published: 23 December 2019

**Abstract:** Green plants provide the foundation for the structure, function, and interactions among organisms in both tropical and temperate zones. To date, many investigations have revealed patterns and mechanisms that generate plant diversity at various scales and from diverse ecological perspectives. However, in the era of climate change, anthropogenic disturbance, and rapid urbanization, new insights are needed to understand how plant species in these forest habitats are changing and adapting. Here, we recognize four themes that link studies from Asia and Europe presented in this Special Issue: (1) genetic analyses of diverse plant species; (2) above- and below-ground forest biodiversity; (3) trait expression and biological mechanisms; and (4) interactions of woody plants within a changing environment. These investigations enlarge our understanding of the origins of diversity, trait variation and heritability, and plant–environment interactions from diverse perspectives.

**Keywords:** climate change; forest biodiversity; plant–environment interactions; plant traits; urbanization

---

## 1. Introduction

Investigations of plants in both little disturbed, more natural environments, as well as in urban areas, are needed where crucial green infrastructure is ever more important for sustaining complex human societies. Recently, numerous studies have addressed the fundamental issues on plant evolution and community phylogenetics via exploring patterns and mechanisms from diverse organismal levels (e.g., molecular, population, species, community, landscape, and ecosystem) [1,2], plant functional traits (e.g., nutrient traits and reproductive traits) [3,4], and interaction of plant species with changing environments (e.g., water, atmosphere, soil, human activities) [5,6].The aim of this Special Issue is to help fill this void in the current research by focusing on diversity, trait, and plant–environment interactions within the context of forest ecosystems. The 19 papers encompassed here can best be linked under four basic themes: (1) genetic analyses of diverse plant species; (2) above- and below-ground forest biodiversity; (3) trait expression and biological mechanisms; and (4) interactions of woody plants within a changing environment.

Together, the guest editors of this Special Issue conceptualized these four themes as a means to progress an open discussion of forest biology, including plants and diverse environments. Our proposal for this Special Issue coincided with the common interests of the ecological and forestry research communities. This Special Issue includes research performed mostly in Asia and Europe with studies originating from Belgium, China, Croatia, India, Poland, and Spain. We could not have hoped to create a more internationally inclusive and relevant Special Issue, and are very proud to present as guest editors this collection of forest biology studies.

## 2. Theme 1: Genetic Analyses of Diverse Plant Species

This theme includes 7 papers investigating the physiological and transcriptome analyses of 6 woody plus 1 herbaceous species from 7 plant families. By studying the differentially expressed genes (DEGs) between yellow-green leaf mutant (yl) and control plants in Birch (Betulaceae), Gang et al. find that 1163 genes and 930 genes differentially express in yl compared with WT and C11. The KEGG pathway enrichment analysis for DEGs reveals that photosynthesis antenna proteins represent the most significant enriched pathway. The expressions of photosynthesis antenna proteins are crucial to the leaf color formation in yl. They also report that Chl accumulation, leaf anatomical structure, photosynthesis, and growth are affected in yl. This study has provided the difference in phenomenal, physiological, and gene expression characteristics in leaves between yl mutant and control plants, and presented a new insight into the mutation underlying the chlorotic leaf phenotype in birch [7].

Kaushik and Kumar report expressed transcripts in the leaves of *Aegle marmelos*, a medicinal and horticultural tree species from Rutaceae. They find that 133,616 contigs are assembled to 46,335 unigenes with minimum and maximum lengths of 201 bp and 14,853 bp, respectively. A total of 482 transcripts are annotated as cytochrome p450s, and 314 transcripts are annotated as glucosyltransferases. They suggest that the monoterpenoid biosynthesis pathway in leaves is predominant [8].

Liu et al. evaluate the genetic diversity of 42 wild individuals from seven populations of *Dalbergiaodorifera*, a semi-deciduous commercially importantand threatened tree species from Fabaceae, indigenous to Hainan Island in tropical China. They find 19 SSR markers harbored 54 alleles across the 42 samples, and the medium genetic diversity level is inferred. Among the 7 wild populations, the expected heterozygosity varies from 0.31 to 0.40. The AMOVA analysis shows that only 3% of genetic variation exists among populations. Moderate population differentiation among the investigated populations is indicated by pairwise Fst. Structure analysis suggests two clusters for the 42 samples. These findings provide a preliminary genetic basis for the conservation, management, and restoration of this endemic species [9].

Lu et al. evaluate the genetic diversity and structure of *Eucalyptus urophylla*, an important commercial tropical plantation species from Myrtaceae. They find that significant deviations from the Hardy–Weinberg equilibrium are recorded at all 16 loci in the populations, revealing reasonably high levels of genetic diversity. The genetic differentiation coefficient reveals low differentiation among pairs of provenances comprising the first cycle population. They also find that the majority of molecular genetic variation exists among individuals rather than among provenances for the first cycle population and among individuals rather than among field trial sources in the third cycle population [10].

Mo, Feng et al. construct four small RNA libraries from the graft union of Pecan, a high-value fruit tree from Juglandaceae. They find that 47 conserved miRNAs belonging to 31 families and 39 novel miRNAs are identified. For the identified miRNAs, 584 target genes are bioinformatically predicted, and 266 of them are annotated. Meanwhile, 29 miRNAs (including 16 conserved and 13 novel miRNAs) are differentially expressed during the graft process. The expression profiles of 12 miRNA are further validated by qRT-PCR. They also find that miRS26 may be involved in callus formation, while miR156, miR160, miR164, miR166, and miRS10 may be associated with vascular bundle formation. These results indicate that the miRNA-mediated gene regulations play important roles in the graft union development of pecan [11].

Wei et al. perform a genome-wide identification and analysis of members of the HbMADS-box gene family associated with floral organ and inflorescence development in *Hevea brasiliensis*, a rubber tree species from Euphorbiaceae. They find 20 MADS-box genes are newly identified in the *H. brasiliensis* genome. Expression profiling reveals that HbMad-box genes are differentially expressed in various tissues, which indicate that HbMad-box genes may exert different functions throughout the life cycle. Additionally, 12 genes are found to be associated with the differentiation of flower buds and may be involved in flower development. All of these floral-enriched HbMADS-box genes are regulated by hormone, salt, cold, high-temperature, and drought stresses. This study demonstrates that HbMad-box genes may be multifunctional regulators, and are mainly involved in the maintenance of floral organ and inflorescence development [12].

In order to understand the function of heat shock transcription factors (Hsfs) in moso bamboo in the family Poaceae, Xie et al. identify 22 non-redundant Hsf genes in the moso bamboo genome. They find members of the PheHsf family can be clustered into three classes, containing stress-, hormone- and development-related cis-acting elements. They also find most PheHsfs participate in rapid shoot growth and flower development in moso bamboo, and PheHsfA1a is expressed mainly during moso bamboo development. Two hub genes are involved in a complex protein interaction network, and five PheHsfAsare predicted to play an important role in flower and shoot development and abiotic stress response of moso bamboo. This study provides an overview of the complexity of the PheHsf gene family and sets a basis for analyzing the functions of PheHsf genes [13].

## 3. Theme 2: Above- and Below-Ground Forest Biodiversity

Within this theme, five papers explore the dimensions of plant DNA barcoding, reproductive biology, and of edible plant and fungalresources in forest diversity. Wu, Li, Liao et al. evaluate the effectiveness of DNA barcoding in identifying 23 mangrove species in Guangdong Province (GP), southern China. They find that the success rates for PCR amplification of *rbc*L, *mat*K, *trn*H-*psb*A, and ITS are 100%, 80.29%, 99.38%, and 97.18%, respectively, and the rates of DNA sequencing are 100%, 75.04%, 94.57%, and 83.35%, respectively. These results suggest that both *rbc*L and *trn*H-*psb*A are universal in mangrove species in the sampled sites. The highest success rate for species identification is 84.48% for *trn*H-*psb*A, followed by *rbc*L (82.16%), ITS (66.48%), and *mat*K (65.09%), which increases to 91.25% with the addition of *rbc*L. They suggest that *rbc*L and *trn*H-*psb*A are the most suitable DNA barcode fragments for species identification in mangrove plants, and the combination of *mat*K + *rbc*L + *trn*H-*psb*A + ITS is optimal when constructing the phylogenetic tree in mangrove communities [14]. In addition, in GP, Liao et al. obtain diverse datasets of edible plants and macro-fungi from field collections, historical publications, and community surveys across seven cities. This work is guided on "Observation Methodology for Long-term Forest Ecosystem Research" of National Standards of the People's Republic of China (GB/T 33027-2016). They find that at least 100 plant species (with 64 plant species producing fruit) and 20 macro-fungi are commonly used as edible forest products in subtropical GP. There are 55 and 57 species providing edible parts in summer and autumn, respectively. Many edible plants have multiple uses. They suggest that edible plants and macro-fungi can enrich the food supply for residents in rural and urban areas by acting as supplemental resources to support the increasing demand for food in the era of rapid urbanization and global change [15].

Mao et al. describe the flowering phenology pattern of *Cyclocarya paliurus*, a monoecious species with a heterodichogamous mating system, in a juvenile plantation at the individual and population levels for 5 consecutive years. They find that four flowering phenotypes and strongly skewed ratios of protandry/protogyny and male/female occur in the juvenile population. Sexual type and ratio change significantly with the growth of the population over the years, showing an increasing monoecious group and a decreasing unisexual group, as well as a tendency for the sexual ratio to move towards equilibrium. Two flowering phases and bimodality in gender are displayed, as in other heterodichogamous species, thereby verifying the presence of heterodichogamy in *C. paliurus* [16].

In order to detect the reason for impaired cone maturation in the Pinaceae, Mo, Xu et al. compare transcriptome libraries of *Pinus massoniana* and Z pine (a natural introgression hybrid) cones at seven successive growth stages. They find that several cones indeed relate to reproductive processes. At every growth stage, these genes are expressed at a higher level in *P. massoniana* than in the Z pine. These data provide insights into understanding which molecular mechanisms are altered between *P. massoniana* and the Z pine that might cause changes in the reproductive process [17].

To examine the diversity and antimicrobial activities of endophytic fungi in *Litsea cubeba*, a medicinal plant from Lauraceae, Wu, Yang et al. obtain 970 isolates from the root, stem, leaf, and fruit segments. They find that the fungal endophytes belonged to the phylum Ascomycota and can be classified into 3 taxonomic classes, 9 orders, 12 families, and 17 genera. *Colletotrichum boninense* is the dominant species. For the antimicrobial activities, 17 isolates could inhibit the growth of plant pathogenic fungi, while the extracts of 6endophytes show antimicrobial activity to all the tested pathogenic fungi [18].

## 4. Theme 3: Trait Expression and Biological Mechanisms

Theme three includes three papers that investigate the physiological and morphological traits and the possible biological mechanisms that generate and maintain particular patterns in trait expression. In order to interpret the patterns of genetic variation of photosynthesis and the relationships with growth traits within gene resources of teak (*Tectonagrandis*), a commercially important tree species in the plant family Lamiaceae found in tropical regions, Huang et al. measure gas exchange, chlorophyll fluorescence parameters, growth traits of plants in nursery, and field trials for 20 teak clones originating from different countries. They report abundant genetic variation in gas exchange, chlorophyll fluorescence, and growth among the teak clones. The measured traits are found to have generally high heritability. The net photosynthetic rate, seedling height, and individual volume of wood are significantly correlated with each other, and seedling height is significantly correlated with plant height [19].

Zheng et al. screen creeping genes in crape myrtle (*Lagerstroemia indica*) in the plant family Lythraceae, which has significant good polymorphisms. They detect two SSR markers, with genetic distances of 23.49 centimorgan and 25.86 cM from the loci controlling the plant opening angle trait and the branching angle trait, respectively. The accuracy rate for phenotypic verification is 76.51% and 74.14%, respectively, which provides basic information for the molecular marker-assisted selective breeding and cloning of the creeping gene to improve architecture diversity in the breeding of crape myrtle [20].

To describe the spatial arrangement of shoot tissues (rectangular vs. cylindrical) and allometric relationships in two contrasting species of the *Polygonatum* from plant family Asparagaceae, Tulik et al. measure the mass and length of the aerial shoots of the individual plants. They find that both species differ significantly with respect to the length, diameter, and thickness of the outer zone of parenchyma. Allometric relationships are stronger for *P. multiflorum* [21].

## 5. Theme 4: Interactions of Woody Plants within a Changing Environment

The four papers addressing this theme describe the investigations of adaptations and responses of plants to the changing environment and provide feedback on forest management at specific sites. To determine the impact of drought on leaf phenology of *Quercus robur*, an economically and ecologically important tree species in the plant family Fagaceae, and spring frost susceptibility in nine provenances, Čehulić et al. expose one-year-old saplings to experimental drought, re-water, and score leaf phenology and frost injury. They find that leaf phenology from most provenances is significantly influenced by the drought treatment. Drought induces a carry-over effect on flushing phenology. In contrast to flushing, autumn leaf phenology is unambiguously delayed following the drought treatments for all studied provenances. This higher susceptibility to spring frost is most likely caused by the advanced flushing phenology, which results from the drought treatment in the previous year [22].

To explore the effect of heritable phenotypic plasticity in the adaptation of woody species to a quickly changing environment, Vander Mijnsbrugge and Janssens observe the timing of bud burst, flower opening, leaf senescence, and leaf fall in two successive years in a common garden of *Crataegus monogyna* from Rosaceae. They find a strong auto-correlation is present among the spring phenophases as well as among the autumnal phenophases, with spring phenophases being negatively correlated with fall phenophases. The strongest between-provenance differentiation is found for the timing of bud burst in spring. Warmer spring temperatures advance the timing of bud burst. However, advancement is non-linear among the provenances. It can be hypothesized that non-local provenances display larger temporal phenotypic plastic responses in the timing of their spring phenophases compared to local provenances when temperatures in the common garden deviate more from their home-sites [23].

The impacts of lime application, understory removal, and their interactive effects on soil microbial communities are tested by Wan et al., who conduct a lime application experiment combined with understory removal in a subtropical plantation of *Eucalyptus* (Myrtaceae). They find that lime application significantly decreases both fungal and bacterial phospholipid fatty acids (PLFAs). Understory removal reduces the fungal PLFAs but has no effect on the bacterial PLFAs. Changes in soil microbial communities caused by the lime application are mainly attributed to increases in soil pH and $NO_3^-$-N contents, while changes caused by understory removal are mainly due to the indirect effects on soil microclimate and the decreased soil-dissolved carbon contents. Furthermore, both lime application and understory removal significantly reduce the litter decomposition rates, which may impact the microbe-mediated soil ecological process. They suggest that lime applications may not be suitable for the management of subtropical *Eucalyptus* plantations [24].

Wu, Li, Chen et al. examine the effects of two native phosphate solubilizing bacteria (PSB), and a mixture of both strains on the growth of seedlings of *Camellia oleifera* (Theaceae). They report a significant promotion of the growth of *C. oleifera* plants by three inoculation treatments. All the PSB inoculation treatments can improve the leaf N and P content and have positive effects on the available N, P, and K content of the rhizosphere soil. A co-inoculation of the two native PSB strains causes a synergistic effect and achieves the best benefit. PSB can convert the insoluble phosphates into plant-available forms and may have the potential for use in sustainable agricultural practices [25].

## 6. Summary and Future Directions

We are pleased to present this Special Issue and believe that many of the studies included here from across the world will make a lasting contribution to biology, ecology, and forestry at diverse scales in the era of changing climate and rapid urbanization. All of the case studies here highlight the important role of emerging techniques, new methods, and novel theories to promote the development of forest biology and plant ecology. We expect that subsequent contributions to this field might consider plant biology from the perspectives of morphology, genetics, trait function, and plant–environment interactions with biotic and abiotic factors [1,2,5,26–30]. Such studies may provide novel insights and new knowledge on quantitative evaluation and description of interactions of plants with animals and microbes, both in natural and urban environments, including terrestrial and aquatic systems.

**Author Contributions:** N.P. and W.J.K., proposed and guest-edited the Special Issue and wrote this editorial together. All authors have read and agreed to the published version of the manuscript.

**Funding:** This work was supported by the National Natural Science Foundation of China (31570594), Fundamental Research Funds of CAF (CAFYBB2017QB002), Research Funds of Guangdong Academy of Sciences (2020GDASYL-20200401001), and CFERN & BEIJING TECHNO SOLUTIONS Award Funds on excellent academic achievements.

**Acknowledgments:** We would like to acknowledge the contributions made by the authors and all reviewers of the 19 manuscripts in this *Genetic and Morphological Variation in Tropical and Temperate Plant Species* Special Issue.

**Conflicts of Interest:** The authors declare no conflict of interest.

## References

1. Meineke, E.K.; Davies, T.J.; Daru, B.H.; Davis, C.C. Biological Collections for Understanding Biodiversity in the Anthropocene. *Philos. Trans. R. Soc. B Biol. Sci.* **2019**, *374*, 20170386. [CrossRef] [PubMed]
2. Leebens-Mack, J.H.; Barker, M.S.; Carpenter, E.J.; Deyholos, M.K.; Gitzendanner, M.A.; Graham, S.W.; Grosse, I.; Li, Z.; Melkonian, M.; Mirarab, S.; et al. One Thousand Plant Transcriptomes and the Phylogenomics of Green Plants. *Nature* **2019**, *574*, 679–685.
3. He, N.; Liu, C.; Piao, S.; Sack, L.; Xu, L.; Luo, Y.; He, J.; Han, X.; Zhou, G.; Zhou, X.; et al. Ecosystem Traits Linking Functional Traits to Macroecology. *Trends Ecol. Evol.* **2019**, *34*, 200–210. [CrossRef] [PubMed]

4. Faucon, M.-P.; Houben, D.; Lambers, H. Plant Functional Traits: Soil and Ecosystem Services. *Trends Plant Sci.* **2017**, *22*, 385–394. [CrossRef] [PubMed]
5. Horváth, Z.; Ptacnik, R.; Vad, C.F.; Chase, J.M. Habitat Loss over Six Decades Accelerates Regional and Local Biodiversity Loss via Changing Landscape Connectance. *Ecol. Lett.* **2019**, *22*, 1019–1027. [CrossRef] [PubMed]
6. Pei, N.; Wang, C.; Sun, R.; Xu, X.; He, Q.; Shi, X.; Gu, L.; Jin, J.; Liao, J.; Li, J.; et al. Towards an Integrated Research Approach for Urban Forestry: The Case of China. *Urban For. Urban Green.* **2019**, *46*, 126472. [CrossRef]
7. Gang, H.; Liu, G.; Chen, S.; Jiang, J. Physiological and Transcriptome Analysis of a Yellow-Green Leaf Mutant in Birch (*Betula platyphylla* × *B. pendula*). *Forests* **2019**, *10*, 120. [CrossRef]
8. Kaushik, P.; Kumar, S. Transcriptome Analysis of Bael (*Aegle marmelos* (L.) Corr.) a Member of Family Rutaceae. *Forests* **2018**, *9*, 450. [CrossRef]
9. Liu, F.; Hong, Z.; Xu, D.; Jia, H.; Zhang, N.; Liu, X.; Yang, Z.; Lu, M. Genetic Diversity of the Endangered *Dalbergia odorifera* Revealed by SSR Markers. *Forests* **2019**, *10*, 225. [CrossRef]
10. Lu, W.; Arnold, R.J.; Zhang, L.; Luo, J. Genetic Diversity and Structure through Three Cycles of a *Eucalyptus urophylla* S.T. Blake Breeding Program. *Forests* **2018**, *9*, 372. [CrossRef]
11. Mo, Z.; Feng, G.; Su, W.; Liu, Z.; Peng, F. Identification of miRNAs Associated with Graft Union Development in Pecan [*Carya illinoinensis* (Wangenh.) K. Koch]. *Forests* **2018**, *9*, 472. [CrossRef]
12. Wei, M.; Wang, Y.; Pan, R.; Li, W. Genome-Wide Identification and Characterization of MADS-box Family Genes Related to Floral Organ Development and Stress Resistance in *Hevea brasiliensis* Müll. Arg. *Forests* **2018**, *9*, 304. [CrossRef]
13. Xie, L.; Li, X.; Hou, D.; Cheng, Z.; Liu, J.; Li, J.; Mu, S.; Gao, J. Genome-Wide Analysis and Expression Profiling of the Heat Shock Factor Gene Family in *Phyllostachys edulis* during Development and in Response to Abiotic Stresses. *Forests* **2019**, *10*, 100. [CrossRef]
14. Wu, F.; Li, M.; Liao, B.; Shi, X.; Xu, Y. DNA Barcoding Analysis and Phylogenetic Relation of Mangroves in Guangdong Province, China. *Forests* **2019**, *10*, 56. [CrossRef]
15. Liao, J.; Zhang, L.; Liu, Y.; Li, Q.; Chen, D.; Zhang, Q.; Li, J. Diversity and Utilization of Edible Plants and Macro-Fungi in Subtropical Guangdong Province, Southern China. *Forests* **2018**, *9*, 666. [CrossRef]
16. Mao, X.; Fu, X.-X.; Huang, P.; Chen, X.-L.; Qu, Y.-Q. Heterodichogamy, Pollen Viability, and Seed Set in a Population of Polyploidy *Cyclocarya Paliurus* (Batal) Iljinskaja (Juglandaceae). *Forests* **2019**, *10*, 347. [CrossRef]
17. Mo, J.; Xu, J.; Cao, Y.; Yang, L.; Yin, T.; Hua, H.; Zhao, H.; Guo, Z.; Yang, J.; Shi, J. *Pinus massoniana* Introgression Hybrids Display Differential Expression of Reproductive Genes. *Forests* **2019**, *10*, 230. [CrossRef]
18. Wu, F.; Yang, D.; Zhang, L.; Chen, Y.; Hu, X.; Li, L.; Liang, J. Diversity Estimation and Antimicrobial Activity of Culturable Endophytic Fungi from *Litsea cubeba* (Lour.) Pers. in China. *Forests* **2019**, *10*, 33. [CrossRef]
19. Huang, G.; Liang, K.; Zhou, Z.; Yang, G. Muralidharan EM Variation in Photosynthetic Traits and Correlation with Growth in Teak (*Tectona grandis* Linn.) Clones. *Forests* **2019**, *10*, 44. [CrossRef]
20. Zheng, T.; Qin, B.; Li, S.; Cai, M.; Pan, H.; Wang, J.; Cheng, T.; Zhang, Q. Screening of Applicable SSR Molecular Markers Linked to Creeping Trait in Crape Myrtle. *Forests* **2019**, *10*, 429. [CrossRef]
21. Tulik, M.; Karczewski, J.; Szeliga, N.; Jura-Morawiec, J.; Jarzyna, I. Morphological Characteristics and Allometric Relationships of Shoot in Two Undergrowth Plants: *Polygonatum odoratum* and *Polygonatum multiflorum*. *Forests* **2018**, *9*, 783. [CrossRef]
22. Čehulić, I.; Sever, K.; Katičić Bogdan, I.; Jazbec, A.; Škvorc, Ž.; Bogdan, S. Drought Impact on Leaf Phenology and Spring Frost Susceptibility in a *Quercus robur* L. Provenance Trial. *Forests* **2019**, *10*, 50. [CrossRef]
23. Vander Mijnsbrugge, K.; Janssens, A. Differentiation and Non-Linear Responses in Temporal Phenotypic Plasticity of Seasonal Phenophases in a Common Garden of *Crataegus monogyna* Jacq. *Forests* **2019**, *10*, 293. [CrossRef]
24. Wan, S.; Liu, Z.; Chen, Y.; Zhao, J.; Ying, Q.; Liu, J. Effects of Lime Application and Understory Removal on Soil Microbial Communities in Subtropical *Eucalyptus* L'Hér. Plantations. *Forests* **2019**, *10*, 338. [CrossRef]
25. Wu, F.; Li, J.; Chen, Y.; Zhang, L.; Zhang, Y.; Wang, S.; Shi, X.; Li, L.; Liang, J. Effects of Phosphate Solubilizing Bacteria on the Growth, Photosynthesis, and Nutrient Uptake of *Camellia oleifera* Abel. *Forests* **2019**, *10*, 348. [CrossRef]
26. Dornelas, M.; Gotelli, N.J.; Shimadzu, H.; Moyes, F.; Magurran, A.E.; McGill, B.J. A Balance of Winners and Losers in the Anthropocene. *Ecol. Lett.* **2019**, *22*, 847–854. [CrossRef]

27. McCallen, E.; Knott, J.; Nunez-Mir, G.; Taylor, B.; Jo, I.; Fei, S. Trends in Ecology: Shifts in Ecological Research Themes over the Past Four Decades. *Front. Ecol. Environ.* **2019**, *17*, 109–116. [CrossRef]
28. Brancalion, P.H.S.; Niamir, A.; Broadbent, E.; Crouzeilles, R.; Barros, F.S.M.; Almeyda Zambrano, A.M.; Baccini, A.; Aronson, J.; Goetz, S.; Reid, J.L.; et al. Global Restoration Opportunities in Tropical Rainforest Landscapes. *Sci. Adv.* **2019**, *5*, eaav3223. [CrossRef]
29. Zhou, W.; Fisher, B.; Pickett, S.T. Cities are hungry for actionable ecological knowledge. *Front. Ecol. Environ.* **2019**, *17*, 135. [CrossRef]
30. Maes, M.J.A.; Jones, K.E.; Toledano, M.B.; Milligan, B. Mapping Synergies and Trade-offs between Urban Ecosystems and the Sustainable Development Goals. *Environ. Sci. Policy* **2019**, *93*, 181–188. [CrossRef]

*forests*

MDPI

*Article*

# Physiological and Transcriptome Analysis of a Yellow-Green Leaf Mutant in Birch (*Betula platyphylla* × *B. Pendula*)

**Huixin Gang, Guifeng Liu, Su Chen and Jing Jiang ***

State Key Laboratory of Tree Genetics and Breeding, Northeast Forestry University, 26 Hexing Road, Harbin 150040, China; gang_em@163.com (H.G.); liuguifeng@126.com (G.L.); chensunefu@163.com (S.C.)

* Correspondence: jiangjing@nefu.edu.cn; Tel.: +86-139-4602-6246

Received: 19 December 2018; Accepted: 1 February 2019; Published: 2 February 2019

**Abstract:** Chlorophyll (Chl)-deficient mutants are ideal materials for the study of Chl biosynthesis, chloroplast development, and photosynthesis. Although the genes encoding key enzymes related to Chl biosynthesis have been well-characterized in herbaceous plants, rice (*Oryza sativa* L.), Arabidopsis (*Arabidopsis thaliana*), and maize (*Zea mays* L.), yellow-green leaf mutants have not yet been fully studied in tree species. In this work, we explored the molecular mechanism of the leaf color formation in a yellow-green leaf mutant (*yl*). We investigated the differentially expressed genes (DEGs) between *yl* and control plants (wild type birch (WT) and *BpCCR1* overexpression line 11, (C11)) by transcriptome sequencing. Approximately 1163 genes (874 down-regulated and 289 up-regulated) and 930 genes (755 down-regulated and 175 up-regulated) were found to be differentially expressed in *yl* compared with WT and C11, respectively. Kyoto Encyclopedia of Genes and Genomes (KEGG) pathway enrichment analysis for DEGs revealed that photosynthesis antenna proteins represent the most significant enriched pathway. The expressions of photosynthesis antenna proteins are crucial to the leaf color formation in *yl*. We also found that Chl accumulate, leaf anatomical structure, photosynthesis, and growth were affected in *yl*. Taken together, our results not only provide the difference of phenomenal, physiological, and gene expression characteristics in leaves between *yl* mutant and control plants, but also provide a new insight into the mutation underlying the chlorotic leaf phenotype in birch.

**Keywords:** yellow-green leaf mutant; transcriptome; antenna protein; photosynthesis; birch

## 1. Introduction

Birch (*Betula*), a member of the tall deciduous tree family of Betulaceae, contains approximately 20 taxa birch species. Birch is an ecologically important tree species native to parts of China, Siberia, Korea, Japan, and Russia that has been introduced to many northern areas of the world [1]. This species grows fast and has a high tolerance that allows it to be used for revegetation and reforestation [2]. Birch also plays an important role in the forestry industry as a source of timber, fuelwood, plywood, pulpwood, and furniture [3–6]. Recently, there has been an increased interest in plants with colored leaves. *Betula pendula* 'Purple Rain', an intraspecific variety of *B. pendula*, has been used as a decorative plant for its purple leaves [7]. Previously, a yellow-green leaf mutant (*yl*) was derived from cinnamoyl-CoA reductase (*BpCCR1*)-overexpressing transgenic birch plants. The *yl* mutant displayed a distinct yellow-green phenotype, while the leaves of all the other *BpCCR1*-overexpressing lines were a normal color, the same as the wild type birch (WT). One of the *BpCCR1*-overexpressing lines, C11 with green leaves, was used as control birch [8]. The *yl* mutant is a very valuable resource because a yellow-green leaf is one of the most popular traits in landscape greening.

Leaf color variation is one of the most common mutated traits as it is easily discovered in higher plants. Up to now, many Chl-deficient mutants have been identified in herbaceous and woody plants, including Arabidopsis [9], rice [10], maize [11], wheat (*Triticum aestivum* L.) [12], cotton (*Gossypium barbadense* L.) [13], and tea (*Camellia sinensis* (L.) O. Kuntze) [14]. It has been reported that a yellow-green leaf mutant (*siygl1*) of foxtail millet (*Setaria italic* L.) isolated following ethyl methanesulfonate (EMS) treatment was due to the loss function of the *SiYGL1* gene. The identification of the *SiYGL1* gene that encodes Mg-chelatase ATPase subunit D facilitated the understanding of the biological processes of chlorophyll (Chl) biosynthesis in millet [15]. Another yellow-green leaf mutant (*ygl8*) was proved to be controlled by the *Ygl8* gene, which encodes a chloroplast-targeted uridine monophosphate (UMP) kinase and affects chloroplast development in rice [16]. Therefore, Chl-deficient mutants are valuable genetic materials for exploring the molecular mechanisms of Chl biosynthesis and regulation, chloroplast development, plastid-to-nucleus signal transduction, and photosynthesis.

Next-generation sequencing (NGS) technologies have been considered as powerful tools for advanced research in many areas. Examples are genome and transcriptome sequencing of animals, plants, and microbes with high-throughput, high-speed, and high-accuracy sequencing data [17–19]. Wang, using transcriptome sequencing, analyzed a Chl-deficient chlorina tea plant culticar and reported the molecular mechanisms of the chlorine tea phenotype [14]. Study of a *Lagerstroemia indica* yellow leaf mutant using transcriptome analysis revealed the formation pathway of a yellow leaf mutant and discovered novel candidate genes related to leaf color [20]. The development of NGS technologies has increased the rate and efficiency of gene discovery and permitted a deeper understanding of the gene expression network.

In this work, we performed transcriptome sequencing for *yl* with a yellow-green leaf phenotype, *BpCCR1* overexpression line 11(C11), and wild type birch (WT) with a normal green leaf phenotype to analyze gene expression in these plants and elucidate the molecular mechanisms related to the different phenotypes. In addition, comparative physiological studies were conducted to investigate the phenotypic differences between *yl* and control (WT and C11) plants. This study improved our understanding of the yellow-green phenotype in birch.

## 2. Materials and Methods

### 2.1. Materials

A birch (*Betula platyphylla* × *B. pendula*) Chl-deficient mutant *yl*, *BpCCR1* overexpression line 11 (C11), and wild type birch (WT) were used as the experimental materials. The *yl* mutant was derived from *BpCCR1* overexpression lines. All the plants were grown in the pots with dimensions of 8 × 8 cm and substrata of 9 cm under natural conditions and were well watered at the birch breeding base, Harbin, China. Mature leaves from the new stems were collected in the spring. Immediately after harvest, samples were frozen in liquid nitrogen and stored at −80 °C for RNA extraction.

### 2.2. Methods

#### 2.2.1. Measurement of Growth Traits and Pigment Content

Forty plants of each line (WT, C11, and *yl*) were used for the measurement of plant height. Each value was the average of the measurements.

Fresh leaves of WT, C11, and *yl* plants (from first to sixth leaf) were collected during the growing season and used for the measurement of pigment contents, according to the method of Lichtenthaler [21]. The first leaf was the youngest and the sixth leaf was the oldest on the main stem. Chl and carotenoid (Car) were extracted with 80% acetone at 4 °C for 24 h in the dark, and then calculated from the absorbance at 470 nm, 646 nm, and 663 nm in a Vis-UV spectrophotometer (TU-1901, Persee, China). Contents of Chl a (mg/g), Chl b (mg/g), and Car (mg/g) were calculated as follows:

$$C_{\text{Chl a}}\ (\text{mg/L}) = 12.21\ A_{663} - 2.81\ A_{646} \tag{1}$$

$$C_{\text{Chl b}}\ (\text{mg/L}) = 20.13\ A_{646} - 5.03\ A_{663} \tag{2}$$

$$C_{\text{Car}}\ (\text{mg/L}) = 4.37\ A_{470} + 2.11\ A_{663} - 9.10\ A_{646} \tag{3}$$

$$\text{Chl a (mg/g)} = C_{\text{Chl a}}\ (\text{mg/L}) \times V\ (\text{L}) \ /\ M_{\text{fresh}}\ (\text{g}) \tag{4}$$

$$\text{Chl b (mg/g)} = C_{\text{Chl b}}\ (\text{mg/L}) \times V\ (\text{L}) \ /\ M_{\text{fresh}}\ (\text{g}) \tag{5}$$

$$\text{Car (mg/g)} = C_{\text{car}}\ (\text{mg/L}) \times V\ (\text{L}) \ /\ M_{\text{fresh}}\ (\text{g}) \tag{6}$$

$$\text{Chl a/b ratio} = \text{Chl a} \ /\ \text{Chl b} \tag{7}$$

### 2.2.2. Light Microscopy

The fourth leaves of WT, C11, and *yl* were used as samples and fixed in FAA (formaldehyde, glacial acetic acid and 50% ethyl alcohol, V:V:V = 1:1:18) for 24 h, dehydrated in a graded ethanol series and xylene, and then embedded in paraffin wax. Sections (10 μm thick) were stained with safranine and fast green dyes. The cell structures of the samples were examined and photographed using an Olympus DP26 digital camera (Olympus, Tokyo, Japan). Five leaf anatomical features, including lamina thickness (LT), adaxial epidermis thickness (UE), abaxial epidermis thickness (LE), palisade parenchyma thickness (PT), and spongy parenchyma thickness (ST), were examined with the cellSens Entry software. Additionally, the palisade parenchyma/mesophyll ratio was calculated. Four positions of each section were measured and each value of leaf anatomical feature was the average of 20 measurements from five individual plants.

### 2.2.3. Leaf Gas-Exchange Measurement

The net photosynthetic rate (Pn) of the fourth leaves from WT, C11, and *yl* was measured using an Li-6400 portable photosynthesis system (LI-COR Inc, Lincoln, NE, USA) at 9:00–11:00 am on sunny days. $CO_2$ concentration was controlled at 400 μmol $mol^{-1}$. Relative air humidity was about 50% and leaf temperature was about 28 °C. The default red/blue LED light source (LI6400-02B) was chosen as the light source. The photosynthetic curves were made against the light intensity of 2000, 1800, 1500, 1200, 1000, 800, 600, 400, 200, 100, 50, 20, and 0 μmol photos $m^{-2}$ $s^{-1}$. About 15 min light adaptation was applied to leaves at an initial light step before $CO_2$ measurement and then the values were recorded when they were stable at each light step. The measurements of WT, C11, and *yl* were made under the same conditions (including time interval of illumination). The averaged values of each light step for each plant were used in the light-response curve.

### 2.2.4. RNA Extraction, Library Construction, and RNA-seq

The fourth leaves of WT, C11, and *yl* were used as samples. Total RNAs were extracted from leaf samples using the CTAB (cetyltrimethylammonium bromide) method [22]. A summary of the procedure was as follows: The samples were individually milled in a mortar with liquid nitrogen and then incubated with 2% CTAB (added 2% β-mercaptoethanol) at 65 °C in a water-bath for 5 min. The samples were centrifuged at 13,400 g for 10 min and an equal volume of chloroform was added to the supernatant. A half volume of ethyl alcohol and 0.8 times volume of 5 mol/L LiCl were then added to the supernatant after centrifuging at 13,400 g, 4 °C for 10 min. After standing for 10 min, the samples were centrifuged at 13,400 g, 4 °C for 20 min. Then, the precipitates were washed in 70% ethanol and dried. RNAs were dissolved with diethylpyrocarbonate (DEPC)-treated water and treated with DNaseI. RNA quality, purity, and integrity were detected by 1% agrose gel electrophoresis, a NanoDrop2000 microvolume spectrophotometer (Thermo, Waltham, MA, USA) and an Agilent 2100 Bioanalyzer (Agilent, Palo Alto, CA, USA), respectively. RNAs from three independent replicates were mixed by equal volume. Poly (A) mRNA was enriched using Oligo (dT)-magnetic beads and cleaved into short fragments with fragmentation buffer. These short fragments were used as templates to synthesize the first-strand cDNA. Second-strand cDNA was synthesized using buffer, dNTPs, RNaseH, and DNA polymerase I. Purified cDNA was used as a template for PCR amplification and library construction. Lastly, the library was sequenced on an Illumina HiSeq™ 2500 platform by Biomarker Technology Company (Beijing, China). CTAB, β-mercaptoethanol, chloroform, ethyl alcohol, LiCl, and DEPC were bought from Sigma-Aldrich, St. Louis, MO, USA. DNaseI, dNTPs, RNaseH, and DNA polymerase I were bought from Promega, Madison, WI, USA.

#### 2.2.5. Gene Annotation and DEG Analysis

To elucidate the reason for the different phenotype observed in *yl*, we explored the gene expression of WT, C11, and *yl* at the molecular level. After removing adapters and low-quality sequences, we generated an average of approximately 3.46 Gb RNA-seq data with 91.28% Q30 bases and 46.28% GC content for each sample in the transcriptome sequencing. More than 82.21% of the clean reads were mapped to the birch reference genome [23] using TopHat2 [24]. Gene functions were annotated using the Nr, Swiss-Prot, Kyoto Encyclopedia of Genes and Genomes (KEGG), Eukaryotic Orthologous Groups (KOG), Clusters of Orthologous Groups (COG) and Gene Ontology (GO) databases. The expression level of each gene was calculated using FPKM (Fragments per kilobase per million mapped reads). Differentially expressed genes (DEGs) between each two-sample comparison were defined with fold change ≥2 and FDR (false discovery rate) <0.01 as a threshold, according to the statistical analysis performed by EBSeq. The percentages of DEGs in Go classification were calculated as follows:

$$\text{Percentages of genes} = (\text{Number}_{\text{DEGs in a specific term}}) / (\text{Number}_{\text{all DEGs}}) \times 100\% \quad (8)$$

KEGG pathway terms with corrected enrichment $p$ values less than 0.05 (Fisher's exact test) were considered to be significantly enriched. The genes involved in photosynthesis-antenna proteins were extracted according to the functional annotation information of the genes.

#### 2.2.6. RNA Extraction and Quantitative RT-PCR

Total RNAs were extracted from the functional leaves of the WT, C11, and *yl* lines as described in 2.1.4 and treated with DNaseI (Promega, Madison, WI, USA). cDNA was synthesized from 1 μg RNA of WT, C11, and *yl* using a ReverTreAce® qPCR RT Kit (Toyobo, Osaka, Japan), according to the manufacturers' instructions, respectively. The procedure was as follows: RNA was incubated with 5x RT Master Mix at 37 °C for 15 min, 50 °C for 5 min, and 98 °C for 5 min, and then diluted 10-fold with nuclease-free water. The quantitative (q)RT-PCR was performed on a 7500 real-time PCR system (Applied Biosystems, Darmstadt, Germany) using SYBR® Green PCR master mix (Toyobo, Osaka, Japan). Each qRT-PCR reaction (total 20 μL) contained 2 μL of cDNA, 10 μL of 2× SYBR Green Real-time PCR Master mix, 0.5 μL of each PCR primer, and 7 μL of nuclease-free water. qRT-PCR conditions were as follows: 95 °C for 30 s followed by 45 cycles of 95 °C for 15 s, 60 °C for 45 s, and a final extension at 72 °C for 30 s. The results were calculated using the $2^{-\Delta\Delta Ct}$ method [25], and *Bp18S rRNA* was selected as an internal control gene [26]. Primer sequences are listed in Table 1.

#### 2.2.7. Statistical Analysis

Data were analyzed using SPSS statistics software, version 19.0 (International Business Machines, Armonk, NY, USA). Differences between the means of each line on leaf anatomical and plant height were determined using one-way analysis of variance and the Duncan multiple comparison procedure. A $p$ value less than 0.05 was considered statistically significant and labeled with a different letter. The same letter represented that they were not significantly different. The correlations between pigment contents and Pn, stomatal conductance (Gs) and Pn, Gs and transpiration rate (Tr), and Tr and Pn were evaluated using Pearson's correlation coefficients. The relationship between Tr and Pn was tested for linear, exponential, and logarithmic functions and the best fit regressions were selected.

Correlation coefficients ($R^2$) and equations were obtained from nonlinear regression analysis of Tr and Pn using Origin Pro software, version 8.1 (OriginLab, Northampton, MA, USA).

**Table 1.** The primer sequences used in qRT-PCR.

| Gene | Forward Primer (5′-3′) | Reverse Primer (5′-3′) |
|---|---|---|
| Bpev01.c0080.g0006.m0001 | GGCTCAATCCAGCATGGTTGC | CCACACCTGCATGCATTGCAC |
| Bpev01.c0577.g0002.m0001 | CGTTCATGGTGGACCTGAGCC | AGCTAAAGTGAGGGACTTTGTCGAT |
| Bpev01.c0167.g0013.m0001 | CACAACATAGCCAGCCACCTTC | GTCGGTGCTACCCAAGGACTC |
| Bpev01.c0243.g0056.m0001 | CAACGGACGTTATGCCATGTTGG | TCCCAGCCGGCGGAATTAC |
| Bpev01.c0264.g0036.m0001 | CGGTGAGGCCGTGTGGTT | CGCCCATCAGGATGACCTGT |
| Bpev01.c1040.g0049.m0001 | CCCGAGACATTTGCCAAGAACC | GCCTTGAACCAGACTGCCTCTC |
| Bpev01.c0190.g0044.m0001 | CTTGGCCGGCGATGTGATC | GCCAACATGGCCCACCTC |
| Bpev01.c0621.g0012.m0001 | GCAATAGGCCTTGCCTCCTTCATAG | CGAGTACCCGTCTTCTCATTCGC |
| Bpev01.c0984.g0005.m0001 | GGGAGACTAAGGTACAAGCAGTGG | CCAGCTGCTCAATTGCTTCAGAG |
| Bpev01.c0401.g0006.m0001 | GGGAAGGCAAGGCTAGTGCAG | GCCTTCAACAGCAAGGCAAGT |
| Bpev01.c0894.g0003.m0001 | CCTCCAACAGGGAGTGGCAAC | CTGTCAATCATCCCAGAACAGCTTG |
| Bpev01.c1891.g0003.m0001 | GGCGTTCCGTATATGAGCCTCTTC | GGCTACTGCTGTTTTACCGGTCT |

## 3. Results

### 3.1. Pigment Contents and Leaf Anatomical Structure of Chl-Deficient Mutant yl

Previously, we transformed the *BpCCR1* gene to a hybrid birch, WT (*Betula platyphylla* × *B. Pendula*) by *Agrobacterium tumefaciens*. We obtained 19 *BpCCR1*-overexpressing transgenic lines. Among them, a transgenic line (*yl*) displayed a yellow-green leaf phenotype, which was distinct from other transgenic lines, including C11. During the growth season, the *yl* mutant exhibited yellow-green leaves, while C11 and WT exhibited green leaves (Figure 1a,b). In order to investigate the difference, we measured the pigment contents and leaf anatomical structure of *yl*. We measured the Chl and Car contents from the first leaf to the sixth leaf. The result showed that the pigment contents, including Chl a, Chl b, and Car, were increased from the first leaf to sixth leaf in all samples (Figure 1c). However, all pigment levels were lower in the leaves of *yl* than those of WT and C11. Chl a in *yl* was decreased by 44–65% of WT, and 41–57% of C11. Chl b in *yl* was decreased by 62–79% of WT, and 61–73% of C11. Similarly, Car in *yl* was decreased by 40–55% compared with WT, and 30–54% compared with C11. The ratio of Chl a to Chl b was increased in all leaves of *yl*.

We then investigated the leaf anatomical structure of WT, C11, and *yl*, as leaves are important organs for photosynthesis, and their anatomical structures could affect photosynthetic and physiological activities. A significant reduction in the lamina thickness of *yl* leaves compared with the leaves of WT and C11 was observed in transverse sections of the leaf blades (Figure 2a). Also, adaxial epidermis, abaxial epidermis, palisade parenchyma, and spongy parenchyma were reduced in thickness in *yl* leaves (Figure 2b). However, there was no difference in the ratio of palisade to spongy parenchyma.

### 3.2. Changes in Photosynthesis and Growth

To examine the effect of low pigment contents on photosynthesis in the *yl* mutant, we measured the photosynthetic rate of WT, C11, and *yl*. The result showed that *yl* had a lower net photosynthetic rate and a lower transpiration rate than WT and C11 at all light intensities determined (Figure 3a,b).

We then investigated the correlations between pigment contents, transpiration rate, stomatal conductance, and Pn. The results showed that the Pn values had a strong relationship with total Chl, Chl a, Chl b, and Car under high light intensity ($p < 0.05$). The correlations between pigment contents and Pn were decreased under low light intensity, revealing that not all pigments participated in photosynthesis under low light intensity (Table S1). In addition, we found that Pn and Tr values displayed a significant positive correlation ($p < 0.01$) with Gs in WT, C11, and *yl* plants. Similarly, there was also a highly significant correlation ($p < 0.01$) between Pn and Tr in WT, C11, and *yl* plants (Table 2).

Then, we analyzed the scatter plots between Pn and Tr and found that the exponential decay function was the best fit. The curve could be divided into two stages. In the first stage, the Pn value increased with the rising Tr and the curve was likely to be linear. Gs may be the primary limiting factor in this stage. In the second stage, Pn increased slowly or mantained invariability along with the increase of Tr. Gs may not be the primary limiting factor in this stage (Figure 4a–c).

To explore the effect of low photosynthesis on growth, we measured the height of WT, C11, and *yl* plants. The heights of one-year-old WT and C11 were 36.7 cm and 37.0 cm, while the *yl* mutant was 30.9 cm, about 84% and 83% of WT and C11, respectively. The result revealed that *yl* grew slower compared to WT and C11 (Figure 3c).

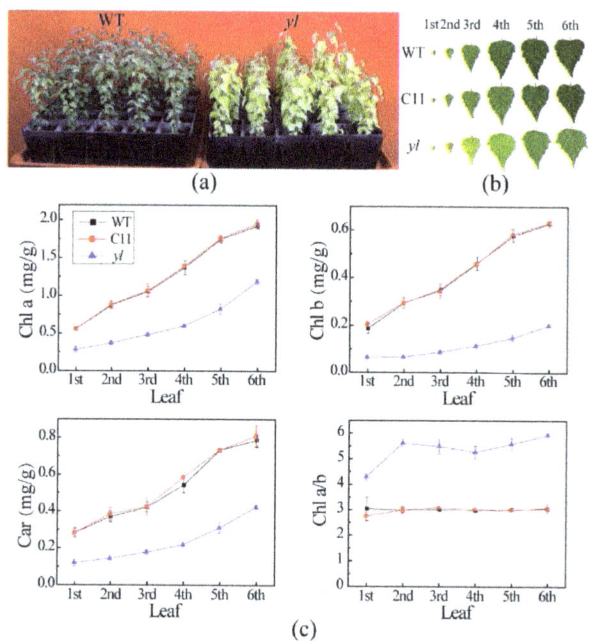


**Figure 1.** Growth performance and pigment content of wild type birch (WT), *BpCCR1* overexpression line 11 (C11) and yellow-green leaf mutant (*yl*) lines. (**a**) Growth performance of one-year-old WT and *yl* plants. (**b**) The leaves from first to sixth of WT, C11, and *yl* lines. (**c**) Chl a, Chl b, Car, and Chl a/b in first to sixth leaves of WT, C11, and *yl* lines. Error bars represent the standard deviation (SD) of three independent experiments.

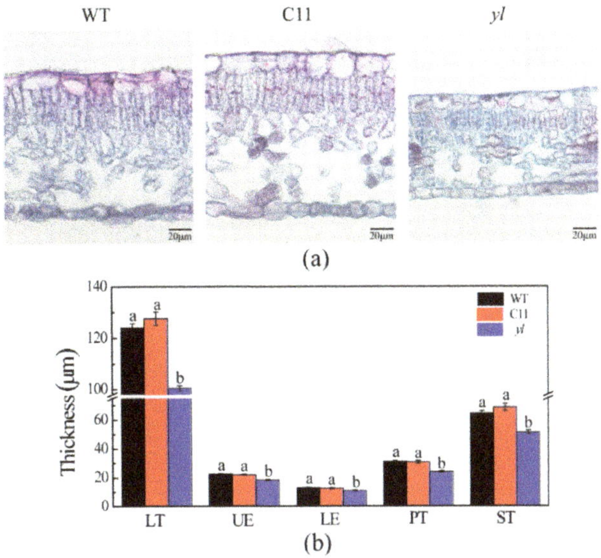


**Figure 2.** Leaf anatomical characteristics of WT, C11, and *yl*. (**a**) Leaf transections of WT, C11, and *yl* lines. (**b**) Leaf anatomical structure of WT, C11, and *yl*. Lamina thickness, LT. Adaxial epidermis thickness, UE. Abaxial epidermis thickness, LE. Palisade parenchyma thickness, PT. Spongy parenchyma thickness, ST. Error bars represent the standard deviation (SD) of 20 measurements from five individual plants.

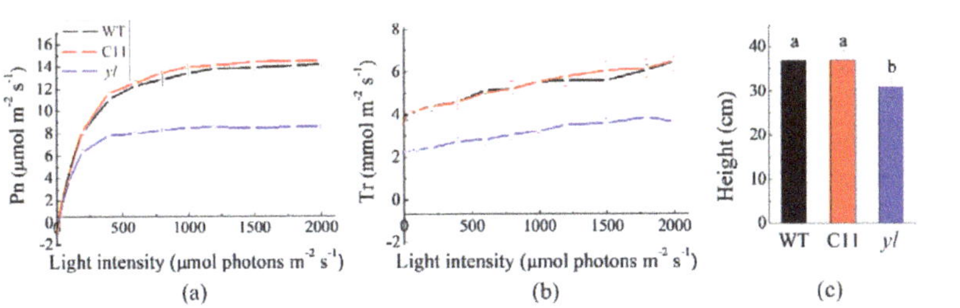


**Figure 3.** Photosynthetic and growth performance of WT, C11, and *yl* lines. Photosynthetic rate (**a**) and transpiration rate (**b**) of WT, C11, and *yl*. Error bars represent the SD of three measurements. (**c**) Plant height of WT, C11, and *yl*. Error bars represent the SD of 40 measurements.

**Table 2.** The correlation analysis of transpiration rate (Tr), stomatal conductance (Gs), and Pn in WT, C11, and *yl*. ** Correlation is significant at the 0.01 level (2-tailed).

| Correlation | Gs in WT (mol $m^{-2}$ $s^{-1}$) | Tr in WT (mmol $m^{-2}$ $s^{-1}$) | Gs in C11 (mol $m^{-2}$ $s^{-1}$) | Tr in C11 (mmol $m^{-2}$ $s^{-1}$) | Gs in *yl* (mol $m^{-2}$ $s^{-1}$) | Tr in *yl* (mmol $m^{-2}$ $s^{-1}$) |
|---|---|---|---|---|---|---|
| Tr in WT (mmol $m^{-2}$ $s^{-1}$) | 0.987 ** | - | - | - | - | - |
| Pn in WT (μmol $m^{-2}$ $s^{-1}$) | 0.860 ** | 0.915 ** | - | - | - | - |
| Tr in C11 (mmol $m^{-2}$ $s^{-1}$) | - | - | 0.986 ** | - | - | - |
| Pn in C11 (μmol $m^{-2}$ $s^{-1}$) | - | - | 0.856 ** | 0.895 ** | - | - |
| Tr in *yl* (mmol $m^{-2}$ $s^{-1}$) | - | - | - | - | 0.991 ** | - |
| Pn in *yl* (μmol $m^{-2}$ $s^{-1}$) | - | - | - | - | 0.773 ** | 0.815 ** |

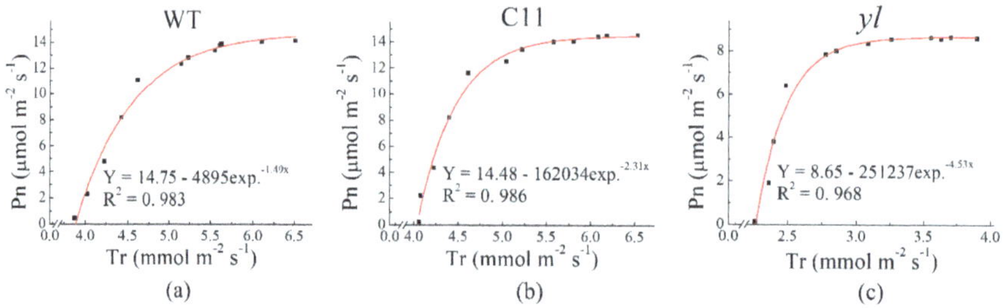


**Figure 4.** The scatter plots between Pn and Tr in WT, C11, and *yl*.

## 3.3. Differently Expressed Genes between yl and Control Plants

In order to generate DEGs, two transcriptome comparisons were carried-out, including C11 vs. *yl* and WT vs. *yl*. As a result, 1163 genes (874 down-regulated and 289 up-regulated) were found to be differentially expressed in *yl* compared with C11. 930 genes, including 755 down-regulated and 175 up-regulated that were differentially expressed in *yl* compared with WT (Figure 5a).

GO categories were then assigned to evaluate the potential functions of these DEGs according to the biological process, cellular component, and molecular function ontology. In the biological process, the DEGs were classified into twenty categories, and the most three overrepresented terms were cellular process, single-organism process, and metabolic process. For cell components, there are many DEGs involved in cells, cell parts, and organelles. For molecular functions, most of the DEGs were participates in binding and catalytic activity, as shown in Figure 5b.

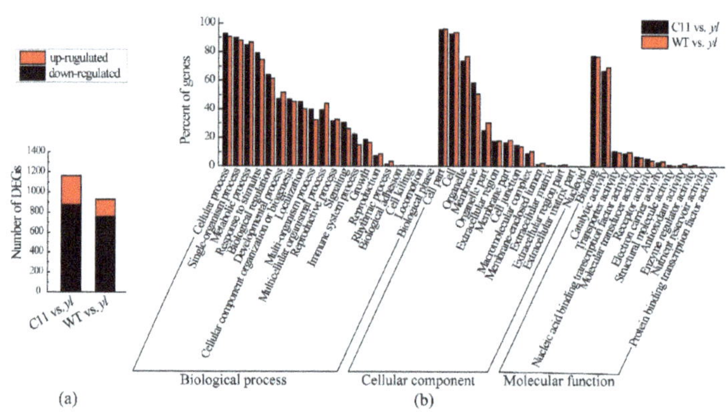


**Figure 5.** Differentially expressed genes (DEGs) analysis based on Gene ontology. (**a**) Number of DEGs in C11 vs. *yl* and WT vs. *yl*. (**b**) Gene ontology classification of DEGs in C11 vs. *yl* and WT vs. *yl*.

## 3.4. KEGG Pathway Analysis of DEGs

The DEGs were further subjected to KEGG pathway analysis to identify the enriched biological pathways. The DEGs of C11 vs. *yl* and WT vs. *yl* were mapped to 81 and 85 KEGG pathways, respectively. Among them, two pathways, including photosynthesis-antenna proteins and phenylalanine metabolism, were considered significantly enriched at a cut-off P-value $< 0.05$ and FDR $< 0.05$ in both C11 vs. *yl* and WT vs. *yl*. Photosynthesis-antenna proteins represented the most significantly enriched pathway in the DEGs of *yl* compared to WT and C11 (Figure 6a,b).

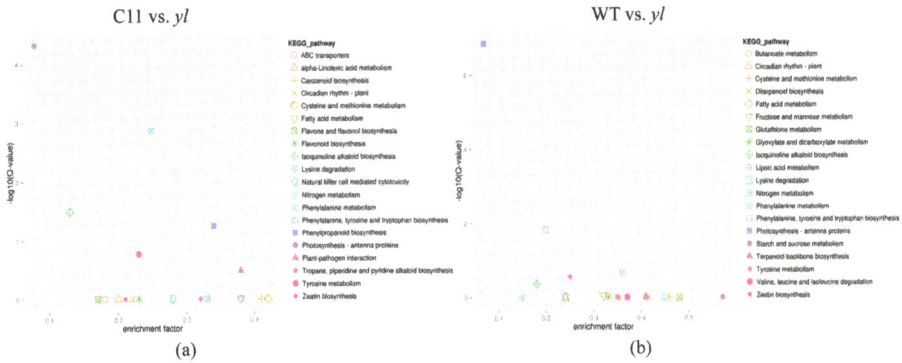


**Figure 6.** KEGG-based pathway enrichment of DEGs in C11 vs. *yl* and WT vs. *yl*.

We then explored the photosynthesis-antenna proteins pathway in more detail. A total of twenty-one *Lhc* (light-harvesting complex) genes were found to be related to the antenna proteins in the birch genome. However, seven of these genes were differently expressed in *yl* compared to control plants and six displayed low or undetectable levels of expression in all samples. One gene (Bpev01.c0243.g0056.m0001) involved in light-harvesting the Chl a/b binding protein Lhca3 of Photosystem I, three genes (Bpev01.c0362.g0012.m0001, Bpev01.c0264.g0036.m0001, Bpev01.c1767.g0010.m0001) involved in light-harvesting the Chl a/b binding protein Lhcb1, and one gene (Bpev01.c1040.g0049.m0001) involved in light-harvesting the Chl a/b binding protein Lhcb2 of Photosystem II showed significantly reduced transcript levels in *yl*. Two genes (Bpev01.c0190.g0044.m0001, Bpev01.c0841.g0007.m0001) were involved in light-harvesting the Chl a/b binding protein Lhcb4 of Photosystem II, and Bpev01.c0190.g0044.m0001 was down-regulated and Bpev01.c0841.g0007.m0001 was up-regulated (Figure 7a,b). These results suggested that the changes in the photosynthesis-antenna proteins pathway were important to the unique phenotype of *yl*.

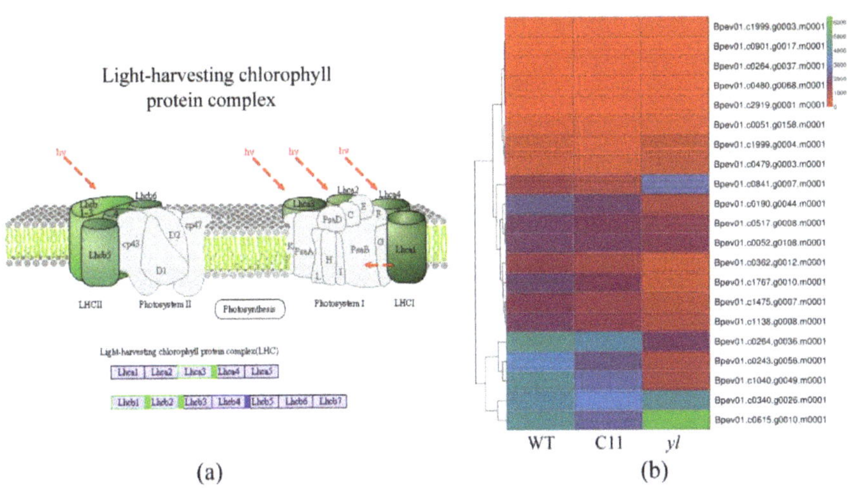


**Figure 7.** Expression pattern of genes involved in light-harvesting Chl complex in WT, C11, and *yl*. (**a**) DEGs in the pathway of light-harvesting Chl complex. Green box represents down-regulated genes. Blue box represents down-regulated and up-regulated genes. (**b**) Expression of all genes involved in light-harvesting Chl complex in WT, C11, and *yl*.

### 3.5. qRT-PCR Verification of RNA-seq

To test the reliability of RNA-Seq, we selected 12 functionally important and representative genes for validation using qRT-PCR, including two non-DEGs, five down-regulated genes, and five up-regulated genes (Table 3). The expression of all these genes obtained via qRT-PCR showed a similar pattern to that detected by transcriptome sequencing (Figure 8).

**Table 3.** The genes used for qRT-PCR.

| Gene | C11 and WT vs. *yl* |
|---|---|
| Bpev01.c0080.g0006.m0001 | normal |
| Bpev01.c0577.g0002.m0001 | normal |
| Bpev01.c0167.g0013.m0001 | down-regulated |
| Bpev01.c0243.g0056.m0001 | down-regulated |
| Bpev01.c0264.g0036.m0001 | down-regulated |
| Bpev01.c1040.g0049.m0001 | down-regulated |
| Bpev01.c0190.g0044.m0001 | down-regulated |
| Bpev01.c0621.g0012.m0001 | up-regulated |
| Bpev01.c0984.g0005.m0001 | up-regulated |
| Bpev01.c0401.g0006.m0001 | up-regulated |
| Bpev01.c0894.g0003.m0001 | up-regulated |
| Bpev01.c1891.g0003.m0001 | up-regulated |

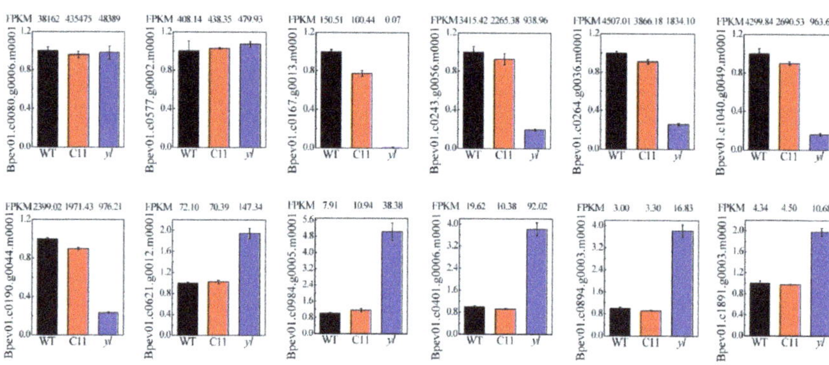

**Figure 8.** Quantificational real-time PCR verification of RNA-seq. Error bars represent the SD of three measurements.

## 4. Discussion

In this study, we reported a Chl-deficient mutant *yl* that produced yellow-green leaves in birch. The mutant was isolated from *BpCCR1* transgenic lines in birch plant breeding. Physiological analysis and gene expression characterization of *yl* were performed to investigate the difference between *yl* and control plants (WT and C11).

Chl and Car are the main pigments that trap light energy in leaf tissue. It has been demonstrated that the leaves of Chl-deficient mutants always contain less Chl and Car. Some mutants also showed a change in the ratio of Chl a/b [16,27,28]. Consistent with the previous research, pigment analysis of *yl* showed that Chl a, Chl b, and Car were reduced in young leaf or mature leaf specimens, compared to WT and C11. We also found that the ratio of Chl a/b was increased in the *yl* plants (Figure 1). Chl b is thought to be essential to the stability of the light-harvesting Chl a/b protein complex [29]. Thus, the decreased Chl content and increased Chl a/b ratio in *yl* indicated that there might be fewer light-harvesting antenna complexes than in the WT and C11 controls. There are numerous Chl-deficient mutants that have shown reduced amounts of light-harvesting proteins (LHC) in the thylakoid membranes of the chloroplast [30,31]. Andersson reported that the absence of Lhcb1 and

Lhcb2 proteins in Arabidopsis showed reduced Chl levels and increased Chl a/b [32]. According to transcriptomic analysis, DEGs of *yl* compared with WT and C11 were both significantly enriched in photosynthesis-antenna proteins (Figures 6 and 7). These results revealed that the expression change of genes involved in photosynthesis-antenna proteins plays an important role in the formation of the yellow-green leaf phenotype in *yl*.

Green plants absorb light energy to convert $CO_2$ and water into carbohydrates and oxygen through photosynthesis. Photosynthesis is the key process that provides energy for catabolic processes and growth in plants. It has been reported that photosynthesis could be influenced by many environmental factors, such as intensity, spectrum and duration of illumination, mechanical wounding, and heating [33,34]. These stimulations would induce the generation and propagation of variation potential [35]. Then, electron flow connected to pH would be changed and the light-harvesting complex would transfer to photosystem II (PSII). As a result, photosynthesis would be decreased in plants [36]. Photosynthesis is a complex process. The changes in pigment contents, stomata conductance, and gene expressions may also affect photosynthesis [37]. Due to the reduced pigment contents, most chlorina mutants have a poorer photosynthetic performance than that observed in wild type specimens [9,38]. However, this is not true for all chlorine mutants. For example, a chlorina rice mutant *Huangyu B* was found to have a higher photosynthetic efficiency than its wild type [39], and the photosynthetic rate of a Chl-deficient mutant *siygl1* in foxtail millet was even higher than that of Yugu1 plants during the reproductive growth stage [15]. In this study, *yl* showed a reduced pigment, including Chl a, Chl b, and Car contents (Figure 1). The photosynthetic rate, stomata conductance, and transpiration rate of the *yl* line were lower than WT and C11 under all light intensities set (Figure 3). Additionally, the measurement of photosynthesis in WT, C11, and *yl* was performed under the same conditions. We also found that there was a significant positive correlation between stomata conductance, pigment contents, and photosynthetic rate. RNA-seq results showed that many genes related to photosynthesis-antenna proteins were down-regulated in *yl* compared to WT and C11 (Figure 7). As a result, the energy absorbed, trapped, and transferred in photosystem I (PSI) and PSII would probably be affected. Taken together, the decreased photosynthesis in *yl* was probably mainly due to the low expression level of antenna protein genes, and reduced photosynthetic pigment contents. The difference in Pn may underlie the retarded growth in one-year-old *yl* mutant plants (Figures 1 and 3).

The molecular mechanism of leaf color mutation is complex. Mutation of genes related to chloroplast development, blocked in photosynthetic pigment biosynthesis, blocked in chloroplast protein transport, and blocked in phytochrome regulation would lead to the formation of a yellow-green mutant [26–31]. Studies have shown that plentiful genes are related to the yellow-green leaf phenotype in plants. Examples are cytokinin-responsive gata transcription factor1, *Cga1* [40]; chaperone protein ClpC, *ClpC1* [41]; signal recognition particle 43 kDa protein, *cpSRP43* [42]; chloroplast Signal recognition particle subunit, *cpSRP54* [43]; metallo-beta-lactamase, *GRY79* [44]; nuclear transcription factor Y, *HAP3A* [45]; NADPH-dependent thioredoxin reductase C, *NTRC* [46]; protein stay-green, *SGR* [47]; and YbeY endoribonuclease, *YbeY* [38]. However, the expression of all these genes did not change in the *yl* transcriptome (Table 4).

**Table 4.** The expression of well-known genes related to leaf color in the transcriptional level of a *yl* mutant.

| Gene | ID | Annotation | C11 and WT vs. *yl* | Mutant phenotype |
|---|---|---|---|---|
| Cga1 | Bpev01.c0051.g0036.m0001 | Cytokinin-responsive gata transcription factor1 | normal | yellow leaf |
| ClpC1 | Bpev01.c1202.g0038.m0001 | Chaperone protein ClpC | normal | chlorotic leaf |
| cpSRP43 | Bpev01.c1171.g0012.m0001 | Signal recognition particle 43 kDa protein | normal | yellow-green leaf |
| cpSRP54 | Bpev01.c1238.g0002.m0001 | Chloroplast signal recognition particle subunit | normal | yellow-green leaf |
| GRY79 | Bpev01.c0787.g0006.m0001 | Metallo-beta-lactamase | normal | yellow turn green |
| HAP3A | Bpev01.c0288.g0025.m0001 | Nuclear transcription factor Y | normal | pale-green leaf |
| NTRC | Bpev01.c0029.g0129.m0001 | NADPH-dependent thioredoxin reductase C | normal | light-green leaf |
| SGR | Bpev01.c0717.g0015.m0001 | Protein stay-green | normal | stay-green |
| YbeY | Bpev01.c1489.g0003.m0001 | YbeY endoribonuclease | normal | pale-green leaf |

Plastid-to-nucleus retrograde signaling is considered to coordinate nuclear gene expression. Nott has summarized three independent retrograde signaling pathways from previous studies, including signals generated by Mg-Protoporphyrin IX (Mg-Proto IX), chloroplast gene expression, and the redox state of photosynthetic electron transport components [48]. One possibly important function of retrograde signaling is to regulate the biosynthesis of Chl with the expression of genes for nuclear-encoded Chl-binding proteins, such as the Lhca and Lhcb proteins. Here, we found that the transcription of Chl-binding protein genes (*Lhc* gene family) was down-regulated via RNA-seq and qRT-PCR analysis (Figures 7 and 8). Members of the *Golden2-like* (*GLK*) gene family have been reported to regulate chloroplast development in diverse plant species [49,50]. *GLK* genes are sensitive to retrograde signaling from the chloroplast, and they could then operate downstream of genes for plastid retrograde signaling [51]. The expression of *GLK* (Bp023762) was only 0.7% of WT and 0.9% of C11 (Figure 8). In addition, the *PSRP1* gene (BP012524), an encoding ribosomal-binding factor (plastid-specific ribosomal protein 1) that inhibits plastid translation by blocking tRNA-binding sites on ribosomes, was upregulated in the *yl* mutant. These results suggest that plastid-to-nucleus retrograde signaling triggered in *yl* may regulate nuclear gene expression.

## 5. Conclusions

In this study, about 1163 DEGs and 930 DEGs were obtained in *yl* compared with WT and C11, respectively. The DEGs related to the photosynthesis antenna proteins pathway were significantly enriched. In addition, the physiological characteristics analysis showed that the yellow-green leaf mutant *yl* had reduced amounts of Chl, an increased Chl a/b value, and reduced leaf anatomical compared to control plants. Based on these results, we can conclude that the expression of genes involved in the photosynthesis antenna proteins pathway might be responsible for the lower pigment contents and ratio of Chl a/b, resulting in yellow-green leaves in *yl*.

**Supplementary Materials:** The following are available online at http://www.mdpi.com/1999-4907/10/2/120/s1, Table S1: The correlation analysis between pigment contents and Pn.

**Author Contributions:** J.J. and S.C. designed the experiments; H.G. performed the experiments; S.C. analyzed the data; H.G. wrote the manuscript; J.J., G.L., and S.C. revised the manuscript.

**Funding:** This research was funded by the National Natural Science Foundation of China (No. 31570647) and the 111 Project (B16010).

**Acknowledgments:** We are grateful to the transgenic material *yl* generated by Rui Wei and Wenbo Zhang.

**Conflicts of Interest:** The authors declare no conflict of interest.

## References

1. Zyryanova, O.A.; Terazawa, M.; Koike, T.; Zyryanov, V.I. White birch trees as resource species of Russia: Their distribution, ecophysiological features, multiple utilizations. *Eurasian J. For. Res.* **2010**, *13*, 25–40.
2. Dulamsuren, C.; Hauck, M.; Bader, M.K.-F.; Osokhjargal, D.; Oyungerel, S.; Nyambayar, S.; Runge, M.; Leuschner, C. Water relations and photosynthetic performance in Larix sibirica growing in the forest-steppe ecotone of northern Mongolia. *Tree Physiol.* **2008**, *29*, 99–110. [CrossRef]
3. Bekhta, P.; Hiziroglu, S.; Shepelyuk, O. Properties of plywood manufactured from compressed veneer as building material. *Mater. Des.* **2009**, *30*, 947–953. [CrossRef]
4. Liu, X.; Wang, Q.; Chen, P.; Song, F.; Guan, M.; Jin, L.; Wang, Y.; Yang, C. Four Novel Cellulose Synthase (CESA) Genes from Birch (Betula platyphylla Suk.) Involved in Primary and Secondary Cell Wall Biosynthesis. *IJMS* **2012**, *13*, 12195–12212. [CrossRef] [PubMed]
5. Borrega, M.; Tolonen, L.K.; Bardot, F.; Testova, L.; Sixta, H. Potential of hot water extraction of birch wood to produce high-purity dissolving pulp after alkaline pulping. *Bioresour. Technol.* **2013**, *135*, 665–671. [CrossRef] [PubMed]
6. Rubert-Nason, K.F.; Holeski, L.M.; Couture, J.J.; Gusse, A.; Undersander, D.J.; Lindroth, R.L. Rapid phytochemical analysis of birch (Betula) and poplar (Populus) foliage by near-infrared reflectance spectroscopy. *Anal. Bioanal. Chem.* **2012**, *405*, 1333–1344. [CrossRef] [PubMed]

7. Lin, L.; Mu, H.; Jiang, J.; Liu, G. Transcriptomic analysis of purple leaf determination in birch. *Gene* **2013**, *526*, 251–258. [CrossRef] [PubMed]
8. Zhang, W.B.; Wei, R.; Chen, S.; Jiang, J.; Li, H.Y.; Huang, H.J.; Yang, G.; Wang, S.; Wei, H.R.; Liu, G.F. Functional characterization of CCR in birch (Betula platyphylla x Betula pendula) through overexpression and suppression analysis. *Physiol. Plant.* **2015**, *154*, 283–296. [CrossRef]
9. Córdoba, J.; Molina-Cano, J.-L.; Martínez-Carrasco, R.; Morcuende, R.; Pérez, P. Functional and transcriptional characterization of a barley mutant with impaired photosynthesis. *Plant Sci.* **2016**, *244*, 19–30. [CrossRef]
10. Liu, J.; Wang, J.; Yao, X.; Dong, X.; Chen, W.; Li, Z.K. Fine mapping and photosynthetic characteristics of the lower chlorophyllb1 mutant in rice (Oryza sativa L.). *Plant Breed.* **2015**, *134*, 661–667. [CrossRef]
11. Zhong, X.M.; Sun, S.F.; Li, F.H.; Wang, J.; Shi, Z.S. Photosynthesis of a yellow-green mutant line in maize. *Photosynthetica* **2015**, *53*, 499–505. [CrossRef]
12. Li, N.; Jia, J.; Xia, C.; Liu, X.; Kong, X. Characterization and mapping of novel chlorophyll deficient mutant genes in durum wheat. *Breed. Sci.* **2013**, *63*, 169–175. [CrossRef] [PubMed]
13. Karaca, M.; Saha, S.; Callahan, F.E.; Jenkins, J.N.; Read, J.J.; Percy, R.G. Molecular and cytological characterization of a cytoplasmic-specific mutant in pima cotton (Gossypium barbadense L.). *Euphytica* **2004**, *139*, 187–197. [CrossRef]
14. Wang, L.; Yue, C.; Cao, H.; Zhou, Y.; Zeng, J.; Yang, Y.; Wang, X. Biochemical and transcriptome analyses of a novel chlorophyll-deficient chlorina tea plant cultivar. *BMC Plant Biol.* **2014**, *14*, 239. [CrossRef]
15. Li, W.; Tang, S.; Zhang, S.; Shan, J.; Tang, C.; Chen, Q.; Jia, G.; Han, Y.; Zhi, H.; Diao, X. Gene mapping and functional analysis of the novel leaf color gene SiYGL1 in foxtail millet [Setaria italica (L.) P. Beauv]. *Physiol. Plant.* **2016**, *157*, 24–37. [CrossRef] [PubMed]
16. Zhu, X.; Guo, S.; Wang, Z.; Du, Q.; Xing, Y.; Zhang, T.; Shen, W.; Sang, X.; Ling, Y.; He, G. Map-based cloning and functional analysis of YGL8, which controls leaf colour in rice (Oryza sativa). *BMC Plant Biol.* **2016**, *16*, 134. [CrossRef] [PubMed]
17. Cao, J.; Schneeberger, K.; Ossowski, S.; Günther, T.; Bender, S.; Fitz, J.; Koenig, D.; Lanz, C.; Stegle, O.; Lippert, C.; et al. Whole-genome sequencing of multiple Arabidopsis thaliana populations. *Nat. Genet.* **2011**, *43*, 956–963. [CrossRef]
18. Cai, Q.; Li, B.; Lin, F.; Huang, P.; Guo, W.; Zheng, Y. De novo sequencing and sssembly analysis of transcriptome in *Pinus bungeana* Zucc. ex Endl. *Forests* **2018**, *9*, 156. [CrossRef]
19. Du, M.; Ding, G.; Cai, Q. The Transcriptomic Responses of Pinus massoniana to Drought Stress. *Forests* **2018**, *9*, 326. [CrossRef]
20. Li, Y.; Zhang, Z.; Wang, P.; Wang, S.; Ma, L.; Li, L.; Yang, R.; Ma, Y.; Wang, Q. Comprehensive transcriptome analysis discovers novel candidate genes related to leaf color in a Lagerstroemia indica yellow leaf mutant. *Genes Genom.* **2015**, *37*, 851–863. [CrossRef]
21. Lichtenthaler, H.K.; Wellburn, A.R. Determinations of total carotenoids and chlorophylls a and b of leaf extracts in different solvents. *Biochem. Soc. Trans.* **1983**, *11*, 591–592. [CrossRef]
22. Chang, S.; Puryear, J.; Cairney, J. A simple and efficient method for isolating RNA from pine trees. *Plant Mol. Biol. Report.* **1993**, *11*, 113–116. [CrossRef]
23. Salojärvi, J.; Smolander, O.-P.; Nieminen, K.; Rajaraman, S.; Safronov, O.; Safdari, P.; Lamminmäki, A.; Immanen, J.; Lan, T.; Tanskanen, J.; et al. Genome sequencing and population genomic analyses provide insights into the adaptive landscape of silver birch. *Nat. Genet.* **2017**, *49*, 904–912. [CrossRef] [PubMed]
24. Dobin, A.; Gingeras, T.R. Comment on "TopHat2: Accurate alignment of transcriptomes in the presence of insertions, deletions and gene fusions" by Kim et al. *Bioinformatics* **2013**, *14*, 000851.
25. Livak, K.J.; Schmittgen, T.D. Analysis of Relative Gene Expression Data Using Real-Time Quantitative PCR and the $2-\Delta\Delta CT$ Method. *Methods* **2001**, *25*, 402–408. [CrossRef] [PubMed]
26. Huang, H.; Wang, S.; Jiang, J.; Liu, G.; Li, H.; Chen, S.; Xu, H. Overexpression of BpAP1 induces early flowering and produces dwarfism in Betula platyphylla x Betula pendula. *Physiol. Plant.* **2014**, *151*, 495–506. [CrossRef] [PubMed]
27. Ma, X.; Sun, X.; Li, C.; Huan, R.; Sun, C.; Wang, Y.; Xiao, F.; Wang, Q.; Chen, P.; Ma, F.; et al. Map-based cloning and characterization of the novel yellow-green leaf gene ys83 in rice (Oryza sativa). *Plant Physiol. Biochem.* **2017**, *111*, 1–9. [CrossRef]

28. Dong, H.; Fei, G.-L.; Wu, C.-Y.; Wu, F.-Q.; Sun, Y.-Y.; Chen, M.-J.; Ren, Y.-L.; Zhou, K.-N.; Cheng, Z.-J.; Wang, J.-L.; et al. A rice virescent-yellow Leaf mutant reveals new insights into the role and sssembly of plastid caseinolytic protease in higher plants. *Plant Physiol.* **2013**, *162*, 1867–1880. [CrossRef]
29. Tanaka, R.; Koshino-Kimura, Y.; Sawa, S.; Ishiguro, S.; Okada, K.; Tanaka, A. Overexpression of chlorophyllide an oxygenase (CAO) enlarges the antenna size of photosystem II in Arabidopsis thaliana. *Plant J.* **2002**, *26*, 365–373. [CrossRef]
30. Highkin, H.R.; Boardman, N.K.; Goodchild, D.J. Photosynthetic Studies on a Pea-mutant Deficient in Chlorophyll. *Plant Physiol.* **1969**, *44*, 1310–1320. [CrossRef]
31. Polle, J.E.W.; Benemann, J.R.; Tanaka, A.; Melis, A. Photosynthetic apparatus organization and function in the wild type and a chlorophyll b -less mutant of Chlamydomonas reinhardtii. Dependence on carbon source. *Planta* **2000**, *211*, 335–344. [CrossRef] [PubMed]
32. Andersson, J.; Wentworth, M.; Walters, R.G.; Howard, C.A.; Ruban, A.; Horton, P.; Jansson, S. Absence of the Lhcb1 and Lhcb2 proteins of the light-harvesting complex of photosystem II—Effects on photosynthesis, grana stacking and fitness. *Plant J.* **2003**, *35*, 350–361. [PubMed]
33. Maxwell, K.; Johnson, G.N. Chlorophyll fluorescence—A practical guide. *J. Exp. Bot.* **2000**, *51*, 659–668. [CrossRef] [PubMed]
34. Müller, P.; Li, X.P.; Niyogi, K.K. Non-photochemical quenching. A response to excess light energy. *Plant Physiol.* **2001**, *125*, 1558–1566. [CrossRef] [PubMed]
35. Sukhov, V.; Surova, L.; Sherstneva, O.; Katicheva, L.; Vodeneev, V. Variation potential influence on photosynthetic cyclic electron flow in pea. *Front. Plant Sci.* **2015**, *5*, 766. [CrossRef] [PubMed]
36. Sukhova, E.; Mudrilov, M.; Vodeneev, V.; Sukhov, V. Influence of the variation potential on photosynthetic flows of light energy and electrons in pea. *Photosynth. Res.* **2017**, *136*, 215–228. [CrossRef]
37. Cheng, J.; Fan, P.; Liang, Z.; Wang, Y.; Niu, N.; Li, W.; Li, S. Accumulation of End Products in Source Leaves Affects Photosynthetic Rate in Peach via Alteration of Stomatal Conductance and Photosynthetic Efficiency. *J. Amer. Soc. Hort. Sci.* **2009**, *134*, 667–676. [CrossRef]
38. Liu, J.; Zhou, W.; Liu, G.; Yang, C.; Sun, Y.; Wu, W.; Cao, S.; Wang, C.; Hai, G.; Wang, Z.; et al. The Conserved Endoribonuclease YbeY Is Required for Chloroplast Ribosomal RNA Processing in Arabidopsis. *Plant Physiol.* **2015**, *168*, 205–221. [CrossRef]
39. Zhou, X.-S.; Wu, D.-X.; Shen, S.-Q.; Sun, J.-W.; Shu, Q.-Y. High photosynthetic efficiency of a rice (Oryza sativa L.) xantha mutant. *Photosynthetica* **2006**, *44*, 316–319. [CrossRef]
40. Hudson, D.; Guevara, D.R.; Hand, A.J.; Xu, Z.; Hao, L.; Chen, X.; Zhu, T.; Bi, Y.-M.; Rothstein, S.J. Rice Cytokinin GATA Transcription Factor1 Regulates Chloroplast Development and Plant Architecture. *Plant Physiol.* **2013**, *162*, 132–144. [CrossRef]
41. Sjögren, L.L.; Macdonald, T.M.; Sutinen, S.; Clarke, A.K. Inactivation of the clpC1 Gene Encoding a Chloroplast Hsp100 Molecular Chaperone Causes Growth Retardation, Leaf Chlorosis, Lower Photosynthetic Activity, and a Specific Reduction in Photosystem Content. *Plant Physiol.* **2004**, *136*, 4114–4126. [CrossRef] [PubMed]
42. Lv, X.-G.; Shi, Y.-F.; Xu, X.; Wei, Y.-L.; Wang, H.-M.; Zhang, X.-B.; Wu, J.-L. Oryza sativa Chloroplast Signal Recognition Particle 43 (OscpSRP43) Is Required for Chloroplast Development and Photosynthesis. *PLoS ONE* **2015**, *10*, e0143249. [CrossRef] [PubMed]
43. Yu, B.; Gruber, M.Y.; Khachatourians, G.G.; Zhou, R.; Epp, D.J.; Hegedus, D.D.; Parkin, I.A.P.; Welsch, R.; Hannoufa, A. Arabidopsis cpSRP54 regulates carotenoid accumulation in Arabidopsis and Brassica napus. *J. Exp. Bot.* **2012**, *63*, 5189–5202. [CrossRef] [PubMed]
44. Wan, C.M.; Li, C.M.; Ma, X.Z.; Wang, Y.; Sun, C.H.; Huang, R.; Zhong, P.; Gao, Z.Y.; Chen, D.; Xu, Z.J.; et al. GRY79 encoding a putative metallo-beta-lactamase-trihelix chimera is involved in chloroplast development at early seedling stage of rice. *Plant Cell Rep.* **2015**, *34*, 1353–1363. [CrossRef] [PubMed]
45. Miyoshi, K.; Ito, Y.; Serizawa, A.; Kurata, N. OsHAP3genes regulate chloroplast biogenesis in rice. *Plant J.* **2003**, *36*, 532–540. [CrossRef] [PubMed]
46. Perez-Ruiz, J.M. Rice NTRC Is a High-Efficiency Redox System for Chloroplast Protection against Oxidative Damage. *Plant Cell* **2006**, *18*, 2356–2368. [CrossRef] [PubMed]
47. Jiang, H.; Li, M.; Liang, N.; Yan, H.; Wei, Y.; Xu, X.; Liu, J.; Xu, Z.; Chen, F.; Wu, G.; et al. Molecular cloning and function analysis of the stay green gene in rice. *Plant J.* **2007**, *52*, 197–209. [CrossRef]

48. Nott, A.; Jung, H.-S.; Koussevitzky, S.; Chory, J. Plastid-to-nucleus retrograde signaling. *Annu. Rev. Plant Biol.* **2006**, *57*, 739–759. [CrossRef]
49. Rossini, L.; Cribb, L.; Martin, D.J.; Langdale, J.A. The Maize Golden2 Gene Defines a Novel Class of Transcriptional Regulators in Plants. *Plant Cell* **2001**, *13*, 1231. [CrossRef]
50. Yasumura, Y.; Moylan, E.C.; Langdale, J.A. A Conserved Transcription Factor Mediates Nuclear Control of Organelle Biogenesis in Anciently Diverged Land Plants. *Plant Cell* **2005**, *17*, 1894–1907. [CrossRef]
51. Wang, P.; Korkaric, M.; Waters, M.T.; Capper, R.G.; Saunders, N.J.; Langdale, J.A. GLK transcription factors coordinate expression of the photosynthetic apparatus in Arabidopsis. *Plant Cell* **2009**, *21*, 1109–1128.

*forests*

MDPI

*Article*

# Transcriptome Analysis of Bael (*Aegle marmelos* (L.) Corr.) a Member of Family Rutaceae

**Prashant Kaushik [1] and Shashi Kumar [2,*]**

1 Instituto de Conservación y Mejora de la Agrodiversidad Valenciana, Universitat Politècnica de València, 46022 Valencia, Spain; prakau@doctor.upv.es

2 International Center for Genetic Engineering and Biotechnology, Aruna Asaf Ali Marg, New Delhi 110 067, India

* Correspondence: skrhode@icgeb.res.in; Tel.: +91-11-26741358

Received: 13 June 2018; Accepted: 10 July 2018; Published: 26 July 2018

**Abstract:** *Aegle marmelos* (L.) Corr. is a medicinally and horticulturally important tree member of the family Rutaceae. It is native to India, where it is also known as Bael. Despite its importance, the genomic resources of this plant are scarce. This study presented the first-ever report of expressed transcripts in the leaves of *Aegle marmelos*. A total of 133,616 contigs were assembled to 46,335 unigenes with minimum and maximum lengths of 201 bp and 14,853 bp, respectively. There were 7002 transcription factors and 94,479 simple sequence repeat (SSR) markers. The *A. marmelos* transcripts were also annotated based on information from other members of Rutaceae; namely *Citrus clementina* and *Citrus sinensis*. A total of 482 transcripts were annotated as cytochrome p450s (CYPs), and 314 transcripts were annotated as glucosyltransferases (GTs). In the *A. marmelos* leaves, the monoterpenoid biosynthesis pathway was predominant. This study provides an important genomic resource along with useful information about *A. marmelos*.

**Keywords:** *Aegle marmelos* (L.) Corr.; transcripts; transcriptome assembly; simple sequence repeats; transcription factors; cytochrome p450; glycotransferases; metabolic pathway

---

## 1. Introduction

*Aegle marmelos* (L.) Corr. (2 n = 18) or Bael is an underexploited member of family Rutaceae. Believed to be native to the Indian subcontinent, it is well distributed throughout the tropical and subtropical belts of southeast Asia [1,2]. Botanically, *A. marmelos* is a deciduous tree stretching up to 10 m in height that flowers during the months of May–June [3,4]. It is also commonly grown as a horticultural plant in India, and its fruits are processed as juice or candies, as well as eaten fresh. During the past few decades, a spike in its cultivation as a horticulture plant has been attributed to its medicinal properties, along with a hardy nature that allows it to be cultivated on marginal lands with acidic or alkaline soils [5,6].

The traditional medicine system of Ayurveda in India routinely uses every part of *A. marmelos* as a therapy for medical conditions [7,8]. The leaves are most easily accessible, and are therefore more regularly used for the treatments than any other plant part. *A. maremelos* leaves are used to treat jaundice and help in wound-healing when applied as a paste on a wound surface [9]. Moreover, *A. marmelos* leaf extracts have been proved to be a better cure for gastrointestinal and hematopoietic damage than its fruits [10]. The leaf extract of *A. marmelos* is used as a medication against a number of chronic diseases such as diabetes, pancreatic cancer, and arthritis [11–14]. All of these medicinal properties of *A. marmelos* leaves are attributed to various phytochemicals present in the leaves such as aeglin, rutin, γ-sitosterole, β-sitosterol, eugenol, marmesinin, glycoside, skimmianine, etc. Broadly, these phytochemicals can be divided into three main classes: alkaloids, phenylpropanoids, and terpenoids [15,16]. However, there is no genomic data-based information about the pathways of these important metabolic compounds that are present in the *A. marmelos* leaf. The information regarding the biosynthetic pathways and the encoding enzymes present in the *A. marmelos* leaf will be highly useful for the functional genomics in *A. marmelos* via transgenics and metabolic engineering approaches. Furthermore, *A. marmelos* leaf extract is used for the green synthesis of gold and silver nanoparticles [17,18].

The sum total of all of the transcripts captured in the cell of an individual organism is called its transcriptome [19]. There are two ways to capture the expressed transcripts: either by microarray, which is limited to predefined sequences, or by performing RNA-Seq using second-generation sequencing technologies [20]. This kind of sequencing has revolutionized the understanding of non-model organisms, and has evolved as one of the first choices of methods to apply to gene discovery and the expression profiling of non-model organisms [21,22]. The availability of well-defined computational tools, along with a well-applied methodology, has further demonstrated the effectiveness of de novo transcriptome assemblies in organisms even without a reference genome [23,24].

Genomic resources in A. marmelos are scarce compared with other members of Rutaceae, such as *Citrus sinensis* (Sweet Orange) and *Citrus clementina* (Clementine), both having well-annotated genomes [25,26]. Moreover, the unavailability of molecular markers based on the genomic information has further decelerated the molecular breeding efforts in *A. marmelos*. Earlier, a diversity study was carried out using only 12 random amplification of polymorphic DNA (RAPDs) [1]. This limitation can be overcome by developing an appropriate resource of genomic information-based molecular markers using a next-generation sequencing (NGS)-based approach such as transcriptomics [20,22]. To the best of our knowledge, this is the first detailed report on the transcriptome of this medically important plant. Moreover, only six expressed sequence tag (ESTs) are available in the National Center for Biotechnology Information (NCBI) database (accessed on 25 May 2018) [27]. An investigation into the leaf transcriptome of *A. marmelos* can help answer key questions regarding various aspects related to genes and their gene function, via the pathways involved in the metabolic compound formation. Therefore, we used RNA sequencing followed by the de novo transcriptome assembly of *A. marmelos* leaves to identify the transcription factors, simple sequence repeats (SSRs), and transcripts related to important metabolic pathways in the leaves of *A. marmelos*. Also, the information regarding cytochrome P450s (CYPs) and glucosyltransferases (GTs) extant in the leaf of *A. marmelos* was also accomplished.

## 2. Materials and Methods

### 2.1. RNA Isolation and Sequencing

Young and tender leaves from three mature and healthy plants of *A. marmelos* variety "Kaghzi" (~five years old) were collected from the Government Garden Nursery (coordinates at 29°58′06.9″ N 76°52′50.8″ E) in Haryana, India. The sampled leaf tissues were stored in RNAlater (Life Technologies, Carlsbad, CA, USA) till further use. RNA was extracted with a TRIZOL reagent (Life Technologies Corporation, Carlsbad, CA, USA) based RNA extraction protocol for plant leaves [28,29]. The quality of the extracted RNA was checked on a 1% formaldehyde denaturing agarose gel, and further quantified using a Nanodrop ND-1000 spectrophotometer (Nanodrop Technologies, Montchanin, DE, USA). A pooled sample of RNA from three selected plants was used for a single cDNA library preparation. The library was prepared with a TruSeq RNA Library Prep Kit v2 from Illumina® (Illumina Inc., San Diego, CA, USA), and the library quantification was done using a Qubit Fluorometer (Qubit™ dsDNA HS Assay Kit, Life Technologies Corporation, Carlsbad, California, USA) and Agilent D1000 ScreenTape system (Agilent Technologies, Santa Clara, CA, USA) The library was further sequenced on the Illumina HiSeq 2500 (2 × 150 bp) platform (Illumina, Dedham, MA, USA).

### 2.2. De Novo Assembly and Identification Coding Sequences

The cleaned reads were assembled using Trinity software (version 2.4.0) and TransDecoder v. 3.0.1 (http://transdecoder.sourceforge.net/) [30] was used to identify candidate coding regions within the generated transcripts and look for the open reading frames (ORF) that were at least 100 amino acids long in order to decrease the chances of false positives.

### 2.3. Gene Function Annotation

The transcripts with ORFs were annotated with BLASTX (default parameters, *e*-value cut-off $10^{-5}$) by resemblance counter to NR (non-redundant protein sequences database of NCBI), protein family (Pfam), Kyoto Encyclopedia of Genes and Genomes (KEGG) (http://www.genome.jp/kegg/) [31], *e*-value cut-off of $1 \times 10^{-5}$, and cluster of orthologous groups (COG) (https://www.ncbi.nlm.nih.gov/COG/) [32]. The gVolantes server (https://gvolante.riken.jp) [33] was used for the assessment of the completeness of transcriptome assembly via BUSCO_v3 selecting the plant ortholog set. Only transcripts pertaining to plant species were extracted and used for gene ontology. Pfam annotation was done with Hmmerscan, while Blast2Go was used for Gene Ontology (GO) annotation [34,35]. KEGG orthologies were estimated using the KEGG Automated Annotation Server (KAAS) by means of single-directional best hit method (http://www.genome.jp/kegg/kaas/) [36].

### 2.4. Identification of Transcription Factors

Transcription factors families present in the leaves of A. marmelos were identified by searching coding sequences identified by TransDecoder against the plant transcription factor database (PlnTFDB) (http://plntfdb.bio.uni-potsdam.de/v3.0/downloads.php) [37] with an e-value cut-off of $< 1 \times 10^{-10}$.

### 2.5. Identification of Simple Sequence Repeats (SSRs)

The presence of SSRs was determined by using MIcroSAtelliteidentification tool v1.0 (MISA) (http://pgrc.ipk-gatersleben.de/misa/) [38]. Briefly, the transcripts were checked 10 times for monorepeats, six times for direpeats, and five times for tri/tetra/penta/hexarepeats.

## 3. Results

### 3.1. De Novo Assembly, Gene Prediction, and Functional Annotation

RNA-Seq targeting expressed coding sequences has been used successfully in many medicinal and non-model plant species that do not have a reference genome (e.g., *Prosopis cineraria* L. [39], *Andrographis paniculata* Burm.f. [40], *Phyllanthus emblica* L. [41], *Picrorhiza kurroa* [42], and *Azadirachta indica* Royle ex Benth. [43]). Moreover, being a tree, *A. marmelos* can have a large genome size, which further restricts genome sequencing efforts [44].

The pooled RNA sample of *A. marmelos* leaves with RIN values around 8.0 generated a total of 115.92 million paired reads of high quality (Phred score > 30). Trinity assembler was used for the assembly, and after trimming of adapters, there was a total of 133,616 contigs (only from reads of 200 bp and above in length) clustered into 46,345 unigenes (Table 1). The raw data that was obtained as a result of sequencing was submitted to NCBI BioProject (PRJNA433585). The assembly completeness report from gVolante estimated that the transcriptome assembly was 90.15% complete (Figure S1). We scrutinized for an open reading frame that was at least 100 amino acids long in order to decrease the chances of false positives during open reading frames (ORF) predictions. The annotated transcripts with ORFs are listed in Table 2. A total of 90,525 transcripts were annotated to GO terms (Table S1). The transcripts related to plant species were extracted and used for gene ontology (Table 2).

**Table 1.** Assembly statistics of the leaf transcriptome.

| *A. marmelos* Assembly Statistics | |
|---|---|
| Number of assembled sequences | 133,616 |
| Total length (nt) | 225,969,847 |
| Range of coding sequence length (nt) | 201–14,853 |
| Mean sequence length (nt) | 1691 |
| Median sequence length (nt) | 1395 |
| N50 sequence length (nt) | 2544 |
| L50 sequence count | 30,077 |
| Number of sequences > 1K (nt) | 81,171 (60.7%) |
| Number of sequences > 10K (nt) | 96 (0.1%) |
| Base composition (%) | A: 29.75 |
| | T: 29.75 |
| | G: 20.36 |
| | C: 20.14 |
| Guanine-Cytosine content (%) | 40.50 |

**Table 2.** Annotation summary of *A. marmelos* leaf transcripts. COG: cluster of orthologous groups, GO: gene ontology, KEGG: Kyoto Encyclopedia of Genes and Genomes, ORFs: open reading frames, Pfam: protein family.

| Parameters | Numbers |
|---|---|
| Total Transcripts | 133,616 |
| Total ORFs | 165,230 |
| Transcript BLASTX (Plant Species) | 126,101 |
| Transcript annotated with GO terms | 90,525 |
| Transcript BLASTP | 82,445 |
| Transcript BLASTX against Pfam | 67,451 |
| Transcript annotations against KEGG | 83,221 |
| Transcript annotated with COG | 81,636 |

### 3.2. GO Annotation

In total, 600,642 Gene Ontology (GO) terms were mapped to the *A. marmelos* leaf contigs belonging to all the three possible classes, i.e., biological process (227,921 transcripts), cellular component (188,465 transcripts), and molecular function (184,25 transcripts) in the GO database (Figure 1). The breakdown of the proteins associated with the various biological process, cellular components, and molecular functions is illustrated in Figure 2. The "integral component of membrane" (GO: 0016021) associated with various cellular components, "transcription DNA-templated" (GO: 00006351) associated with biological processes and "ATP binding" (GO: 00005524) associated with molecular function were the most mapped terms in their respective categories (Figure 2.).

The GO terms primarily define three categories of functions: namely, the biological, cellular, and molecular functions for a gene product. This is achieved by associating a gene with their ontologies [45,46]. Earlier studies have pointed out a higher metabolic activity in the leaves of *A. marmelos*, which is because of the presence of phytochemicals such as alkaloids, flavonoids, and phenols [47,48]. We have identified a number of GO terms in the leaves of *A. marmelos*; this information could lead to the identification of important pathways of metabolic compounds in *A. marmelos* [49].

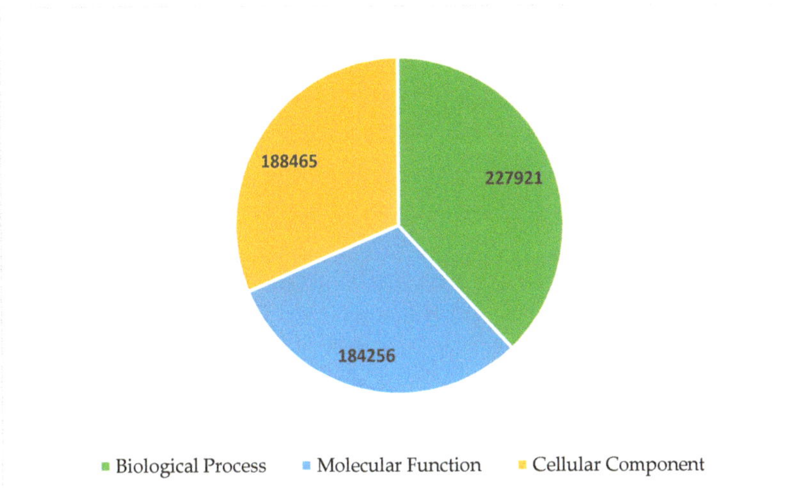


**Figure 1.** Genes associated with the biological process, cellular components, and molecular functions in the *A. marmelos* leaf transcriptome assembly.

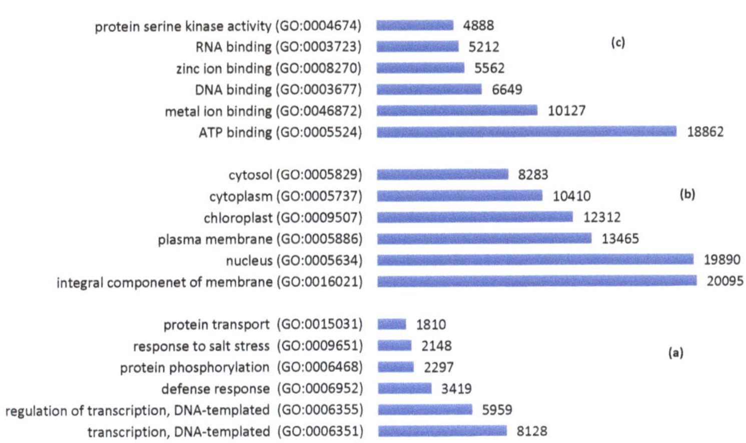


**Figure 2.** Gene Ontology (GO) classification of *A. marmelos* transcripts. GO term are divided in three main categories: biological process (**a**), cellular component (**b**), and molecular function (**c**).

### 3.3. Citrus Database Annotation

The *A. marmelos* transcripts were also annotated via Phytozome (https://phytozome.jgi.doe.gov/) with reference to the *Citrus clementina* and *Citrus sinensis* genomes. This resulted in the mapping of 78.44% of the transcripts to the *Citrus clementine*, and 79.85% to the *Citrus sinensis* genome (Table 3). An almost similar number of transcripts were annotated with GO terms and KEGG annotation, respectively (Tables S2 and S3). However, recently, an extensive amount of relatedness was observed within the members of genus Citrus of family Rutaceae, which was based on the study performed by using the whole genome sequences of 60 members in the Citrus genus; the authors even pointed out the need for reformulation of the genus [50].

**Table 3.** Annotation summary of *A. marmelos* leaf transcripts with Citrus *sinensis* and *Citrus clementine* genome.

| Genome | *Citrus clementina* | *Citrus sinensis* |
|---|---|---|
| No. of transcripts annotated | 98,917 (78.44) | 100,701 (79.85) |
| No. of transcripts annotated by Gene Ontology | 98,184 | 100,053 |
| No. of transcripts annotated by KEGG | 48,184 | 48,881 |

### 3.4. *Simple Sequence Repeats (SSRs) Prediction*

Simple sequence repeats (SSRs), or short tandem repeats or microsatellites, are short repeat motifs that show length polymorphism due to the insertion or deletion mutations of one or more repeat types [51]. We analyzed for the abundance of SSRs of annotated plant transcripts for *A. marmelos* leaf transcripts using the MISA tool, and the predicted SSRs statistics are shown in Figure 3. There were 58,354 transcripts that contained SSRs, and among these, 23,034 had more than one SSRs (Table S4). In total, 94,479 SSRs were identified, of which 65.27% were monorepeats, 19.78% were direpeats, and 13.40% were trirepeats (Figure 3). Tetra, penta and hexarepeats made up 1.01%, 0.26% and 0.24% of the total, respectively (Figure 3). However, out of a total of 94,479 identified SSRs, 11,400 (12.06%) were related to the compound formation.

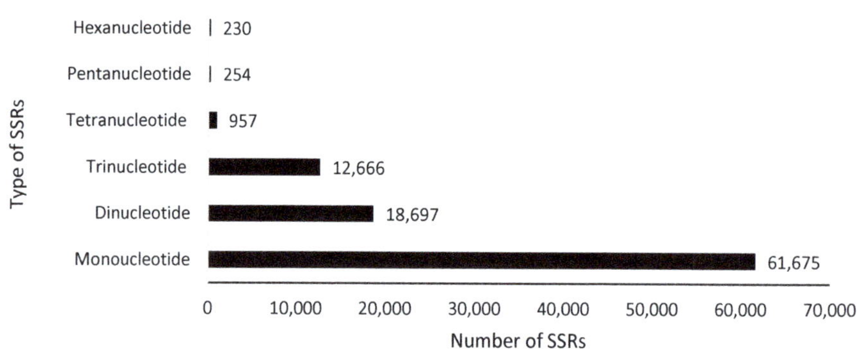


**Figure 3.** Simple sequence repeats (SSRs) classes identified in the leaf transcripts of *A. marmelos*.

SSRs are codominant markers that are well dispersed throughout plant genomes. SSRs are popularly used for marker-assisted selection, fingerprinting, diversity assessment, and quantitative trait loci (QTLs) identification [52]. Routinely, SSRs are identified in the medicinal plants via transcriptome assemblies, because they are more robust and can also be transferred among different species within the same genus. These identified SSRs can also be used for the marker-assisted breeding in *A. marmelos* i.e., to breed this tree for a particular environment or condition. Otherwise, until recently, only diversity-related studies were conducted in *A. marmelos* using universal primers, and researchers were even limited to only 12 RAPDs and 16 universal ISSRs to access diversity among their *A. marmelos* genotypes collection [1,53]. Furthermore, these genomic information-based SSRs can help to identify and differentiate between homozygous and heterozygous individuals. SSRs are also commonly used for the map-based cloning of genes; a close association between genes and their SSRs is crucial in the context of genotyping and haplotyping [51,52].

### 3.5. *Transcriptional Factors Identification*

Gene expression patterns are regulated by transcription factors that in turn determine the different biological process [54]. A total of 7002 transcription factors were retrieved from the PlnTFDB.

The 52 transcription factors were unique to the *A. marmelos* leaves; although these were out of a total 6122 that were extant above 100 in the unigenes (Table S5). The most abundant were Auxin response factors (ARFs) (717), myeloblastosis (MYB-related) (562), a basic domain/leucine zipper (bZIP) (437), and basic helix–loop–helix (bHLH) (417), whereas HB-Other (132) and CAMATA (109) were the least abundant (Figure 4).

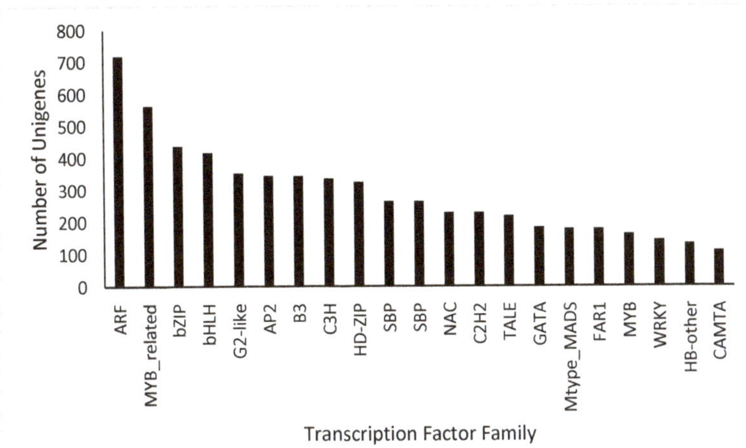


**Figure 4.** Top 21 families of transcription factors identified in the *A. marmelos* leaf.

Auxin is the plant hormone that regulates the different plant processes from growth to senescence. Auxin response factors are necessary for the plant to response to auxin stimuli; they channelize the response via auxin response DNA elements that are present in the primary auxin response genes. ARFs switch the auxin response gene on and off via their transcriptional activation domain or transcriptional repression domain [55,56]. MYB-related transcription factors play many roles like protection against biotic and abiotic stresses. MYB transcription factors also regulate the metabolism of the phenylpropanoid pathway, and are well studied with respect to the regulation of primary and secondary metabolism in the plant [57,58]. Likewise, bZIP and bHLH transcription factors are also involved in the metabolic biosynthesis in plants, especially by activation of phenylpropanoid genes [59,60].

### 3.6. Transcripts Encoding Cytochrome p450s (CYPs) and Glucosyltransferase (GTs)

CYPs help in the primary and secondary metabolism of plants by catalyzing monooxygenation reactions. These cytochromes assist in the diversification of metabolic pathways in plants. Currently, these are potential targets for metabolic engineering for the overproduction of metabolites of interest [61]. There were 477 transcripts in total that were annotated cytochrome p450s (Table S6). Considering their vital role inmetabolic pathways, we further analyzed the abundance of SSRs annotated within these cytochrome p450 transcripts (Table 4). Among the 128 identified SSRs, 85 were with monorepeats, seven were with direpeats, and 36 were with trirepeats (Table S7).

**Table 4.** Prediction of simple sequence repeats (SSRs) for the annotated transcripts with cytochrome p450s (CYPs) and glucosyltransferase (GTs).

| Description | (CYPs) | (GTs) |
|---|---|---|
| Total number of annotated transcripts encoding | 477 | 314 |
| Total size of annotated transcripts (bp) | 1,111,258 | 778,664 |
| Number of SSRs identified | 128 | 247 |
| Number of annotated transcripts containing SSR | 94 | 165 |
| Number of annotated transcripts containing more than 1 SSR | 30 | 64 |
| Number of SSRs present in the compound formation | 5 | 16 |

The last step in the production of plant secondary metabolites is glycosylation, which is carried out by glycotransferases (GTs) [62–64]. A total of 314 transcripts were annotated as glucosyltransferase (Table S8). We analyzed the abundance of SSRs that were present in these transcripts (Table S9), and among the 247 identified, 109 were monorepeats, 79 were direpeats, 58 were trirepeats, and only one was identified as a tetrarepeat (Table 4). The SSRs that were identified as using CYPs and GTs can be of immense potential for identifying genetic diversity among different *A. marmelos* accessions with divergent metabolic profiles.

### *3.7. Identification of Biosynthetic Pathways in A. Marmelos Leaf*

*A. marmelos* leaves are used for the treatment of several medical conditions in Ayurveda and Yunani medicine systems [65]. The transcripts with the highest fragments per kilobase per million mapped reads (FPKM) values were extracted from an annotation file along with the Kyoto Encyclopedia of Genes and Genomes (KEGG) ID and sorted from the RNA-Seq by Expectation Maximization (RSEM) file that was obtained from the assembly for transcript quantification. Using the KEGG ID, pathways were identified (Table S10). RSEM is commonly used to obtain information regarding transcript abundance from the RNA-Seq data of an organism, even without a reference genome [66]. The pathway analysis identified that monoterpenoid biosynthesis and thiamine pathways were the two most expressed pathways present in the *A. marmelos* leaves (Figure 5).

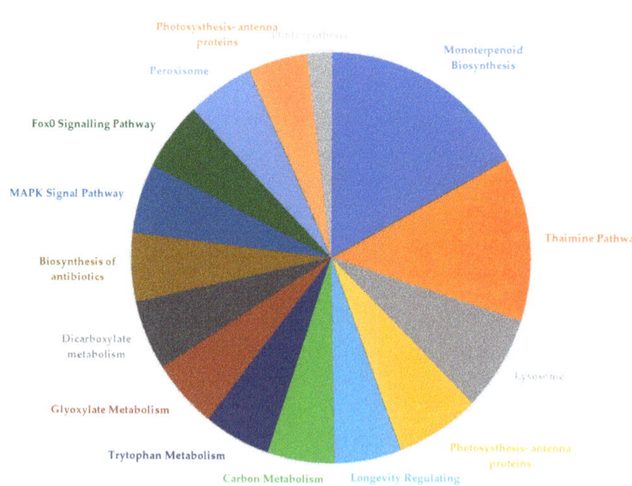


**Figure 5.** The top 15 pathways in the *A. marmelos* leaf.

*A. marmelos* leaves have been reported to contain monoterpenoids as the principal metabolites in the leaves in levels as high as 93.9% [67,68]. Moreover, this monoterpenoid content is not affected by the geographic location of the plant, as it remains unaffected by changes in altitude, unlike many other metabolites [69]. Thiamine is naturally produced in the plants as a sulfur-comprising and water-soluble compound. *A. marmelos* contains thiamine, although thiamine concentration is higher in the fruits than the leaves, and is among the fruits with the highest thiamine content [70,71]. Detailed information regarding the routes, reactions, and encoded enzymes of the two most expressed pathways identified in the leaf transcriptome of *A. marmelos* is provided in Figures 6 and 7. This detailed information generated regarding the monoterpenoid and thiamine biosynthetic pathway will be useful for further genetic analysis of the production of these important metabolites and their pathway engineering.

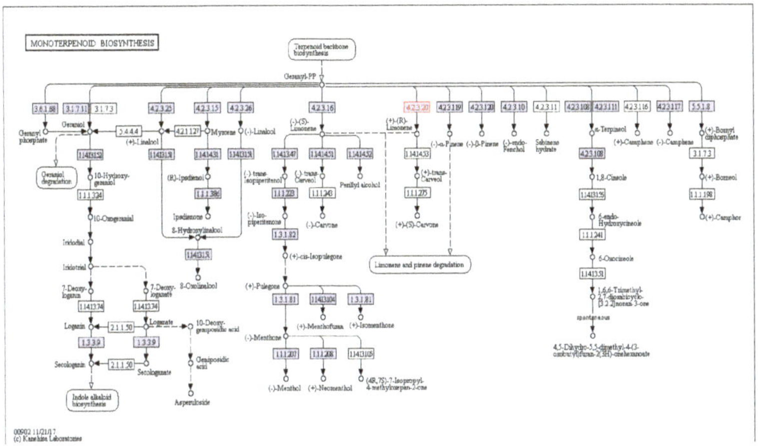


**Figure 6.** *A. marmelos* leaf transcriptome encoded enzymes (highlighted) involved in the monoterpenoid biosynthetic pathway identified in the leaf.

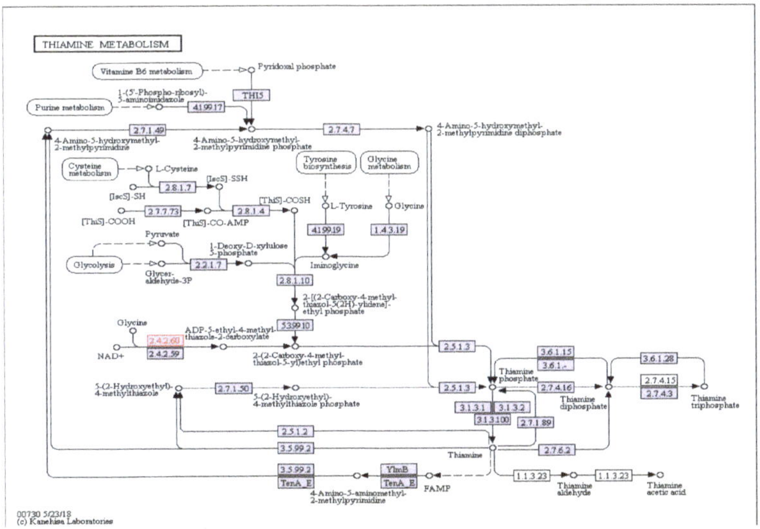


**Figure 7.** *A. marmelos* leaf transcriptome encoded enzymes (highlighted) involved in the thiamine biosynthetic pathway identified in the leaf.

## 4. Conclusions

In cases of underexploited plant species, there is often not enough genomic information available to proceed with their genetic improvement, and subsequently transfer important genes from them to cultivated crops. Transcriptome assembly is a cost-effective alternative to genome sequencing for obtaining the information of expressed genes and assisting in the more effective development of underexploited crops and medicinal plants. RNA-Seq shines a light on genes and their functions, as well as the pathways that are present, and can subsequently lead to evolutionary studies via molecular markers. We have successfully performed the first de novo transcriptome assembly of *A. marmelos*, which is a plant with religious, medicinal, and horticultural importance. It is the first-ever information about this plant, which will be of immense value for evolutionary studies and represents the development of a valuable resource for *A. marmelos*. Also, once a transcriptome reference is available, anchored-based transcriptome assemblies and different types of evolutionary studies can be performed within family Rutaceae involving genus *Aegle*.

**Supplementary Materials:** Supplementary materials can be found at http://www.mdpi.com/1999-4907/9/8/450/s1. Figure S1. Completeness assessment of *A. maremelos* leaf transcriptome assembly using gVolante server. Table S1. *A. marmelos* detailed annotation results. Table S2. Detail list of *A. maremlos* transcripts annotation with *Citrus sinensis* genome. Table S3. Detail result of *A. maremlos* transcripts annotation against *Citrus clementina* genome. Table S4. Detailed of SSRs in the *A. marmelos* leaf transcriptome. Table S5. Transcriptional factors family identified in *A. marmelos* leaf. Table S6. Detail list of cytochrome 450s (CYPs) annotation result. Table S7. SSRs identified in the cytochromes P450s transcripts. Table S8. Detail list of glycosyltransferases (GTs) annotation result. Table S9. SSRs identified in Glycosyltransferases (GTs) transcripts. Table S10. FPKM based top KEGG IDS in *A. maremlos* leaf transcriptome.

**Author Contributions:** P.K. and S.K. conceived and designed the project. P.K. performed the experiments. P.K. analyzed the data. P.K. and S.K. wrote the paper. Both authors read and approved the final manuscript.

**Funding:** This research received no external funding.

**Acknowledgments:** We thank the anonymous reviewers for their careful reading of the manuscript and providing the insightful suggestions. P.K. would like to thank Bengaluru Genomics Centre Pvt. Limited, Bengaluru, India for letting him use their facility.

**Conflicts of Interest:** The authors declare no conflict of interest.

## Abbreviations

| | |
|---|---|
| BLAST | Basic Local Alignment Search Tool |
| COG | Cluster of Orthologous Groups |
| CYP | Cytochrome p450 |
| FPKM | Fragments per kilobase of exon per million fragments mapped |
| GO | Gene Ontology |
| GT | Glucosyltransferase |
| KEGG | Kyoto Encyclopedia of Genes and Genomes |
| KO | Kyoto Encyclopedia of Genes and Genomes Orthology |
| KOG | euKaryotic Ortholog Groups |
| MISA | MIcroSAtelliteidentification |
| NCBI | National Center of Biotechnology Information |
| NR | Non-redundant |
| Nt | Nucleotide |
| ORF | Open reading frame |
| Pfam | Protein family |
| PlnTFDB | Plant Transcription Factor DataBase |
| RNA-Seq | Ribonucleic acid sequencing |
| RSEM | RNA-Seq expression estimation by expectation maximization |
| SSR | Simple Sequence Repeat |

## References


1. Nayak, D.; Singh, D.R.; Sabarinathan, P.; Singh, S.; Nayak, T. Random amplified polymorphic DNA (RAPD) markers reveal genetic diversity in bael (*Aegle marmelos* Correa) genotypes of Andaman Islands, India. *Afr. J. Biotechnol.* **2013**, *12*. [CrossRef]
2. Ajithkumar, D.; Seeni, S. Rapid clonal multiplication through in vitro axillary shoot proliferation of *Aegle marmelos* (L.) Corr., a medicinal tree. *Plant Cell Rep.* **1998**, *17*, 422–426. [CrossRef]
3. Kishore, K.; Mahanti, K.K.; Samant, D. Phenological growth stages of bael (*Aegle marmelos*) according to the extended Biologische Bundesantalt, Bundessortenamt und Chemische Industrie scale. *Ann. App. Biol.* **2017**, *170*, 425–433. [CrossRef]
4. Singhal, V.; Salwan, A.; Kumar, P.; Kaur, J. Erratum to: Phenology, pollination and breeding system of *Aegle marmelos* (Linn.) Correa (Rutaceae) from India. *New For.* **2011**, *42*, 85–100. [CrossRef]
5. Hiwale, S. *Sustainable Horticulture in Semiarid Dry Lands*; Springer: New Delhi, India, 2015; ISBN 978-81-322-2243-9.
6. Khare, C.P. *Indian Medicinal Plants: An Illustrated Dictionary*; Springer-Verlag: New York, NY, USA, 2007; ISBN 978-0-387-70637-5.
7. Bhardwaj, R.L.; Nandal, U. Nutritional and therapeutic potential of bael (*Aegle marmelos* Corr.) fruit juice: A review. *Nutr. Food Sci.* **2015**, *45*, 895–919. [CrossRef]
8. Gutiérrez, S.P.; Sánchez, M.A.Z.; González, C.P.; García, L.A. Antidiarrhoeal activity of different plants used in traditional medicine. *Afr. J. Biotechnol.* **2007**, *6*, 2988–2994. [CrossRef]
9. Arul, V.; Miyazaki, S.; Dhananjayan, R. Studies on the anti-inflammatory, antipyretic and analgesic properties of the leaves of *Aegle marmelos* Corr. *J. Ethnopharmacol.* **2005**, *96*, 159–163. [CrossRef] [PubMed]
10. Singh, P.; Dutta, S.R.; Guha, D. Gastric mucosal protection by aegle marmelos against gastric mucosal damage: Role of enterochromaffin cell and serotonin. *Saudi J. Gastroenterol.* **2015**, *21*, 35. [CrossRef] [PubMed]
11. Manandhar, B.; Paudel, K.R.; Sharma, B.; Karki, R. Phytochemical profile and pharmacological activity of *Aegle marmelos* Linn. *J. Integr. Med.* **2018**, *16*, 153–163. [CrossRef] [PubMed]
12. Baliga, M.S.; Thilakchand, K.R.; Rai, M.P.; Rao, S.; Venkatesh, P. *Aegle marmelos* (L.) Correa (Bael) and its phytochemicals in the treatment and prevention of cancer. *Integr. Cancer Ther.* **2013**, *12*, 187–196. [CrossRef] [PubMed]
13. Shankarananth, V.; Balakrishnan, N.; Suresh, D.; Sureshpandian, G.; Edwin, E.; Sheeja, E. Analgesic activity of methanol extract of *Aegle marmelos* leaves. *Fitoterapia* **2007**, *78*, 258–259. [CrossRef] [PubMed]
14. Nigam, V.; Nambiar, V.S. *Aegle Marmelos* Leaf Juice As A Complementary Therapy To Control Type 2 Diabetes-Randomised Controlled Trial In Gujarat, India (in press). *Adv. Integr. Med.* **2018**. [CrossRef]
15. Laphookhieo, S.; Phungpanya, C.; Tantapakul, C.; Techa, S.; Tha-in, S.; Narmdorkmai, W. Chemical Constituents from *Aegle marmelos*. *J. Braz. Chem. Soc.* **2011**, *22*, 176–178. [CrossRef]
16. Maity, P.; Hansda, D.; Bandyopadhyay, U.; Mishra, D.K. Biological activities of crude extracts and chemical constituents of Bael, *Aegle marmelos* (L.) Corr. *Indian J. Exp. Biol.* **2009**, *47*, 849–861. [PubMed]
17. Jha, A.K.; Prasad, K. Biosynthesis of Gold Nanoparticles Using Bael (*Aegle marmelos*) Leaf: Mythology Meets Technology. *Int. J. Green Nanotechnol. Biomed.* **2011**, *3*, 92–97. [CrossRef]
18. Rao, K.J.; Paria, S. *Aegle marmelos* Leaf Extract and Plant Surfactants Mediated Green Synthesis of Au and Ag Nanoparticles by Optimizing Process Parameters Using Taguchi Method. *ACS Sustain. Chem. Eng.* **2015**, *3*, 483–491. [CrossRef]
19. Hancock, J.M.; Zvelebil, M.J. (Eds.) *Concise Encyclopaedia of Bioinformatics and Computational Biology*, 2nd ed.; John Wiley & Sons: New York, NY, USA, 2014; Available online: https://www.wiley.com/enus/Concise/Encyclopaedia/of/Bioinformatics/and/Computational+Biology%2C+2nd+Edition-p-9780470978719 (accessed on 6 May 2018).
20. Martin, J.A.; Wang, Z. Next-generation transcriptome assembly. *Nat. Rev. Genet.* **2011**, *12*, 671–682. [CrossRef] [PubMed]
21. Papanicolaou, A.; Stierli, R.; ffrench-Constant, R.H.; Heckel, D.G. Next generation transcriptomes for next generation genomes using est2assembly. *BMC Bioinformatics* **2009**, *10*, 447. [CrossRef] [PubMed]
22. Morin, P.A.; Foote, A.D.; Hill, C.M.; Simon-Bouhet, B.; Lang, A.R.; Louis, M. SNP Discovery from Single and Multiplex Genome Assemblies of Non-model Organisms. Methods in Molecular Biology. In *Next Generation Sequencing*; Humana Press: New York, NY, USA, 2018; pp. 113–144. ISBN 978-1-4939-7512-9.

23. Grabherr, M.G.; Haas, B.J.; Yassour, M.; Levin, J.Z.; Thompson, D.A.; Amit, I.; Adiconis, X.; Fan, L.; Raychowdhury, R.; Zeng, Q.; et al. Full-length transcriptome assembly from RNA-Seq data without a reference genome. *Nat. Biotechnol.* **2011**, *29*, 644–652. [CrossRef] [PubMed]
24. Geniza, M.; Jaiswal, P. Tools for building de novo transcriptome assembly. *Curr. Plant Biol.* **2017**, *11*, 41–45. [CrossRef]
25. Xu, Q.; Chen, L.L.; Ruan, X.; Chen, D.; Zhu, A.; Chen, C.; Bertrand, D.; Jiao, W.B.; Hao, B.H.; Lyon, M.P.; et al. The draft genome of sweet orange (*Citrus sinensis*). *Nat. Genet.* **2013**, *45*, 59–66. [CrossRef] [PubMed]
26. Terol, J.; Naranjo, M.A.; Ollitrault, P.; Talon, M. Development of genomic resources for Citrus clementina: Characterization of three deep-coverage BAC libraries and analysis of 46,000 BAC end sequences. *BMC Genom.* **2008**, *9*, 423. [CrossRef] [PubMed]
27. National Center for Biotechnology Information (NCBI) Database. Available online: https://www.ncbi.nlm.nih.gov/nucest/?term=Aegle+marmelos (accessed on 25 May 2018).
28. Portillo, M.; Fenoll, C.; Escobar, C. Evaluation of different RNA extraction methods for small quantities of plant tissue: Combined effects of reagent type and homogenization procedure on RNA quality-integrity and yield. *Physiol. Plant.* **2006**, *128*, 1–7. [CrossRef]
29. Rio, D.C.; Ares, M.; Hannon, G.J.; Nilsen, T.W. Purification of RNA using TRIzol (TRI reagent). *Cold Spring Harb. Protoc.* **2010**, *2010*, pdb–prot5439. [CrossRef] [PubMed]
30. Haas, B.J.; Papanicolaou, A.; Yassour, M.; Grabherr, M.; Blood, P.D.; Bowden, J.; Couger, M.B.; Eccles, D.; Li, B.; Lieber, M.; et al. De novo transcript sequence reconstruction from RNA-seq using the Trinity platform for reference generation and analysis. *Nat. Protoc.* **2013**, *8*, 1494–1512. [CrossRef] [PubMed]
31. Kanehisa, M.; Goto, S.; Sato, Y.; Furumichi, M.; Tanabe, M. KEGG for integration and interpretation of large-scale molecular data sets. *Nucleic Acids Res.* **2011**, *40*, 109–114. [CrossRef] [PubMed]
32. Tatusov, R.L.; Galperin, M.Y.; Natale, D.A.; Koonin, E.V. The COG database: A tool for genome-scale analysis of protein functions and evolution. *Nucleic Acids Res.* **2000**, *28*, 33–36. [CrossRef] [PubMed]
33. Nishimura, O.; Hara, Y.; Kuraku, S. gVolante for standardizing completeness assessment of genome and transcriptome assemblies. *Bioinformatics* **2017**, *33*, 3635–3637. [CrossRef] [PubMed]
34. Conesa, A.; Götz, S. Blast2GO: A comprehensive suite for functional analysis in plant genomics. *Int. J. Plant Genomics* **2008**, *2008*, 619832. [CrossRef] [PubMed]
35. Finn, R.D.; Clements, J.; Eddy, S.R. HMMER web server: Interactive sequence similarity searching. *Nucleic Acids Res.* **2011**, *39*, W29–W37. [CrossRef] [PubMed]
36. Moriya, Y.; Itoh, M.; Okuda, S.; Yoshizawa, A.; Kanehisa, M. KAAS: An automatic genome annotation and pathway reconstruction server. *Nucleic Acids Res* **2007**, *35*, W182–W185. [CrossRef] [PubMed]
37. Pérez-Rodríguez, P.; Riaño-Pachón, D.; Corrêa, L.; Rensing, S.; Kersten, B.; Mueller-Roeber, B. PlnTFDB: Updated content and new features of the plant transcription factor database. *Nucleic Acids Res.* **2010**, D822–D827. [CrossRef] [PubMed]
38. Thiel, T.; Michalek, W.; Varshney, R.K.; Graner, A. Exploiting EST databases for the development and characterization of gene-derived SSR-markers in barley (Hordeum vulgare L.). *Theor. Appl. Genet.* **2013**, *106*, 411–422. [CrossRef] [PubMed]
39. Rai, M.K.; Shekhawat, J.K.; Kataria, V.; Shekhawat, N.S. De novo assembly of leaf transcriptome, functional annotation and genomic resources development in Prosopis cineraria, a multipurpose tree of Indian Thar Desert. *Plant Gene* **2017**, *12*, 88–97. [CrossRef]
40. Cherukupalli, N.; Divate, M.; Mittapelli, S.R.; Khareedu, V.R.; Vudem, D.R. De novo Assembly of Leaf Transcriptome in the Medicinal Plant Andrographis paniculata. *Front. Plant Sci.* **2016**, *7*, 1203. [CrossRef] [PubMed]
41. Kumar, A.; Kumar, S.; Bains, S.; Vaidya, V.; Singh, B.; Kaur, R.; Kaur, J.; Singh, K. De novo Transcriptome Analysis Revealed Genes Involved in Flavonoid and Vitamin C Biosynthesis in *Phyllanthus emblica* (L.). *Front. Plant Sci.* **2016**, *7*, 481. [CrossRef] [PubMed]
42. Kumar, V.; Sharma, N.; Sood, H.; Chauhan, R.S. Exogenous feeding of immediate precursors reveals synergistic effect on picroside-I biosynthesis in shoot cultures of Picrorhiza kurroa Royle ex Benth. *Sci. Rep.* **2016**, *6*, 29750. [CrossRef] [PubMed]
43. Krishnan, N.M.; Pattnaik, S.; Jain, P.; Gaur, P.; Choudhary, R.; Vaidyanathan, S.; Deepak, S.; Hariharan, A.K.; Krishna, P.B.; Nair, J.; et al. A draft of the genome and four transcriptomes of a medicinal and pesticidal angiosperm Azadirachta indica. *BMC Genom.* **2012**, *13*, 464. [CrossRef] [PubMed]

44. Schatz, M.C.; Witkowski, J.; McCombie, W.R. Current challenges in de novo plant genome sequencing and assembly. *Genome Biol.* **2012**, *13*, 243. [CrossRef] [PubMed]
45. Blake, J.A. Ten Quick Tips for Using the Gene Ontology. *PLOS Comput. Biol.* **2013**, *9*, e1003343. [CrossRef] [PubMed]
46. Rhee, S.Y.; Wood, V.; Dolinski, K.; Draghici, S. Use and misuse of the gene ontology annotations. *Nat. Rev. Genet.* **2008**, *9*, 509–515. [CrossRef] [PubMed]
47. Mujeeb, F.; Bajpai, P.; Pathak, N. Phytochemical Evaluation, Antimicrobial Activity, and Determination of Bioactive Components from Leaves of *Aegle marmelos*. *BioMed Res. Int.* **2014**, *2014*, 1–11. Available online: https://www.hindawi.com/journals/bmri/2014/497606/ (accessed on 12 May 2018). [CrossRef] [PubMed]
48. Pati, R.; Muthukumar, M. Genetic Transformation of Bael (*Aegle marmelos* Corr.). In *Biotechnology of Neglected and Underutilized Crops*; Springer: Dordrecht, The Netherlands, 2013; pp. 343–365. ISBN 978-94-007-5499-7.
49. Terol, J.; Tadeo, F.; Ventimilla, D.; Talon, M. An RNA-Seq-based reference transcriptome for Citrus. *Plant Biotechnol. J.* **2016**, *14*, 938–950. [CrossRef] [PubMed]
50. Wu, G.A.; Terol, J.; Ibanez, V.; López-García, A.; Pérez-Román, E.; Borredá, C.; Domingo, C.; Tadeo, F.R.; Carbonell-Caballero, J.; Alonso, R.; et al. Genomics of the origin and evolution of Citrus. *Nature* **2018**, *554*, 311–316. [CrossRef] [PubMed]
51. Nadeem, M.A.; Nawaz, M.A.; Shahid, M.Q.; Doğan, Y.; Comertpay, G.; Yıldız, M.; Hatipoğlu, R.; Ahmad, F.; Alsaleh, A.; Labhane, N.; et al. DNA molecular markers in plant breeding: Current status and recent advancements in genomic selection and genome editing. *Biotechnol. Biotec. Eq.* **2018**, *32*, 261–285. [CrossRef]
52. Vieira, M.L.C.; Santini, L.; Diniz, A.L.; de Freitas Munhoz, C. Microsatellite markers: What they mean and why they are so useful. *Genet. Mol. Biol.* **2016**, *39*, 312–328. [CrossRef] [PubMed]
53. Mujeeb, F.; Bajpai, P.; Pathak, N.; Verma, S.R. Genetic Diversity Analysis of Medicinally Important Horticultural Crop *Aegle marmelos* by ISSR Markers. *Methods Mol. Biol.* **2017**, *1620*, 195–211. [CrossRef] [PubMed]
54. Liu, L.; White, M.J.; MacRae, T.H. Transcription factors and their genes in higher plants. *Eur. J. Biochem.* **1999**, *262*, 247–257. [CrossRef] [PubMed]
55. Boer, D.R.; Freire-Rios, A.; van den Berg, W.A.M.; Saaki, T.; Manfield, I.W.; Kepinski, S.; López-Vidrieo, I.; Franco-Zorrilla, J.M.; de Vries, S.C.; Solano, R.; et al. Structural basis for DNA binding specificity by the auxin-dependent ARF transcription factors. *Cell* **2014**, *156*, 577–589. [CrossRef] [PubMed]
56. Finet, C.; Berne-Dedieu, A.; Scutt, C.P.; Marlétaz, F. Evolution of the ARF gene family in land plants: Old domains, new tricks. *Mol. Biol. Evol.* **2013**, *30*, 45–56. [CrossRef] [PubMed]
57. Ambawat, S.; Sharma, P.; Yadav, N.R.; Yadav, R.C. MYB transcription factor genes as regulators for plant responses: An overview. *Physiol. Mol. Biol. Plants* **2013**, *19*, 307–321. [CrossRef] [PubMed]
58. Baldoni, E.; Genga, A.; Cominelli, E. Plant MYB Transcription Factors: Their Role in Drought Response Mechanisms. *Int. J. Mol. Sci.* **2015**, *16*, 15811–15851. [CrossRef] [PubMed]
59. Hartmann, U.; Sagasser, M.; Mehrtens, F.; Stracke, R.; Weisshaar, B. Differential combinatorial interactions of cis-acting elements recognized by R2R3-MYB, BZIP, and BHLH factors control light-responsive and tissue-specific activation of phenylpropanoid biosynthesis genes. *Plant Mol. Biol.* **2005**, *57*, 155–171. [CrossRef] [PubMed]
60. Baxevanis, A.D.; Vinson, C.R. Interactions of coiled coils in transcription factors: Where is the specificity? *Curr. Opin. Genet. Dev.* **1993**, *3*, 278–285. [CrossRef]
61. Kumar, S. Engineering Cytochrome P450 Biocatalysts for Biotechnology, Medicine, and Bioremediation. *Expert Opin. Drug Metab. Toxicol.* **2010**, *6*, 115–131. [CrossRef] [PubMed]
62. Chang, C.W.T. Predictable Enzymatic Glycosylation. *Chem. Biol.* **2009**, *16*, 579–580. [CrossRef] [PubMed]
63. Keegstra, K.; Raikhel, N. Plant glycosyltransferases. *Curr. Opin. Plant Biol.* **2001**, *4*, 219–224. [CrossRef]
64. Vogt, T.; Jones, P. Glycosyltransferases in plant natural product synthesis: Characterization of a supergene family. *Trends Plant Sci.* **2000**, *5*, 380–386. [CrossRef]
65. Rishabha, M.; Ajay, K.; Anupama, S.; Gt, K. Pharmacological Screening, Ayurvedic values and Commercial Utility of *Aegle Marmelos*. *Int. J. Drug Dev. Res.* **2012**, *4*, 28–37.
66. Li, B.; Dewey, C.N. RSEM: Accurate transcript quantification from RNA-Seq data with or without a reference genome. *BMC Bioinform.* **2011**, *12*, 323. [CrossRef] [PubMed]

67. Kaur, H.P.; Garg, S.N.; Sashidhara, K.V.; Yadav, A.; Naqvi, A.A.; Khanuja, S.P.S. Chemical Composition of the Essential Oil of the Twigs and Leaves of *Aegle marmelos* (L.) Correa. *J. Essent. Oil Res.* **2006**, *18*, 288–289. [CrossRef]
68. Raju, P.M.; Agarwal, S.S.; Ali, M.; Velasco-Negueruela, A.; Pérez-Alonso, M.J. Chemical Composition of the Leaf Oil of *Aegle marmelos* (L.) Correa. *J. Essent. Oil Res.* **1999**, *11*, 311–313. [CrossRef]
69. Verma, R.S.; Padalia, R.C.; Chauhan, A. Essential oil composition of *Aegle marmelos* (L.) Correa: Chemotypic and seasonal variations. *J. Sci. Food Agric.* **2013**, *94*, 1904–1913. [CrossRef] [PubMed]
70. Tarwadi, K.; Agte, V. Antioxidant and micronutrient potential of common fruits available in the Indian subcontinent. *Int. J. Food Sci. Nutr.* **2007**, *58*, 341–349. [CrossRef] [PubMed]
71. Baliga, M.S.; Bhat, H.P.; Joseph, N.; Fazal, F. Phytochemistry and medicinal uses of the bael fruit (*Aegle marmelos* Correa): A. concise review. *Food Res. Int.* **2011**, *44*, 1768–1775. [CrossRef]

*forests*

Article

# Genetic Diversity of the Endangered *Dalbergia odorifera* Revealed by SSR Markers

**Fumei Liu [1,2,3], Zhou Hong [1], Daping Xu [1], Hongyan Jia [2], Ningnan Zhang [1], Xiaojin Liu [1], Zengjiang Yang [1] and Mengzhu Lu [3,*]**

1 State Key Laboratory of Tree Genetics and Breeding, Research Institute of Tropical Forestry, Chinese Academy of Forestry, Longdong, Guangzhou 510520, China; liufumei1115@163.com (F.L.); hzhou1981@caf.ac.cn (Z.H.); gzfsrd@163.com (D.X.); ningnanzhang@126.com (N.Z.); xjliucaf@163.com (X.L.); yzengjiang@126.com (Z.Y.)
2 The Experimental Centre of Tropical Forestry, Chinese Academy of Forestry, Pingxiang 532600, China; rlzxjhy@163.com
3 State Key Laboratory of Tree Genetics and Breeding, Research Institute of Forestry, Chinese Academy of Forestry, Beijing 100091, China
* Correspondence: lumz@caf.ac.cn; Tel.: +86-0106-288-9606

Received: 1 February 2019; Accepted: 25 February 2019; Published: 3 March 2019

**Abstract:** *Dalbergia odorifera* T. Chen (Fabaceae) is a semi-deciduous tree species indigenous to Hainan Island in China. Due to its precious heartwood "Hualimu (Chinese)" and Chinese medicinal components "Jiangxiang", *D. odorifera* is seriously threatened of long-term overexploitation and has been listed on the IUCN (International Union for Conservation of Nature's) red list since 1998. Therefore, the elucidation of its genetic diversity is imperative for conservation and breeding purposes. In this study, we evaluated the genetic diversity of 42 wild *D. odorifera* trees from seven populations covering its whole native distribution. In total, 19 SSR (simple sequence repeat) markers harbored 54 alleles across the 42 samples, and the medium genetic diversity level was inferred by Nei's gene diversity (0.36), observed (0.28) and expected heterozygosity (0.37). Among the seven wild populations, the expected heterozygosity (He) varied from 0.31 (HNQS) to 0.40 (HNCJ). The analysis of molecular variance (AMOVA) showed that only 3% genetic variation existed among populations. Moderate population differentiations among the investigated populations were indicated by pairwise $F_{st}$ (0.042–0.115). Structure analysis suggested two clusters for the 42 samples. Moreover, the seven populations were clearly distinguished into two clusters from both the principal coordinate analysis (PCoA) and neighbor-joining (NJ) analysis. Populations from Haikou city (HNHK), Baisha autonomous county (HNBS), Ledong autonomous county (HNLD), and Dongfang city (HNDF) comprised cluster I, while cluster II comprised the populations from Wenchang city and Sansha city (HNQS), Changjiang autonomous county (HNCJ), and Wuzhisan city (HNWZS). The findings of this study provide a preliminary genetic basis for the conservation, management, and restoration of this endemic species.

**Keywords:** *Dalbergia odorifera* T. Chen; genetic diversity; population structure; EST-SSR marker; microsatellite marker; rosewood; conservation

## 1. Introduction

*Dalbergia odorifera* T. Chen, formerly named *Dalbergia hainanensis* Merr. et Chun, is endemic to Hainan province, southern China. It is a semi-deciduous perennial tree species (diploid) of predominant outcrossing in the Fabaceae family and one of the most valuable timber species in China. *Dalbergia odorifera* is restricted to relatively narrow tropical geographic areas in Hainan Island at altitudes below 600 m. Obviously, the development of *D. odorifera* plantations is required to alleviate the demand for this valuable wood, but so far no breeding systems have been established. In the 1950s, it was introduced to the subtropical areas of Guangdong, Guangxi, and Fujian provinces in China [1]. Following several decades, the introduced trees now exhibit a satisfactory growth performance, and have even formed valuable heartwood at most sites [1].

The heartwood of this species, locally known as "Hualimu" or "Huanghuli" (Chinese name), takes more than 50 years to mature. It is one of the most precious fragrant rosewoods with a high value on the furniture and craft markets (especially for luxury furniture and crafts) in China. As a source of traditional Chinese medicine, it is also known as "Jiangxiang", and contains a series of chemical components, such as flavonoid [2], phenolic [3], and sesquiterpene derivatives [4–6], which play important roles in the pharmaceutical industry for treatment of cardiovascular diseases [7], cancer, diabetes [8], blood disorders, ischemia, swelling, and rheumatic pain [9,10]. Due to its high medicinal and commercial value, *D. odorifera* has been overexploited for a long time and has been listed on the IUCN (International Union for Conservation of Nature's) red list by World Conservation Monitoring Centre (WCMC) since 1998 [11]. As a result, the species became rare, only limited numbers of individuals are found in parts of their original habitat, which was highly fragmented in the remaining forests in Hainan Island [12]. Therefore, a comprehensive survey is urgently needed to obtain information on the levels and patterns of genetic variation for *D. odorifera*. Such information is imperative for establishing an effective strategy for conservation and breeding purposes.

Molecular markers are often used to elucidate genetic variation in tree species [13]. However, in *D. odorifera* there are very few studies conducted using DNA molecular markers [12,14]. Compared to other molecular markers, microsatellite (simple sequence repeat, SSR) markers are the ideal choices for studying the genetic composition of wild populations because of their co-dominant character and high variability [15,16]. The use of microsatellite markers to analyze the genetic diversity of *D. odorifera* can provide an invaluable means for conservation and protection of this endangered species. Moreover, the developments of SSR markers have been innovated by next-generation sequencing based on transcriptomes (RNA-seq), especially for species without a reference genome [17–19]. This approach has been applied for SSR identification, development, and association studies in many tree species [20–22]. In the present study, we applied this approach and developed 19 polymorphic SSR markers specific for *D. odorifera*. The main objectives of this study were to use these developed SSR markers to evaluate the genetic diversity of wild *D. odorifera* populations, and find out the causes for the endangered and fragmented status of this species. The findings of this study will provide useful genetic information for conservation and breeding strategies in *D. odorifera*.

## 2. Materials and Methods

### 2.1. Plant Materials and DNA Extraction Materials and Methods

In total, 42 wild individuals representing seven *D. odorifera* populations were sampled from the whole Hainan Island of China (Figure 1, Table 1). We sampled all the trees with a diameter at breast height (DBH) larger than eight cm, and the 42 individuals were the last remaining resources. Ten leaves were collected from each individual and sealed in plastic bags with desiccants. Total genomic DNA was extracted for each sample using the Hi-DNAsecure Plant Kit (Tiangen, Beijing, China) according to the manufacturer's instructions. The quality and quantity of DNAs were determined by NanoDrop 2000 (Thermo Scientific, Wilmington, DE, USA).

**Table 1.** Geographical location of seven investigated *D. odorifera* populations in Hainan Island of China.

| Population | Size | Origin | Status | Altitude (m) | Latitude (N) | Longitude (E) |
|---|---|---|---|---|---|---|
| HNQS | 7 | Wenchang city and Sansha city | Wild, Qiong mountain | 20–80 | 19°37′–19°53′ | 110°19′–110°23′ |
| HNHK | 3 | Haikou city | Wild, strand plain | 10–80 | 20°04′–20°32′ | 110°21′–109°21′ |
| HNBS | 5 | Baisha autonomous county | Wild, Li mu mountain | 150–250 | 19°12′–19°23′ | 109°7′–109°24′ |
| HNDF | 5 | Dongfang city | Wild, strand plain | 5–100 | 19°6′–19°17′ | 108°37′–108°39′ |
| HNLD | 7 | Ledong autonomous county | Wild, Jianfeng mountain | 35–80 | 18°34′–18°42′ | 108°49′–109°08′ |
| HNCJ | 5 | Changjiang autonomous county | Wild, Bawang mountains | 125–185 | 19°4′–19°08′ | 109°03′–109°12′ |
| HNWZS | 10 | Wuzhisan city | Wild, Wuzhisan mountain | 60–150 | 18°48′–18°53′ | 109°30′–109°67′ |

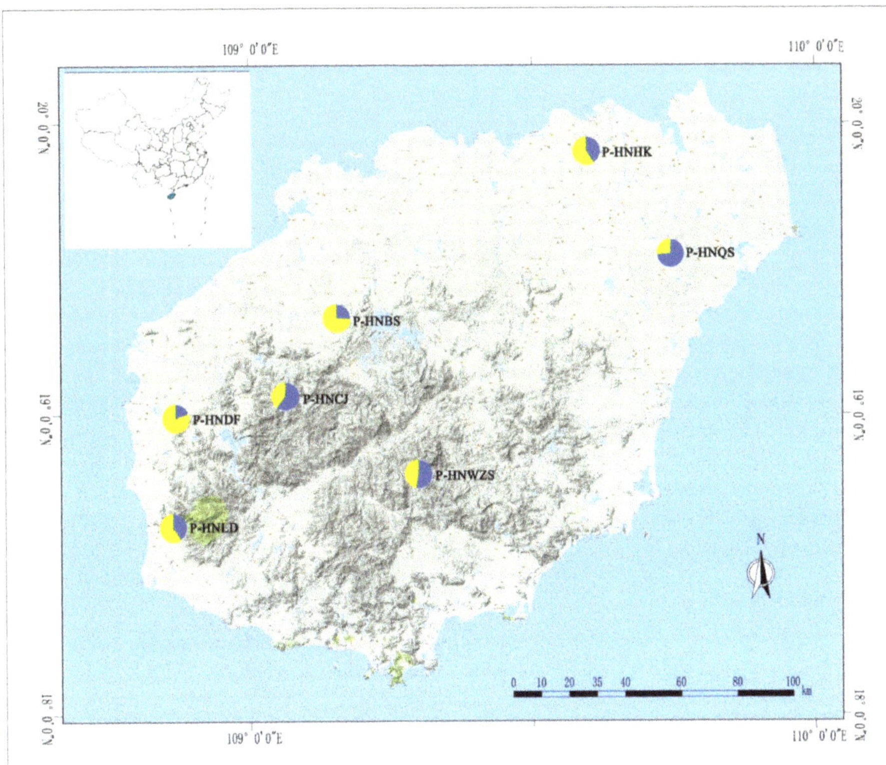


**Figure 1.** Geographic location of seven investigated *D. odorifera* populations collected from Hainan Island in China. P-HNQS—population of Wenchang city and Sansha city in Hainan province, P-HNHK—population of Haikou city, P-HNBS—population of Baisha city, P-HNDF—population of Dongfang city, P-HNLD—population of Ledong autonomous county, P-HNCJ—population of Changjiang autonomous county, P-HNWZS—population of Wuzhisan city. The pie charts estimated genetic structure of the seven populations based on STRUCTURE analysis with cluster number of two, in each chart, different color represents a different cluster accounted in each population.

### 2.2. RNA Sequencing and Data Deposition

To develop protocols, three leaves from three trees (H27, H98, and H100) were collected from three different populations of Haikou city (HNHK), Dongfang city (HNDF), and Changjiang autonomous county (HNCJ), respectively, and immediately put into liquid nitrogen. RNA extraction and sequencing were done by Beijing Novogene Biological Information Technology Co., Ltd., Beijing, China (http://www.novogene.com/). The sequence data were deposited in the database of SRA (Sequence Read Archive) at the National Center for Biotechnology Information (NCBI, https://www.ncbi.nlm.nih.gov/), under accession number SRP175426, and SRR8398210, SRR8398212, and SRR8398211 were the three biosample accession numbers for H27, H98, and H100, respectively [23].

### 2.3. SSR Identification and Marker Development

The software of MISA (MIcro SAtellite; http://pgrc.ipk-gatersleben.de/misa) was employed to detect, locate, and identify SSR loci. The minimum number of motifs used to select the SSR was ten for mono-nucleotide repeats, and six for di-nucleotide motifs, five for tri-, tetra-, penta-, and hexa-nucleotide repeats. Primers were designed using Primer 3.0 software [24] using default settings with the following criteria: Predicted primer lengths of 18–24 bases, GC content of 40%–60%, annealing temperature of 56–62 °C, and predicted product sizes of 150–300 bp.

### 2.4. Validation of SSR Marker by PCR and Capillary Electrophoresis

Subsequently, DNAs from three samples (randomly selected from 42 individuals) were used to validate the 192 randomly selected SSR loci (exclude mononucleotide repeats) with the designed primers. PCR reactions were performed in 15 μL final volume, containing 10.25 μL water, 1.5 μL 10 x DNA polymerase buffer, 1.5 μL $MgCl_2$ (25 mM), 0.3 μL dNTPs (10 mM each), 0.15 μL of each primer at 10 μM, 0.3 μL Taq polymerase at 5 units/μL (TaqUBA), and 1 μL of genomic DNA (40–50 ng). Totally 35 cycles of 94 °C for 15 s, appropriate annealing temperature for 15 s, and 72 °C for 30 s were performed, following the pre-denaturation at 94 °C for 3 min. PCR products of clear, stable, and specific bands with an expected length (100–350 bp) were considered as successful PCR amplifications. All the PCR reactions were repeated at least once. Finally, 22 SSR markers were randomly selected from the successful ones and used to analyze 42 samples. Their diluted PCR products mixed with 12.5 Hi-Di formamide and 0.25 μL size standard (Shanghai Generay Biotech Co., Ltd., Shanghai, China) were separated by capillary electrophoresis, and genotyped with an ABI 3730 Genetic Analyzer (Applied Biosystem, Foster, CA, USA) at Shanghai Generay Biotech Co., Ltd., Shanghai, China (http://www.generay.com.cn). Peak identification and fragment sizing were done using Gene Mapper v4.0 (Applied Biosystems, Foster, CA, USA) with default settings.

### 2.5. Statistical Analysis

The frequency of null alleles (FNA) and scoring errors were estimated using the Micro-checker software 2.2.3 [25]. POPGENE v1.3.1 software [26] was used to estimate the following genetic diversity parameters: Allele frequency, observed number of alleles (Na), effective number of alleles (Ne), expected and observed heterozygosities (He and Ho, respectively), Nei's gene diversity (GD), the percentage of polymorphic loci (PPB), and Wright's fixation index (F) and gene flow ($N_m$). The polymorphism information content (PIC) was calculated for each locus using the online program PICcalc [27]. F-statistics, including inbreeding coefficient within individuals ($F_{IS}$), genetic differentiation among populations ($F_{ST}$), were computed using GenAlEx version 6.5, so were the pairwise $F_{st}$, pairwise $G'_{ST}$ (Hedrick's standardized genetic differentiation index, adjusted for bias) [28]. Hardy-Weinberg equilibrium (HWE) was evaluated using chi-squared tests for each population at individual loci [26]. The Ewens-Watterson test for neutrality at each locus was performed using POPGENE v1.3.1 [26]. Hierarchical analyses of molecular variance (AMOVA) were conducted using GenAlEx version 6.5 [28].

The genetic structure of the investigated populations was analyzed using STRUCTURE 2.0 [29]. The number of discontinuous K was estimated from one to seven with 20 replicates, both length of burn-in period and value of MCMC (Markov chain Monte Carlo) were set to 100,000 times [30]. The true value of clusters (K) were harvested online (http:taylor0.biology.ucla.edu/struct_harvest/) according to the highest mean of estimated lnP(D) (log probability of data) and lnP(D)-derived delta K value [31]. Repeated sampling analysis and the genetic structural plot were performed by CLUMPAK [32]. To summarize the patterns of variation in the multi-locus dataset, principal coordinate analysis (PCoA) was performed using GenAlEx version 6.5 software based on pairwise $G'_{ST}$ matrix. Next, Mantel tests were carried out between matrixes of pairwise $G'_{ST}$ and geographic and genetic distance (Nei's unbiased genetic distance) using GenAlEx version 6.5 software, respectively. Additionally, a Neighbor-Joining (NJ) tree based on Nei's unbiased genetic distance was drawn in MEGAX [33].

## 3. Results

### 3.1. Distribution of SSR Loci in D. odorifera

In total, 35,774 potential SSR loci were identified and distributed in 26,880 unigenes, of which 6629 (24.7%) contained more than one SSR locus (Table S1). The SSR loci distributed in the leaf transcriptome were of a frequency of 1/2.18 kb. According to the unevenly distributed prediction (Figure S1), mono-nucleotide repeat motifs were the most frequent (21,623, 60.44%), followed by di- (7612, 21.28%) and tri-nucleotide (6112 or 17.09%) repeat motifs. These three motifs represented 98.81% in all, whereas only 40 and 14 penta- and hexa-nucleotide repeat motifs were found, respectively.

### 3.2. Development of Polymorphic SSR Markers

We randomly selected 192 SSR loci and designed primers to test the specificity of amplification for three samples and the informative nature of these SSR markers. Of these, 104 pairs of primers (54.2%) either did not give any amplification products or gave unexpected products, while 88 (45.8%) produced clear amplicons with the expected size of 100–350 bp. Next, 22 of the 88 primers were randomly selected for polymorphism detection and 19 (86.4%) showed polymorphism (Table 2). Further information on these validated 88 SSR markers, including ID of cDNA sequence, SSR type, repeat motif, position in template sequence, primer sequence, annealing temperature, and expected amplicon length (for developing alternative primers if desired) is available in Table S2. Among the polymorphic SSR loci, three (15.79%) were confirmed to locate in coding sequences (CDSs), six (31.58%) in 5′-untranslated regions (5′UTRs), and three (15.79%) in 3′-untranslated regions (3′UTRs).

### 3.3. Polymorphism of 19 SSR Loci

In total, 19 SSR loci harbored 54 alleles across the 42 *D. odorifera* samples (File S1), the number of alleles detected per locus was in a range of two to five, with an allele frequency range of 0.01–0.99 (Table 3, Table S3). The largest number of alleles (five) was detected at locus S21, which also harbored the largest effective number of alleles (Ne, 2.79), expected heterozygosity (He, 0.65), Nei's gene diversity (GD, 0.64), and polymorphic information contents (PIC, 0.60). In terms of the overall PIC, both S09 and S21 were highly informative with PIC values higher than 0.50, while S02, S12, S23, S26, and S27 were less informative with PIC values smaller than 0.25, and the remaining 12 loci were moderately informative with PIC values between 0.25 and 0.50. The average of Wright's fixation index (F) was 0.16, ranging from −0.19 (S22) to 0.44 (S24). Furthermore, null alleles were found at loci S04, S09, S21, S24, and S29. Six loci (S04, S08, S09, S24, S27, and S29) showed significant deviations from the Hardy-Weinberg equilibrium across the 42 *D. odorifera* individuals. Additionally, all the 19 SSR loci were selectively neutral according to the Ewens-Watterson test for neutrality (Table S4).

**Table 2.** Characteristics of 19 SSR (simple sequence repeat) markers developed for *D. odorifera*.

| Locus | ID | Repeat Motif | Forward Primer (5′-3′) | Reverse Primer (3′-5′) | Product Size (bp) | SSR Position | Tm (°C) |
|---|---|---|---|---|---|---|---|
| S01 | c102105_g1 | (ATA)6 | AGTCCCGCCCACAAAATCAT | CTGGTCAGTCATTCCCCCAC | 252–258 | Unknown | 60 |
| S02 | c11754_g1 | (AAG)6 | GGTCCCTGACTCACTGAAGC | CAACCTCTCTCTGCAGAACCA | 269–275 | 5′UTR | 60 |
| S03 | c1204_g1 | (ATA)6 | GCACGTGGTCAAAGCAATCA | ATGAGCCCCTTCTGCACTTC | 263–266 | Unknown | 60 |
| S04 | c25868_g1 | (GAT)6 | GCTGTGGAGTCACGTTCTCA | TCCCCACAGAATCACAAGCC | 278–287 | 3′UTR | 60 |
| S07 | c29390_g1 | (CCT)6 | GCCAATGACATAATGGGCGG | TGCAGAGAGTCAGGAGCTCT | 226–229 | CDS | 60 |
| S08 | c31361_g1 | (TGT)6 | GGAAGAGAAATGGAGGGTAGCT | TGCCAGACAACCAGAATGCT | 305–320 | Unknown | 60 |
| S09 | c33497_g1 | (CAT)6 | ACCCTCCTCCTCCACCTTTT | ACCGGCTTCAGTGATTGGTT | 230–239 | 5′UTR | 60 |
| S10 | c40172_g1 | (TGC)7 | CACGTACCCAACCGTCAAGA | TCCGACGACCACCTAATCCT | 261–273 | 5′UTR | 60 |
| S11 | c40452_g1 | (ATC)6 | AAAAAGCGAGGACTACGGCA | TGGAGAAGCAGTGCTCGTTT | 224–227 | Unknown | 60 |
| S12 | c34672_g2 | (GAT)7 | GGTGAACAAGCTGGAGTGGA | AAGCCCAGCATCTAAACCCC | 259–271 | CDS | 60 |
| S21 | c56000_g1 | (TCCC)6 | GAGCCTTGAGTTCACCTCCC | TTGGGTGTGAGATTGAGGGC | 238–250 | 5′UTR | 60 |
| S22 | c53146_g1 | (TC)10 | CCACCGATCTTAACCTCCGG | ACTACAAGTGCGTGTGACCC | 254–260 | Unknown | 60 |
| S23 | c56684_g2 | (TC)10 | TGGCGTTGACTTCCAGCATT | GAGCAGTGTCAGCATGATGC | 278–280 | 3′UTR | 60 |
| S24 | c59001_g1 | (CTA)7 | GCTGCAAATGCCAGTGCTTA | CGCTGTTGTCAGTGCATTGG | 232–241 | Unknown | 60 |
| S26 | c60831_g5 | (GTT)7 | CCAATCCCACCAGTGAGGAG | GCAGCACCTCTGAGACAAGT | 241–247 | CDS | 60 |
| S27 | c49315_g1 | (TAC)7 | GAACCTTTCCTTCTGCGCCT | CCTATGAAGCGTGTGCATGC | 266–272 | 3′UTR | 60 |
| S28 | c63495_g1 | (TAC)7 | ACAGCATTTGTGTTTGTGCA | CAGCTGCGCTCTCATTCCTA | 246–249 | Unknown | 60 |
| S29 | c57231_g1 | (TAT)7 | TCCCCGTTCCTCTCTCTCAG | GGACTGTCACATGGCACTCA | 150–168 | 5′UTR | 60 |
| S30 | c48304_g3 | (TGT)7 | TGCCTTGATCCGCTGAGATC | TCCCAAAATCGATGCAAAGCA | 246–258 | 5′UTR | 59 |

**Table 3.** Diversity statistics of the 19 SSR loci across 42 *D. odorifera* samples.

| Locus | Na | Ne | Ho | He | GD | PIC | $F_{ST}$ | $N_m$ | F | FNA | $P_{HWE}$ [a] |
|---|---|---|---|---|---|---|---|---|---|---|---|
| S01 | 3 | 1.99 | 0.38 | 0.50 | 0.50 | 0.41 | 0.08 | 2.79 | 0.23 | 0.08 | 0.108 ns |
| S02 | 2 | 1.05 | 0.05 | 0.05 | 0.05 | 0.05 | 0.07 | 3.58 | −0.02 | 0.00 | 0.874 ns |
| S03 | 2 | 1.80 | 0.38 | 0.45 | 0.44 | 0.35 | 0.16 | 1.30 | 0.14 | 0.04 | 0.355 ns |
| S04 | 3 | 1.71 | 0.29 | 0.42 | 0.42 | 0.37 | 0.11 | 2.08 | 0.31 | 0.09 * | 0.002 ** |
| S07 | 2 | 1.83 | 0.36 | 0.46 | 0.45 | 0.35 | 0.12 | 1.85 | 0.21 | 0.07 | 0.173 ns |
| S08 | 4 | 1.38 | 0.24 | 0.28 | 0.27 | 0.26 | 0.05 | 4.68 | 0.13 | 0.03 | 0.001 ** |
| S09 | 3 | 2.57 | 0.36 | 0.62 | 0.61 | 0.54 | 0.17 | 1.22 | 0.42 | 0.16 * | 0.001 *** |
| S10 | 3 | 1.97 | 0.48 | 0.50 | 0.49 | 0.44 | 0.09 | 2.65 | 0.03 | 0.01 | 0.412 ns |
| S11 | 2 | 1.75 | 0.38 | 0.43 | 0.43 | 0.34 | 0.14 | 1.48 | 0.11 | 0.03 | 0.481 ns |
| S12 | 2 | 1.21 | 0.14 | 0.17 | 0.17 | 0.16 | 0.13 | 1.70 | 0.17 | 0.03 | 0.268 ns |
| S21 | 5 | 2.79 | 0.45 | 0.65 | 0.64 | 0.60 | 0.10 | 2.20 | 0.30 | 0.12 * | 0.067 ns |
| S22 | 4 | 1.56 | 0.43 | 0.36 | 0.36 | 0.33 | 0.11 | 1.96 | −0.19 | −0.05 | 0.793 ns |
| S23 | 2 | 1.02 | 0.02 | 0.02 | 0.02 | 0.02 | 0.06 | 3.79 | −0.01 | 0.00 | 0.938 ns |
| S24 | 3 | 2.05 | 0.29 | 0.52 | 0.51 | 0.46 | 0.22 | 0.87 | 0.44 | 0.15 * | 0.002 ** |
| S26 | 2 | 1.02 | 0.02 | 0.02 | 0.02 | 0.02 | 0.09 | 2.63 | −0.01 | 0.00 | 0.938 ns |
| S27 | 3 | 1.27 | 0.21 | 0.21 | 0.21 | 0.19 | 0.09 | 2.42 | −0.01 | 0.00 | 0.032 * |
| S28 | 2 | 1.80 | 0.26 | 0.42 | 0.44 | 0.35 | 0.10 | 2.19 | 0.25 | 0.08 | 0.105 ns |
| S29 | 4 | 1.71 | 0.26 | 0.42 | 0.42 | 0.35 | 0.12 | 1.75 | 0.37 | 0.11 * | 0.000 *** |
| S30 | 3 | 1.81 | 0.33 | 0.45 | 0.45 | 0.37 | 0.03 | 7.77 | 0.25 | 0.08 | 0.135 ns |
| Mean | 2.84 | 1.70 | 0.28 | 0.37 | 0.36 | 0.31 | 0.11 | 2.58 | 0.16 | 0.03 | |

Na—observed number of alleles, Ne—effective number of alleles, Ho—observed heterozygosity, He—expected heterozygosity, GD—Nei's gene diversity, PIC—polymorphic information content, $F_{ST}$—genetic differentiation coefficient, $N_m$—Gene flow, estimated from Fst, $N_m = [(1/Fst) - 1]/4$, F Wright's (1978) fixation index, FNA—frequency of null alleles, * $p < 0.05$, likely contained null alleles, $P_{HWE}$ [a] *p*-value for deviation from Hardy-Weinberg equilibrium: ns not significant, * $p < 0.05$, ** $p < 0.01$, *** $p < 0.001$.

### 3.4. Genetic Diversity in D. odorifera

Among the seven populations investigated, the number of polymorphic loci varied from 13 to 17, along with the percentage of polymorphic loci (PPB) from 68.42% to 89.47% (Table 4). Presenting the largest PPB, population HNLD also had the largest alleles number of 46 (Alleles), whereas HNHK, with the smallest PPB, had the smallest number of 35 (Alleles). In total, eight private alleles appeared among the investigated populations, of which, three appeared in HNDF, two in HNLD, and one in HNQS, HNBS and HNWZS, respectively. The observed heterozygosity (Ho) ranged from 0.24 (HNWZS) to 0.38 (HNDF) and expected heterozygosity (He) from 0.31 (HNQS) to 0.40 (HNCJ), with an average of 0.28 and 0.37, respectively. At the population and species level, the expected (He) heterozygosity was 0.36 and 0.37, respectively. Additionally, population HNCJ possessing the highest genetic diversity level (He, 0.40) also showed the largest value of Nei's gene diversity (0.36).

**Table 4.** Summary of different *D. odorifera* population diversity statistics averaged over the 19 SSR loci.

| Population | Size | Alleles | Na | Ne | Np | Ho | He | GD | $F_{IS}$ | PPB % |
|---|---|---|---|---|---|---|---|---|---|---|
| HNQS | 7 | 40 | 2.11 | 1.52 | 1 | 0.25 | 0.31 | 0.28 | 0.09 | 73.68 |
| HNHK | 3 | 35 | 1.84 | 1.54 | 0 | 0.26 | 0.34 | 0.28 | 0.00 | 68.42 |
| HNBS | 5 | 40 | 2.11 | 1.69 | 1 | 0.31 | 0.36 | 0.32 | 0.02 | 78.95 |
| HNDF | 5 | 41 | 2.16 | 1.59 | 3 | 0.38 | 0.37 | 0.33 | −0.11 | 89.47 |
| HNLD | 7 | 46 | 2.42 | 1.64 | 2 | 0.26 | 0.38 | 0.35 | 0.20 | 89.47 |
| HNCJ | 5 | 40 | 2.11 | 1.77 | 0 | 0.36 | 0.40 | 0.36 | −0.03 | 78.95 |
| HNWZS | 10 | 44 | 2.32 | 1.63 | 1 | 0.24 | 0.34 | 0.33 | 0.22 | 78.95 |
| Mean | 5.25 | 40 | 2.15 | 1.63 | - | 0.29 | 0.36 | 0.32 | 0.06 | 79.70 |
| Total [a] | 42 [b] | 56 | 2.84 | 1.70 | - | 0.28 | 0.37 | 0.36 | 0.21 | 100.00 |

Population see Table 1, Size—number of sampled individuals, Alleles—total number of detected alleles, Na—observed mean number of alleles, Ne—mean effective number of alleles, Np—number of private alleles, Ho—observed heterozygosity, He—expected heterozygosity, GD—Nei's gene diversity, $F_{IS}$—inbreeding coefficient, PPB %—the percentage of polymorphic loci, [a] diversity indices averaged over the 19 loci across all *D. odorifera* populations, [b] total number of sampled individuals.

Both AMOVA and pairwise $F_{st}$ analysis were performed to investigate the genetic variations among these populations. The AMOVA analysis was conducted without grouping the investigated populations (population HNHK was not in this analysis for individuals below five). The result showed that only 3% of the total genetic variation occurred among populations, and 20% of the within population variation was due to the heterozygosity of the individuals within each population (Table 5). The overall $F_{ST}$ was very small (0.03, Table 5), the overall gene flow was 2.58 ($N_m$) estimated among all these populations (Table 3). Furthermore, the pairwise $F_{st}$ ranged from 0.042 to 0.115 (Table 6). The highest level appeared between populations HNQS and HNDF (0.115), whereas the lowest appeared between HNLD and HNCJ (0.042).

**Table 5.** Analysis of molecular variance (AMOVA) for six populations of *D. odorifera*.

| Source | *d.f.* | Sum of Square | Mean of Square | Variance Components | Percentage of Variation | $F_{ST}$ | $F_{IS}$ |
|---|---|---|---|---|---|---|---|
| Among populations | 5 | 27.864 | 5.573 | 0.111 | 3% | | |
| Within populations | | | | | | | |
| Among Individuals | 33 | 136.893 | 4.148 | 0.715 | 20% | | |
| Within Individuals | 39 | 106.000 | 2.718 | 2.718 | 77% | | |
| Total | 77 | 270.756 | | 3.544 | 100% | 0.03 * | 0.21 *** |

*d.f.* degrees of freedom, population HNHK was not in this analysis for individuals below five, $F_{ST}$ and $F_{IS}$ is based on standard permutation across the full data set, * $p < 0.05$, *** $p < 0.001$.

**Table 6.** Pairwise Genetic Differentiation Index ($F_{st}$) between the seven populations.

| Population | HNQS | HNHK | HNBS | HNDF | HNLD | HNCJ | HNWZS |
|---|---|---|---|---|---|---|---|
| HNQS | | | | | | | |
| HNHK | 0.088 | | | | | | |
| HNBS | 0.095 | 0.055 | | | | | |
| HNDF | 0.115 | 0.090 | 0.062 | | | | |
| HNLD | 0.074 | 0.055 | 0.043 | 0.051 | | | |
| HNCJ | 0.071 | 0.100 | 0.087 | 0.090 | 0.042 | | |
| HNWZS | 0.061 | 0.078 | 0.051 | 0.051 | 0.044 | 0.054 | |

Population see Table 1.

### 3.5. Population Structure of *D. odorifera*

An admixture model-based approach was implemented to evaluate the population structure of the 42 *D. odorifera* individuals. The optimum cluster number (K) of the investigated populations was two, with the largest values of both ln P(D) (log probability of data, −978) and delta K (15) harvested from the STRUCTURE HARVESTER website (Figure 2a,b). Based on K of two, a graphic representation of estimated membership coefficients of each individual was exhibited in Figure 2c. Each color showed the proportion of membership of each individual, represented by a vertical line, to the two clusters. The individual with the probability higher than a score of 0.75 was considered a pure one, and lower than 0.75 an admixture one. In this analysis, the yellow cluster included 17 individuals with 14 pure and 3 admixture ones, while the blue cluster included 25 individuals with 14 pure and 11 admixture ones. However, only HNBS and HNDF entirely consisted of individuals from the blue cluster, other populations consisted of individuals from both clusters.

The pairwise $G'_{ST}$ matrix was used for the principal coordinate analysis (PCoA). The first and second axis explained 63.25% and 22.13% of the variance within the molecular data, respectively (Figure 3a). Two clusters were clearly distinguished by PCoA analysis: Populations from HNHK, HNBS, HNLD and HNDF were grouped as cluster I, the other three populations (HNQS, HNCJ and HNWZS) grouped as cluster II. Moreover, the NJ (Neighbor-joining) dendrogram tree showed similar results based on Nei's unbiased genetic distance among the investigated populations (Figure 3b).

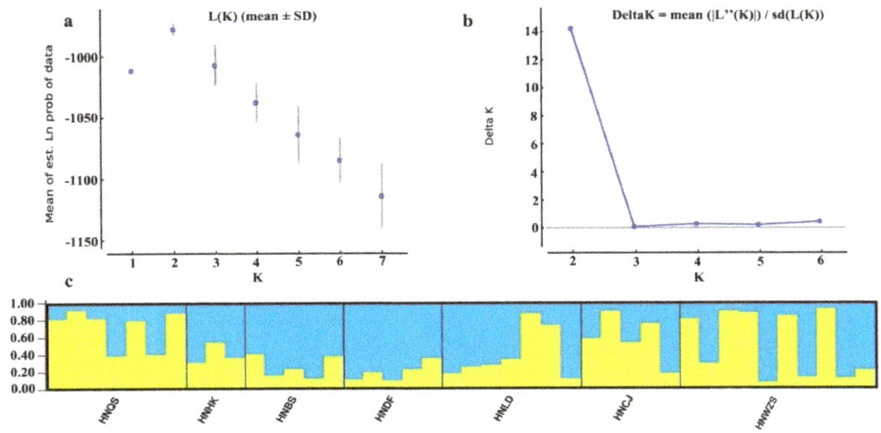


**Figure 2.** Results of STRUCTURE analysis for 42 *D. odorifera* individuals based on microsatellite data. (**a**) Estimation of population using mean of estimated lnP(D) (log probability of data) with cluster number (K) ranged from one to seven. (**b**) Estimation of population using lnP(D)-derived delta K with cluster number (K) ranged from one to seven. (**c**) Estimated genetic structure of the seven populations based on STRUCTURE analysis with cluster number (K) of two. In each plot, different color represents a different cluster and black segments separate the populations.

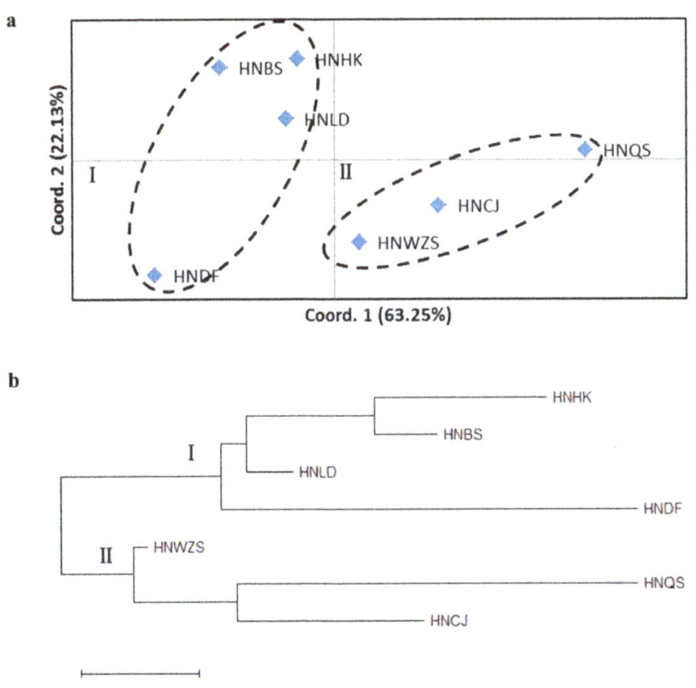


**Figure 3.** Relationships among the seven wild *D. odorifera* populations in Hainan Island. (**a**) Principal coordinate analysis (PCoA) based on pairwise $G'_{ST}$ (Hedrick's standardized Gst, analog of Fst, adjusted for bias), Coord.1 (63.25%): The first principal coordinate, explained 63.25% of variation; Coord.2 (22.13%): The second principal coordinate, explained 22.13% of variation. (**b**) Neighbour-joining (NJ) tree based on Nei's unbiased genetic distance among seven populations of *D. odorifera* in Hainan Island.

Subsequently, Mantel tests between the matrixes of pairwise $G'_{ST}$ and geographic distance (Figure 4a) and genetic distance (Figure 4b) were carried out, respectively. The results showed that genetic differentiations among the investigated populations were more attributed to the genetic distance (91.2%, Figure 4b) rather than to the geographic distance (27.7%, Figure 4a). Hence, there was no clear geographic origin-based structuring, or predominate isolation by distance among the investigated populations.

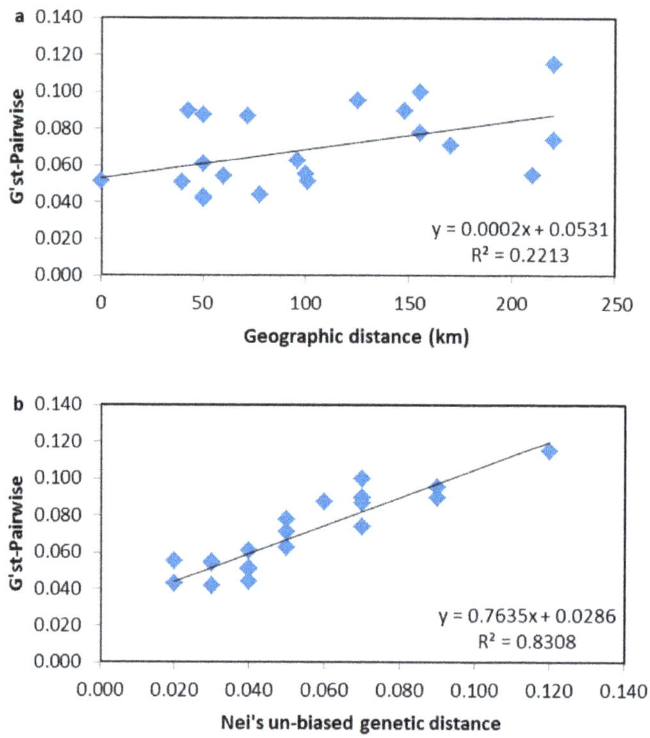


**Figure 4.** Mantel tests for pairwise $G'_{ST}$ matrix correspondence on relationships between geographic and genetic distance for *D. odorifera* populations in Hainan Island. (**a**) Relationships between pairwise $G'_{ST}$ and geographic distance. There is a positive relationship between the two elements (Rxy: 27.7%, $p < 0.05$). (**b**) Relationships between pairwise $G'_{ST}$ and Nei's unbiased genetic distance. There is a positive relationship between the two elements (Rxy: 91.2%, $p < 0.05$).

## 4. Discussion

### *4.1. Development of SSR Marker for D. odorifera*

Measuring levels of genetic diversity within and among populations is essential to understand the adaptability to environments of a species, in particular for conservation studies to explore the causes of rare and/or endangered plant species [34]. However, reports on the genetic diversity of *D. odorifera* is scarce, as a shortage of molecular markers restricted to six RAPD (random amplified polymorphic DNA) [12] and 25 SRAP (sequence-related amplified polymorphism) [14] loci. The use of such dominant markers could give a biased estimation of genetic variation when populations are not in the Hardy-Weinberg equilibrium [35], which may be true for the *D. odorifera* fragmented populations. In the present study, six loci showed significant deviations from the Hardy-Weinberg equilibrium. Therefore, the development of co-dominant SSR markers for *D. odorifera* is of great use for genetic studies. In this study, we have identified 35,774 putative SSR loci from the leaf transcriptome dataset, substantially more than those reported for other legume species such as 5956 in *Prosopis alba* Griseb. [36], 5710 in *Millettia pinnata* (L.) Panigrahi [18], and 7493 in *Mucuna pruriens* (L.) DC. [37]. In addition, the dominant repeats and motif types of SSR also vary among the transcriptomes from different species. These differences may be attributable to different genome structure and composition in these species.

The effectiveness and success of SSR development rely considerably on the quality and the accuracy of the sequence data [38]. Therefore, the identified SSR loci need to be further validated. Of the 192 primer pairs selected, 45.8% (88) yielded the expected amplicons for each locus, indicating that no introns presented within the amplified regions. From those validated SSR markers, 22 were randomly selected for polymorphism detection, of which 86.4% (19) exhibited polymorphism among the 42 wild *D. odorifera* trees. The PIC content provides an estimation of the information content of locus. The average PIC value of these newly developed SSR markers is 0.31, which is comparable to or lower than that in other legumes, such as *Vigna umbellata* (Thunb.) Ohwi & Ohashi (0.2898) [39], *Mucuna pruriens* (L.) DC. (0.24) [37], and *Melilotus* species (0.79) [40], but relatively higher than that based on ISSR (inter simple sequence repeat) and RAPD markers in other *Dalbergia* species, such as *Dalbergia cochinchinensis* Pierre ex Laness (ISSR 0.101; RAPD 0.088) [41] and *Dalbergia oliveri* Prain (ISSR 0.147; RAPD 0.116) [42]. Both high and low allelic PIC value markers are useful for genetic diversity to avoid a biased estimation [43,44]. Therefore, these SSR markers developed in our present study appear to be useful for genetic studies of *D. odorifera* populations.

### 4.2. Genetic Diversity of D. odorifera

Genetic diversity is essential to the long-term survival of species and plays an important role in the genetic improvement through breeding programs. However, limited information on genetic diversity of *D. odorifera* is available. Prior to the present study, only one report has been conducted using six RAPD markers, indicating medium genetic diversity level (six populations) inferred by the percentage of polymorphic loci (PPB) of 54.55% and Nei's gene diversity (GD) of 0.21 [12]. Compared to which, our results exhibited a higher genetic diversity level with the higher values of 100% (PPB) and 0.36 (GD) using the 19 newly developed SSR markers. These differences may be due to the different numbers [45] and types of molecular markers [16] used in the studies, or alternatively, due to the different population sizes in the two studies.

Genetic diversity in wild plant species is often related to the geographic range, population size, longevity, mating system, migration, and balancing selection [34,45,46]. Higher genetic diversity is expected to reflect a better adaption to the environments of a species [47]. However, the medium genetic diversity level of *D. odorifera* was indicated by the observed and expected heterozygosity of 0.28 and 0.37, respectively. Most of studies have been concordant with the general trend or prediction that species with narrow or endemic distributions maintain significantly lower levels of genetic diversity than species with widespread distributions [48–51]. Notably, the native habitat of *D. odorifera* is restricted to small regions in Hainan Island. It is no wonder then that the genetic diversity values in the present study are much smaller than those in wide spread tropical tree species such as *Olea europaea* Linn. (12 SSR markers, Ho = 0.75, He = 0.6) [20], *Prunus africana* (Hook.f.) Kalkman (6 SSR markers, Ho = 0.68, He = 0.73) [52], and *Eugenia dysenterica* DC. (9 SSR markers, Ho = 0.545, He = 0.62) [53]. However, the diversity of *D. odorifera* is even lower than in some rare and endemic tree species like *Boswellia papyrifera* (Del. ex Caill.) Hochst (He = 0.69) [54], *D. cochinchinenesis* (He = 0.55), and *D. oliveri* (He = 0.75) [55]. Similar observations were reported for *Ottelia acuminate* (He = 0.35, endemic to southwestern China) [56] and *Dipterocarpus alatus* Roxb. ex G.Don (He = 0.22, endemic to southeastern Vietnam) [57], resulting from the extensive reduction in population sizes caused by human disturbance. Similarly, due to the long-time over-logging for the valuable fragrant heartwood, the distributions of *D. odorifera* populations in Hainan Island have been dramatically reduced in the past thirty years. The present populations were highly fragmented into subpopulations, each composed of only a few individuals and large trees are seldom [12]. This is consistent with the suggestion that the distribution-restricted plant species are associated with a relatively low genetic diversity primarily from over-exploitation of their resources.

In the present study, the mean observed heterozygosity (0.28) was much lower than the mean expected heterozygosity (0.37), and the Wright's (1978) fixation index (F) was up to 0.16 across the 42 wild trees (Table 3), indicating a modest heterozygote deficiency existed within the entire wild distribution range for *D. odorifera*. This result may be attributed to the botanical characteristics of *D. odorifera*, more specifically, to its complicated reproduction system which causes a relatively high inbreeding coefficient (Table 5) [58]. *Dalbergia odorifera* is a predominantly outcrossing species [12]. Its flowers are entomophilous pollinated by small insects and fruits with flattened seedpods are dispersed by wind [59,60], which limits longer distance dispersal. Moreover, *D. odorifera* has the ability of coppice regeneration especially stimulated by trunk injuries [12]. These characters are consistent with a predominantly outcrossing mating system that includes at least some extent of inbreeding. Alternatively, it is due to the small populations in which mating between relatives occurred more frequently than in large populations [61].

### *4.3. Genetic Differentiation and Population Structure*

Woody species with predominately outcrossing tend to have less differentiation among populations and high variation within populations [34]. Similarly, our AMOVA analysis showed that most of the genetic variation was within the investigated populations of *D. odorifera*, while only 3% genetic variation components existed among populations, which is much lower compared to other *Dalbergia* species (0.236, *D. cochinchinensis*; 0.126, *D. oliveri*) [55]. Genetic differentiation among different populations is strongly influenced by gene flow ($N_m$) and genetic drift [62]. For neutral genes, the value of Nm below one indicates that genetic drift is a predominant factor affecting population structure, whereas the value above four indicates that gene flow can replace a genetic drift [48,63]. In the present study, the 19 SSR markers, which were selectively neutral according to the results of the Ewens-Watterson test (Table S4), are excellent for investing the effects of gene flow and genetic drift, showing overall gene flow of 2.58 (Table 3). This relatively high gene flow could curtail parts of the dispersive effects caused by genetic drift, reducing the genetic variation among populations while increasing the diversity within populations. However, the frequent migration indicated by the relatively high gene flow was opposite to the fragmented distributions of the investigated populations, for which, the putative explanation may be that of frequent human actions, primarily due to overexploitation and illegal logging [12]. Similar observations were also found for *Acer miaotaiense* (P. C. Tsoong) [34] and *Plectranthus edulis* (Vatke) Agnew [47]. Additionally, a genetic drift could not be ignored, since the population sizes are so small that any reduction in size could result in genetic drift.

Pairwise $F_{st}$ was in a range of 0.042–0.115 (Table 6), suggested that moderate population differentiation were found among these wild populations. The highest level of genetic differentiation (0.115) was found between populations HNDF and HNQS, and the distance between them was about 220 km, which was matching with the indication that long-term isolation may limit the level of gene flow between two populations [34]. However, the genetic differentiation level between HNLD and HNHK was only 0.055, the distance between them was also 220 km, which was opposite to the indication. An admixture model-based approach was implemented to evaluate the population structure, and suggested two clusters were the best for the 42 *D. odorifera* trees. Similar results were generated from both neighbor-joining and PCoA analysis. They all distinguished the investigated populations into two clusters: cluster I consisted of populations from Haikou city (HNHK), Baisha autonomous county (HNBS), Ledong autonomous county (HNLD), and Dongfang city (HNDF); while cluster II consisted of populations from Wenchang city and Sansha city (HNQS), Changjiang autonomous county (HNCJ), and Wuzhisan city (HNWZS). Moreover, genetic differentiation among the investigated populations showed positive relationships with both geographic distance (27.7%, Figure 4a) and genetic distance (91.2%, Figure 4b) distinguished by the Mantel tests, which was attributed more to the genetic distance rather than the geographic distance. Across all these analyses, no clear geographic origin-based structuring or predominant signs of "isolation by distance" were found, the present population structure of *D. odorifera* was more likely to be inferred by human activities.

### 4.4. Conservation

The main goal of conservation is to establish a suitable strategy for maintaining current genetic diversity and ensuring the long-term evolution of an endangered species [64]. The current state of *D. odorifera* is: Medium genetic diversity level along with modest heterozygosity deficiency, low genetic differentiation, and really small population size. This pattern mainly results from extensive human activities, primarily due to the over-logging. Many necessary approaches have been implemented by the Chinese government: (1) *Dalbergia odorifera* has been promoted to a second-grade state-protected species and it is forbidden to exploit natural resources; (2) national parks and sanctuaries have been established for in situ conservation covering almost every habitat in Hainan Island, such as Hainan Jianfengling national reservation, Bawangling national park, and Wangning Botany Park, etc. However, the population size of *D. odorifera* is still decreasing due to illegal-logging. Therefore, impactful ex situ conservation strategies should be the best choice, to avoid the loss of genetic diversity due to illegal logging and increase the variability of progenies by "outcrossing" the trees available.

## 5. Conclusions

The present study provides an initial assessment on genetic diversity and structure of *D. odorifera* conducted using 19 SSR markers. Medium genetic diversity at the species level and low genetic differentiations among populations were found in this endangered endemic tree species. This pattern of genetic variation may be primarily caused by extensive human activities, and this information could be used in the establishment of conservation strategies of this endangered species. In addition, the large number of SSR loci may serve as tools for assisting breeding programs in future studies.

**Supplementary Materials:** The following are available online at http://www.mdpi.com/1999-4907/10/3/225/s1, Figure S1: SSR motifs distribution, Table S1: Summary of SSR identified from the transcriptome. Table S2: Details of 88 validate SSR markers. Table S3: Allele frequency distribution across 42 *Dalbergia odorifera*. Table S4: The Ewens-Watterson test for neutrality across 19 microsatellite loci in *Dalbergia odorifera*. File S1: Raw data. Zip, contains three files: file a: 1-Allele-PDF, Allele reports, captured all the peaks in **.pdf. (** locus code, S01-S30); file b: 2-Raw and Statistical data-Excel, the 42 samples are identified to the "SAMPLE" labeled in the Allele reports. Scored raw data in "Raw-S01-S30. xlsx", then corrected the wrong captures and defined the allele series as integers to generate "S01-S30.xlsx" for statistics; file c: 3-Status of 42 wild *D. odorifera* trees.

**Author Contributions:** Data curation, Z.H. and N.Z.; formal analysis, F.L., H.J.; funding acquisition, Z.H., D.X. and N.Z.; investigation, N.Z. and Z.Y.; methodology, Z.H., D.X., X.L., Z.Y. and M.L.; project administration, Z.H. and D.X.; resources, X.L., H.J.; supervision, D.X.; writing—original draft, F.L.; writing—review & editing, F.L. and M.L.

**Funding:** This research was funded by Research Funds for the Central Non-profit Research Institution of Chinese Academy of Forestry (CAFYBB2017ZX001-4), National Natural Science Foundation of China (31500537), and Science Innovation Projects of Guangdong Province (2016KJCX009).

**Acknowledgments:** The authors are very grateful to Szmidt, A.E. (Department of Biology, Kyushu University) for the helpful comments on this manuscript.

**Conflicts of Interest:** The authors declare no conflict of interest.

## References

1. Liu, X.; Xu, D.; Yang, Z.; Zhang, N. Geographic variations in seed germination of *Dalbergia odorifera* T. Chen in response to temperature. *Ind. Crop. Prod.* **2017**, *102*, 45–50. [CrossRef]
2. Lee, D.S.; Kim, K.S.; Ko, W.; Li, B.; Keo, S.; Jeong, G.S.; Oh, H.; Kim, Y.C. The neoflavonoid latifolin isolated from MeOH extract of *Dalbergia odorifera* attenuates inflammatory responses by inhibiting NF-kappaB activation via Nrf2-mediated heme oxygenase-1 expression. *Phytother. Res.* **2014**, *28*, 1216–1223. [CrossRef] [PubMed]

3. Wang, H.; Dong, W.H.; Zuo, W.J.; Wang, H.; Zhong, H.M.; Mei, W.L.; Dai, H.F. Three new phenolic compounds from *Dalbergia odorifera*. *J. Asian Nat. Prod. Res.* **2014**, *16*, 1109–1118. [CrossRef] [PubMed]
4. Meng, H.; Chen, D.L.; Yang, Y.; Liu, Y.Y.; Wei, J.H. Sesquiterpenoids with cytotoxicity from heartwood of *Dalbergia odorifera*. *J. Asian Nat. Prod. Res.* **2018**. [CrossRef] [PubMed]
5. Tao, Y.; Wang, Y. Bioactive sesquiterpenes isolated from the essential oil of *Dalbergia odorifera* T. Chen. *Fitoterapia* **2010**, *81*, 393–396. [CrossRef] [PubMed]
6. Wang, H.; Dong, W.H.; Zuo, W.J.; Liu, S.; Zhong, H.M.; Mei, W.L.; Dai, H.F. Five new sesquiterpenoids from *Dalbergia odorifera*. *Fitoterapia* **2014**, *95*, 16–21. [CrossRef] [PubMed]
7. Fan, Z.M.; Wang, D.Y.; Yang, J.M.; Lin, Z.X.; Lin, Y.X.; Yang, A.L.; Fan, H.; Cao, M.; Yuan, S.Y.; Liu, Z.J.; et al. *Dalbergia odorifera* extract promotes angiogenesis through upregulation of VEGFRs and PI3K/MAPK signaling pathways. *J. Ethnopharmacol.* **2017**, *204*, 132–141. [CrossRef] [PubMed]
8. Lee, D.S.; Li, B.; Keo, S.; Kim, K.S.; Jeong, G.S.; Oh, H.; Kim, Y.C. Inhibitory effect of 9-hydroxy-6,7-dimethoxydalbergiquinol from *Dalbergia odorifera* on the NF-kappaB-related neuroinflammatory response in lipopolysaccharide-stimulated mouse BV2 microglial cells is mediated by heme oxygenase-1. *Int. Immunopharmacol.* **2013**, *17*, 828–835. [CrossRef] [PubMed]
9. Choi, H.S.; Park, J.A.; Hwang, J.S.; Ham, S.A.; Yoo, T.; Lee, W.J.; Paek, K.S.; Shin, H.C.; Lee, C.H.; Seo, H.G. A *Dalbergia odorifera* extract improves the survival of endotoxemia model mice by inhibiting HMGB1 release. *BMC Complement. Altern. Med.* **2017**, *17*, 212. [CrossRef] [PubMed]
10. Wang, W.; Weng, X.C.; Cheng, D.L. Antioxidant activities of natural phenolic components from *Dalbergia odorifera* T. Chen. *Food Chem.* **2000**, *71*, 45–49. [CrossRef]
11. Dalbergia Cochinchinensis. The IUCN Red List of Threatened Species. Available online: http://www.iucnredlist.org/details/32398/0 (accessed on 18 July 2018).
12. Yang, Q.X.; Feng, J.D.; Wei, J.H.; Li, R.T.; He, M.J. Genetic diversity of China's endangered medicinal plant *Dalbergia odorifera*. *World Sci. Technol. Mod. Tradit. Chin. Med. Mater. Med.* **2007**, *9*, 73–79. Available online: http://kns.cnki.net/KCMS/detail/detail.aspx?dbcode=CJFQ&dbname=CJFD2007&filename=SJKX200702016&v=MTgxNzJGeUhuVmJ2TE5pZkFkckc0SHRiTXJZOUVZb1I4ZVgxTHV4WVM3RGgxVDNxVHJXTTFGckNVUkxLZVplZHE= (accessed on 20 June 2018).
13. Nybom, H. Comparison of different nuclear DNA markers for estimating intraspecific genetic diversity in plants. *Mol. Ecol.* **2004**, *13*, 1143–1155. [CrossRef] [PubMed]
14. Yang, Y.; Meng, H.; Wu, Y.; Chen, B.; Gan, B.C. Primer screening of SRAP molecular marker in *Dalbergia odorifera*. *Acta Agric. Jiangxi.* **2011**, *23*, 29–31.
15. Fregene, M.A.; Suarez, M.; Mkumbira, J.; Kulembeka, H.; Ndedya, E.; Kulaya, A.; Mitchel, S.; Gullberg, U.; Rosling, H.; Dixon, A.G.O.; et al. Simple sequence repeat marker diversity in cassava landraces: Genetic diversity and differentiation in an asexually propagated crop. *Theor. Appl. Genet.* **2003**, *107*, 1083–1093. [CrossRef] [PubMed]
16. Powell, W.; Morgante, M.; Andre, C.; Hanafey, M.; Vogel, J.; Tingey, S.; Rafalski, A. The comparison of RFLP, RAPD, AFLP and SSR (microsatellite) markers for germplasm analysis. *Mol. Breed.* **1996**, *2*, 225–238. [CrossRef]
17. Dai, F.; Tang, C.; Wang, Z.; Luo, G.; He, L.; Yao, L. De novo assembly, gene annotation, and marker development of mulberry (*Morus atropurpurea*) transcriptome. *Tree. Genet. Genomes.* **2015**, *11*, 26. [CrossRef]
18. Huang, J.; Guo, X.; Hao, X.; Zhang, W.; Chen, S.; Huang, R.; Gresshoff, P.M.; Zheng, Y. De novo sequencing and characterization of seed transcriptome of the tree legume *Millettia pinnata* for gene discovery and SSR marker development. *Mol. Breed.* **2016**, *36*, 75. [CrossRef]
19. Taheri, S.; Lee Abdullah, T.; Yusop, M.; Hanafi, M.; Sahebi, M.; Azizi, P.; Shamshiri, R. Mining and development of novel SSR markers using next generation sequencing (NGS) data in plants. *Molecules* **2018**, *23*, 399. Available online: http://www.mdpi.com/1420-3049/23/2/399 (accessed on 2 January 2019). [CrossRef] [PubMed]
20. Dervishi, A.; Jakše, J.; Ismaili, H.; Javornik, B.; Štajner, N. Comparative assessment of genetic diversity in *Albanian olive* (*Olea europaea* L.) using SSRs from anonymous and transcribed genomic regions. *Tree Genet. Genomes* **2018**, 14. [CrossRef]
21. Dong, M.; Wang, Z.; He, Q.; Zhao, J.; Fan, Z.; Zhang, J. Development of EST-SSR markers in *Larix principis-rupprechtii* Mayr and evaluation of their polymorphism and cross-species amplification. *Trees* **2018**. [CrossRef]

22. Li, N.; Zheng, Y.; Ding, H.; Li, H.; Peng, H.; Jiang, B.; Li, H. Development and validation of SSR markers based on transcriptome sequencing of Casuarina equisetifolia. *Trees-Struct. Funct.* **2018**, *32*, 41–49. [CrossRef]
23. Liu, F.-M.; Hong, Z.; Yang, Z.-J.; Zhang, N.-N.; Liu, X.-J.; Xu, D.-P. De novo transcriptome analysis of *Dalbergia odorifera* and transferability of SSR markers developed from the transcriptome. *Forests* **2019**, *10*, 98. Available online: http://www.mdpi.com/1999-4907/10/2/98 (accessed on 28 January 2019). [CrossRef]
24. Untergasser, A.; Cutcutache, I.; Koressaar, T.; Ye, J.; Faircloth, B.C.; Remm, M.; Rozen, S.G. Primer3–new capabilities and interfaces. *Nucleic Acids Res.* **2012**, *40*, e115. [CrossRef] [PubMed]
25. Van Oosterhout, C.; Hutchinson, W.F.; Wills, D.P.M.; Shipley, P. Micro-checker: Software for identifying and correcting genotyping errors in microsatellite data. *Mol. Ecol. Notes* **2004**, *4*, 535–538. [CrossRef]
26. *Popgene*, Version 1.32; The User-Friendly Shareware for Population Genetic Analysis Molecular Biology and Biotechnology Center. University of Alberta Edmonton. Available online: http://www.ualberta.ca/~{}fyeh (accessed on 23 November 2017).
27. Nagy, S.; Poczai, P.; Cernák, I.; Gorji, A.M.; Hegedűs, G.; Taller, J. PICcalc: An online program to calculate polymorphic information content for molecular genetic studies. *Biochem. Genet.* **2012**, *50*, 670–672. [CrossRef] [PubMed]
28. Peakall, R.; Smouse, P.E. GenAlEx 6.5: Genetic analysis in Excel. Population genetic software for teaching and research—An update. *Bioinformatics* **2012**, *28*, 2537–2539. [CrossRef] [PubMed]
29. Evanno, G.; Regnaut, S.; Goudet, J. Detecting the number of clusters of individuals using the software structure: A simulation study. *Mol. Ecol.* **2010**, *14*, 2611–2620. [CrossRef] [PubMed]
30. Porras-Hurtado, L.; Ruiz, Y.; Santos, C.; Phillips, C.; Carracedo, A.; Lareu, M.V. An overview of STRUCTURE: Applications, parameter settings, and supporting software. *Front. Genet.* **2013**, *4*, 98. [CrossRef] [PubMed]
31. Earl, D.A.; Vonholdt, B.M. STRUCTURE HARVESTER: A website and program for visualizing STRUCTURE output and implementing the Evanno method. *Conserv. Genet. Resour.* **2012**, *4*, 359–361. [CrossRef]
32. Kopelman, N.M.; Jonathan, M.; Mattias, J.; Rosenberg, N.A.; Itay, M. Clumpak: A program for identifying clustering modes and packaging population structure inferences across K. *Mol. Ecol. Resour.* **2015**, *15*, 1179–1191. [CrossRef] [PubMed]
33. Kumar, S.; Stecher, G.; Li, M.; Knyaz, C.; Tamura, K. MEGA X: Molecular evolutionary genetics analysis across computing platforms. *Mol. Biol. Evol.* **2018**, *35*, 1547–1549. [CrossRef] [PubMed]
34. Li, X.; Li, M.; Hou, L.; Zhang, Z.; Pang, X.; Li, Y. De novo transcriptome assembly and population genetic analyses for an endangered Chinese endemic *Acer miaotaiense* (Aceraceae). *Genes* **2018**, *9*, 378. [CrossRef] [PubMed]
35. Szmidt, A.E.; Wang, X.R.; Lu, M.Z. Empirical assessment of allozyme and RAPD variation in *Pinus sylvestris* (L.) using haploid tissue analysis. *Heredity* **1996**, *76*, 412–420. [CrossRef]
36. Torales, S.L.; Rivarola, M.; Pomponio, M.F.; Gonzalez, S.; Acuña, C.V.; Fernández, P.; Lauenstein, D.L.; Verga, A.R.; Hopp, H.E.; Paniego, N.B.; et al. De novo assembly and characterization of leaf transcriptome for the development of functional molecular markers of the extremophile multipurpose tree species *Prosopis alba*. *BMC. Genom.* **2013**, *14*, 705. [CrossRef] [PubMed]
37. Sathyanarayana, N.; Pittala, R.K.; Tripathi, P.K.; Chopra, R.; Singh, H.R.; Belamkar, V.; Bhardwaj, P.K.; Doyle, J.J.; Egan, A.N. Transcriptomic resources for the medicinal legume *Mucuna pruriens*: De novo transcriptome assembly, annotation, identification and validation of EST-SSR markers. *BMC Genom.* **2017**, *18*, 409. [CrossRef] [PubMed]
38. Liu, S.; Liu, H.; Wu, A.; Hou, Y.; An, Y.; Wei, C. Construction of fingerprinting for tea plant (*Camellia sinensis*) accessions using new genomic SSR markers. *Mol. Breed.* **2017**, *37*, 93. [CrossRef]
39. Chen, H.; Chen, X.; Tian, J.; Yang, Y.; Liu, Z.; Hao, X.; Wang, L.; Wang, S.; Liang, J.; Zhang, L.; et al. Development of gene-based SSR markers in rice bean (*Vigna umbellata* L.) based on transcriptome data. *PLoS ONE* **2016**, *11*, e0151040. [CrossRef] [PubMed]
40. Yan, Z.; Wu, F.; Luo, K.; Zhao, Y.; Yan, Q.; Zhang, Y.; Wang, Y.; Zhang, J. Cross-species transferability of EST-SSR markers developed from the transcriptome of *Melilotus* and their application to population genetics research. *Sci. Rep.-UK* **2017**, *7*, 17959. [CrossRef] [PubMed]
41. Vu Thi Thu, H. Genetic diversity among endangered rare *Dalbergia cochinchinensis* (Fabaceae) genotypes in Vietnam revealed by random amplified polymorphic DNA (RAPD) and inter simple sequence repeats (ISSR) markers. *Afr. J. Biotechnol.* **2012**, *11*. [CrossRef]

42. Phong, D.T.; Hien, V.T.; Thanh, T.T.; Tang, D.V. Comparison of RAPD and ISSR markers for assessment of genetic diversity among endangered rare *Dalbergia oliveri* (Fabaceae) genotypes in Vietnam. *Genet. Mol. Resh.* **2011**, *10*, 2382–2393. [CrossRef] [PubMed]
43. Liu, S.; An, Y.; Li, F.; Li, S.; Liu, L.; Zhou, Q.; Zhao, S.; Wei, C. Genome-wide identification of simple sequence repeats and development of polymorphic SSR markers for genetic studies in tea plant (*Camellia sinensis*). *Mol. Breed.* **2018**, *38*, 59. [CrossRef]
44. Väli, Ü.; Einarsson, A.; Waits, L.; Ellegren, H. To what extent do microsatellite markers reflect genome-wide genetic diversity in natural populations? *Mol. Ecol.* **2008**, *17*, 3808–3817. [CrossRef] [PubMed]
45. Ferrer, M.M.; Eguiarte, L.E.; Montana, C. Genetic structure and outcrossing rates in *Flourensia cernua* (Asteraceae) growing at different densities in the South-western Chihuahuan Desert. *Ann. Bot.* **2004**, *94*, 419–426. [CrossRef] [PubMed]
46. White, T.L.; Adams, W.T.; Neale, D.B. *Forest Genetics*; CABI Publishing: Boston, MA, USA, 2007; pp. 149–186.
47. Gadissa, F.; Tesfaye, K.; Dagne, K.; Geleta, M. Genetic diversity and population structure analyses of *Plectranthus edulis* (Vatke) Agnew collections from diverse agro-ecologies in Ethiopia using newly developed EST-SSRs marker system. *BMC Genet.* **2018**, *19*, 92. [CrossRef] [PubMed]
48. Kang, S.S.; Chung, M.G. Genetic variation and population structure in Korean endemic species: IV. *Hemerocallis hakuunensis (Liliaceae). J. Plant. Res.* **1997**, *110*, 209–217. [CrossRef]
49. Zhong, T.; Zhao, G.; Lou, Y.; Lin, X.; Guo, X. Genetic diversity analysis of *Sinojackia microcarpa*, a rare tree species endemic in China, based on simple sequence repeat markers. *J. For. Res.* **2018**. [CrossRef]
50. Gichira, A.W.; Li, Z.-Z.; Saina, J.K.; Hu, G.-W.; Gituru, R.W.; Wang, Q.-F.; Chen, J.-M. Demographic history and population genetic structure of *Hagenia abyssinica* (Rosaceae), a tropical tree endemic to the Ethiopian highlands and eastern African mountains. *Tree Genet. Genomes* **2017**, *13*. [CrossRef]
51. Hamrick, J.L.; Godt, M.J.W.; Sherman-Broyles, S.L. Factors influencing levels of genetic diversity in woody plant species. *New For.* **1992**, *6*, 95–124. [CrossRef]
52. Mihretie, Z.; Schueler, S.; Konrad, H.; Bekele, E.; Geburek, T. Patterns of genetic diversity of *Prunus africana* in Ethiopia: Hot spot but not point of origin for range-wide diversity. *Tree Genet. Genomes* **2015**, *11*, 118. [CrossRef]
53. Boaventura-Novaes, C.R.D.; Novaes, E.; Mota, E.E.S.; Telles, M.P.C.; Coelho, A.S.G.; Chaves, L.J. Genetic drift and uniform selection shape evolution of most traits in *Eugenia dysenterica* DC. (Myrtaceae). *Tree Genet. Genomes* **2018**, *14*, 76. [CrossRef]
54. Addisalem, A.B.; Duminil, J.; Wouters, D.; Bongers, F.; Smulders, M.J.M. Fine-scale spatial genetic structure in the frankincense tree *Boswellia papyrifera* (Del.) Hochst. and implications for conservation. *Tree Genet. Genomes* **2016**, *12*, 86. [CrossRef]
55. Hartvig, I.; So, T.; Changtragoon, S.; Tran, H.T.; Bouamanivong, S.; Theilade, I.; Kjær, E.D.; Nielsen, L.R. Population genetic structure of the endemic rosewoods *Dalbergia cochinchinensis* and *D. oliveri* at a regional scale reflects the Indochinese landscape and life-history traits. *Ecol. Evol.* **2018**, *8*, 530–545. [CrossRef] [PubMed]
56. Zhai, S.H.; Yin, G.S.; Yang, X.H. Population genetics of the endangered and wild edible plant *Ottelia acuminata* in southwestern China using novel SSR markers. *Biochem. Genet.* **2018**, *56*, 235–254. [CrossRef] [PubMed]
57. Vu, D.-D.; Bui, T.T.-X.; Nguyen, M.-D.; Shah, S.N.M.; Vu, D.-G.; Zhang, Y.; Nguyen, M.-T.; Huang, X.-H. Genetic diversity and conservation of two threatened dipterocarps (Dipterocarpaceae) in southeast Vietnam. *J. For. Res.* **2018**. [CrossRef]
58. Mutegi, E.; Snow, A.A.; Rajkumar, M.; Pasquet, R.; Ponniah, H.; Daunay, M.-C.; Davidar, P. Genetic diversity and population structure of wild/weedy eggplant (*Solanum insanum*, Solanaceae) in southern India: Implications for conservation. *Am. J. Bot.* **2015**, *102*, 140–148. [CrossRef] [PubMed]
59. Institute of Botany, the Chinese Academy of Sciences. *Flora Reipublicae Popularis Sinicae*; Science Press: Beijing, China, 1994; Volume 40, p. 114.
60. The, S.N. A review on the medicinal plant *Dalbergia odorifera* species. Phytochemistry and biological activity. *Evid-Based. Complement. Altern.* **2017**. [CrossRef]
61. Moritsuka, E.; Chhang, P.; Tagane, S.; Toyama, H.; Sokh, H.; Yahara, T.; Tachida, H. Genetic variation and population structure of a threatened timber tree *Dalbergia cochinchinensis* in Cambodia. *Tree Genet. Genomes* **2017**, *13*, 115. [CrossRef]

62. Schaal, B.A.; Hayworth, D.A.; Olsen, K.M.; Rauscher, J.T.; Smith, W.A. Phylogeographic studies in plants: Problems and prospects. *Mol. Ecol.* **1998**, *7*, 465–474. [CrossRef]
63. Slatkin, M. Gene flow and the geographic structure of natural populations. *Science* **1987**, *236*, 787–792. [CrossRef] [PubMed]
64. Forest, F.; Grenyer, R.; Rouget, M.; Davies, T.J.; Cowling, R.M.; Faith, D.P.; Balmford, A.; Manning, J.C.; Proches, S.; van der Bank, M.; et al. Preserving the evolutionary potential of floras in biodiversity hotspots. *Nature* **2007**, *445*, 757–760. [CrossRef] [PubMed]

forests

MDPI

Article

# Genetic Diversity and Structure through Three Cycles of a *Eucalyptus urophylla* S.T.Blake Breeding Program

**Wanhong Lu [1], Roger J. Arnold [1,*], Lei Zhang [2] and Jianzhong Luo [1]**

1 China Eucalypt Research Centre, Chinese Academy of Forestry, 30 Mid Renmin Dadao, Zhanjiang 524022, China; luwanhong@outlook.com (W.L.); Luojz69@hotmail.com (J.L.)

2 Forest Science Institute, Dongmen State Forest Farm, Fusui 532108, China; zhanglei506@163.com

* Correspondence: roger.arnold@y7mail.com; Tel.: +86-158-1170-6013

Received: 17 April 2018; Accepted: 15 June 2018; Published: 21 June 2018

**Abstract:** *Eucalyptus urophylla* S.T.Blake is an important commercial tropical plantation species worldwide. In China, a breeding program for this species has progressed through three cycles but genetic diversity and structure in the breeding populations are uncertain. A sampling of field trials from these populations was carried out to evaluate their genetic diversity and structure using 16 microsatellite loci. Significant deviations from Hardy-Weinberg equilibrium were recorded at all 16 loci in the populations. Overall expected and observed heterozygosity ($H_e$ and $H_o$) estimates of 0.87 and 0.59 respectively for the first cycle population, and 0.88 and 0.60 respectively for the third cycle population, revealed reasonably high levels of genetic diversity. The genetic differentiation coefficient ($F_{st}$) revealed low differentiation among pairs of provenances (from the species' native range) comprising the first cycle population (range: 0.012–0.108), and AMOVA results showed that the majority of molecular genetic variation existed among individuals rather than among provenances for the first cycle population and among individuals rather than among field trial sources in the third cycle population. Levels of genetic diversity appeared to remain unchanged from the first to third cycle populations, and the results indicate prospects for maintaining if not increasing diversity through recurrent breeding. Likely effects of artificial directional selection, prior to sampling, on both populations examined are discussed along with implications for future *E. urophylla* breeding.

**Keywords:** microsatellite locus; Hardy-Weinberg equilibrium; genetic differentiation; breeding population; artificial selection

## 1. Introduction

*Eucalyptus urophylla* S.T.Blake is a tall forest tree that has a natural distribution spanning seven of the Lesser Sunda Islands in eastern Indonesia, where it is mostly found growing on volcanically derived soils, and it also extends into East Timor. Across this natural range, the species can vary from a tall forest tree up to 45 m high to a shrub like form of less than 2 m [1,2]. Cross-pollination in the species is mostly effected by insects and birds [3] and though self-compatible, it is predominantly outcrossing but with a mixed mating system in natural stands [4]. On lower slopes it often co-occurs with *Eucalyptus alba* Reinw. ex Blume in mosaic stands [1] and it was only in 1977 that *E. urophylla* was described as a species separate from *E. alba* [3].

As an exotic forest plantation species *E. urophylla* is now one of the most commercially important hardwood species worldwide. Both the pure species and hybrid varieties involving this species (most commonly with *Eucalyptus grandis* W.Hill ex Maiden) are the foundation of substantial areas of commercial plantations in tropical and warmer subtropical regions for the production of pulpwood, fuelwood, poles, veneer logs and even saw logs [2,5,6]. It was first introduced to China in 1971 [7] and by mid-1990s the species and its hybrids had become leading genetic material for commercial plantations established in tropical and warmer sub-tropical areas of southern China [8,9]. Today in this country, there are over three million hectares of plantations established with hybrid varieties of this species, and this resource provides livelihoods for hundreds of thousands of people [10].

Following the phenotypic diversity observed in the species for growth and stem-form through its natural range, high levels of genetic diversity have been recorded across this range through designed genetic field trials. Phenotypic observations/measurements on quantitative traits have been carried out in various statistically designed provenance/family cum progeny field trials of this species to examine genetic variability and diversity in adaptive and economically important growth, stem form and wood quality traits. Examples of such work with *E. urophylla* are provided by Hodge and Dvorak [11], who reported results for 65 provenances originating from the seven Indonesian islands where the species occurs naturally, that were tested in a series of 125 provenance/progeny trials planted in five countries, and by Kien et al. [12] who reported results on 144 families, representing 9 provenances, tested in two field trials located in northern Vietnam.

Genetic variation and relationships among and within natural populations of *E. urophylla* have also been examined in a number of molecular genetic studies. House and Bell [4] examined material from across the species' full natural range by using isozymes and found that most of the genetic diversity of the species was attributed to variation within populations. In contrast, they found that genetic differences between populations, at least for the isozymes examined, appeared to be small, while no striking patterns relating to geography were detected. Similarly, Payn et al. [13] and Payn et al. [14] investigated genetic diversity and geographical distribution of chloroplast DNA variations in the species and found moderate to high levels of genetic diversity throughout its geographic range ($H_e$ = 0.70–0.78). However, the latter of these studies also found relatively low genetic differentiation among populations ($F_{st}$ = 0.03), which the authors took to indicate low levels of recurrent gene flow among the Indonesian islands of the species occurrence.

In a separate study Tripiana et al. [15] used 10 microsatellite markers to study 360 seedlings, representing 49 provenances (referred to by them as "subpopulations"), spanning *E. urophylla*'s natural range in Indonesia and East Timor, which they grouped into 17 "natural populations", in order to assess the species' genetic diversity and structure. They found that microsatellite heterozygosity was moderate to high within populations ($H_o$ = 0.51–0.72) based on the loci they examined. They also found that the index of fixation was significantly different from zero for all populations ($F_{IS}$ = 0.13–0.31), whilst the differentiation among populations was low ($F_{st}$ = 0.04) and not significantly different from zero, due to extensive gene flow across the species' natural range via pollen flow. They also suggested that the $F_{IS}$ values observed might have been due to a Wahlund effect. The latter effect arises when two genetically distinct groups are (inadvertently or intentionally) lumped into a single sampling unit, either because they co-occur but rarely interbreed, or because the spatial scale chosen for sampling is larger than the true scale of a population (or of a subpopulation, depending on the definitions used for these terms). This Wahlund effect (substructure within populations) can lead to heterozygote deficits and deviations from Hardy-Weinberg equilibrium (HWE) [16,17].

However, despite the convincing evidence presented by such molecular genetic studies on *E. urophylla*, it is noteworthy that House and Bell [4] saw a paradox in that an apparent lack of, or at least relative minor, differentiation between populations of *E. urophylla* from its natural range found in isozyme, and subsequently also in molecular genetics studies, seemed somewhat contradictory compared to a high degree of population differentiation for morphological/adaptive traits. Pryor et al.'s [1] detailed study on morphological features from 23 populations across *E. urophylla*'s natural range supported the identification of three separate taxa from within the species: *E. urophylla*, *Eucalyptus orophila* L.D.Pryor and *Eucalyptus wetarensis* L.D. Pryor. Indeed House and Bell's [4] study identified a large degree of allelic diversity between the genetic material originating from the island of Wetar and that originating from other islands; the former populations having been classified as *E. wetarensis* by Pryor et al. [1]. Although such division into separate species has generally not yet been accepted either by eucalypt growers or researchers, a recently published taxonomic classification of *Eucalyptus* species does identify *E. urophylla*, *E. orophila* and *E. wetarensis* as separate species [18].

Though studies such as those cited above [1,4,11,12,14,15] have provided excellent insights into the natural populations of *E. urophylla*, today most commercial growers of the species no longer focus on genetic material collected directly from natural populations. Worldwide, many *E. urophylla* improvement programs have already captured in their breeding populations diverse genetic material originating from the species' natural range, and out of the initial breeding populations most of them have by now progressed through several cycles of selection and breeding.

During the 1980s and 1990s genetically diverse seedlots of *E. urophylla* were imported to China for the establishment of a base population to support ongoing genetic improvement [9,19]. With this material a first cycle breeding population of *E. urophylla* was established as a series of field trials during the period 1988 to 1998 and included over 400 open-pollinated families representing 30 provenances from the species' natural range along with families and seedlots from various planted stands/exotic seed sources. Then in 2004 a second cycle breeding population of *E. urophylla* was established in China as a single field trial, with open-pollinated (OP) seed collected from selected plus-trees of the first cycle population (i.e., OP families). In 2006, a Chinese cooperative tree improvement program was initiated involving commercial growers and a number of government research institutes. Through this cooperative program, a third cycle breeding population for *E. urophylla* was established in 2010, with material selected from both the first and the second cycle populations.

How much genetic diversity is currently present in the third cycle breeding population of *E. urophylla* in China, and how this has been affected by the preceding cycles of selection, are critical questions. Maintaining broad genetic diversity through the successive cycles of breeding is essential for achieving genetic gains from both the current and future cycles of this species. Variation is needed in key economic traits, so that artificial selection can ultimately result in heritable genetic improvements [2]. The genetic diversity—quantifiable, unquantifiable and/or "cryptic"—will serve as the primary basis for adaptation to future biotic and abiotic challenges, and selection for traits not currently seen or linked to economic values, e.g. adaptation to future climatic shifts and/or emergence of new pathogens [15].

Thus, the present study was initiated in order to examine the first and third cycle *E. urophylla* breeding populations in China, using molecular genetic markers with the following specific objective: To evaluate the potential loss of genetic diversity through breeding cycles by comparing the genetic diversity, being assessed by molecular markers, between the founding (first cycle) and the descendent (third cycle) breeding populations. The parameters estimated in this study will also provide a benchmark for comparison with future, successive breeding populations.

## 2. Materials and Methods

### 2.1. Plant Material

The natural stand origins of some of the genetic materials included in the breeding populations and examined in this study are known to be on the island of Wetar, Indonesia, and thus could be classified as *E. wetarensis* according to Pryor et al. [1] and Nicole [18]. However, for the purpose of the study reported here, the taxonomic classification of *E. urophylla* according to Brooker [20], is followed and all material involved is referred to as *E. urophylla*.

Samples were obtained from the first cycle breeding population of *E. urophylla* in China; see Table 1, Figure 1, and Supplementary Material Table S1 and Figure S1. In mid-2016 three out of the five field trials comprising this population (T46, T77 and T94) were sampled. Regarding the two other trials of the same cycle that were not sampled, trial T54 was terminated some years ago and therefore not available for sampling, and trial TJJ was just a duplicate of trial T94.

**Table 1.** Details of the samples obtained from 3 field trials (T46, T77 and T94) of the first cycle breeding population of *E. urophylla*.

| Trial (year) | Provenance Origins of Families in Trial * (Locations of the Mothers of Trees Sampled) | | | | No. Families Sampled (in Trial) |
|---|---|---|---|---|---|
| | Provenance Geographic Location ** | Island, Country | Latitude & Longitude | Altitude (m a.s.l.) | |
| T46 (1988) | 1. Mt Lewotobi I | Flores, Ind. | 8°31′ S, 122°45′ E | 398 | 10 (14) |
| | 2. Mt Egon *** | Flores, Ind. | 8°38′ S,122°27′ E | 450 | 14 (17) |
| | 3. Ulanu River | Alor, Ind. | 8°20′ S,124°27′ E | 500 | 9 (11) |
| subtotals | 3 provenances | | | | 33 (42) |
| T77 (1992) | 4. Mandiri | Flores, Ind. | 8°45′ S,122°50′ E | 410 | 9 (15) |
| | 5. Mt Lewotobi II | Flores, Ind. | 8°32′ S,122°48′ E | 375 | 25 (30) |
| | 6. Wukoh | Flores, Ind. | 8°35′ S,122°48′ E | 375 | 9 (14) |
| | 2. Mt Egon *** | Flores, Ind. | 8°38′ S,122°27′ E | 450 | 22 (31) |
| | 7. Bangat | Flores, Ind. | 8°38′ S,122°27′ E | 330 | 10 (14) |
| | 8. Wairteban | Flores, Ind. | 8°38′ S,122°27′ E | 525 | 10 (13) |
| | 9. Iling Gele | Flores, Ind. | 8°37′ S,122°27′ E | 600 | 7 (14) |
| | 10. Andalan | Flores, Ind. | 8°36′ S,122°28′ E | 725 | 10 (13) |
| | 11. Jawaghar | Flores, Ind. | 8°36′ S,122°28′ E | 725 | 10 (14) |
| | 12. DMFF I | China | 22°23′ N,107°30′ E | 200 | 23 (29) |
| subtotals | 10 provenances | | | | 135 (190) |
| T94 (1998) | 13. Waikui | Alor, Ind. | 8°14′ S,124°44′ E | 540 | 2 (7) |
| | 14. Arnau | Wetar, Ind. | 7°49′ S, 126°10′ E | 315 | 3 (7) |
| | 15. Uhak | Wetar, Ind. | 7°39′ S, 126°29′ E | 350 | 5 (27) |
| | 16. Apui | Alor, Ind. | 8°17′ S, 124°40′ E | 1115 | 2 (19) |
| | 17. Piritumas | Alor, Ind. | 8°19′ S, 124°31′ E | 355 | 4 (24) |
| | 18. Ilwaki | Wetar, Ind. | 7°52′ S, 126°27′ E | 515 | 5 (35) |
| | 19. Telemar | Wetar, Ind. | 7°54′ S, 125°58′ E | 180 | 1 (3) |
| | 20. Carbubu | Wetar, Ind. | 7°56′ S, 125°53′ E | 175 | 1 (3) |
| | 21. Dalaki | Pantar, Ind. | 8°31′ S, 124°5′ E | 440 | 2 (8) |
| | 22. DMFF II | China | 22°23′ N,107°30′ E | 200 | 9 (24) |
| subtotals | 10 provenances | | | | 34 (157) |
| Totals | 22 provenances/ seed sources | | | | 202 (389) families |

Note: * Key to abbreviations: Mt = Mount; DMFF = Dongmen State Forest Farm, Guangxi, China; Ind. = Indonesia. ** numerical ID's of provenances relate to locations indicated by numbered blue dots in Figure 2; *** The Mt Egon provenance was included in both T46 and T77 trials, but was represented by different families each trial (i.e., families in the 2 trials were mutually exclusive).

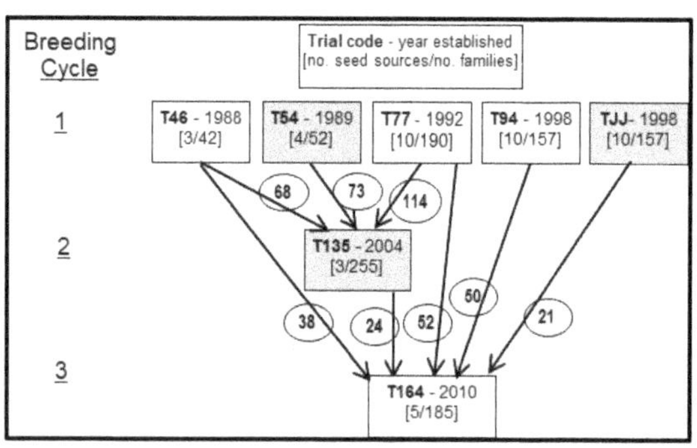


**Figure 1.** Development of three successive cycles of *Eucalyptus urophylla* S.T.Blake breeding in China. Trials shaded in grey were not sampled for this study; circled numbers indicate number of families contributing to the succeeding trial indicated by the associated arrow.

From the first cycle breeding population, 202 families were sampled; Table 1. Of these families, 170 were first generation progeny from mother trees originating from 20 natural stand provenances (i.e., provenances from the species' native range) from 4 Indonesian Islands (see Figure 2), and 32 families were progeny from plus-trees selected in earlier trials at Dongmen Forest Farm in China and presumed to be second generation genetic material. Unfortunately, the origins of the latter material are now not known as clear records are not available. The provenance categorization across the native range of the species was based on designations provided originally by CSIRO

(Commonwealth Scientific and Industrial Research Organization)'s Australian Tree Seed Centre, which was the supplier of the seedlots.

The single trial that comprised the second cycle breeding population (T135) had been intensively thinned and then suffered wind/typhoon damage over 5 years ago and was unsuitable for sampling. From the descendant third cycle breeding population, 125 families were sampled from one field trial (T164) that (initially) included almost a comprehensive set of the seed sources and families of third cycle breeding population, planted in 2010; Table 2 and Supplementary Material Table S1, Table S2 and Figure S1. Though this third cycle population comprised multiple field trials, each contained about the same set of families so just one trial was sampled for this study. Of the third cycle families sampled: 91 were progeny from mother trees selected in 4 trials of the first cycle breeding population (their mother trees represented 20 provenances from 4 Indonesian Islands of the species' natural range); 20 were progeny from mother trees selected in the first cycle breeding population which themselves were progeny of plus-trees selected in earlier trials at Dongmen Forest Farm; 14 were progeny from mother trees selected in the trial (T135), which comprised the second cycle breeding population, that were maternal descendants of first cycle families not related (maternally) to other families sampled from the third cycle population.

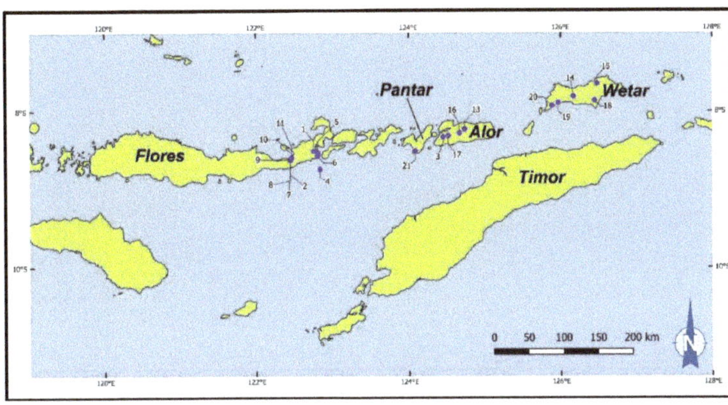


**Figure 2.** Geographic locations of *E. urophylla* natural stand provenances in Indonesia from which the genetic material included in the trials of the first cycle breeding population originated—the numbers linked to blue dots indicate the locations of the numbered provenances listed in Table 1.

Whilst each cycle of the breeding population might 'nominally' be considered as a generation, there is some variation among the families comprising each cycle with respect to the number of generations, at least on the maternal sides, and their descent from mother trees in natural stands in Indonesia and East Timor. The number of families sampled from the first and third cycles were a compromise between: (1) balancing the number of samples to represent the provenances/seed sources and their respective sizes (i.e., number of families from them) in each population; (2) trees available for sampling (considerations mostly for first cycle populations where thinning and some typhoon damage had reduced representation); and, (3) resources available for this study.

From each family sampled in each breeding cycle, fresh leaf tissue was collected from one tree (i.e., samples obtained from one tree per family, each family having originated from a different mother tree) for DNA extraction, in accordance with the methodologies used by Payn et al. [13], Payn et al. [14] and Tripiana et al. [15]. The one tree sampled per family in each cycle, was selected randomly, being the first tree found within each target family of the respective trials, starting from replicate 1 and working methodically through the replicates in numerical sequence.

**Table 2.** Details of the samples obtained from one of the field trials (T164), planted in 2010, of the third cycle breeding population of *E. urophylla*.

| Mothers of Trees Sampled | | Origins of Grandmothers of Families Sampled * | | |
|---|---|---|---|---|
| **Trial ID** | **No. Families Sampled (in Trial)** | **No. Families Sampled (in Trial)** | **Provenance Geographic Location **** | **Island, Country** |
| T46 | 20 (38) | 6 (11) | 2. Mt Egon | Flores, Ind. ** |
| | | 6 (14) | 1. Mt Lewotobi I | Flores, Ind. |
| | | 5 (5) | 3. Ulanu River | Alor, Ind. |
| | | 3 (9) | 23. DMFF III | China |
| T77 | 43 (52) | 1 (1) | 4. Mandiri | Flores, Ind. |
| | | 20 (21) | 5. Mt Lewotobi II | Flores, Ind. |
| | | 9 (10) | 2. Mt Egon | Flores, Ind. |
| | | 2 (2) | 8. Wairteban | Flores, Ind. |
| | | 2 (2) | 9. Iling Gele | Flores, Ind. |
| | | 1 (1) | 10. Andalan | Flores, Ind. |
| | | 8 (12) | 12. DMFF I | China |
| T94 | 32 (50) | 1 (7) | 13. Waikui | Alor, Ind. |
| | | 1 (7) | 14. Arnau | Wetar, Ind. |
| | | 1 (27) | 15. Uhak | Wetar, Ind. |
| | | 1 (19) | 16. Apui | Alor, Ind. |
| | | 2 (24) | 17. Piritumas | Alor, Ind. |
| | | 2 (35) | 18. Ilwaki | Wetar, Ind. |
| | | 1 (3) | 19. Telemar | Wetar, Ind. |
| | | 1 (3) | 20. Carbubu | Wetar, Ind. |
| | | 9 (32) | 22. DMFF II | China |
| TJJ | 16 (21) | 0 (7) | 13. Waikui | Alor, Ind. |
| | | 2 (7) | 14. Arnau | Wetar, Ind. |
| | | 5 (27) | 15. Uhak | Wetar, Ind. |
| | | 0 (19) | 16. Apui | Alor, Ind. |
| | | 0 (24) | 17. Piritumas | Alor, Ind. |
| | | 9 (35) | 18. Ilwaki | Wetar, Ind. |
| | | 0 (3) | 19. Telemar | Wetar, Ind. |
| | | 0 (3) | 20. Carbubu | Wetar, Ind. |
| | | 0 (32) | 22. DMFF II | China |
| T135 | 14 (24) | 6 (22) | Trial T77 | China |
| | | 4 (14) | Trial T54 | China |
| | | 4 (9) | Trial T46 | China |
| 125 (185) families | | Grandmothers represent 22 provenances/seed sources | | |

Note: * Key to abbreviations: Mt = Mount; DMFF = Dongmen State Forest Farm, Guangxi, China; Ind. = Indonesia. ** numerical ID's of provenances relate to locations indicated by numbered blue dots in Figure 2.

## 2.2. DNA Isolation and Microsatellite PCR Amplification

Total genomic DNA was extracted from 300 mg of fresh leaf (from each tree sampled) using the modified cetyltrimethyl ammonium bromide (CTAB) method, following the methodology described by Wang [21]. DNA quality and quantity were determined by applying agarose gel electrophoresis and spectrophotometery, using a NanoDrop 2000 spectrophotometer (Thermo Fisher Scientific Inc., Waltham, MA, USA).

Sixteen microsatellite loci were used in the study, all of which had been previously described and used in *Eucalyptus* species (Table 3). These loci were selected from among a total of 608 published microsatellite loci, as described by He et al. [22] and Brondani et al. [23], for their high polymorphism and allelic frequency differences between pools selected for genotyping a 'discovery' population following procedures described by Wang et al. [24].

**Table 3.** Details of 16 microsatellite loci examined in this study.

| No. | Locus | Reference | No. | Locus | Reference |
|---|---|---|---|---|---|
| 1 | EUCeSSR805 | [22] | 9 | Embra165 | [23] |
| 2 | Embra57 | [23] | 10 | Embra186 | [23] |
| 3 | Embra78 | [23] | 11 | Embra192 | [23] |
| 4 | Embra86 | [23] | 12 | Embra198 | [23] |
| 5 | Embra94 | [23] | 13 | Embra226 | [23] |
| 6 | Embra119 | [23] | 14 | Embra333 | [23] |
| 7 | Embra130 | [23] | 15 | Embra368 | [23] |
| 8 | Embra145 | [23] | 16 | Embra377 | [23] |

Note: EUCeSSR805 is an expressed sequence tag marker while all others are neutral markers across the whole genome.

Polymerase chain reactions (PCRs) were performed in a total volume of 10 μL, following a touchdown PCR procedure, as described by Li and Gan [25]. This procedure involved: Incubation at 94 °C for 1 min; 20 cycles of incubation at 94 °C for 20 s, 66 °C for 30 s with a decrease of 0.5 °C per cycle, and 72 °C for 1 min; 25 cycles of 94 °C for 20 s, 56 °C for 30 s, 72 °C for 30 s; then, a final extension at 72 °C for 10 min. Fluorescein-12-dUTP1 mM aqueous solution (MBI Fermentas Inc., Burlington, ON, Canada) was added to facilitate subsequent detection of PCR products using an Applied Biosystems 3130xl Series Genetic Analyzer (Applied Biosystems, Foster City, CA, USA).

### 2.3. Statistical Analyses

GenePop v4.2 [26] software was used to test the breeding populations studied for Hardy-Weinberg equilibrium, heterozygote deficits and for heterozygote excesses based on the Markov chain method using 10,000 dememorizations, 20 batches and 5000 iterations per batch. The above provided probability test parameters for Hardy-Weinberg equilibrium ($P_{HWE}$) separately for each one of the 16 microsatellite loci. Analyses with this software also provided estimates of the frequencies of null alleles at each locus.

The polymorphism information content (PIC) for every microsatellite locus was estimated separately for each breeding population by using the software PowerMarker v3.25 [27]. The number of alleles ($N_a$), effective number of alleles ($N_e$), Shannon's Information Index (I), observed heterozygosity ($H_o$), expected heterozygosity ($H_e$), and Wright's fixation index ($F_{IS}$) were also calculated separately for each locus (by population) using the software GenAlEx v6.4.1 [28], and then averaged across loci for each population. The same software was also used to determine counts of the number of private alleles ($N_{pa}$) for each provenance and seed source [29]. Meanwhile, FSTAT v2.9.3.2 [30] software was used to evaluate genetic differentiation index ($F_{st}$) values on a pairwise basis between provenances of the first cycle breeding population.

### 2.4. Population Genetic Structure

Analyses of molecular variance (AMOVA) were carried out separately for the first and third cycle populations using GenAlEx v6.4.1 to apportion genetic variance within each population, based on 999 permutations. For the first cycle population, data from provenances represented by four or fewer families was omitted (i.e., 5 provenances omitted), and across the other 15 provenances the genetic variance was apportioned into that attributed to variation among provenances, where 'provenance' refers to both natural stand provenances and exotic seed sources (i.e., collections from trials at Dongmen), and to variation among individuals within provenances. As origins of two provenances included in the first cycle—DMFF I and DMFF II—were somewhat uncertain, the AMOVA was then repeated for the first cycle population without these two provenances. For the third cycle population, genetic variance was attributed to variation among seed sources, with each trial contributing material designated as a separate seed source, and to variation among individuals within seed sources.

Nei's genetic distance between the 15 provenances that contributed to the first cycle population (represented by ≥5 or more families in our sample), was estimated using Powermarker v3.25. These estimates were then used to create a neighbor-joining (NJ) dendrogram by applying cluster analysis using the unweighted pair group method with arithmetic means (UPGMA) and a bootstrap resampling number of 1000. MEGA v7 software [31] was used to edit the UPGMA-NJ dendrogram.

The number of genetically homogeneous clusters (K) in both the first and third cycle populations, were estimated using the software STRUCTURE v2.3.4 [32], which uses a Bayesian model-based clustering method that does not require prior information on either the number of sampling sites or the locations from which the individuals were sampled. The program parameters in STRUCTURE v2.3.4 were set as recommended by Pritchard et al. [33], including the assumption of admixture among populations and correlated allele frequencies. A burn-in period of 50,000 iterations was followed by 100,000 iterations of the Markov Chain Monte Carlo model (MCMC). The model was run for a range of K values varying from 2 to 16 and with 5 replications in each iteration. The optimal K value supported by the data was assessed according to the recommendations of Evanno et al. [34], whereby the statistic ΔK was calculated based on the rate of change in the log probability of data between successive K values. The optimal K value was determined with the highest ΔK method, and this was carried out using the software STRUCTURE Harvester v0.6 [35].

## 3. Results

### 3.1. Microsatellite Loci Diversity and Polymorphism

In total 459 and 428 alleles were identified across the 16 microsatellite loci examined in the first and third cycle populations respectively (Table 4). The average number of alleles per locus ($N_a$) and the average effective number alleles per locus ($N_e$) were similar in both populations at 28.7 and 10.6 respectively for the first cycle, and at 26.8 and 10.4 respectively for the third cycle. Averaged across loci, both the polymorphism information content (PIC) and Shannon's information index (I) values for the third cycle population, being 0.87 and 2.58 respectively, showed little change from the first cycle population (0.86 and 2.56 respectively). Similarly, the values of average observed heterozygosity ($H_o$) and average expected heterozygosity ($H_e$) remained similar from the first to third cycle population, as did the values of Wright's fixation index ($F_{IS}$) (0.32 and 0.31 respectively) (Table 4); values which suggest an excess of homozygotes in both populations.

**Table 4.** Loci genetic diversity indices for the first and third cycle breeding populations of *E. urophylla*; except for 'N' and '$N_a$ total' the numbers for each trait in each population represent means across the 16 loci analyzed with standard deviations of these means given in brackets.

| Population | N | $N_a$ Total | $N_a$ | $N_e$ | PIC | I | $H_o$ | $H_e$ | $F_{IS}$ |
|---|---|---|---|---|---|---|---|---|---|
| 1st | 202 | 459 | 28.7 (2.7) | 10.6 (1.8) | 0.86 (0.02) | 2.56 (0.14) | 0.59 (0.06) | 0.87 (0.02) | 0.32 (0.07) |
| 3rd | 125 | 428 | 26.8 (2.0) | 10.4 (1.5) | 0.87 (0.02) | 2.58 (0.12) | 0.60 (0.06) | 0.88 (0.01) | 0.31 (0.07) |

Note: N, number of trees and families sampled; $N_a$ Total, total number of alleles observed across all loci; $N_a$, observed number of alleles per locus averaged across the 16 loci for each population; $N_e$, effective number of alleles per locus averaged across loci; PIC, polymorphism information content, averaged across loci; I, Shannon's information index, averaged across loci; $H_o$, observed heterozygosity averaged across loci; $H_e$, expected heterozygosity, averaged across loci; $F_{IS}$, Wright's fixation index averaged across loci.

### 3.2. Hardy-Weinberg Equilibrium and Null Alleles

Both the first and third cycle populations showed marked deviations from Hardy-Weinberg equilibrium; at all 16 loci analyzed, the probabilities of Hardy–Weinberg equilibrium were non-significant (probability test parameters for Hardy-Weinberg equilibrium—$P_{HWE}$—were less than

0.01 at each of the 16 loci). These results are in agreement with the excess of homozygotes suggested by some of the indices presented for both populations in Table 4.

Together with the indications of deviations from equilibrium at each locus, relatively high estimated frequencies of null alleles (from over 0.10 up to 0.36) were found at some loci in both populations studied (9 loci in both the first and third cycle populations). Although such null alleles were likely to have biased homozygote frequencies and hence the magnitudes of departure from the Hardy–Weinberg equilibrium, it is worth noting that all loci with low frequencies of null alleles (<0.05) also showed a departure from Hardy–Weinberg equilibrium.

### 3.3. Population Diversity and Variation

The portion of the first cycle population sampled in this study comprised predominantly first generation progeny (170 out of 202 families, representing 20 provenances) from the species' native range. The other families sampled (32 out of 202) were from mother trees selected at Dongmen Forest Farm (DMFF) and assumed to be second generation progeny. Among the 15 provenances represented by ≥5 families, $N_a$ by provenance (over all loci) ranged from 5.7 (Ilwaki) to 16.0 (Mt Egon) and $N_e$ ranged from 3.8 (Ulanu River) to 8.3 (Mt Egon) and Shannon's information index (I) ranged from 1.44 (Ulanu River) to 2.25 (Mt Egon) (Table 5). Three of these 15 provenances—Andalan, Bangat and Wukoh—had no private alleles (i.e., $N_{pa}$ = 0) whilst in the other 12 of these provenances $N_{pa}$ varied from 2 (Jawaghar and Mt Lewotobi I) up to 18 (Mt Egon). Expected heterozygosity ($H_e$) by provenance ranged from 0.71 (Ulanu River) to 0.88 (Andalan) and observed heterozygosity ($H_o$) ranged from 0.33 (Ulanu River) to 0.69 (Jawaghar). The average of Wright's fixation index ($F_{IS}$) across the 15 provenances was 0.28, and by provenance this parameter ranged from 0.13 (Jawaghar) to 0.54 (Ulanu River) indicating significant heterozygotic deficits and excesses of homozygotes for most provenances.

**Table 5.** Genetic diversity indices for the provenances comprising the first cycle breeding population of *E. urophylla*—results presented only for the 15 provenances which had 5 or more families sampled.

| Provenance | N | $N_a$ | $N_e$ | I | $N_{pa}$ | $H_o$ | $H_e$ | $F_{IS}$ |
|---|---|---|---|---|---|---|---|---|
| Andalan, Flores | 10 | 9.4 | 7.2 | 2.03 | 0 | 0.62 | 0.88 | 0.26 |
| Bangat, Flores | 10 | 8.9 | 6.5 | 1.96 | 0 | 0.64 | 0.87 | 0.23 |
| DMFF I, China | 18 | 11.9 | 7.4 | 2.14 | 6 | 0.62 | 0.87 | 0.28 |
| DMFF II, China | 14 | 9.2 | 6.0 | 1.89 | 8 | 0.53 | 0.83 | 0.35 |
| Iling Gele, Flores | 7 | 6.9 | 5.4 | 1.74 | 3 | 0.61 | 0.85 | 0.23 |
| Ilwaki, Wetar | 5 | 5.7 | 4.7 | 1.57 | 7 | 0.58 | 0.84 | 0.26 |
| Jawaghar, Flores | 10 | 9.1 | 6.8 | 1.91 | 2 | 0.69 | 0.84 | 0.13 |
| Mandiri, Flores | 9 | 7.4 | 5.4 | 1.74 | 4 | 0.55 | 0.82 | 0.31 |
| Mt Egon, Flores | 36 | 16.0 | 8.3 | 2.25 | 18 | 0.59 | 0.85 | 0.30 |
| Mt Lewotobi I, Flores | 10 | 8.4 | 6.1 | 1.86 | 2 | 0.63 | 0.84 | 0.20 |
| Mt Lewotobi II, Flores | 25 | 13.2 | 8.3 | 2.19 | 14 | 0.66 | 0.86 | 0.23 |
| Uhak, Wetar | 5 | 5.8 | 4.8 | 1.59 | 9 | 0.61 | 0.84 | 0.21 |
| Ulanu River, Alor | 9 | 6.1 | 3.8 | 1.44 | 10 | 0.33 | 0.71 | 0.54 |
| Wairteban, Flores | 10 | 8.6 | 6.2 | 1.89 | 5 | 0.56 | 0.85 | 0.32 |
| Wukoh, Flores | 9 | 7.6 | 5.6 | 1.80 | 0 | 0.54 | 0.84 | 0.33 |
| Average | 12.5 | 8.9 | 6.2 | 1.87 | 5.9 | 0.58 | 0.84 | 0.28 |
| Std. dev. | 8.2 | 2.9 | 1.3 | 0.23 | 5.3 | 0.08 | 0.04 | 0.09 |

Note: N, number of trees and families sampled; $N_a$, number of alleles per locus averaged across the 16 loci; $N_e$, number of effective alleles per locus averaged across loci; I, Shannon's information index, averaged across loci; $N_{pa}$, total number of private alleles across all 16 loci; $H_o$, observed heterozygosity averaged across loci; $H_e$, expected heterozygosity averaged across loci; $F_{IS}$, Wright's fixation index averaged across loci; Mt: Mount; Std. dev.: standard deviation.

For all the above genetic diversity indices, the two second generation seed sources were not exceptional; though DMFF I indices values were generally above those of DMFF II, both of them were only slightly above average. However, differences between these two sources may have been affected somewhat by the sample sizes; 23 families sampled represented DMFF I, but only 9 families represented DMFF II.

An AMOVA of hierarchical genetic variance across the 15 provenances of the first cycle breeding population represented by 5 or more families, revealed that only 3.4% of the total variance was attributed to variation among provenances, whilst most of the variance was due to variation among individuals within provenances (Table 6). When the AMOVA for the first cycle population was repeated after the two provenances of uncertain origins were removed—DMFF I and DMFF II—it yielded almost identical results to the previous one, indicating that those two sources, despite being potentially homogeneous with respect to provenance origins, had little effect on the AMOVA estimate regarding the level of among provenance variation. A separate AMOVA, based on the same loci, was carried out for the third cycle population and revealed similar results; genetic variation among individuals within seed sources accounted for almost 99% of the total molecular variance, whilst only 1% was attributed to variation among the seed sources (i.e., the five field trials of the first and second cycles from which the families were sourced).

**Table 6.** Analyses of molecular variance (AMOVA) based on 16 microsatellite loci across: 15 provenances of the first cycle breeding population of *E. urophylla* that were represented by 5 or more families; and, across the five field trial sources that contributed to the third cycle breeding population of *E. urophylla*.

| Source of Variation | Degrees of Freedom | Sum of Squares | Variance Components [1] | Percentage of Variation |
|---|---|---|---|---|
| 1st cycle population | | | | |
| Among provenances | 14 | 174.9 | 0.24 * | 3.4 |
| Among Individuals within provenances | 359 | 2423.5 | 6.75 * | 96.6 |
| Total | 373 | 2598.4 | 6.99 | 100 |
| 3rd cycle population | | | | |
| Among seed sources | 4 | 85.7 | 0.13 * | 1 |
| Among Individuals within seed sources | 120 | 2212.4 | 18.44 * | 99 |
| Total | 124 | 2298.1 | 18.56 | 100 |

[1] Significance levels of variance components were based on 999 permutations; * indicates significant at $p < 0.001$.

### 3.4. Genetic Structure of Populations

The average genetic differentiation index values ($F_{st}$) for all pairwise provenance comparisons in the first cycle population was 0.044 and between any two provenances this index ranged from 0.012 (Egon vs. DMFF I) to 0.108 (Ulanu River vs. Jawaghar) (Table 7), suggesting a low degree of genetic differentiation among these *E. urophylla* provenances. The highest value achieved for this index was between two provenances from geographically distant origins; Ulanu River (Alor) and Jawaghar (Flores) which have a straight line distance of separation of approximately 220 km. Similarly, the $F_{st}$ values of Ulanu River vs. Iling Gele, Ulanu River vs. Mandiri, and Ulanu River vs. Wukoh were also higher than other pairwise provenance comparisons and in each of these pairs the provenances had straight line distances of separation varying from around 184 to 222 km.

**Table 7.** Pairwise genetic differentiation indices ($F_{st}$) for the provenances comprising the first cycle breeding population of *E. urophylla*—grey shading.

| Provenance | Andalan | Apui | Arnau | Bangat | Carbubu | Dalaki | DMFF I | DMFF II | Mt Egon | Iling Gele | Ilwaki | Jawag-har | Lewotobi I | Lewotobi II | Man-diri | Pirit-umas | Tele-mar | Uhak | Ulanu R | Waikui | Wair-teban |
|---|---|---|---|---|---|---|---|---|---|---|---|---|---|---|---|---|---|---|---|---|---|
| Apui | 0.032 | | | | | | | | | | | | | | | | | | | | |
| Arnau | 0.030 | 0.042 | | | | | | | | | | | | | | | | | | | |
| Bangat | 0.024 | 0.034 | 0.034 | | | | | | | | | | | | | | | | | | |
| Carbubu | 0.035 | 0.044 | 0.046 | 0.041 | | | | | | | | | | | | | | | | | |
| Dalaki | 0.036 | 0.046 | 0.034 | 0.041 | 0.047 | | | | | | | | | | | | | | | | |
| DMFF I | 0.018 | 0.024 | 0.027 | 0.022 | 0.029 | 0.030 | | | | | | | | | | | | | | | |
| DMFF II | 0.028 | 0.040 | 0.022 | 0.033 | 0.033 | 0.026 | 0.023 | | | | | | | | | | | | | | |
| Mt Egon | 0.018 | 0.031 | 0.027 | 0.024 | 0.027 | 0.029 | 0.012 | 0.018 | | | | | | | | | | | | | |
| Iling Gele | 0.031 | 0.049 | 0.034 | 0.033 | 0.049 | 0.041 | 0.032 | 0.038 | 0.030 | | | | | | | | | | | | |
| Ilwaki | 0.037 | 0.040 | 0.050 | 0.047 | 0.046 | 0.055 | 0.034 | 0.042 | 0.030 | 0.059 | | | | | | | | | | | |
| Jawaghar | 0.037 | 0.044 | 0.045 | 0.047 | 0.040 | 0.045 | 0.035 | 0.039 | 0.024 | 0.047 | 0.050 | | | | | | | | | | |
| LewotobiI | 0.037 | 0.038 | 0.042 | 0.037 | 0.041 | 0.043 | 0.031 | 0.035 | 0.027 | 0.057 | 0.048 | 0.036 | | | | | | | | | |
| LewotobiII | 0.026 | 0.031 | 0.030 | 0.027 | 0.031 | 0.029 | 0.022 | 0.026 | 0.018 | 0.031 | 0.039 | 0.020 | 0.026 | | | | | | | | |
| Mandiri | 0.046 | 0.043 | 0.054 | 0.049 | 0.054 | 0.053 | 0.039 | 0.051 | 0.037 | 0.066 | 0.044 | 0.062 | 0.045 | 0.043 | | | | | | | |
| Piritumas | 0.036 | 0.042 | 0.054 | 0.046 | 0.039 | 0.056 | 0.021 | 0.031 | 0.024 | 0.054 | 0.047 | 0.040 | 0.045 | 0.039 | 0.054 | | | | | | |
| Telemar | 0.043 | 0.046 | 0.049 | 0.054 | 0.058 | 0.047 | 0.032 | 0.044 | 0.035 | 0.057 | 0.049 | 0.052 | 0.054 | 0.042 | 0.038 | 0.044 | | | | | |
| Uhak | 0.043 | 0.045 | 0.049 | 0.040 | 0.045 | 0.049 | 0.034 | 0.035 | 0.028 | 0.057 | 0.041 | 0.059 | 0.043 | 0.042 | 0.047 | 0.051 | 0.055 | | | | |
| Ulanu R | 0.086 | 0.093 | 0.096 | 0.081 | 0.105 | 0.075 | 0.067 | 0.069 | 0.068 | 0.101 | 0.096 | 0.108 | 0.085 | 0.088 | 0.097 | 0.082 | 0.091 | 0.070 | | | |
| Waikui | 0.036 | 0.038 | 0.050 | 0.044 | 0.043 | 0.043 | 0.030 | 0.037 | 0.024 | 0.057 | 0.041 | 0.045 | 0.038 | 0.037 | 0.035 | 0.043 | 0.044 | 0.037 | 0.083 | | |
| Wairteban | 0.040 | 0.040 | 0.043 | 0.048 | 0.046 | 0.036 | 0.029 | 0.034 | 0.026 | 0.053 | 0.047 | 0.047 | 0.042 | 0.041 | 0.042 | 0.038 | 0.027 | 0.048 | 0.079 | 0.022 | |
| Wukoh | 0.035 | 0.044 | 0.042 | 0.031 | 0.040 | 0.047 | 0.032 | 0.036 | 0.028 | 0.041 | 0.035 | 0.051 | 0.051 | 0.035 | 0.056 | 0.050 | 0.058 | 0.031 | 0.095 | 0.043 | 0.050 |

### 3.5. Structure

Relationships between the 15 provenances from the first cycle population (that had 5 or more families sampled for this study) are summarized in an unrooted neighbor-joining dendrogram, which is based on Nei's genetic distance estimates, as shown in Figure 3. While the dendrogram generally does not show a strong connection to provenance geographic origins, it must be noted that many clusters and nodes identified in this are poorly supported (i.e. bootstrap values <50—see [36]). On account of this there can be little confidence in the patterns and genetic associations observed in the dendrogram. Of the two seed sources originating from previous trials (of exotic genetic material), DMFF I was closest to the provenances of Anadalan, Bangat and Iling Gele from the island of Flores, whilst DMFF II was closest to the provenance Ulanu from Alor Island, but these associations are dubious and uncertain on account of low bootstrap values associated with their branches.

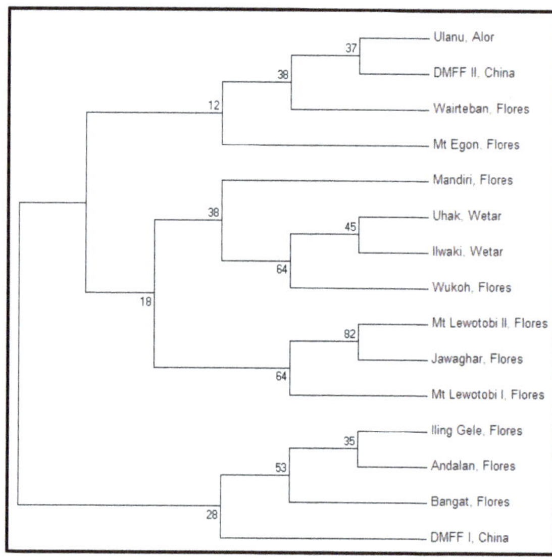


**Figure 3.** Unrooted neighbor joining (NJ) dendrogram for the 15 *E. urophylla* provenances (from which 5 or more families were sampled) represented in the first cycle breeding population in China, numbers placed at the head of branches are bootstrap values (based on 1000 iterations).

Bayesian cluster analysis performed using STRUCTURE software on the first cycle population (all provenances and families sampled) initially suggested the existence of 12 genetically homogenous clusters within this population; log-likelihoods of the number of clusters plateaued at K = 12. However, methodology of Evanno et al. [34] strongly supported K = 4 as the correct number of clusters within this population. As STRUCTURE has been found to work "extremely well for inferring the number of clusters" even with $F_{st}$ values down as low as 0.02 [37], there can be reasonable confidence in these cluster numbers; in the present study only 5 of the 231 relevant pair-wise provenance comparisons (first cycle population) had $F_{st}$ values of less than 0.02.

Whilst the 12 clusters indicated by the initial analysis showed no discernible associations with geographic origins, the subsequently indicated 4 clusters aligned somewhat weakly with geographic provenance origins (Figure 4a). Individuals originating from earlier trials in China (i.e., DMFF I and DMFF II) were predominantly allocated to cluster 2, individuals from Flores Island were predominantly allocated to cluster 4 and those from Alor and Wetar were predominantly allocated to cluster 3. The members of cluster 1 were mixed, with only 34% or less of individuals originating from any particular island/seed source being allocated to this cluster. However, it must be noted that some individuals might have been assigned to the wrong cluster, as $F_{st}$ needs to be over 0.05 to achieve an assignment accuracy of 95% or more [37]; in the present study, the average of $F_{st}$ across all relevant pair wise provenance comparisons was 0.044.

Similar cluster analyses conducted on the third cycle population suggested that it could be separated into just three genetically homogeneous clusters, i.e., K = 3 from the methodology of Evanno et al. [34] (Figure 4b). Weak patterns were also evident in the membership of these three clusters, with progeny from trials E94, TJJ and E135 being predominantly assigned to cluster 1; the former two of these trials comprised predominantly provenances from the islands of Wetar and Alor, whilst the latter trial itself comprised the second cycle population (and was predominantly second generation trees). Progeny from trials E46 and E77 were predominantly allocated to cluster 3; both of these trials contained predominantly progeny from provenances originating from the island of Flores. For cluster 2, membership was mixed with only 33% or less of individuals from any source (field trial) being allocated to this cluster.

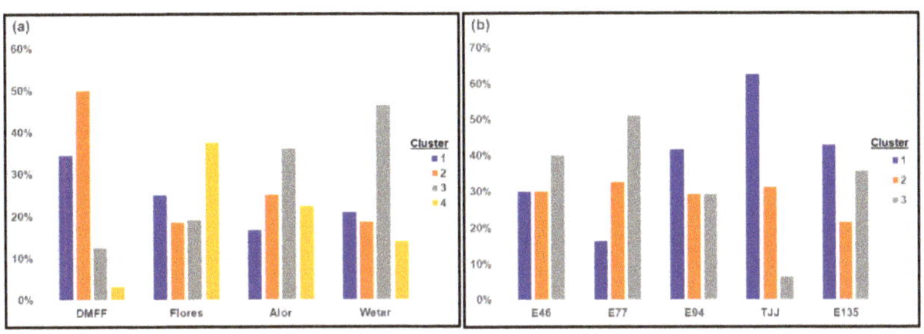


**Figure 4.** Proportions from each source group belonging to genetically homogenous clusters, determined by Bayesian cluster analyses with number of clusters determined by methods of Evanno et al. [34], for: (**a**) the first cycle population (presented by source groups comprising island/country of origin); and, (**b**) the third cycle population (presented by source groups comprising field trials of origin) of *E. urophylla*. Categories on X-axes represent geographic origins (parent locations), and these are for: (**a**) DMFF = Dongmen Forest Farm, China; Flores = Flores Island, Indonesia; Alor = Alor Island, Indonesia; Wetar = Wetar Island, Indonesia; for (**b**) E46, E77, E94 and TJJ are first cycle family trials, and E135 is the second cycle family trial.

## 4. Discussion

### 4.1. First Cycle Population

The first cycle breeding population, which included over 25 *E. urophylla* provenances originating from its natural distribution and 5 or possibly more exotic seed sources (though only 20 provenances and 2 exotic seed sources sampled in this study), was established to provide a foundation for long term breeding through recurrent cycles of selection and inter-mating. The intentionally broad selection of provenances, representing 5 of the 7 islands where the species grows naturally in Indonesia and East Timor, was anticipated to provide broad genetic diversity and hence a solid foundation for long term genetic gains.

The diversity indices estimated in this study for the provenances included in the first cycle population, $N_a$ = 5.7–16.0 alleles/locus and average $H_o$ and $H_e$ of 0.58 and 0.84 respectively, indicate that it harbors reasonable levels of genetic diversity. While these results provide clear insights on the breeding population, it must be emphasized that they do not necessarily reflect the population genetic parameters of the 20 provenances of the natural range, nor when considered collectively do they reflect the genetic diversity within the species' entire natural range. Five of the provenances were represented by four or fewer families (thus DNA samples were only collected from four or less trees per provenance), and the trees that were sampled as representatives of those provenances had been subject to intensive artificial selection well before the sampling for this study was carried out.

Despite the above limitations, the genetic diversity parameters estimated for the first cycle breeding population were similar to those found by previous studies carried out on natural populations of this species not subjected to prior selection. The values estimated were also higher than those estimated for a number of other *Eucalyptus* species. For example, Jones et al. [38] obtained a $H_e$ estimate of 0.62 from an *E. globulus* study that included 158 trees from four natural populations of the species, and Elliot and Byrne [39] reported $H_e$ estimates, by population, in *E. occidentalis* ranging from 0.30 to 0.41.

The relatively low genetic differentiation observed for the first cycle population, as indicated by the pairwise $F_{st}$ values that ranged from 0.008 to 0.108, was not unexpected. Similar values (for $F_{st}$) have been reported for natural populations of a range of other *Eucalyptus* species; i.e., $F_{st}$ = 0.03 in *E. populnea* [40], $F_{st}$ = 0.045 in *E. marginata* [41], $F_{st}$ = 0.044–0.065 in *E. camaldulensis* [42], and $F_{st}$ = 0.08 in *E. globulus* [38]. In the case of the species involved in the current study, House and Bell [4] suggested that pollen flow among populations and even among islands, mediated by birds and bats, may have contributed to the low differentiation between its geographically disparate populations.

The diversity indices estimated in this study from material sampled from the first cycle population (Table 5) were generally slightly lower than those found for "populations" of *E. urophylla*'s native range by two previous studies, though most "populations" in those previous studies encompassed multiple natural stand provenances. Payn et al.'s [13] study of 357 families from 19 populations (encompassing 45 natural stand provenances), representing all 7 islands, reported average number of alleles ($N_a$) per locus per population of 7.7 to 12.0, expected heterozygosities ($H_e$) per population of 0.44 to 0.90 and observed heterozygosities ($H_o$) per population of 0.44 to 0.78. Tripiana et al.'s [15] study of 17 populations (encompassing 49 natural stand provenances) found $H_o$'s per population of 0.51 to 0.72 and average $N_a$'s per marker loci, by population, of 5.2 to 10.6. As well as sampling differences arising from examination of populations (each comprising multiple provenances) vs. individual provenances, the lower indices of the present study might also be due to the use of different microsatellite markers than what were used by the earlier studies and/or the effects of artificial selection (prior to sampling) on the provenances examined in this present study.

Notable differences between this current study and both Payn et al.'s [13] and Tripiana et al.'s [15] studies regarding the indices estimated, were for Wright's fixation index ($F_{IS}$). In the current study the estimates for this index for the first cycle population ranged from 0.13 to 0.54 at a provenance level (while when averaged across the 15 provenances represented by $\geq$5 families it was 0.28), and were generally higher than those reported by Payn et al. [13], who found values (per population) of just 0.017 to 0.150 and a little higher than those of Tripiana et al. [15], who reported values (per population) of 0.13 to 0.31. It is noteworthy that the latter authors suggested that the fixation index ($F_{IS}$) values they reported could have been overestimated, as their DNA samples were extracted from non-selected seedlings that might have also included some seedlings originating from selfed seed. Given that intensive selection for growth and form had been a factor in the breeding populations sampled in this study, such a factor is quite unlikely to have contributed to our results.

Several factors might have contributed to the elevated fixation index values of the current study, compared to natural stand populations not subjected to artificial selection. In the first cycle population a number of provenances had relatively high and positive fixation index ($F_{IS}$) values (>0.30), suggesting possible inbreeding; the provenance with the highest $F_{IS}$ value (Ulanu River, $F_{IS}$ = 0.54) had the lowest observed heterozygosity ($H_o$ = 0.33).

The positive and relatively high fixation index values ($F_{IS}$) found in the first cycle population of this study indicate a marked deviation from Hardy-Weinberg expectations, corroborating the Hardy-Weinberg test results obtained by individual loci. The latter result was not unexpected; both Tripiana et al. [15] and Faria et al. [43] had previously observed deviations from the Hardy-Weinberg equilibrium in natural stand genetic material of this species, a situation they attributed to an excess of homozygotes across most microsatellite loci they examined. At least three factors may have contributed to this situation in the first cycle population studied here. Firstly, the presence of directional selection: a key requirement for equilibrium to be reached is the absence of directional selection [44], yet when sampled for this study, prior heavy selective thinning had already been performed on the population resulting in less than 15% of the trees originally planted (and 75% of families) remaining. But while such selection may have contributed to deviation from the Hardy-Weinberg equilibrium, it's hard to explain how the selection for traits of relatively low heritabilities—Kien et al. [12] and Hodge and Dvorak [11] reported within provenance narrow sense heritabilities for growth traits of mostly less than 0.25—could have resulted in marked selective pressure on the alleles of the 16 microsatellite loci examined in this study, as most or all of these loci were likely neutral for traits of selection (growth and form). Secondly, the Wahlund effect; such an effect was suggested by Tripiana et al. [15] based on results of their study on 17 "natural populations" of *E. urophylla* which included 49 provenances. Their aggregation of provenances resulted in their "populations" (within the total population they studied) having originated from wide geographical ranges and each likely "consisted of several possibly differentiated subpopulations". But, for the present study, any Wahlund effect could not have come from lumping of geographically distinct groups into single sampling units: we examined heterozygosity (in the first cycle) by provenance (Table 5) and all but two of the provenances were natural stand provenances. But rather than being due to lumping, it's possible that variable levels of inbreeding and/or hybrid introgression within provenances might have resulted in some unknown substructuring within provenances resulting in an apparent Wahlund effect. On all of the islands of *E. urophylla*'s natural range except Timor, the species and *E. alba* are sympatric at elevations between 400 and 800 m, and occasional natural hybrids between these two species have been recorded in their natural ranges [3,45]. On the island of Timor the two species occasionally co-occur [45]. Dvorak et al. [46] suggested that natural introgression with *E. alba*, may have had great influence on the genetic architecture of *E. urophylla*, as natural hybridization and introgression is often apparent in field trials. Indeed, Hodge and Dvorak [11] noted that some *E. urophylla* provenances in their extensive trials had up to 50% of progeny being white-barked trees, which they considered as indicating high levels of introgression with *E. alba*. Bark characteristics of the two 'pure' species are distinct, with *E. alba* having a smooth white bark and *E. urophylla* having rough brown fibrous bark that varies from a short basal stocking through to covering the trunk and extending to small branches [3,46]. Besides such introgression, inbreeding might also have been a factor contributing to an apparent Wahlund effect in the first cycle *E. urophylla* population (as well as directly contributing to a violation of Hardy-Weinberg Equilibrium through creating homozygotic excesses). While inbreeding due to selfing was unlikely, as House and Bell [4] found *E. urophylla* to be predominantly outcrossing (at least in natural stands) with mean multi-locus outcrossing rates (t) of around 0.90 and low variations between individual trees in outcrossing rates, the species' mixed mating system could have contributed to inbreeding. The latter authors had found high levels of Wright's fixation index ($F_{IS}$), as is the case in the current study, a situation they attributed to breeding among close relatives (individuals with high coancestry) being prevalent in natural stands that had resulted in a lower level of inbreeding compared to actual selfing. Thirdly, the amplification failure of certain alleles at individual loci resulted in some null alleles and likely led to some heterozygotes being genotyped as homozygotes [17]. We estimated that null alleles occurred at frequencies (per locus) of 0.00 to 0.36 across the 16 microsatellite loci examined in the first cycle population. However, it must be noted that even at the seven loci where the frequencies of null alleles were low (below 0.05), significant deviations from the Hardy–Weinberg Equilibrium still occurred.

Out of the total molecular genetic variance recorded in the first cycle population, the vast majority (96.6%) was associated with variation among individuals within provenances. Despite the artificial selection that this population had been subject to, the percentage of variance attributed to variation among individuals was almost the same as that reported by Payn et al. [13], whose AMOVA showed that variation among unselected individuals within populations (most of which incorporated a range of geographically proximal provenances) also accounted for around 96.6% of the total molecular variance.

Even so, artificial selection probably played some role in the generally poor support (low bootstrap values), and hence lack of clear patterns, in the topology of the dendrogram illustrating genetic distances and clustering of the provenances, that was generated for the first cycle population. This result was contrary to results obtained by Payn et al. [13], whose dendrogram generated for 19 *E. urophylla* 'populations' of the species' native range coincided to a large extent with the geographic origins; their populations were generally clustered by island of origin. The latter authors identified two genetically homogenous groups (clusters), with strong geographic patterns; one of the clusters was clearly dominated by populations from the western islands (i.e., Flores, Pantar, Alor, Lomblen and Adanara) and the other one by populations from the eastern islands (Wetar and Timor). A number of factors in addition to artificial selection likely contributed to the differences between Payn et al.'s [13] cluster analysis results and those obtained from the present study (for the first cycle population), including sampling differences and loci differences, besides the fact that alignment of genetic differentiation with geographic origins is generally weak for *E. urophylla* for a variety of traits [4,13].

### 4.2. Third Cycle Population

The third cycle *E. urophylla* population involved in this study included families selected from the populations of both preceding cycles. This strategy was implemented in order to capture more genetic diversity than would have likely been available from just the single somewhat limited trial that comprised the second cycle population (it omitted selections from two of the first cycle trials). Indeed, the genetic diversity indices estimated for the third cycle population, with the total number of alleles ($N_a$) being 26.8, observed heterozygosity ($H_o$) being 0.60 and expected heterozygosity ($H_e$) being 0.88, indicate that it does indeed contain a reasonable amount of genetic diversity which is very close to what was found in the first cycle population. This result was obtained even though the third population (and the sample obtained from it) was smaller, with respect to the number of families, than the first cycle population (and the sample obtained from it).

Departure from Hardy-Weinberg Equilibrium persisted in the third cycle too. This was indicated by the expected and observed heterozygosity estimates, along with the value estimated for Wright's fixation index ($F_{IS}$), and the Hardy-Weinberg equilibrium probability test parameters for each one of the 16 microsatellite loci. But, had panmixia been achieved among the individuals selected for retention in the trials of the first and second cycle populations, the descendent third cycle population would have been expected to have a somewhat closer approximation to equilibrium. But this was not the case and we need to consider the reasons for this outcome.

One factor contributing to the third cycle population's deviation from Hardy-Weinberg equilibrium would have been artificial selection. Parents of all families included were superior trees selected within the field trials comprising the first and second cycle populations. Then, within the third cycle population, additional selection had been carried out prior to sampling: at the time of sampling for this study only 25% of trees originally planted remained in the trials, representing 158 out of the 195 families initially included.

Could the Wahlund effect, whatever its actual cause, also have persisted through to the third cycle and contributed to the observed departure from Equilibrium? This population comprised families that were second and third generation descendants from a wide range of provenances that exhibited generally limited genetic differentiation, while a range of factors discussed above could have contributed to a substructuring of the population groups examined into gene pools differing in their allele frequencies. Substructuring might have arisen due to provenance origins, with effects persisting through the cycles on account of inadequate panmixia, and/or due to some individuals having originated through hybridization with *E. alba*, and/or having varying levels of inbreeding.

Panmixia might not been achieved in the first and second cycle populations due to a combination of differences in phenology associated with geographic ancestral origins (i.e., between islands of origin and even between provenances within islands) and/or spatial separation of the separate field trials comprising the first cycle population. Swain et al. [47] found that ancestral provenance origins had a significant influence on genetic parameters into at least the second generation of an *E. nitens* breeding population (i.e., grand-maternal provenances effects were evident). Differences in phenology could have been a factor in maintenance of provenances effects through generations of open pollinated breeding.

Differences between the field trials of origin (i.e., immediate parents) in regards to average effective outcrossing rates between the selected parental plus-trees may also have been a contributor to substructuring. It is well known that outcrossing rates can affect the genetic quality of seed from *Eucalyptus* species [48], and despite that the overall genetic diversity of the third cycle population seemed reasonable (compared to the first cycle population), outcrossing rates weren't examined specifically.

Also, even into the third cycle population differential introgression (from *E. alba*) might still be leading to some substructuring. Parent tree selection in the first and second cycles was based on growth and form, irrespective of bark characteristics and/or leaf morphology, and thus some of the parent trees selected could well have carried significant amounts of *E. alba* alleles. In an intensive, well advanced breeding program (>3 generations) for *E. urophylla* in Indonesia, it was found that even after several generations of selection as *E. urophylla*, distinct *E. alba* traits were still present in some individuals [49].

Results from the cluster analyses of the third cycle population, when considered together with those from the first cycle population, accord somewhat with results from the studies of House and Bell [4] and Payn et al. [13], in that populations from the island Wetar exhibited a level of genetic differentiation from populations on Flores and closely adjacent islands, i.e., on Payn et al.'s [13] "western islands". However, the cluster analysis together with the AMOVA carried out in the current study also generally concur with House and Bell's [4] conclusions in that genetic differences and differentiation between populations originating across *E. urophylla's* natural range are generally small.

While the observed heterozygosity was around 30% below the expected heterzogosity in the third cycle population, there was almost no change in the fixation index estimated for the first cycle population. This lack of change suggests no change in genetic variability between the studied populations. Reasons for the discrepancy between the observed and expected heterozygosity are uncertain, and even if inbreeding might have been expected to be a factor contributing to such discrepancies in the first cycle population (as it mainly comprised natural stand progeny), in the third cycle population lower rates of inbreeding were expected due to the crossing between unrelated individuals in the preceding one to two generations.

### 4.3. Practical Implications

Domesticated populations of many forest tree species have been found to show relatively little reduction in overall expected heterozygosity compared to the native distributions from which they were derived [50]. The current study shows that this also holds for the *E. urophylla* breeding populations in China, at least up to the third cycle. Lefevre [50] also suggested that apparent genetic diversity in the breeding population of a domesticated tree species might be increased, relative to the original populations, due to crossing between differentiated native populations. However, results from House and Bell [4], Tripiana et al. [15] and Payn et al. [13], as well as the current study, suggest that the differentiation between the native populations of *E. urophylla* is relatively minor. Payn et al. [13] found that significant gene flow among *E. urophylla*'s geographically separate island populations has had continuous influence on the genetic diversity of the species in its natural range. Thus, crossing, within a diverse breeding population of this species, between geographically disparate native provenances, might be unlikely to result in much change to genetic diversity. Indeed the results obtained from the third cycle population of this current study support this.

Even so, evidence from other studies attest to real prospects for increasing diversity somewhat through future *E. urophylla* breeding cycles in China. Across 4 generations of *E. urophylla* breeding in Brazil, Pigato and Lopes [51] found genetic distances between individuals to be markedly higher in their third and fourth generations than their first and second generations, which they took as indicating an increase in genetic variability with the advance of their program. In an *E. regnans* program in New Zealand that had progressed through three cycles of breeding, Suontama et al. [52] found that some of their third cycle field trials provided the largest heritabilities and additive genetic variances (for height at age 3 years).

In order to better understand the genetic architecture of the current and future cycles of the *E. urophylla* breeding program in China, it would be of value in the future assessments to score trees selected (as parents for a subsequent cycle) for *E. alba* characteristics. Such characteristics include large and roundish leaves with a blunt tip and/or smooth white bark on the mature tree stems, compared to *E. urophylla* with broad-lanceolate leaves narrowing abruptly to a short point and a variable stocking of rough, sub-fibrous bark [3,45]. Such phenotypic data would enable some examination of the potential proportion of individuals that potentially express some level of introgression, and hence enable a better understanding of sources of variation within the breeding population.

The question of whether panmixia can be achieved in the *E. urophylla* breeding populations studied is also of critical importance; results from the current study suggest that it has not been achieved in at least the first and/or second cycles. An assessment of phenology to understand any temporal differences on account of ancestral geographical origins (i.e., island and provenances within island origins) is needed to understand if there is asynchronous flowering within the third cycle population, a process that might be creating unwanted substructure within the population. If real substructures and/or barriers to free interbreeding do exist, then measures such as controlled pollinations and/or sublining of the main population might be required to manage barriers to panmixia and prevent unwanted substructures limiting the potential for future genetic gains.

That null alleles were recorded at relatively high frequencies at some of the loci examined in both the first and third cycle populations, indicates a possible shortcoming in the current study's methodology. Such null alleles might be biasing some parameter estimates as they can result in inflated heterozygote deficits, and decreased estimates of $H_o$, $H_e$, and even genetic diversity [53]. For all or at least a random subset of the samples examined in the current study, no repeated marker amplifications were carried out, so estimation of error rates on locus scoring was not possible. Although this shortcoming is common among the majority of published studies on microsatellites in eucalypts, as well as in other plants (see [17]), it would be better if it was avoided. In any future work on *E. urophylla* and/or other species, we would undertake repeated marker amplification of at least a random subset (10–15%) of samples, to enable error rates to be calculated from the number of inconsistent genotypes between the first and second amplification attempts [54].

## 5. Conclusions

On account of a number of factors, including the origin of genetic materials used to develop the founding (first cycle) population for the *E. urophylla* breeding program in China, deficits of heterozygotes were found for the 16 microsatellite loci examined in both the founding (first cycle) and the descendant (third cycle) breeding populations of *E. urophylla*. Even so, the high allelic diversity observed in the founding population was maintained through cycles of intensive selection into the descendant third cycle population.

Most of the genetic variation within the two populations examined in this study existed among individuals, rather than between provenances or seed sources. This finding reaffirms that the number of (unrelated) individuals included in the populations was the key to capturing adequate genetic variation, rather than the number of seed sources/provenances represented by such individuals. Similarly, the results suggested that maintaining a high number of unrelated individuals in descendant populations should contribute to maintenance of genetic variation.

In general, the level of genetic diversity was maintained through the successive cycles observed in the current study. This indicates good prospects for maintaining if not increasing diversity through future descendant cycles of breeding *E. urophylla*.

## 6. Data Archiving

The data obtained in the course of this study, consisting of genotypic data for 16 microsatellite loci from 202 individuals of the first cycle population and 125 individuals from the third cycle population along with the necessary pedigree data, will be submitted to an appropriate online data repository upon article acceptance.

**Supplementary Materials:** The following are available online at http://www.mdpi.com/1999-4907/9/7/372/s1, Table S1: Details of the *E. urophylla* trials comprising the first, second and third cycle breeding populations for this species in China. Table S2: Additional details on Trial T164 (third cycle breeding population). Figure S1: Locations of the *E. urophylla* trials comprising the first, second and third cycle breeding populations for this species in China.

**Author Contributions:** W.L. and J.L. conceived and designed the study; W.L. and L.Z. organized and conducted the trial sampling and managed the samples; W.L. carried out the laboratory analyses; W.L., L.Z., J.L. and R.J.A. collated, managed and analyzed data; W.L. and R.J.A. wrote the paper.

**Funding:** This research was funded by Fundamental Research Funds of Chinese Academy of Forestry project [project number CAFYBB2017MA022], National Natural Science Foundation of China (project number 31700599).

**Acknowledgments:** We are grateful to Lan Jun from Dongmen State Forest Farm for the assistance in accessing the field trials sampled in this study, and to Paul Macdonell from Queensland, Australia, for preparation of Figure 1. Two anonymous reviewers provided numerous suggestions for significantly improving the content of this paper and we greatly appreciate their guidance.

**Conflicts of Interest:** The authors declare no conflict of interest.

## References

1. Pryor, L.D.; Williams, E.R.; Gunn, B.V. A morphometric analysis of *Eucalyptus urophylla* and some related taxa with descriptions of two new species. *Aust. Syst. Bot.* **1995**, *8*, 57–70. [CrossRef]
2. Eldridge, K.; Davidson, J.; Harwood, C.; Van Wyk, G. *Eucalypt Domestication and Breeding*; Oxford University Press: Oxford, UK, 1993; ISBN 019854149X9780198541493.
3. Centre for Agriculture and Biosciences International. Eucalyptus urophylla. In *The Forestry Compendium: Global Module*; CAB International: Oxon, UK, 2000.
4. House, A.P.N.; Bell, J.C. Isozyme variation and mating system in *Eucalyptus urophylla* S.T. Blake. *Silvae Genetica* **1994**, *43*, 167–176.
5. Pepe, B.; Surata, K.; Suhartono, F.; Sipayung, M.; Purwanto, A.; Dvorak, W.S. Conservation status of natural populations of *Eucalyptus urophylla* in Indonesia and international efforts to protect dwindling genepools. *For. Gen. Res.* **2004**, *31*, 62–64.
6. Arnold, R.J.; Xie, Y.J.; Midgley, S.J.; Luo, J.Z.; Chen, X.F. Emergence and rise of eucalypt veneer production in China. *Int. For. Rev.* **2013**, *15*, 33–47. [CrossRef]
7. Turnbull, J.W. Eucalypts in China. *Aust. For.* **1981**, *44*, 222–234. [CrossRef]
8. Wei, J. Research achievements of the Dongmen Eucalyptus demonstration project, Guangxi State Forest Farm Company, Guangxi, China. In Proceedings of the Regional Expert Consultation on Eucalyptus, Bangkok, Thailand, 4–8 October 1996; Kashio, M., White, K., Eds.; FAO Regional Office for Asia and the Pacific: Bangkok, Thailand, 1996; Volume II, pp. 33–45.
9. Turnbull, J.W. *Development of Sustainable Forestry Plantations in China: A Review*; ACIAR Impact Assessment Series Report No. 45; Australian Centre for International Agricultural Research: Canberra, Australia, 2007; ISBN 1832-1879.
10. Xie, Y.; Arnold, R.J.; Wu, Z.; Chen, S.; Du, A.; Luo, J. Advances in eucalypt research in China. *Front. Agric. Sci. Eng.* **2017**, *4*, 380–390. [CrossRef]
11. Hodge, G.R.; Dvorak, W.S. Provenance variation and within-provenance genetic parameters in *Eucalyptus urophylla*, across 125 test sites in Brazil, Colombia, Mexico, South Africa and Venezuela. *Tree Genet. Genomes* **2015**, *11*, 57. [CrossRef]
12. Kien, N.D.; Jansson, G.; Harwood, C.; Thinh, H.H. Genetic control of growth and form in *Eucalyptus urophylla* in northern Vietnam. *J. Trop. For. Sci.* **2009**, *21*, 50–65.
13. Payn, K.G.; Dvorak, W.S.; Janse, B.J.H.; Myburg, A.A. Microsatellite diversity and genetic structure of the commercially important tropical tree species *Eucalyptus urophylla*, endemic to seven islands in eastern Indonesia. *Tree Genet. Genomes* **2008**, *4*, 519–530. [CrossRef]

14. Payn, K.G.; Dvorak, W.S.; Myburg, A.A. Chloroplast DNA phylogeography reveals the island colonisation route of *Eucalyptus urophylla* (Myrtaceae). *Aust. J. Bot.* **2007**, *55*, 673–683. [CrossRef]
15. Tripiana, V.T.; Bourgeoi, S.M.; Verhaegen, D.V.; Vigneron, P.V.; Bouvet, J.M.M. Combining microsatellites, growth, and adaptive traits for managing in situ genetic resources of *Eucalyptus urophylla*. *Can. J. For. Res.* **2007**, *37*, 773–785. [CrossRef]
16. Wahlund, S. Zusammensetzung von population und correlation serscheinung vom standpunkt der vererbungslehre aus betrachtet. *Hereditas* **1928**, *11*, 65–106. [CrossRef]
17. Selkoeobe, K.A.; Toonen, R.J. Microsatellites for ecologists: A practical guide to using and evaluating microsatellite markers. *Ecol. Lett.* **2006**, *9*, 615–629. [CrossRef] [PubMed]
18. Nicolle, D. Classification of the Eucalypts (Angophora, Corymbia and Eucalyptus) Version 3. Available online: http://www.dn.com.au/Classification-Of-The-Eucalypts.pdf (accessed on 19 March 2018).
19. Arnold, R.J.; Luo, J.Z.; Lu, W.H.; Wang, C.B.; Lin, Y. Co-operative Improvement of Key Eucalypt Species in China. In *Scientific Cultivation and Green Development to Enhance the Sustainability of Eucalypt Plantations, Proceedings of IUFRO Eucalypt Conference, Zhanjiang, China, 21–24 October 2015*; IUFRO: Vienna, Austria, 2015; p. 53.
20. Brooker, M.I.H. A new classification of the genus *Eucalyptus* L'Hér. (Myrtaceae). *Aust. Syst. Bot.* **2000**, *13*, 79–148. [CrossRef]
21. Wang, Y. Development of EST-SSR markers and their application to construction of genetic maps in Eucalyptus. Ph.D. Thesis, Nanjing Forestry University, Nanjing, China, 2009.
22. He, X.; Wang, Y.; Li, F.; Weng, Q.; Li, M.; Xu, L.; Shi, J.; Gan, S. Development of 198 novel EST-derived microsatellites in *Eucalyptus* (Myrtaceae). *Am. J. Bot.* **2012**, *99*, e134–e148. [CrossRef] [PubMed]
23. Brondani, R.P.V.; Williams, E.R.; Brondani, C.; Grattapaglia, D. A microsatellite-based consensus linkage map for species of *Eucalyptus* and a novel set of 230 microsatellite markers for the genus. *BMC Plant Biol.* **2006**, *6*, 20. [CrossRef] [PubMed]
24. Wang, Y.H.; Bible, P.; Loganantharaj, R.; Upadhyaya, H.D. Identification of SSR markers associated with height using pool-based genome-wide association mapping in sorghum. *Mol. Breed.* **2012**, *30*, 281–292. [CrossRef]
25. Li, F.; Gan, S. An optimised protocol for fluorescent-dUTP based SSR genotyping and its application to genetic mapping in *Eucalyptus*. *Silvae Genetica* **2011**, *60*, 18–25. [CrossRef]
26. Rousset, F. Genepop'007: A complete reimplementation of the Genepop software for Windows and Linux. *Mol. Ecol. Resour.* **2008**, *8*, 103–106. [CrossRef] [PubMed]
27. Liu, K.; Muse, S.V. PowerMarker: An integrated analysis environment for genetic marker analysis. *Bioinformatics* **2005**, *21*, 2128–2129. [CrossRef] [PubMed]
28. Peakall, R.; Smouse, P. GenAlEx 6: Genetic analysis in Excel. Population genetic software for teaching and research. *Mol. Ecol. Notes* **2006**, *6*, 288–295. [CrossRef]
29. Song, Z.J.; Yang, H.Y.; Weng, Q.J.; Zhou, C.P.; Li, F.G.; Li, M.; Lu, W.H.; Luo, J.Z.; Gan, S.M. Genetic Diversity and Selective Loci in *Eucalyptus tereticornis* Populations. *Sci. Silvae Sin.* **2016**, *52*, 39–47. [CrossRef]
30. FSTAT v2.9.3.2. Available online: http://www2.unil.ch/popgen/softwares/fstat.htm (accessed on 14 October 2016).
31. Kumar, S.; Stecher, G.; Tamura, K. MEGA7: Molecular Evolutionary Genetics Analysis Version 7.0 for Bigger Datasets. *Mol. Biol. Evol.* **2016**, *33*, 1870–1874. [CrossRef] [PubMed]
32. Pritchard, J.K.; Stephens, M.; Donnelly, P. Inference of population structure using multilocus genotype data. *Genetics* **2000**, *155*, 945–959. [PubMed]
33. Pritchard, J.K.; Wen, X.; Falush, D. Documentation for structure software: Version 2.3. *J. Pediatr. Surg.* **2010**, *41*, 55–63.
34. Evanno, G.; Regnaut, S.; Goudet, J. Detecting the number of clusters of individuals using the software STRUCTURE: A simulation study. *Mol. Ecol.* **2005**, *14*, 2611–2620. [CrossRef] [PubMed]
35. Earl, D.A.; von Holdt, B.M. STRUCTURE HARVESTER: A website and program for visualizing STRUCTURE output and implementing the Evanno method. *Conserv. Genet. Resour.* **2012**, *4*, 359–361. [CrossRef]
36. Hillis, D.M.; Bull, J.J. An empirical test of bootstrapping as a method for assessing confidence in phylogenetic analysis. *Syst. Biol.* **1993**, *42*, 82–192. [CrossRef]
37. Latch, E.K.; Dharmarajan, G.; Glaubitz, J.C.; Rhodes, O.E. Relative performance of Bayesian clustering software for inferring population substructure and individual assignment at low levels of population differentiation. *Conserv. Genet.* **2006**, *7*, 295–302. [CrossRef]

38. Jones, R.C.; Steane, D.A.; Potts, B.M.; Vaillancourt, R.E. Microsatellite and morphological analysis of *Eucalyptus globulus* populations. *Can. J. For. Res.* **2002**, *32*, 59–66. [CrossRef]
39. Elliott, C.; Byrne, M. Genetic diversity within and between natural populations of *Eucalyptus occidentalis* (Myrtaceae). *Silvae Genetica* **2003**, *52*, 169–173.
40. Holman, J.E.; Hughes, J.M.; Fensham, R.J. A morphological cline in *Eucalyptus*: A genetic perspective. *Mol. Ecol.* **2003**, *12*, 3013–3025. [CrossRef] [PubMed]
41. Wheeler, M.A.; Byrne, M.; McComb, J.A. Little genetic differentiation within the dominant forest tree, *Eucalyptus marginata* (Myrtaceae) of south-western Australia. *Silvae Genetica* **2003**, *52*, 254–259.
42. Butcher, P.A.; Otero, A.; McDonald, M.W.; Moran, G.F. Nuclear RFLP variation in *Eucalyptus camaldulensis* Dehnh. from northern Australia. *Heredity* **2002**, *88*, 402–412. [CrossRef] [PubMed]
43. Faria, D.A.; Mamani, E.M.C.; Pappas, M.R.; Pappas, G.J.; Grattapaglia, D. A selected set of EST-derived microsatellites, polymorphic and transferable across 6 species of *Eucalyptus*. *J. Heredity* **2010**, *101*, 512–520. [CrossRef] [PubMed]
44. Hartl, D.L.; Clarke, A.G. *Principles of Population Genetics*; Sinauer Associates: Sunderland, MA, USA, 2007; ISBN 9780878933068.
45. Martin, B.; Cossalter, C. Les *Eucalyptus* des îles de la Sonde. *Bois For. Trop.* **1975**, *164*, 3–14.
46. Dvorak, W.S.; Hodge, G.R.; Payn, K.G. The conservation and breeding of *Eucalyptus urophylla*: A case study to better protect important populations and improve productivity. *South For.* **2008**, *70*, 77–85. [CrossRef]
47. Swain, T.L.; Verryn, S.D.; Laing, M.D. An investigation of assumptions made in estimating genetic parameters and predicting genetic gain in a *Eucalyptus nitens* breeding programme in South Africa. *New For.* **2014**, *46*, 7–21. [CrossRef]
48. Burgess, I.P.; Williams, E.R.; Bell, J.C.; Harwood, C.E.; Owen, J.V. The effect of outcrossing rate on the growth of selected families of *Eucalyptus grandis*. *Silvae Genetica* **1996**, *45*, 97–100.
49. Clegg, P.A.; Toba Pulp Lestari, Medan, Sumatra, Indonesia. Personal communication, 2018.
50. Lefevre, F. Human impacts on forest genetic resources in the temperate zone: An updated review. *For. Ecol. Manag.* **2004**, *197*, 257–271. [CrossRef]
51. Pigato, S.M.P.C.; Lopes, C.R. The evaluation of genetic variability in four generations of *Eucalyptus urophylla* S.T. Blake by RAPD marker. *Sci. For.* **2001**, *60*, 119–133.
52. Suontama, M.; Low, C.B.; Stovold, G.T.; Miller, M.A.; Fleet, K.R.; Li, Y.; Dungey, H.S. Genetic parameters and genetic gains across three breeding cycles for growth and form traits of *Eucalyptus regnans* in New Zealand. *Tree Genet. Genomes* **2015**, *11*, 133. [CrossRef]
53. Paetkau, D.; Waits, L.P.; Clarkson, P.L. An empirical evaluation of genetic distance statistics using microsatellite data from bear (*Ursidae*) populations. *Genetics* **1997**, *147*, 1943–1957. [PubMed]
54. Hoffman, J.I.; Amos, W. Microsatellite genotyping errors: Detection approaches, common sources and consequences for paternal exclusion. *Mol. Ecol.* **2005**, *14*, 599–612. [CrossRef] [PubMed]

*forests*

MDPI

*Article*

# Identification of miRNAs Associated with Graft Union Development in Pecan [*Carya illinoinensis* (Wangenh.) K. Koch]

**Zhenghai Mo [1,2,3], Gang Feng [1], Wenchuan Su [1], Zhuangzhuang Liu [1] and Fangren Peng [1,*]**

1 Co-Innovation Center for the Sustainable Forestry in Southern China, Nanjing Forestry University, Nanjing 210037, China; mozhenghai@yeah.net (Z.M.); m15895892025@163.com (G.F.); swcsdau@163.com (W.S.); wlclekd@126.com (Z.L.)

2 Institute of Botany, Jiangsu Province and Chinese Academy of Sciences, Nanjing 210014, China

3 Green Universe Pecan Science and Technology Co., Ltd., 38 Muxuyuan Street, Nanjing 210007, China

* Correspondence: frpeng@njfu.edu.cn; Tel.: +86-025-8542-7995

Received: 23 May 2018; Accepted: 31 July 2018; Published: 3 August 2018

**Abstract:** Pecan [*Carya illinoinensis* (Wangenh.) K. Koch] is a high-value fruit tree with a long juvenile period. The fruiting process of pecan seedlings can be largely accelerated through grafting. As non-coding small RNAs, plant miRNAs participate in various biological processes through negative regulation of gene expression. To reveal the roles of miRNAs in the graft union development of pecan, four small RNA libraries were constructed from the graft union at days 0, 8, 15, and 30 after grafting. A total of 47 conserved miRNAs belonging to 31 families and 39 novel miRNAs were identified. For identified miRNAs, 584 target genes were bioinformatically predicted, and 266 of them were annotated; 29 miRNAs (including 16 conserved and 13 novel miRNAs) were differentially expressed during the graft process. The expression profiles of 12 miRNA were further validated by quantitative reverse transcription PCR (qRT-PCR). In addition, qRT-PCR revealed that the expression levels of 3 target genes were negatively correlated with their corresponding miRNAs. We found that miRS26 might be involved in callus formation; miR156, miR160, miR164, miR166, and miRS10 might be associated with vascular bundle formation. These results indicate that the miRNA-mediated gene regulations play important roles in the graft union development of pecan.

**Keywords:** grafting; pecan; miRNA; graft union; sequencing

## 1. Introduction

Grafting, as an asexual propagation technology, has been applied extensively in fruit trees to aid the adaptation of scion cultivars to potentially disadvantageous soil and climatic conditions, avoid the juvenile period, increase productivity, and improve quality [1]. Successful grafting is a complicated process that involves the initial adhesion of rootstock and scion, callus formation, and vascular connection at the graft union [2]. It has been reported that phytohormones (especially auxin) and antioxidant enzymes are important players during graft union development [3–6]. At the molecular level, a successful graft is controlled by numerous genes in plants, especially for the genes involved in hormone signaling. cDNA amplified fragment length polymorphism (AFLP) analysis of graft union in hickory [*Carya tomentosa* (Lam.) Nutt.] indicated that some genes related to signal transduction, metabolism, auxin transportation, wound response, cell cycle, and cell wall synthesis were responsive to grafting [7]. In *Arabidopsis*, genes involved in hormone signaling, wounding, and cellular debris clearing were induced during graft union development [8]. In grapevine, graft union formation activated the differential expression of genes participated in secondary metabolism, cell wall modification, and signaling [9]. Transcriptomic analysis of graft union in *Litchi chinensis* Sonn. revealed that 9 unigenes annotated in auxin signaling had higher expression levels in the compatible grafts compared with the incompatible ones [10].

miRNAs, a category of non-coding RNAs with approximate 22 nucleotides (nt), are critical regulatory molecules of gene expression; they induce either post-transcriptional degradation or translational inhibition of their target mRNAs [11]. In plants, miRNAs bind to their target mRNA sequences with perfect or near-perfect complementarity, and negatively regulate gene expression mainly via targeted cleavage [12]. The binding sites of plant miRNAs are almost exclusively located within the open reading frames of their target genes [13]. In recent years, with the development of second-generation sequencing technology, miRNAs could be identified from non-model plants [14,15]. Numerous studies have suggested that miRNAs play regulatory roles in plant resistance to biotic and abiotic stresses, such as cold [16], heat [17], and virus infection [18]. In addition, miRNAs have been confirmed to participate in various development processes [19–21]. In grafted plants, miRNA has been reported to be involved in the regulation of scion and rootstock interaction. In watermelon cultivation, grafting is commonly used to increase resistance to environmental stresses. With high-throughput sequencing, Liu et al. [22] found that miRNAs would differentially express in grafted watermelon to regulate plant adaptation to stresses. Li et al. [23] identified grafted-responsive miRNAs in cucumber/pumpkin, pumpkin/cucumber heterografts, and found that miRNAs were involved in regulating physiological process of hetrografts. Khaldun et al. [24] investigated the expression profiles of miRNAs within a distant grafting of tomato/goji, and the result showed that when compared with tomato autografts, tomato/goji heterografts had 43 and 163 differently expressed miRNAs in shoot and fruit, respectively. Although mounting evidence indicates the involvement of miRNAs in scion-rootstock interactions, there was only one published report concerning the functions of miRNAs which participate in the graft process, which was presented in hickory [25].

Pecan [*Carya illinoinensis* (Wangenh.) K. Koch] is an economic nut tree which belongs to the family Juglandaceae and genus *Carya*. It has been widely planted in China in recent years. As a woody plant, the juvenile phase of pecan seedlings is very long, lasting about 10 years. To accelerate the fruit bearing process, grafting is widely used in the cultivation of pecan, by which, trees can begin to produce fruit in 5–7 years. In industrial pecan cultivation, grafting success rate is very low; 75% grafting success is considered to be good [26]. Nowadays, in China, using the graft technique of patch budding can sometimes achieve a grafting success of 90% for some cultivars of pecan, such as 'Pawnee', 'Stuart', and 'Shaoxing'. However, a low grafting success rate still exists in some cultivars, such as 'Mahan' and 'Jinhua'. To improve the graft survival rate of industrial pecan, a better understanding of the mechanism associated with the graft union development is needed. In our previous studies, morphological, proteomic, and transcriptomic analyses have been conducted in the graft process of pecan [27,28]. In this work, we investigated miRNA expression during the graft process of pecan using RNA-sequencing technology. Four small RNA libraries from the graft union collected at different time points (days 0, 8, 15, and 30 after grafting) were constructed, and the differentially expressed miRNAs were analyzed. Subsequently, the potential roles of these miRNAs and their target genes were discussed.

## 2. Materials and Methods

### 2.1. Plant Materials

Pecan homograft was performed through patch grafting in August at the experimental farm at Nanjing Forestry University. The pecan cultivar 'Pawnee' was used as scion, and one-year-old seedlings propagated from pecan seeds were used as rootstock. Based on our morphological observation of graft union development, samples from the graft unions (approximately 5 mm in length, the budding segment that includes the tissues of scion, and the developing xylem of rootstock) were collected at day 0 (ungrafted materials, and were used as control), day 8 (the stage of initial callus proliferation), day 15 (the stage of massive callus proliferation along with cambium establishment), and day 30 (the stage of vascular bundles formation). For each sample, three different graft unions were pooled and frozen in liquid nitrogen immediately, and then stored at −80 °C until required for use.

### 2.2. RNA Extraction and Deep Sequencing of Small RNA

Total RNA was isolated from graft unions at four time points using the trizol reagent (Invitrogen, Carlsbad, CA, USA), following the manufacturer's instructions, and then digested with RNA-free DNase I (Takara, Kyoto, Japan) to degrade genomic DNA. Sequencing libraries were constructed by NEBNext® UltraTM small RNA Sample Library Prep Kit for Illumina® (NEB, Boston, MA, USA) according to the protocol. Briefly, approximately 1.5 μg RNA was ligated to 5′ and 3′ adapter by T4 RNA ligase for each of the samples. Next, reverse transcription synthetic first chain and PCR amplification was conducted. The resulting PCR products were subjected to polyacrylamide gel electrophoresis, and the 140–160 bp fragments were screened for sequencing. The sequencing raw data was deposited in the NCBI Sequence Read Archive (SRA) with the accession number SRP131300.

### 2.3. Sequence Analysis and Target Prediction of Pecan miRNA

Following sequencing, raw reads of the four libraries were processed through in-house Perl scripts. In this step, clean reads were obtained by removing low-quality reads and trimming adapter sequences. Reads smaller than 18 nt or longer than 30 nt were also abandoned. By using Bowtie software, clean reads with 18–30 nt in length were subsequently blasted against the Rfam (http://www.sanger.ac.uk/software/Rfam) and Repbase databases (http://www.girinst.org/) to filter rRNA, tRNA, snRNA, snoRNA, other ncRNA and repeats. The remaining sequences were aligned with the miRBase 21.0 database (http://www.mirbase.org/index.shtml) to identify putative conserved miRNAs, allowing no more than two mismatches. The remaining non-annotated reads were mapped to the pecan graft union development's transcriptome data (accession number SRP118757 and GGRT00000000 in NCBI database) to predict potential novel miRNAs by miRDeep2. The criteria for novel miRNA identification was as follows: (1) miRNA precursors could form hairpin-like structures; (2) miRNA should have a corresponding miRNA * in sequencing data, and the two could form a duplex with 2 nt 3′ overhangs; (3) in miRNA *-deficient cases, candidate miRNAs should derive from multiple and independent libraries [29]. The secondary structures of novel miRNAs were predicted by Randfold software. Putative targets of miRNA were predicted by TargetFinder, and then annotated based on the databases of Nr (NCBI non-redundant protein sequences), Protein family (Pfam) and GO (Gene Ontology). The expression value of putative target genes were obtained from the supplementary materials of our previously published paper (http://www.mdpi.com/2073-4425/9/2/71/s1) [27].

### 2.4. Analysis of Differentially Expressed miRNAs

To calculate the expression levels of miRNAs in four libraries, miRNA counts were first normalized as transcripts per million (TPM) using the following formula: TPM = mapped read count/total reads × 106. Fold changes of miRNA in three comparisons (day 8/day 0, day 15/day 0, and day 30/day 0) were analyzed by IDEG6, and the miRNA were considered to be differentially expressed with the corrected $p$ value ($q$ value) < 0.05 and absolute $\log_2$ fold change >1.

### 2.5. Quantitative Real-Time PCR (qRT-PCR)

To validate the expression profiles of miRNAs, graft unions were collected at days 0, 8, 15, and 30 after grafting, with three biological repetitions. miRNAs were isolated by the Universal Plant microRNA Kit (BioTeke, Beijing, China). The subsequent reverse transcription and real-time PCR were carried out using BioTeke miRNA First Strand cDNA synthesis kit (BioTeke, Beijing, China) and BioTeke miRNA qPCR Detection Kit (BioTeke, Beijing, China), respectively. For target gene detection, total RNAs were extracted from the same samples, as mentioned above. First-strand cDNA synthesis and the following real-time qPCR were conducted by Prime-Script™ II First Strand cDNA synthesis kit (Takara, Dalian, China) and SYBR Premix Ex Taq™ II kit (Takara, Dalian, China), respectively. Primers were designed based on the sequence of corresponding miRNAs and mRNAs, and were available in Table S1. 5.8S rRNA was chosen as an internal reference for miRNA normalization, while the Actin was used as an endogenous reference for mRNA analysis. All qPCR was run in three technical replicates. The relative expression levels of miRNAs and mRNA were calculated using the comparative $2^{-\Delta\Delta Ct}$ method.

To explore tissue-specific expression, miRNAs and total RNAs were extracted from different organs, including wound-induced calluses, xylem, phloem, and leaves. The qPCR primes are listed in Table S1.

## 3. Results

### 3.1. Analysis of Small RNA Sequencing

To identify miRNAs associated with graft union development in pecan, four small RNA libraries were constructed from the graft unions harvested at days 0, 8, 15, and 30 after grafting. Deep sequencing produced 20,691,228, 21,849,708, 22,850,876, and 34,439,863 raw reads for the four libraries, respectively (Table 1). After removing low-quality reads, 17,060,180 (day 0), 19,032,782 (day 8), 19,780,849 (day 15), and 28,632,161 (day 30) clean reads were obtained. Among the clean reads, 6,579,996 (day 0), 6,865,905 (day 8), 6,935,788 (day 15), and 7,815,688 (day 30) reads were unique, and 1,506,852 (day 0), 2,537,261 (day 8), 1,759,479 (day 15), and 2,290,283 (day 30) reads could map to the reference unigene (accession number GGRT00000000 in NCBI database). By aligning to Rfam and Repbase database, clean reads were classified into rRNA, snRNA, snoRNA, tRNA, and repeat-associated sRNA for almost all the libraries, except for day 0, which had no snRNA. The remaining unannotated reads were used for conserved miRNA identification and novel miRNAs predication.

The length distribution of unique clean reads ranging from 18 nt to 30 nt was summarized (Figure 1). We found that the most abundant class was the 24 nt sRNAs, which was consistent with previous studies in hickory [25,30,31]. The second most numerous was 23 nt sRNAs, and the majority of the sRNAs were generally distributed between 21 and 24 nt.

**Table 1.** Analysis of small RNAs from libraries of days 0, 8, 15, and 30 in pecan.

| Libraries | Day 0 | Day 8 | Day 15 | Day 30 |
|---|---|---|---|---|
| Raw reads | 20,691,228 | 21,849,708 | 22,850,876 | 34,439,863 |
| Clean reads | 17,060,180 | 19,032,782 | 19,780,849 | 28,632,161 |
| Unique reads | 6,579,996 | 6,865,905 | 6,935,788 | 7,815,688 |
| rRNA | 7,103,064 | 6,569,083 | 8,616,736 | 12,273,558 |
| snRNA | 0 | 1 | 1 | 54 |
| snoRNA | 4481 | 32,307 | 6889 | 6924 |
| tRNA | 278,816 | 217,321 | 280,911 | 279,464 |
| Repbase | 19,799 | 31,233 | 19,735 | 24,141 |
| Unannotated reads | 9,654,020 | 12,182,837 | 10,856,577 | 16,048,020 |
| Mapped reads | 1,506,852 | 2,537,261 | 1,759,479 | 2,290,283 |

Note: Raw reads, reads generated from squencing platform; Clean reads, reads after quality control; Unique reads, clean reads after clustering; rRNA, ribosomal ribonucleic acid; snRNA, small nuclear ribonucleic acid; tRNA, transfer RNA; Repbase, repeat sequence; Unannotated reads, reads can not align to Rfam and Repbase databases; Mapped reads, the unannotated reads that can map to reference unigenes.

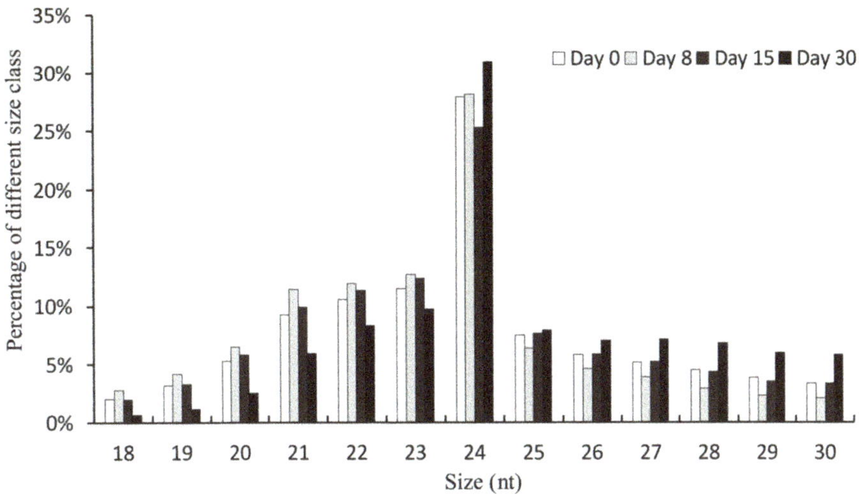


**Figure 1.** Size distribution of sRNAs from the libraries of days 0, 8, 15, and 30 in pecan. For each library, sRNAs were based on the total unique clean reads.

### *3.2. Identification of Conserved miRNAs in Pecan*

To obtain conserved miRNAs in pecan, all unannotated reads in Rfam and Repbase were pooled and used to do a blast against miRbase, allowing two mismatches. Based on miRbase results and hairpin prediction, a total of 47 conserved miRNAs with their corresponding star strands were identified from the four libraries (Table S2). These 47 conserved miRNAs were classified into 31 miRNA families, among them, the miR482 family possessed the maximum members (four), followed by miR166 and miR396, while the remaining families have only one or two members. The 47 miRNAs showed great difference in expression levels, of these, miR159a–b, miR166a–c, and miR319a–b had relatively high expression level, in contrast, members such as miR4998, miR5998, miR6135, miR7504, and miR7717 presented low expression levels.

### *3.3. Identification of Novel miRNAs in Pecan*

To identify novel miRNAs, all the remaining unannotated sRNAs were blasted against our transcriptome data. In total, 39 novel miRNAs corresponding to 39 distinct precursor sequences were obtained from the four libraries (Table S3), and all the precursors of these candidate miRNAs were found to have typical stem-loop structures (Figure S1). Star sequences were detected for all the novel miRNAs, an important evidence of being bona fide miRNAs [29]. The most common base for the first nucleotide of novel miRNAs was Uracil (U), a common pattern observed in other studies [25,32]. The length of these mature miRNAs ranged from 18 nt to 25 nt, and the most common was 24 nt. The range of the minimal free energy (MFE) for these novel miRNA precursors was from −96.9 to −31.8 kcal/mol, with −69.0 kcal/mol on average. The expression level of most novel miRNAs were generally low (TPM < 100), while some miRNAs such as miRS19 and miRS33 presented high level with dynamic TPM > 1000.

### *3.4. Prediction and Functional Annotation of Target Genes of miRNAs*

We searched for putative targets by blasting the miRNAs against our transcriptome sequences with perfect or near-perfect complementarity. As a result, a total of 584 targets were predicted for the 86 miRNAs (with an average of 6.8 targets per miRNA), and 266 of them were annotated (Table S4). For functional classification, these targets were subjected to GO (Gene Ontology) analysis. As shown in Figure 2, targets of miRNAs fell into 17 biological processes, with the three most abundant being metabolic process, cellular process, and single-organism process. Targets in the cellular component category were classified into 11 terms, with the three most frequent being cell, cell part, and organelle. With respect to molecular function, targets were assigned to 10 terms, with the two most frequent being binding and catalytic activity.

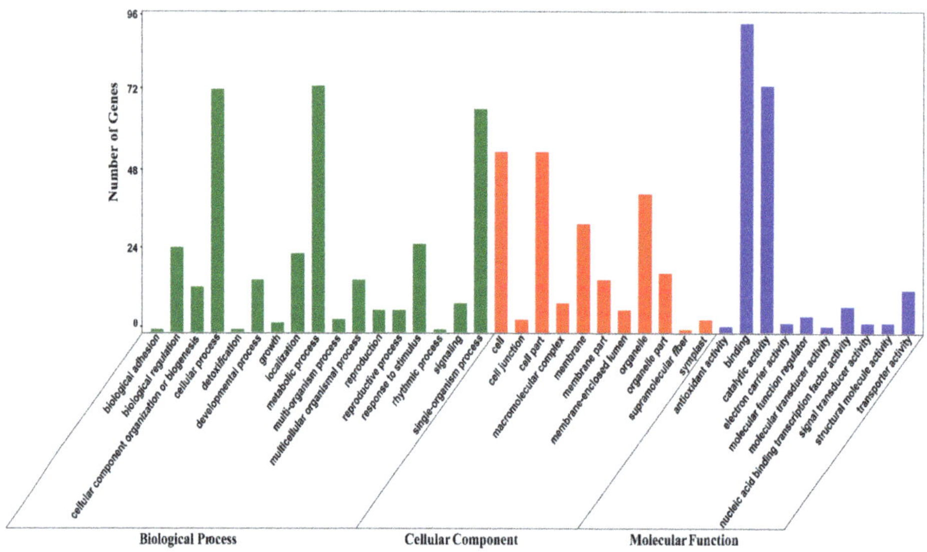


**Figure 2.** GO annotation of targets of identified miRNAs. Targets were functionally categorized by biological process, cellular component and molecular function according to the ontological definitions of the GO terms.

### 3.5. *Differential Expressed miRNAs during the Graft Process of Pecan*

To obtain insight into the possible roles of miRNAs in the graft union development of pecan, differential expressions were analyzed by comparing days 8, 15, 30 to day 0, with the criteria of absolute $\log_2$ fold change >1 and $q$ value < 0.05. In total, 29 miRNAs with 16 conserved and 13 novel were considered to be differentially expressed in the three comparisons (Table 2). Of these, 10 miRNAs were differentially expressed in the comparison of day 8/day 0, with 7 down-regulated and 3 up-regulated. Fourteen differential expressed miRNAs were identified in day 15/day 0 comparison, with 4 down-regulated and 10 up-regulated. In the comparison of day 30/day 0, 23 differential expressed miRNAs were found, with 19 down-regulated and 4 up-regulated. There were 10 miRNAs whose expression changed significantly in two comparisons, and 4 miRNAs changed obviously in three comparisons. We compared the differential expressed value between miRNAs and their targets using our transcriptome data, and found that miRNAs were generally negatively correlated with their corresponding targets (Figure 3).

**Table 2.** Differentially expressed miRNAs during the graft process of pecan.

| miRNA | Mature_Sequence (5′→3′) | TPM | | | | $Log_2$ Fold Change (Treatment vs. Control) | | | Putative Target |
|---|---|---|---|---|---|---|---|---|---|
| | | Day 0 | Day 8 | Day 15 | Day 30 | Day 8/Day 0 | Day 15/Day 0 | Day 30/Day 0 | |
| cil-miR156 | uugacagaagauagagagcac | 1053.64 | 738.88 | 2066.88 | 496.77 | −0.51 | 0.97 | −1.08 | SPL |
| cil-miR160a | ugccuggcucccuguaugcc | 174.90 | 94.66 | 118.25 | 51.99 | −0.89 | −0.56 | −1.75 | ARF |
| cil-miR160b | ugccuggcucccugaaugcc | 191.96 | 103.34 | 185.09 | 66.43 | −0.89 | −0.05 | −1.53 | ARF |
| cil-miR164a | caugugcucuagcucuccagc | 25.59 | 21.04 | 87.41 | 20.22 | −0.28 | 1.77 | −0.34 | NAC |
| cil-miR164b | uggagaagcagggcacgugca | 2337.62 | 1761.75 | 2154.29 | 768.26 | −0.41 | −0.12 | −1.61 | NAC |
| cil-miR166b | ucggaccaggcuucauucccc | 79,283.01 | 33,244.46 | 39,632.89 | 105,566.42 | −1.25 | −1.00 | 0.41 | Class III HD-ZIP |
| cil-miR171b | agguauugauguggcucaauu | 473.50 | 270.84 | 143.96 | 147.30 | −0.81 | −1.72 | −1.68 | |
| cil-miR390 | aagcucaggagggauagcgcc | 1612.45 | 1354.18 | 1933.20 | 704.72 | −0.25 | 0.26 | −1.19 | |
| cil-miR394 | uuggcauucuguccaccucc | 695.32 | 320.80 | 719.81 | 207.95 | −1.12 | 0.05 | −1.74 | F-box only protein 6-like |
| cil-miR396b | guucaauaaagcuguggg aug | 38,707.31 | 95,055.53 | 110,470.14 | 30,583.10 | 1.30 | 1.51 | −0.34 | Serine carboxypeptidase-like |
| cil-miR399 | agggcuucucuccuuuggcagg | 76.78 | 18.41 | 66.84 | 25.99 | −2.06 | −0.20 | −1.56 | |
| cil-miR482a | ggaaugggcuguuuggg auga | 30,943.67 | 36,341.98 | 39,209.03 | 6830.58 | 0.23 | 0.34 | −2.18 | |
| cil-miR482b | aaugggaagauaggaaagaac | 4146.30 | 3168.52 | 3578.48 | 1614.50 | −0.39 | −0.21 | −1.36 | |
| cil-miR482c | uggacaugggugaauugguaag | 16,060.51 | 11,832.64 | 35,229.52 | 6726.61 | −0.44 | 1.13 | −1.26 | |
| cil-miR818 | cacgacgucgguuuauuuaacagg | 25.59 | 12.76 | 41.13 | 92.42 | −1.00 | 0.68 | 1.85 | |
| cil-miR860a | cauaucuuugacuauguacugau | 29.86 | 103.34 | 174.81 | 118.42 | 1.79 | 2.55 | 1.99 | |
| cil-miRS2 | uuuuguaagaucucuguguag | 452.17 | 707.33 | 1007.73 | 872.23 | 0.65 | 1.16 | 0.95 | Syntaxin-131 |
| cil-miRS6 | caaggaaaauaggcuuuugug | 221.82 | 68.37 | 10.28 | 57.76 | −1.70 | −4.43 | −1.94 | ATP synthase gamma chain |
| cil-miRS7 | caucaguuuguggg auuacuuu | 46.92 | 44.70 | 174.81 | 69.32 | −0.07 | 1.90 | 0.56 | |
| cil-miRS8 | aguguggucgggaacccggaacu | 81.05 | 52.59 | 359.90 | 31.77 | −0.62 | 2.15 | −1.35 | |
| cil-miRS9a | ugacagaagagagagagcac | 34.13 | 23.67 | 25.71 | 0.00 | −0.53 | −0.41 | −6.64 | SPL |
| cil-miRS9b | ugacagaagagagagagcac | 34.13 | 23.67 | 25.71 | 0.00 | −0.53 | −0.41 | −6.64 | SPL |
| cil-miRS10 | ugacagaagagagugagcac | 1847.06 | 2987.08 | 3928.10 | 557.42 | 0.69 | 1.09 | −1.73 | Cinnamoyl-CoA reductase 1 |
| cil-miRS18 | caaaaugacuagucaaugau | 81.05 | 42.07 | 97.69 | 28.88 | −0.95 | 0.27 | −1.49 | Esterase-like isoform X1 |
| cil-miRS23 | uguuacuaguuuggcuuugauacu | 46.92 | 2.63 | 5.14 | 0.00 | −4.16 | −3.19 | −6.64 | |
| cil-miRS26 | uuuucguugcuauaaauuggu | 34.13 | 16.38 | 138.82 | 86.65 | −1.06 | 2.02 | 1.34 | cyclin-D1-1 isoform X1 |
| cil-miRS29 | gguggcuggauugaaucc | 230.35 | 199.84 | 534.72 | 5.78 | −0.20 | 1.21 | −5.32 | WD-40 repeat family protein |
| cil-miRS33 | caaaagugauuguauguaacauu | 3911.68 | 3512.98 | 3583.62 | 10,232.88 | −0.16 | −0.13 | 1.39 | Cell division control protein |
| cil-miRS38 | uggcugccaugcaucgucuagc | 4850.14 | 13,539.17 | 7979.60 | 8863.87 | 1.48 | 0.72 | 0.87 | |

Note: SPL, squamosa promoter-binding protein-like; NAC, NAC transcription factor; HD-ZIP, homeobox-leucine zipper; ATP, adenine triphosphate; CoA, coenzyme A; WD, tryptophan-aspartic acid.

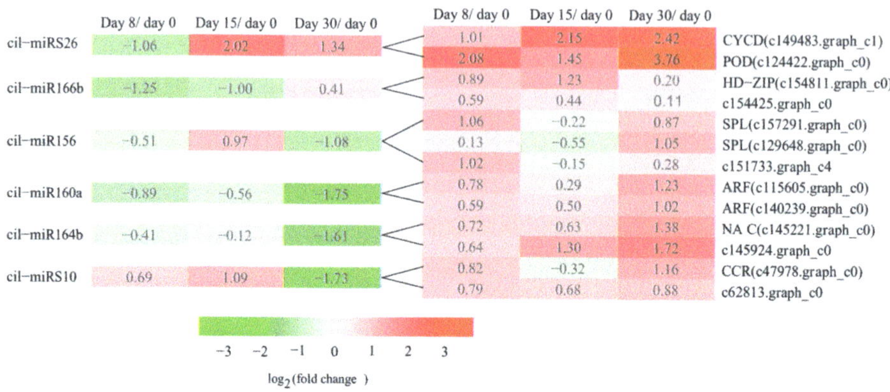


**Figure 3.** Expression profile of some miRNAs and their targets in the graft process. Columns in the heatmap reprent different comparisons (experiment/control: day 8/day 0, day 15/day 0, and day 30/day 0). Comparisons were made to calculate expression changes (fold change). Rows in the heatmap symbolize miRNAs or target genes. The data in the heapmap are the value of $\log_2$ (fold change). Red and green indicate up-regulation and down-regulation respectively.

### 3.6. *Differential Expressed miRNAs during the Graft Process of Pecan*

To validate the dynamic expression of miRNAs at different time points after grafting obtained by sequencing, 12 miRNAs, including 8 conserved and 4 novel, were chosen for qRT-PCR analysis (Figure 4). Results showed that most of the expression profiles of studied miRNAs based on qRT-PCR were similar to those detected by high-throughput sequencing, except miR394. The expression of miR394 at day 30 was down-regulated based on high-throughput sequencing, while it was up-regulated detected by qRT-PCR. Also, for specific time points after grafting, the relative expression level of miRNAs detected by these two methods did not match exactly. For instance, sequencing data indicated that the ratio of miRS23 in day 15/day 0 was 0.11, but it was 0.60 with the corresponding qRT-PCR date. This inconsistency might result from the difference in data normalization protocols of the sequencing data and qRT-PCR. The sequencing was normalized to the global abundance of mapped reads sequenced by illumina, while qRT-PCR was normalized to the level of 5.8S rRNA. A correlation analysis of the fold change of miRNA expression between sequencing and qRT-PCR showed a significant similarity with $R^2 = 0.84$ (Figure S2), confirming the reliability of results obtained by sequencing. Additionally, to further validate the dynamic correlation between miRNAs and their targets, the expression of potential targets were also subjected to qRT-PCR assay. Results showed that all the three targets had an inverse expression profile with their corresponding miRNAs (Figure 5).

### 3.7. *Expression Patterns of miRNAs and Their Targets in Different Tissues of Pecan*

To understand the main roles of miRNAs and their targets, we analyzed the tissue-specific expression profiles of miRNAs and mRNAs in different organs of pecan. Generally, miRNAs and their targets were negatively correlated, and were preferentially expressed in specific tissues (Figure 6). miR156 showed lower expression level in xylem and phloem, while its target had higher expression values in those tissues. miR160, miR164, miR166, and miRS10 exhibited low expression levels in xylem, while their corresponding targets, expect NAC, were highly expressed in xylem. The target of miRS26 displayed the highest expression in callus.

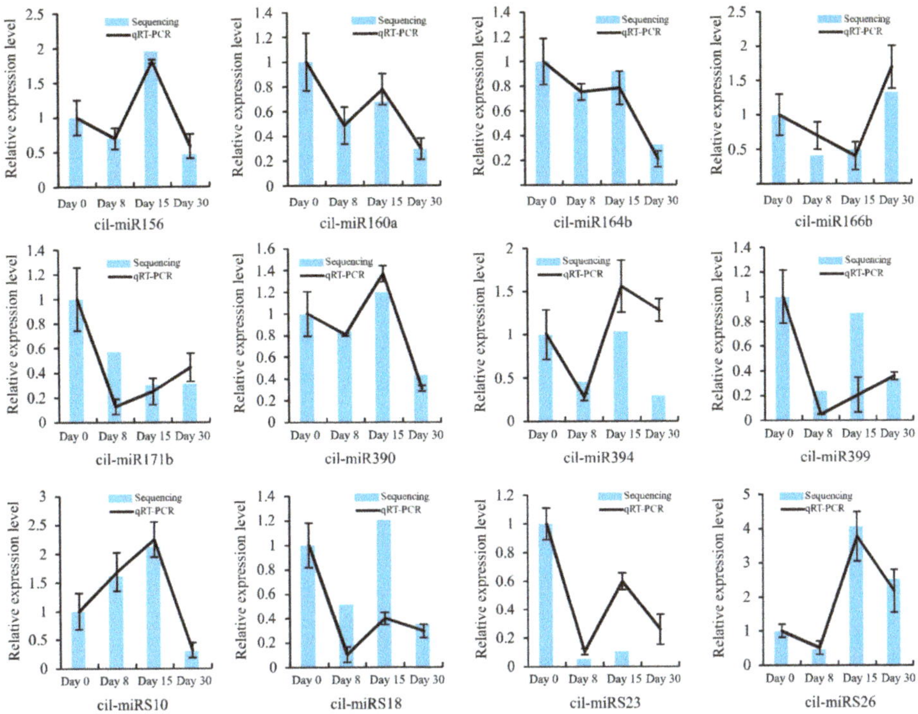


**Figure 4.** qRT-PCR validation of miRNAs in the graft process of pecan. The histograms and lines indicate miRNA expression results obtained by sequencing and qRT-PCR, respectively. The x-axis represents samples collected at different time points after grafting, while the y-axis represents the relative expression level of miRNAs. The expression levels of miRNAs are normalized to the level of 5.8S rRNA. The normalized miRNA levels at day 0 are arbitrarily set to 1. Data from qRT-PCR are means of three replicates and bars represent SE (standard error).

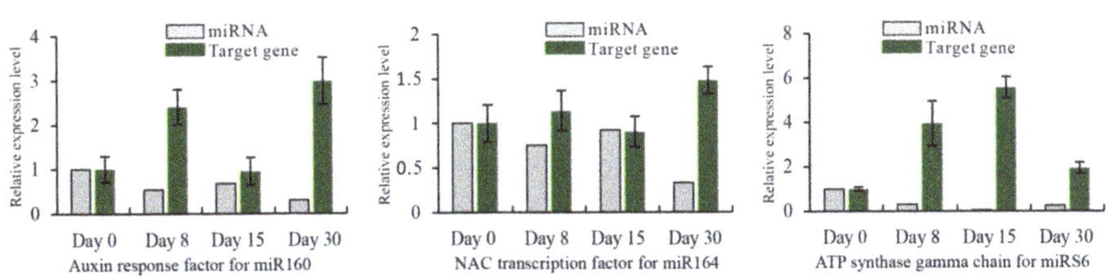


**Figure 5.** The expression of miRNAs and their targets. The relative expression levels of miRNAs and their corresponding target genes are shown in grey and green histograms, respectively. The x-axis represents samples collected at different time points after grafting, while the y-axis represents the relative expression level of miRNAs and their target genes. The expression level of miRNAs and target genes are normalized to the level of 5.8S rRNA and Actin gene. For each miRNA and target gene, the expression levels at day 0 are arbitrarily set to 1. Data from qRT-PCR are means of three replicates and bars represent SE.

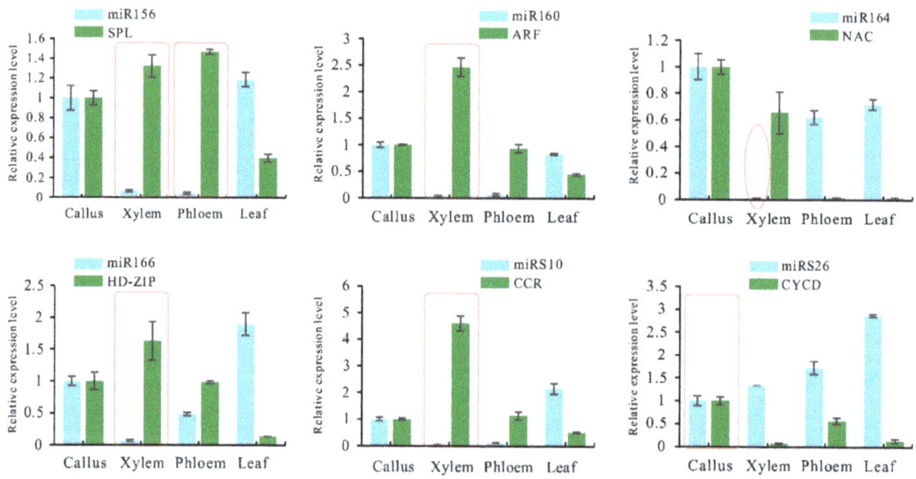


**Figure 6.** Expression of miRNAs and their targets in different tissues of pecan. The relative expression levels of miRNAs and their corresponding target genes are shown in blue and green histograms, respectively. The x-axis represents different tissues collected from pecan, while the y-axis represents the relative expression level of miRNAs and their target genes. The expression level of miRNAs and target gene are normalized to the level of 5.8S rRNA and Actin gene. For each miRNA and target gene, the expression levels at day 0 are arbitrarily set to 1. Data are means of three replicates and bars represent SE. SPL, squamosa promoter-binding protein-like; ARF, auxin response factor; NAC, NAC transcription factor; HD-ZIP, homeobox-leucine zipper; CCR, cinnamoyl-CoA reductase; CYCD, D-type cyclin.

## 4. Discussion

Although grafting has been extensively used in horticulture, our knowledge regarding the molecular mechanism of successful graft remains insufficient. Plant miRNAs are non-coding RNAs that play important roles in various biological processes at post-transcriptional level. In this study, we used high throughput sequencing to explore the conserved and novel miRNAs in pecan, and then analyzed the differentially expressed miRNAs to better understand the function of miRNAs in a successful grafting.

miRNAs are reported to be widely distributed throughout almost all eukaryotes, and some miRNAs are deeply conserved in plant kingdom [33]. In our work, 47 conserved miRNAs belonging to 31 miRNA families were identified. Of those, miRNAs including miR156, miR159, miR160, miR164, miR166, miR167, miR171, miR172, miR390, miR393, miR394, miR396, miR399, and miR403 were confirmed to be well-conserved in both monocot and dicot model plants [33]. We obtained a total of 39 novel miRNAs in the graft process of pecan. Those newly identified miRNAs might be pecan-specific. We detected that the novel miRNAs generally exhibited a lower expression level than the conserved ones, which was consistent with previous literature [34,35].

A total of 16 conserved and 13 novel miRNAs were differentially expressed during the graft processes. Since successful grafting is a developmental processes involving callus formation and vascular bundle formation, the differential expression of a cascade of miRNAs concerning those processes might suggest their involvement in the graft process as well. Previously, it has been reported that miR159, miR169, miR171, and miR172 were identified as being responsive to embryogenic callus formation of *Larix leptolepis* Gordon [36]. miR396 expressed at high level would attenuate cell proliferation in the developing leaves of *Arabidopsis thaliana* [37]. miR166 has been reported to be involved in xylem development of *Acacia mangium* (Willd.) [38]. In *Arabidopsis*, miR166 was found to be involved in vascular development through negatively regulating the expression of *ATHB15*, a *class III homeodomain-leucine zipper* (*HD-ZIP III*) gene [39]. miRNAs including miR156, miR159, miR160, miR172, miR390, and miR482 have been confirmed to participate in the graft process of hickory [25]. Consistent with these previous research, miR156, miR160, miR166, miR171, miR390, miR396, and miR482 showed significantly differential expression for our research, suggesting they might function for graft union development.

The putative target of the differentially expressed miR156, *squamosa promoter-binding protein-like* (*SPL*), encodes a plant-specific transcription factor that functions in multiple biological processes, including plant architecture, leaf development, juvenile-to-adult transition, flower and fruit development, as well as gibberellin (GA) signaling [40]. Among its divergent functions, SPL responses to GA signaling by affecting the genes involved in GAs biosynthesis. Studies have verified that GAs are important regulators in xylem differentiation [41]. In our study, miR156 was significantly down-regulated at day 30, which might induce the up-regulation of SPL during the stage of vasculature formation. It was presumed that the miR156-*SPL* interaction might involve in vascular bundle formation through regulating GA signaling indirectly.

A putative target of miR160 is *auxin response factor* (*ARF*). In the graft process, auxin has been confirmed to be critical in regulating callus formation and vascular development [8,42]. Auxin signaling is transduced via ARFs to regulate the expression of genes containing auxin response elements (AuxREs) in their promoter areas [43]. In *Arabidopsis*, *ARF6* and *ARF8* mutants reduced cell proliferation in response to cutting [42], and *ARF5* mutants showed abnormality in vascular development [44]. In the present study, we hypothesized that the down-regulated miR160a-b at day 30 may induce the accumulation of *ARF*, resulting in vascular connection.

A putative gene targeted by miR164 was the NAC transcription factor, which was in accordance with *Arabidopsis* [45], *Medicago truncatula* Gaertn [46], and *Triticum aestivum* L. [34]. NAC transcription factors are the master regulators in controlling secondary cell wall formation [47], and overexpression NAC1 in *Arabidopsis* was shown to produce thicker stems than the untransformed control plants [48]. Previous studies have reported that secondary cell-wall formation was indispensable for vascular system development [49]; thus, the down-regulated miR164b at day 30 in this work might stimulate *NAC1* expression to regulate vascular development.

miR166b belongs to the miR166 family, and targets the *homeobox-leucine zipper* (*HD-ZIP*) gene. Members of HD-ZIP gene family have been reported to function in various stress conditions, such as drought, salinity, and wounding [50,51]. The class III HD-ZIP gene family plays important roles in vascular bundle development. It was reported to be highly expressed in cambium tissue [52,53]. In *Arabidopsis*, the class III HD-ZIP proteins were also showed to control cambium activity through inducing axial cell elongation and xylem differentiation [54]. Overexpressing a *populus class III HD-ZIP* gene would lead to ectopic formation of vascular cambium within cortical parenchyma in poplar [55]. For a successful grafting, the formation of vascular bundles results from the promotion of vascular cambium activity. In our study, miR166 was down-regulated at day 15, suggesting *class III HD-ZIP* might be up-regulated at the stage of new cambium establishment. It is speculated that the increased *class III HD-ZIP* may stimulate the cambium activity for the subsequent xylem formation. Interestingly, we found that miR166 was significantly down-regulated at day 8 as well, indicating that the initial xylem differentiation might happen before new cambium establishment, as demonstrated by the reports of Pina [2].

A putative target of miRS10, *cinnamoyl-CoA reductase* (*CCR*), is a gene dedicated to monolignol biosynthesis. Down-regulation of *CCR* in poplar exhibited up to 50% reduced lignin level in outer xylem [56]. Since lignin is essential for vascular development, the down-regulated miRS10 at day 30 might induce the up-regulation of *CCR*, and then lead to the lignification of vasculature during the graft process of pecan.

A predicted target of miRS26 was *D-type cyclin* (*CYCD*). CYCD is a critical regulator that promotes the progression of cell cycle by binding to cyclin-dependent kinases A, which plays vital role in cell proliferation [57]. It was found to be induced by auxin and cytokinin [58]. *Arabidopsis* hypocotyl explants of overexpressing *CYCD4* showed faster induction of callus than the control explants on a media with lower auxin concentration [59]. In this study, miRS26 was down-regulated at day 8, a stage during initial callus formation, and then up-regulated at the following time-points, while *CYCD* was up-regulated throughout the grafting process, indicating that a negative correlation between miRS26 and *CYCD* at days 15 and 30 did not exist. However, the expressions of miRS26 and CYCD in different tissues indicated that they were negatively correlated. Considering the negative interaction exiting between miRS26 and *CYCD* during the stage of initial callus formation, and *CYCD* displaying highest expression abundance in callus tissue, we deduced that miRS26 might play a vital role in stimulating callus proliferation during graft union development.

In our study, the tissue-specific expression profiles of miRNAs and their targets might indirectly suggest their specific roles for the graft union development. The low expression of miRNAs in xylem tissues, such as miR156, miR160, miR164, miR166, and miRS10 might be indicative of their specific roles during vascular development. miRS26 showed low expression in callus tissues, suggest its possible involvement in callus formation for a successful graft.

## 5. Conclusions

Our study constructed four sRNA libraries from the graft unions of pecan collected at days 0, 8, 15, and 30 after grafting. We identified a total of 47 conserved miRNAs belonging to 31 families and 39 novel miRNAs. Among them, 29 miRNAs with 16 conserved and 13 novel were differentially expressed in the graft process, suggesting their critical roles in successful grafting. Particularly, for the graft union development of pecan, miRS26 might play an important role in callus formation; miR166, miR156, miR160, miR164, and miRS10 might contribute to vascular bundle formation (Figure 7).

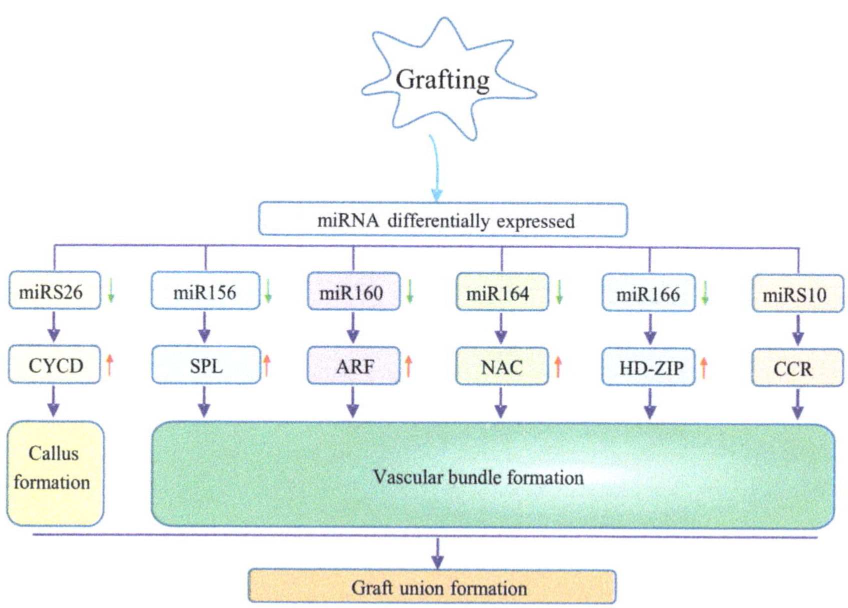


**Figure 7.** Putative regulatory mechanism involving differentially expressed miRNAs and their targets in graft union formation of pecan. The upper arrow indicates upregulation, and the down arrow represents downregulation. CYCD, D-type cyclin; SPL, squamosa promoter-binding protein-like; ARF, auxin response factor; NAC, NAC transcription factor; CCR, cinnamoyl-CoA reductase; HD-ZIP, homeobox-leucine zipper.

**Supplementary Materials:** The following are available online at http://www.mdpi.com/1999-4907/9/8/472/s1, Figure S1: The hairpin structures of novel miRNAs. The mature miRNAs are in red, and the miRNA *s are in blue, Figure S2: Correlation analysis between sequencing and qRT-PCR. Scatter plots show fold-change measured by sequencing and qRT-PCR, Table S1: Primers used in this study, Table S2: Conserved miRNAs identified in pecan, Table S3: Novel miRNAs identified in pecan, Table S4: Target genes of identified miRNAs in pecan.

**Author Contributions:** F.P. conceived and designed the study. Z.M. performed the data analysis and wrote the manuscript. G.F. and W.S. carried out qRT-PCR. Z.L. was involved in sample collection. All authors read and approved the final manuscript.

**Funding:** The authors appreciate the financial support from the SanXin project of Jiangsu province (LYSX(2016)44), the state bureau of forestry 948 project (2015-4-16) and the Priority Academic Program Development of Jiangsu Higher Education Institutions (PAPD).

**Conflicts of Interest:** The authors declare that they have no competing interests.

## References


1. Mudge, K.; Janick, J.; Scofield, S.; Goldschmidt, E.E. A history of grafting. *Hortic. Rev.* **2009**, 437–493.
2. Pina, A.; Errea, P. A review of new advances in mechanism of graft compatibility–incompatibility. *Sci. Hortic.* **2005**, *106*, 1–11. [CrossRef]
3. Aloni, B.; Karni, L.; Deventurero, G.; Levin, Z.; Cohen, R.; Katzir, N.; Lotan-Pompan, M.; Edelstein, M.; Aktas, H.; Turhan, E. Physiological and biochemical changes at the rootstock-scion interface in graft combinations between cucurbita rootstocks and a melon scion. *J. Hortic. Sci. Biotechnol.* **2008**, *83*, 777–783. [CrossRef]
4. Fernández-García, N.; Carvajal, M.; Olmos, E. Graft union formation in tomato plants: Peroxidase and catalase involvement. *Ann. Bot.* **2004**, *93*, 53–60. [CrossRef] [PubMed]
5. Asahina, M.; Iwai, H.; Kikuchi, A.; Yamaguchi, S.; Kamiya, Y.; Kamada, H.; Satoh, S. Gibberellin produced in the cotyledon is required for cell division during tissue reunion in the cortex of cut cucumber and tomato hypocotyls. *Plant Physiol.* **2002**, *129*, 201–210. [CrossRef] [PubMed]
6. Melnyk, C.W.; Schuster, C.; Leyser, O.; Meyerowitz, E.M. A developmental framework for graft formation and vascular reconnection in arabidopsis thaliana. *Curr. Biol.* **2015**, *25*, 1306–1318. [CrossRef] [PubMed]
7. Zheng, B.S.; Chu, H.L.; Jin, S.H.; Huang, Y.J.; Wang, Z.J.; Chen, M.; Huang, J.Q. Cdna-aflp analysis of gene expression in hickory (carya cathayensis) during graft process. *Tree Physiol.* **2009**, *30*, 297–303. [CrossRef] [PubMed]
8. Yin, H.; Yan, B.; Sun, J.; Jia, P.; Zhang, Z.; Yan, X.; Chai, J.; Ren, Z.; Zheng, G.; Liu, H. Graft-union development: A delicate process that involves cell–cell communication between scion and stock for local auxin accumulation. *J. Exp. Bot.* **2012**, *63*, 4219–4232. [CrossRef] [PubMed]
9. Cookson, S.J.; Moreno, M.J.C.; Hevin, C.; Mendome, L.Z.N.; Delrot, S.; Trossatmagnin, C.; Ollat, N. Graft union formation in grapevine induces transcriptional changes related to cell wall modification, wounding, hormone signalling, and secondary metabolism. *J. Exp. Bot.* **2013**, *64*, 2997–3008. [CrossRef] [PubMed]
10. Chen, Z.; Zhao, J.; Hu, F.; Qin, Y.; Wang, X.; Hu, G. Transcriptome changes between compatible and incompatible graft combination of litchi chinensis by digital gene expression profile. *Sci. Rep.* **2017**, *7*, 3954. [CrossRef] [PubMed]
11. Bartel, D.P. Micrornas: Genomics, biogenesis, mechanism, and function. *Cell* **2004**, *116*, 281–297. [CrossRef]
12. Mallory, A.C.; Vaucheret, H. Erratum: Functions of micrornas and related small rnas in plants. *Nat. Genet.* **2006**, *38*, S31. [CrossRef]
13. Millar, A.A.; Waterhouse, P.M. Plant and animal micrornas: Similarities and differences. *Funct. Integr. Genom.* **2005**, *5*, 129–135. [CrossRef] [PubMed]
14. Xu, S.; Jiang, Y.; Wang, N.; Xia, B.; Jiang, Y.; Li, X.; Zhang, Z.; Li, Y.; Wang, R. Identification and differential regulation of micrornas in response to methyl jasmonate treatment in lycoris aurea by deep sequencing. *BMC Genom.* **2016**, *17*, 789. [CrossRef] [PubMed]
15. Liu, F.; Wang, W.; Sun, X.; Liang, Z.; Wang, F. Conserved and novel heat stress-responsive micro rnas were identified by deep sequencing in s accharina japonica (l aminariales, p haeophyta). *Plant Cell Environ.* **2015**, *38*, 1357–1367. [CrossRef] [PubMed]

16. Ferreira, T.H.; Gentile, A.; Vilela, R.D.; Costa, G.G.; Dias, L.I.; Endres, L.; Menossi, M. Micrornas associated with drought response in the bioenergy crop sugarcane (*Saccharum* spp.). *PLoS ONE* **2012**, *7*, e46703. [CrossRef] [PubMed]
17. Li, H.; Hu, T.; Amombo, E.; Fu, J. Genome-wide identification of heat stress-responsive small rnas in tall fescue (*Festuca arundinacea*) by high-throughput sequencing. *J. Plant Physiol.* **2017**, *213*, 157–165. [CrossRef] [PubMed]
18. Wang, Z.; Jiang, D.; Zhang, C.; Tan, H.; Li, Y.; Lv, S.; Hou, X.; Cui, X. Genome-wide identification of turnip mosaic virus -responsive micrornas in non-heading chinese cabbage by high-throughput sequencing. *Gene* **2015**, *571*, 178–187. [CrossRef] [PubMed]
19. Li, T.; Ma, L.; Geng, Y.; Hao, C.; Chen, X.; Zhang, X. Small rna and degradome sequencing reveal complex roles of mirnas and their targets in developing wheat grains. *PLoS ONE* **2015**, *10*, e0139658. [CrossRef] [PubMed]
20. Sun, Y.; Qiu, Y.; Zhang, X.; Chen, X.; Shen, D.; Wang, H.; Li, X. Genome-wide identification of micrornas associated with taproot development in radish (*Raphanus sativus* L.). *Gene* **2015**, *569*, 118–126. [CrossRef] [PubMed]
21. Zhang, S.; Han, S.; Li, W.; Zhou, J.; Li, X.; Qi, L. Mirna regulation in fast- and slow-growing hybrid larix trees. *Trees* **2012**, *26*, 1597–1604. [CrossRef]
22. Liu, N.; Yang, J.; Guo, S.; Xu, Y.; Zhang, M. Genome-wide identification and comparative analysis of conserved and novel micrornas in grafted watermelon by high-throughput sequencing. *PLoS ONE* **2013**, *8*, e57359. [CrossRef] [PubMed]
23. Li, C.; Li, Y.; Bai, L.; Zhang, T.; He, C.; Yan, Y.; Yu, X. Grafting-responsive mirnas in cucumber and pumpkin seedlings identified by high-throughput sequencing at whole genome level. *Physiol. Plant.* **2014**, *151*, 406–422. [CrossRef] [PubMed]
24. Khaldun, A.; Huang, W.; Lv, H.; Liao, S.; Zeng, S.; Wang, Y. Comparative profiling of mirnas and target gene identification in distant-grafting between tomato and lycium (goji berry). *Front. Plant Sci.* **2016**, *7*, 1475. [CrossRef] [PubMed]
25. Sima, X.; Jiang, B.; Fang, J.; He, Y.; Fang, Z.; Saravana Kumar, K.M.; Ren, W.; Qiu, L.; Chen, X.; Zheng, B. Identification by deep sequencing and profiling of conserved and novel hickory micrornas involved in the graft process. *Plant Biotechnol. Rep.* **2015**, *9*, 115–124. [CrossRef]
26. Nesbitt, M.L. Effect of scionwood packing moisture and cut-end scaling on pecan graft success. *Horttechnology* **2002**, *12*, 257–260.
27. Mo, Z.; Feng, G.; Su, W.; Liu, Z.; Peng, F. Transcriptomic analysis provides insights into grafting union development in pecan (*Carya illinoinensis*). *Genes* **2018**, *9*, 71. [CrossRef] [PubMed]
28. Mo, Z.; He, H.; Su, W.; Peng, F. Analysis of differentially accumulated proteins associated with graft union formation in pecan (*Carya illinoensis*). *Sci. Hortic.* **2017**, *224*, 126–134. [CrossRef]
29. Meyers, B.C.; Axtell, M.J.; Bartel, B.; Bartel, D.P.; Baulcombe, D.; Bowman, J.L.; Cao, X.; Carrington, J.C.; Chen, X.; Green, P.J. Criteria for annotation of plant micrornas. *Plant Cell* **2008**, *20*, 3186–3190. [CrossRef] [PubMed]
30. Wang, Z.; Huang, J.; Huang, Y.; Li, Z.; Zheng, B. Discovery and profiling of novel and conserved micrornas during flower development in carya cathayensis via deep sequencing. *Planta* **2012**, *236*, 613–621. [CrossRef] [PubMed]
31. Wang, Z.; Huang, R.; Sun, Z.; Zhang, T.; Huang, J. Identification and profiling of conserved and novel micrornas involved in oil and oleic acid production during embryogenesis in carya cathayensis sarg. *Funct. Integr. Genom.* **2017**, *17*, 1–9. [CrossRef] [PubMed]
32. Chen, L.; Zhang, Y.; Ren, Y.; Xu, J.; Zhang, Z.; Wang, Y. Genome-wide identification of cold-responsive and new micrornas in populus tomentosa by high-throughput sequencing. *Biochem. Biophys. Res. Commun.* **2012**, *417*, 892–896. [CrossRef] [PubMed]
33. Jonesrhoades, M.W.; Bartel, D.P.; Bartel, B. Micrornas and their regulatory roles in plants. *Annu. Rev. Plant Biol.* **2006**, *57*, 19–53. [CrossRef] [PubMed]
34. Ma, X.; Xin, Z.; Wang, Z.; Yang, Q.; Guo, S.; Guo, X.; Cao, L.; Lin, T. Identification and comparative analysis of differentially expressed mirnas in leaves of two wheat (*Triticum aestivum* L.) genotypes during dehydration stress. *BMC Plant Biol.* **2015**, *15*, 21. [CrossRef] [PubMed]

35. Rajagopalan, R.; Vaucheret, H.; Trejo, J.; Bartel, D. A diverse and evolutionarily fluid set of micrornas in arabidopsis thaliana. *Genes Dev.* **2006**, *20*, 3407–3425. [CrossRef] [PubMed]
36. Zhang, S.G.; Zhou, J.; Han, S.Y.; Yang, W.H.; Li, W.F.; Wei, H.L.; Li, X.M.; Qi, L.W. Four abiotic stress-induced mirna families differentially regulated in the embryogenic and non-embryogenic callus tissues of *Larix leptolepis*. *Biochem. Biophys. Res. Commun.* **2010**, *398*, 355–360. [CrossRef] [PubMed]
37. Rodriguez, R.E.; Mecchia, M.A.; Debernardi, J.M.; Schommer, C.; Weigel, D.; Palatnik, J.F. Control of cell proliferation in arabidopsis thaliana by microrna mir396. *Development* **2010**, *137*, 103–112. [CrossRef] [PubMed]
38. Ong, S.S.; Wickneswari, R. Characterization of micrornas expressed during secondary wall biosynthesis in *Acacia mangium*. *PLoS ONE* **2012**, *7*, e49662. [CrossRef] [PubMed]
39. Kim, J.; Jung, J.H.; Reyes, J.L.; Kim, Y.S.; Kim, S.Y.; Chung, K.S.; Kim, J.A.; Lee, M.; Lee, Y.; Narry Kim, V. Microrna-directed cleavage of athb15 mrna regulates vascular development in arabidopsis inflorescence stems. *Plant J.* **2005**, *42*, 84–94. [CrossRef] [PubMed]
40. Chen, X.B.; Zhang, Z.L.; Liu, D.M.; Zhang, K.; Li, A.L.; Mao, L. Squamosa promoter-binding protein-like transcription factors: Star players for plant growth and development. *J. Integr. Plant Biol.* **2010**, *52*, 946–951. [CrossRef] [PubMed]
41. Milhinhos, A.; Miguel, C.M. Hormone interactions in xylem development: A matter of signals. *Plant Cell Rep.* **2013**, *32*, 867–883. [CrossRef] [PubMed]
42. Pitaksaringkarn, W.; Ishiguro, S.; Asahina, M.; Satoh, S. Arf6 and arf8 contribute to tissue reunion in incised arabidopsis inflorescence stems. *Plant Biotechnol.* **2014**, *31*, 49–53. [CrossRef]
43. Li, S.; Xie, Z.; Hu, C.; Zhang, J. A review of auxin response factors (arfs) in plants. *Front. Plant Sci.* **2016**, *7*, 47. [CrossRef] [PubMed]
44. Hardtke, C.S.; Berleth, T. The arabidopsis gene monopteros encodes a transcription factor mediating embryo axis formation and vascular development. *EMBO J.* **1998**, *17*, 1405–1411. [CrossRef] [PubMed]
45. Guo, H.S.; Xie, Q.; Fei, J.F.; Chua, N.H. Microrna directs mrna cleavage of the transcription factor nac1 to downregulate auxin signals for arabidopsis lateral root development. *Plant Cell* **2005**, *17*, 1376–1386. [CrossRef] [PubMed]
46. Zhao, M.; Chen, L.; Wang, T.; Tian, Q.; Zhang, W. Identification of drought-responsive micrornas in medicago truncatula by genome-wide high-throughput sequencing. *BMC Genom.* **2011**, *12*, 367.
47. Zhong, R.; Lee, C.; Zhou, J.; McCarthy, R.L.; Ye, Z.-H. A battery of transcription factors involved in the regulation of secondary cell wall biosynthesis in arabidopsis. *Plant Cell* **2008**, *20*, 2763–2782. [CrossRef] [PubMed]
48. Xie, Q.; Frugis, G.; Colgan, D.; Chua, N. Arabidopsis nac1 transduces auxin signal downstream of tir1 to promote lateral root development. *Genes Dev.* **2000**, *14*, 3024–3036. [CrossRef] [PubMed]
49. Ye, Z.; Zhong, R. Molecular control of wood formation in trees. *J. Exp. Bot.* **2015**, *66*, 4119–4131. [CrossRef] [PubMed]
50. Manavella, P.A.; Dezar, C.A.; Bonaventure, G.; Baldwin, I.T.; Chan, R.L. Hahb4, a sunflower hd-zip protein, integrates signals from the jasmonic acid and ethylene pathways during wounding and biotic stress responses. *Plant J.* **2008**, *56*, 376–388. [CrossRef] [PubMed]
51. Zhao, Y.; Ma, Q.; Jin, X.; Peng, X.; Liu, J.; Deng, L.; Yan, H.; Sheng, L.; Jiang, H.; Cheng, B. A novel maize homeodomain–leucine zipper (hd-zip) i gene, zmhdz10, positively regulates drought and salt tolerance in both rice and arabidopsis. *Plant Cell Physiol.* **2014**, *55*, 1142–1156. [CrossRef] [PubMed]
52. Baima, S.; Possenti, M.; Matteucci, A.E.; Altamura, M.M.; Ruberti, I. The arabidopsis athb-8 hd-zip protein acts as a differentiation-promotingtranscription factor of the vascular meristems. *Plant Physiol.* **2001**, *126*, 643–655. [CrossRef] [PubMed]
53. Schrader, J.; Sandberg, G. A high-resolution transcript profile across the wood-forming meristem of poplar identifies potential regulators of cambial stem cell identity. *Plant Cell* **2004**, *16*, 2278–2292. [CrossRef] [PubMed]
54. Ilegems, M.; Douet, V.; Meylanbettex, M.; Uyttewaal, M.; Brand, L.; Bowman, J.L.; Stieger, P.A. Interplay of auxin, kanadi and class iii hd-zip transcription factors in vascular tissue formation. *Development* **2010**, *137*, 975–984. [CrossRef] [PubMed]
55. Robischon, M.; Du, J.; Miura, E.; Groover, A. The populus class iii hd zip, poprevoluta, influences cambium initiation and patterning of woody stems. *Plant Physiol.* **2011**, *155*, 1214–1225. [CrossRef] [PubMed]

56. Leple, J.; Dauwe, R.; Morreel, K.; Storme, V.; Lapierre, C.; Pollet, B.; Naumann, A.; Kang, K.; Kim, H.; Ruel, K. Downregulation of cinnamoyl-coenzyme a reductase in poplar: Multiple-level phenotyping reveals effects on cell wall polymer metabolism and structure. *Plant Cell* **2007**, *19*, 3669–3691. [CrossRef] [PubMed]
57. Ikeuchi, M.; Sugimoto, K.; Iwase, A. Plant callus: Mechanisms of induction and repression. *Plant Cell* **2013**, *25*, 3159–3173. [CrossRef] [PubMed]
58. Fehér, A.; Magyar, Z. Coordination of cell division and differentiation in plants in comparison to animals. *Acta Biol. Szeged.* **2015**, *59*, 275–289.
59. Kono, A.; Ohno, R.; Umeda-Hara, C.; Uchimiya, H.; Umeda, M. A distinct type of cyclin d, cycd4;2, involved in the activation of cell division in arabidopsis. *Plant Cell Rep.* **2006**, *25*, 540–545. [CrossRef] [PubMed]

*forests*

MDPI

*Article*

# Genome-Wide Identification and Characterization of MADS-box Family Genes Related to Floral Organ Development and Stress Resistance in *Hevea brasiliensis* Müll. Arg.

**Mingming Wei [1], Yajie Wang [2], Ranran Pan [2] and Weiguo Li [1,*]**

[1] Ministry of Agriculture Key Laboratory of Biology and Genetic Resources of Rubber Tree & State Key Laboratory Breeding Base of Cultivation and Physiology for Tropical Crops, Rubber Research Institute, Chinese Academy of Tropical Agricultural Sciences, Danzhou 571737, China; lywmm@catas.cn

[2] Agricultural College, Hainan University, Haikou 570228, China; wangyajiexjs@126.com (Y.W); melody_pr@163.com (R.P)

* Correspondence: anzw@catas.cn or leewg23@163.com; Tel.: +86-0898-2330-1306

Received: 26 April 2018; Accepted: 28 May 2018; Published: 29 May 2018

**Abstract:** Elucidating the genetic mechanisms associated with the transition from the vegetative to reproductive phase in the rubber tree has great importance for both theoretical guidance and practical application to yield genetic improvement. At present, many transcription factors, including those that belong to the MADS-box gene family, have been revealed to have roles in regulating the transition from vegetative growth to reproductive growth. However, to the best of our knowledge, the Mad-box gene family from *H. brasiliensis* Müll. Arg. has not been characterized in detail. To investigate members of the HbMADS-box gene family associated with floral organ and inflorescence development in *H. brasiliensis*, we performed genome-wide identification and analysis of the MADS-box gene family related to flower development in *H. brasiliensis*, and a total of 20 MADS-box genes were newly identified in the *H. brasiliensis* genome. Expression profiling revealed that HbMad-box genes were differentially expressed in various tissues, which indicated that HbMad-box genes may exert different functions throughout the life cycle. Additionally, 12 genes (HbSEP, HbAGL9.1, HbAGL9.2, HbCMB1, HbCMB1-L, HbAGL6, HbAGL8, HbAP1, HbAG, HbDEFL, HbTT16, and HbPADS2) were found to be associated with the differentiation of flower buds and may be involved in flower development in *H. brasiliensis*. All of these floral-enriched HbMADS-box genes were regulated by hormone, salt, cold, high-temperature, and drought stresses. The present study is the first to carry out the genome-wide identification and analysis of the MADS-box gene family related to flower development in *H. brasiliensis*, and 20 new HbMad-box genes were identified in *H. brasiliensis*. Most of the newly identified HbMad-box genes were found to be associated with the differentiation of flower buds and may be involved in flower development in *H. brasiliensis*. Our results demonstrated that HbMad-box genes may be multifunctional regulators that have roles in distinct aspects of development, and are mainly involved in the maintenance of floral organ and inflorescence development.

**Keywords:** *Hevea brasiliensis* Müll. Arg.; HbMad-box genes; conserved domains; gene structures; expression profiles; stress treatments

## 1. Introduction

Natural rubber (NR) is an important industrial and strategic raw material, and has been applied to many aspects of social production [1]. Although more than 2000 plant species in the world are considered to be latex producers, the rubber tree (*Hevea brasiliensis* Müll. Arg.) is the only commercial source of NR because of its high yield and the excellent physical properties of its rubber products [2], and it supplied 92% of the 10.2 million tons of NR consumed worldwide in 2016 [3].

With the rapid development of the world economy, the consumption of NR in major industrial countries is increasing year by year. To meet the growing demand for NR, it is necessary to expand the planting area of rubber trees. However, the rubber tree originates from the Amazon rainforest, so the planting area must be located in sub-tropical to tropical zones [4]. At present, 95% of rubber trees in the world are now mainly cultivated in South-East Asia, so the regions suitable for planting rubber trees are very limited. Therefore, there is an urgent need to improve the rubber yield per area.

To the best of our knowledge, breeding new varieties is one of the most effective approaches to increasing rubber yield per hectare. However, breeding experiments to yield genetic improvement of rubber trees are very inefficient and time-consuming, mainly because of the rubber tree's long life cycle of more than 30–35 years; it is immature for five to eight years until the rubber tree reaches the age of commercial productivity [5,6], and takes more than three decades to breed and select new clones for commercial production [7]. Furthermore, due to the low rate of success for controlled pollinations, genetic improvements of *H. brasiliensis* are very difficult and slow [8]. Therefore, research on the genetic mechanisms that affect the transition from the vegetative to reproductive phase in *H. brasiliensis* can provide insight for producing advantageous genetic improvement methods for controlling reproduction by genetic engineering and accelerate *Hevea* breeding.

In flowering plants, the transition from vegetative growth to reproductive growth is an important developmental process that involves many gene regulatory processes [9]. To date, many researchers have attempted to elucidate the functional genes in association with the transition from the vegetative growth to reproductive growth of plants. It is worth noting that many transcription factors (TFs), including those that belong to the MADS-box gene family, have been demonstrated to have roles in regulating the transition from vegetative growth to reproductive growth [10]. However, to the best our knowledge, the rubber tree Mad-box gene family has not been characterized in detail.

As a floral homeotic gene family, the Mad-box gene family was previously identified and investigated in the model plants *Arabidopsis* Heynh. in Holl & Heynh. and *Nicotiana tabaccum* L. [11–13], and has evolutionarily conserved DNA-binding domains, called the MADS-box [14]. Typically, the MADS-box protein sequences can be divided into four characteristic domains from the N to the C terminus: the MADS-box (M), intervening (I), keratin-like (K), and C-terminal (C) domains [15]. In plants, based on the structural features, MADS-box TFs usually contain two main groups–type I (M-type) and type II (MIKC-type) genes [16]; whereas, the type II genes can be categorized into MIKCc- and MIKC*-type [17]. In previous studies, it has been reported that this superfamily encodes transcriptional regulators that are involved in various processes, including floral organ development [18,19], root development [20], leaf development [21], fruit development and maturation [22–24], and embryonic development [25,26]. In addition to growth and development-related functions, some MADS-box genes also play important roles in response to stress stimuli [27,28]. For instance, MADS-box genes have already been proved to play important roles under low temperature stress in tomato plants [29], while several MADS-box genes have been demonstrated to take part in cold, salt, and drought responses in rice [30]. Furthermore, a few MADS-box genes have been shown to be affected by the application of hormones and they exhibited differential expression in response to cytokinin, gibberellin [31], ethylene [32], and auxin [33] application in other plants.

Despite the fact that the MADS-box gene plays a great role in plant growth and development, only a few MADS-box genes have been identified and characterized in the rubber tree to date [34]. For example, previous studies have found that HbAGL62 is a specific expression in flowers and embryos, and highly expressed in the flower bud differentiation stage, which indicated that HbAGL62 might play an important role in flowering regulation of the rubber tree [35]; MADS27 is highly expressed in the flower buds of rubber trees and may participate in the flowering process of rubber trees [36]; HbMADS1 and HbMADS3 have highly frequent transcriptions in the laticifer cells and somatic embryogenesis, and their transcriptions are induced in the laticifer cells by jamonic acid, which indicates that HbMADS1 and HbMADS3 may be important in natural rubber biosynthesis and somatic embryogenesis in the rubber tree [37]. In the present research, we newly identified 20 MADS-box genes in the rubber tree, and analyzed their gene structure and phylogenetic relationships. To identify differentially expressed patterns of Mad-box genes in various tissues, the 20 MADS-box genes of *H. brasiliensis* were detected and analyzed using real-time quantitative PCR (RT-qPCR). Furthermore, to understand the responses of HbMADS-box genes to various stresses, the expression profiles of 12 floral organ-specific HbMADS-box genes were examined in leaves of *Hevea* seedlings after hormone, salt, cold, high-temperature, and drought stress treatments.

## 2. Materials and Methods

### 2.1. Plant Materials and Treatments

12-year-old rubber tree clones CATAS 7-33-97 were used as the experimental material in this study. The rubber trees were grown under normal field conditions at the Experimental Station of the Rubber Research Institute, the Chinese Academy of Tropical Agricultural Sciences (Danzhou, Hainan, China). The fresh tissues and organs (including roots, stem, stem tips, leaves, labeled bark, xylem, latex, fruits, inflorescence, male flowers, and female flowers) were collected from 12-year-old mature trees of CATAS 7-33-97 during the flowering period. Each sample was harvested on average from five trees, and three biological replicates were taken from each sample (the image of some samples shown in Figure S1). Then, the prepared samples were frozen in liquid nitrogen and transferred to a −80 °C refrigerator for RNA separation.

The tissue culture seedlings of CATAS 7-33-97 were treated with cold, high temperature, and drought stress. Each treatment was set up with three replicates, and each replicate consisted of five seedlings. Under the cold stress condition, the seedlings were grown in a culture incubator set at 5 °C and continuous illumination. For high-temperature stress treatment, the tissue-cultured seedlings were planted at 40 °C and maintained at a relative humidity of 80% in the incubator. Leaf samples of 0, 3, 6, 12, and 24 h treated with low and high temperature stresses were collected for RNA extraction. For drought stress treatments, the tissue-cultured plants were grown in Hoagland nutrient solutions [38] containing 20% PEG6000, and incubated at different times (0 h, 3 h, 6 h, 12 h, 1 day, 3 day, 4 day, and 7 day). Then, the leaf samples of each drought-treated time point were collected for RNA extraction, and samples from untreated plants were used as controls.

For hormone and salt treatments, the tissue-cultured seedlings were treated with abscisic acid (ABA) (200 μmol/L), gibberellin (GA) (100 μmol/L), and high salt (1 M NaCl), respectively. Among these chemicals, ABA and GA were diluted in distilled water that contained 0.05% ($v/v$) ethanol. The diluted ABA, GA, and NaCl solutions were sprayed on the leaves and stems of seedlings until the runoff occurred. For control plants, the distilled water that contained 0.05% ($v/v$) ethanol was sprayed on the leaves and stems of seedlings. Leaf samples were harvested 0, 0.5, 2, 6, 12, 24, and 48 h after treatments. In all treatments, one leaf from each of the five plants was taken and mixed together for RNA extraction.

### 2.2. RNA Isolation and First-Strand cDNA Synthesis

Total RNA was isolated from the collected samples by the described methods [39], and the extracted RNA was digested with DNase (Promega, Madison, WI, USA) to remove genomic DNA contamination. The integrity and concentration of the RNA samples was detected by 1.5% agarose gel electrophoresis and NanoDrop 2000 (Thermo Scientific Inc., Waltham, MA, USA), respectively. Then, the RNA samples were reverse transcribed into First-strand cDNA with the RevertAid™ First Strand cDNA Synthesis Kit (TaKaRa, Shiga, Japan).

### 2.3. Identification and Isolation of Mad-box Genes in H. brasiliensis

Twenty full-length cDNA sequences of *H. brasiliensis* Mad-box genes were obtained from RNA sequencing. The cDNA sequences of these genes were compared with the Transcriptome Shotgun Assembly (TSA) and Expressed Sequence Tags (EST) of *H. brasiliensis* in the NCBI database (http://www.ncbi.nlm.nih.gov/), or searched against the *Hevea* genome database. Then, the NCBI ORF Finder (http://www.ncbi.nlm.nih.gov/gorf/gorf.html) and Softberry (http://linux1.softberry.com/) were used to determine open reading frames (ORFs) of candidate mRNA or genome DNA sequences. In addition, in order to confirm the presence of the Mad-box domain, all the candidate HbMad-box genes were further validated by conserved domain searching using CDD (http://www.ncbi.nlm.nih.gov/Structure/cdd/wrpsb.cgi) and InterProScan (http://www.ebi.ac.uk/interpro/scan.html). After the similarity comparison, ORF Finder, and conserved domain searching, redundant sequences were removed and the sequences of the HbMad-box gene were obtained.

The gene-specific primers used to amplify the corresponding full-length cDNA sequences of HbMad-box genes were designed by Primer 3.0 (http://primer3.ut.ee/). The primer pairs for all HbMad-box genes are listed in Table S1. RT-PCR amplification of HbMad-box genes was conducted using Pyrobest™ DNA polymerase (TaKaRa, Japan) according to the instructions.

### 2.4. Protein Properties and Gene Structure Analysis of HbMad-box Genes

We used the ProtParam tool (http://web.expasy.org/protparam/) to predict the theoretical molecular weight (Mw) and isoelectric point (PI) of HbMad-box gene proteins. To further analyze the structural diversity of HbMADS-box genes, the exon–intron structures of HbMad-box genes were identified by comparing the coding sequence with their corresponding genomic sequence using the FGENESH-C tool (http://linux1.softberry.com), as previously described [40].

### 2.5. Multiple Sequence Alignments and Phylogenetic Analysis

In the present study, amino acid sequence identities and multiple alignments of 26 HbMad-box proteins were calculated using DNAMAN6.0. To examine the evolutionary history and phylogenetic relationships of the HbMADS-box genes, we constructed a phylogenetic tree using MEGA6.0 (http://www.megasoftware.net/) based on multiple sequence alignment of HbMADS-box TFs from *Arabidopsis, Oryza sativa* L., *Vitis vinifera* L., *Jatropha carcas* L., and *H. brasiliensis*.

### 2.6. Quantitative Real-Time PCR (qRT-PCR) Analysis

The real-time quantitative RT-PCR (qRT-PCR) was performed according to the following procedures: 94 °C for 30 s, 94 °C for 5 s, followed by 40 cycles, 60 °C for 15 s, and 72 °C for 10 s. The reaction volume was 20 µL: including 60 ng cDNA per sample, 1×SYBR® Premix Ex Taq™ (TaKaRa, Shiga, Japan), and 0.4 µM per primer. The reaction was carried out in 96-well plates using the CFX96™ Real-Time System (Bio-Rad, Hercules, CA, USA). The 18S rRNA gene (GenBank accession No.: AB268099) was used as a reference gene in the qRT-PCR reaction [41]. After the reaction, we used Bio-Rad CFX Manager Software 3.0 (Bio-Rad, Hercules, CA, USA) to analyze and visualize the data as previously described [42]. All primers used for qRT-PCR analysis were designed by Primer3.0 (http://frodo.wi.mit.edu/primer3). The primer sequences are given in Table S2.

### 2.7. Statistical Analysis

Data and graphical analysis was performed with Sigma Plot 12 software (Systat Software Inc., San Jose, CA, USA). The $2^{-\Delta\Delta CT}$ method was used to calculate the relative expression levels of all HbMad-box genes [43]. The data are represented by the mean ± SD (standard deviation) of the three biological repeats. The statistical significance of the values was determined by a *t*-test.

## 3. Results

### 3.1. Subsection Identification and Characterization of HbMad-box Genes in H. brasiliensis

The 76 candidate HbMADS-box genes were checked by CDS and Inter-ProScan to confirm the existence of the MADS-box domain. After deleting the redundant sequences, a total of 26 non-redundant HbMADS-box genes (designated as HbSEP, HbAGL9.1, HbAGL9.2, HbCMB1, HbCMB1-L, HbAGL6, HbAGL8, HbAP1, HbAGL12, HbAG, HbAGL11, HbAGL15, HbSVP1, HbSVP2, HbTT16, HbDEFL, HbPMADS2, HbAGL30, HbAGL61, HbAGL65, HbMADS1, HbMADS2, HbMADS3, HbMADS4, HbMADS5, and HbMADS27) with complete open reading frames (ORFs) were identified in the rubber tree (Table 1), which included six previously reported HbMADS-box TFs (HbMADS1, HbMADS2, HbMADS3, HbMADS4, HbMADS5, and HbMADS27).

**Table 1.** Characteristics of HbMADS-box genes family in *Hevea brasiliensis* Müll. Arg.

| Gene Name | GenBank Accession No. | ORF Length (bp) | Protein | | |
|---|---|---|---|---|---|
| | | | Length (aa) | Mw (kDa) | PI |
| HbSEP | KY471153 | 732 | 243 | 28.00 | 8.88 |
| HbAGL9.1 | KY471149 | 732 | 243 | 27.77 | 8.21 |
| HbAGL9.2 | KY471165 | 723 | 240 | 27.40 | 8.8 |
| HbCMB1 | KY471159 | 729 | 242 | 27.57 | 8.36 |
| HbCMB1-L | KY471157 | 729 | 242 | 27.87 | 7.6 |
| HbAGL6 | KY471160 | 729 | 242 | 27.63 | 8.73 |
| HbAGL8 | KY471148 | 759 | 252 | 29.16 | 7.72 |
| HbAP1 | KY471161 | 726 | 241 | 28.16 | 8.19 |
| HbMADS3 | GU142915 | 720 | 239 | 27.61 | 9.16 |
| HbAGL12 | KY471152 | 609 | 202 | 23.09 | 6.92 |
| HbAG | KY471158 | 726 | 241 | 27.78 | 9.08 |
| HbAGL11 | KY471166 | 669 | 222 | 25.52 | 9.34 |
| HbMADS2 | GU142914 | 654 | 217 | 24.93 | 9.23 |
| HbAGL15 | KY471151 | 768 | 255 | 29.27 | 6.68 |
| HbMADS27 | KY486278 | 717 | 238 | 27.49 | 9.25 |
| HbSVP1 | KY471154 | 633 | 210 | 23.69 | 6.86 |
| HbSVP2 | KY471156 | 702 | 234 | 26.60 | 6.06 |
| HbMADS4 | KY486276 | 633 | 210 | 23.67 | 6.36 |
| HbTT16 | KY471155 | 705 | 234 | 27.46 | 7.69 |
| HbDEFL | KY471162 | 675 | 224 | 26.23 | 8.43 |
| HbPMADS2 | KY471147 | 639 | 212 | 25.12 | 8.4 |
| HbAGL30 | KY471163 | 1116 | 371 | 41.86 | 7.14 |
| HbAGL61 | KY471164 | 642 | 213 | 24.24 | 8.32 |
| HbAGL65 | KY471150 | 741 | 246 | 28.41 | 8.68 |
| HbMADS5 | KY486277 | 1077 | 358 | 40.74 | 6.38 |
| HbMADS1 | GU142913 | 738 | 245 | 27.78 | 8.34 |

ORF, open reading frame; bp, base pair; aa, amino acids; Mw, molecular weight; PI, isoelectric point.

To confirm the putative HbMADS-box genes, the complete ORF sequences of HbMADS-box genes were isolated through RNA-sequencing and PCR-based approaches. Accurate sequences of HbMADS-box genes were submitted to GenBank, and the accession numbers of HbMADS-box genes are listed in Table 1. The ORF lengths of HbMADS-box genes ranged from 609 bp (HbAGL12) to 1116 bp (HbAGL30), and encoded polypeptides that ranged from 202 to 371 amino acids (Table 1). The corresponding Mw ranged from 23.09 to 41.86 kDa, and the predicted PI varied from 6.06 (HbSVP2) to 9.34 (HbAGL11). The distribution of PI was similar to that of AtMADS-box genes; however, the length and Mw of the HbMADS-box TFs were slightly lower than those of AtMADS-box genes (Table S3).

Pairwise sequence comparisons were performed to check the sequence identities between HbMADS-box proteins. The results showed that the identities between two HbMADS-box genes ranged from 13.67% to 98.57% (Table S4). The average sequence homology between two HbMADS-box genes was 39.99%. The largest sequence identity was observed between HbMADS4 and HbSVP1 (98.5%). HbAGL30 and HbMADS3 showed the least sequence identity (13.67%).

### 3.2. Analysis of Conserved Domains and Structural Features of HbMad-box Proteins

Analysis of conserved domains showed that all of the deduced HbMADS-box proteins contained a highly conserved MADS-box domain of approximately 50 amino acid residues in length, a semi-conservative K domain with an obvious coiled-coil region, a less conservative I domain, and a non-conservative C domain (Figure 1), which indicates that they belonged to the MADS-box TF family. We further analyzed the conserved domain of HbMADS-box proteins, and the results indicated that for amino acids 3 (R), 17 (R), 20 (T), 23–24 (KR), 27 (G), 30–31 (KK), 34 (E), 38–39 (LC), 48 (F), and 52 (G), up to 13 sites were conserved in the MADS-box domains (Figure 1), indicating that the deduced amino acid sequences of HbMADS-box proteins shared significant homology with each other.

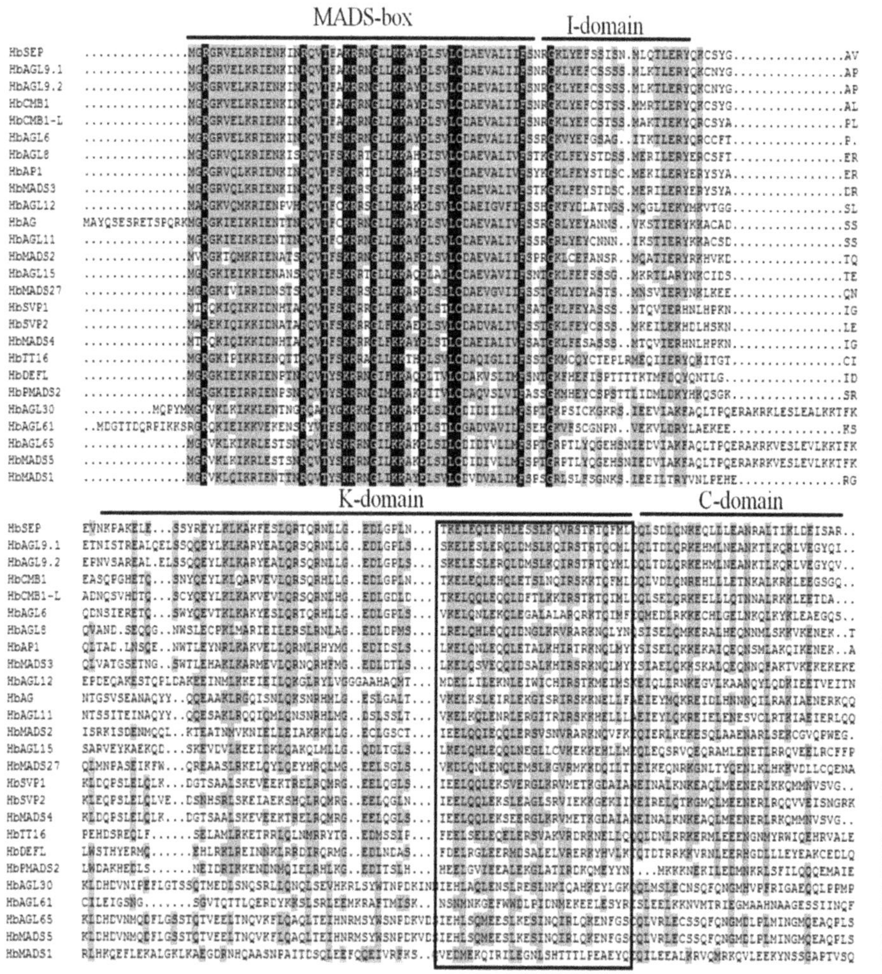


**Figure 1.** Phylogenetic analysis of HbMADS-box genes related proteins. MADS-box, I domain, K domain, and C domain are marked, respectively; coiled-coil domain is boxed.

The conserved motifs and structural characteristics analysis showed that all HbMADS-box genes included four conserved motifs: motifs 1 and 3 belonged to the MADS-box domains, whereas motifs 2 and 4 belonged to the K-box domains. Most HbMADS-box proteins (except HbAGL30 and HbAGL61) contained the MADS-box and K-box domains (Figure S2), which is similar to the previously reported MADS-box conserved motifs in other species [44].

### *3.3. Gene Structures and Sequence Characteristics of HbMad-box Genes*

Sequence analysis revealed that all of the deduced *H. brasiliensis* MADS-box genes belong to the type II (MIKC-type) MADS-box genes. These HbMad-box genes usually contained multiple introns and exons, with a maximum of 11 exons (Figure S3). According to their predicted structures, the 26 candidate HbMADS-box genes could be divided into eight groups (Table S5). The first group of HbMADS-box genes consisted of eight exons interrupted by seven introns, including eight type II HbMADS-box genes (HbSEP, HbAGL9.2, HbCMB1, HbCMB1-L, HbSVP1, HbMADS4, HbAGL65, and HbMADS5). The second and third groups included five type II HbMADS-box genes, which contained six exons interrupted by five introns (HbAGL6, HbMADS2, HbMADS27, HbSVP2, and HbTT16), and seven exons interrupted by six introns (HbAP1, HbMADS3, HbAGL12, HbAG, and HbDEFL), respectively. The fourth and fifth groups consisted of two type II HbMADS-box genes, which contained 10 exons interrupted by nine introns (HbAGL9.1 and HbAGL8), and five exons interrupted by four introns (HbAGL11 and HbPMADS2), respectively. The sixth group included two type II HbMADS-box genes (HbAGL30 and HbMADS1), which contained 11 exons interrupted by 10 introns. The seventh and eighth groups each contained one type II HbMADS-box gene (HbAGL15 and HbAGL61), which consisted of four exons interrupted by three introns, and only one exon, respectively. These results revealed that most of the HbMADS-box genes (69.2% belong to the first, second, and third groups) within the same group shared conserved exon–intron structures, and the other HbMADS-box genes (30.8% of the fourth to eighth groups) with the different exon–intron structures may belong to distinct types of HbMADS-box genes in the rubber tree (Table S5). These characteristics are consistent with the features of MADS-box genes in other flowering plants, such as *Arabidopsis*, *Oryza sativa*, and *Vitis vinifera* [45–47].

### *3.4. Phylogenetic Analysis of HbMad-box TFs*

Based on the phylogenetic tree analysis, the 26 HbMADS-box genes could be divided into three clades (I, II, and III), which contained 11 main branches: SEP, AGL6, FUL/AP1, AGL12, AG, SOC1, AGL15, AGL17, SVP, BS, and AP3/PI subfamilies (Figure 2). Clade I contained 12 type II HbMADS-box genes, and it could be further categorized into two subclades, which contained nine and three members. Clade II contained six type II HbMADS-box genes, and it could also be divided into two subclades, and one subclade contained one member, whereas the other contained five members. Clade III contained eight type II MADS-box genes, which had two subclades, one of which contained three members and the other contained five members.

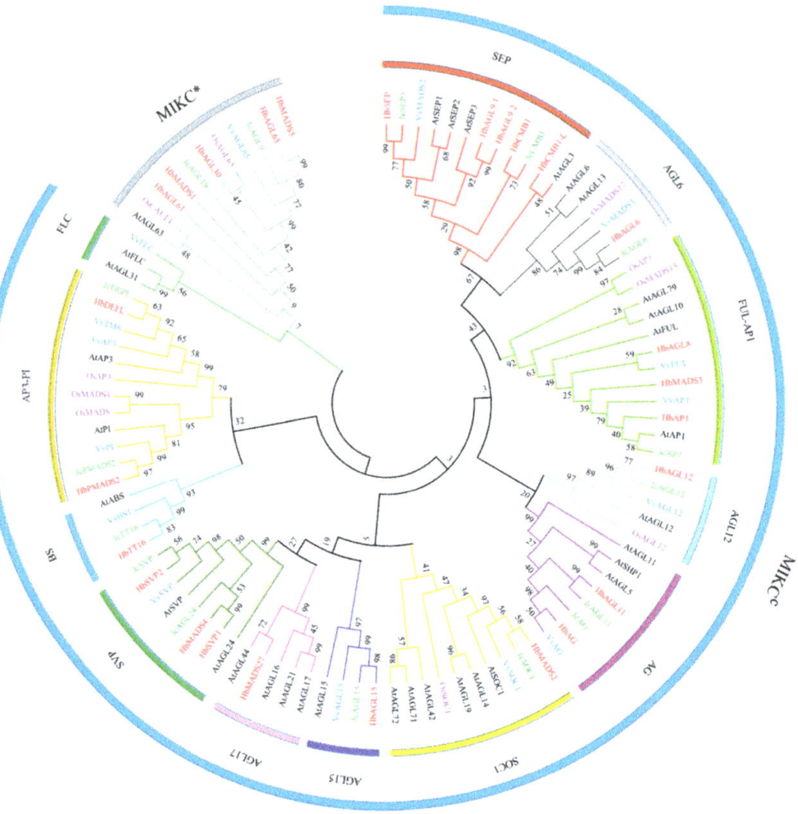


**Figure 2.** Phylogenetic analysis of HbMADS-box genes with other MADS-box genes by MEGA version 6.0. A phylogenetic tree of MADS-box genes was generated by the neighbor-joining (NJ) method. The multiple sequence alignment and construction of phylogenetic tree were performed with MEGA6.06 using the neighbor joining method with 1000 bootstrap replicates. The proteins were clustered and divided into three distinct sub-families. The three sub-families were further divided into 12 sub-groups.

### 3.5. Expression Profiles of HbMad-box Genes in Various Developmental Stages during Rubber Tree Reproductive Development

In the present study, analyzing the expression patterns of HbMADS-box genes in different tissues and organs (including roots, stem, stem tips, leaves, labeled bark, xylem, latex, fruits, inflorescence, male flowers, and female flowers) of *H. brasiliensis* by qRT-PCR, the results showed that most of the deduced HbMADS-box genes might be primarily involved in floral organ differentiation and inflorescence development. For example, HbSEP, HbAGL9.1, HbAGL9.2, HbCMB1, HbCMB1-L, HbAGL6, HbAG, HbDEFL, and HbPADS2 had significantly higher expression in stem tips and floral organs than other tissues; HbAP1 showed the highest expression pattern in stem tips compared to other tissues (Figure 3); HbAGL8 was significantly higher expressed in stem tips and leaves than other tissues (Figure 3); the MADS-box genes HbAGL15, HbTT16, HbMADS2, and HbAG were higher expressed in fruits than other tissues (Figure 3); and the MADS-box genes HbAGL9.2 and HbAGL6 were specifically expressed in floral organs and they were higher expressed in male flowers than other tissues (Figure 3). The higher expression of the HbMad-box genes in floral organs and shoot tips suggests that they may play a specific role in the corresponding tissues. However, in contrast, some MADS-box genes displayed different tissue expression patterns, such as HbAGL15, HbAGL30,

HbAGL61, and HbAGL65, which were almost constitutively expressed in all tested tissues (Figure 3). In addition, we also found that some HbMADS-box genes showed higher expression patterns in other specific tissues; for example, HbMADS1 and HbMADS3 were higher expressed in latex than other tissues (Figure 3), which was consistent with the findings of a previous report that found that HbMADS1 and HbMADS3 were highly expressed in laticifer cells.

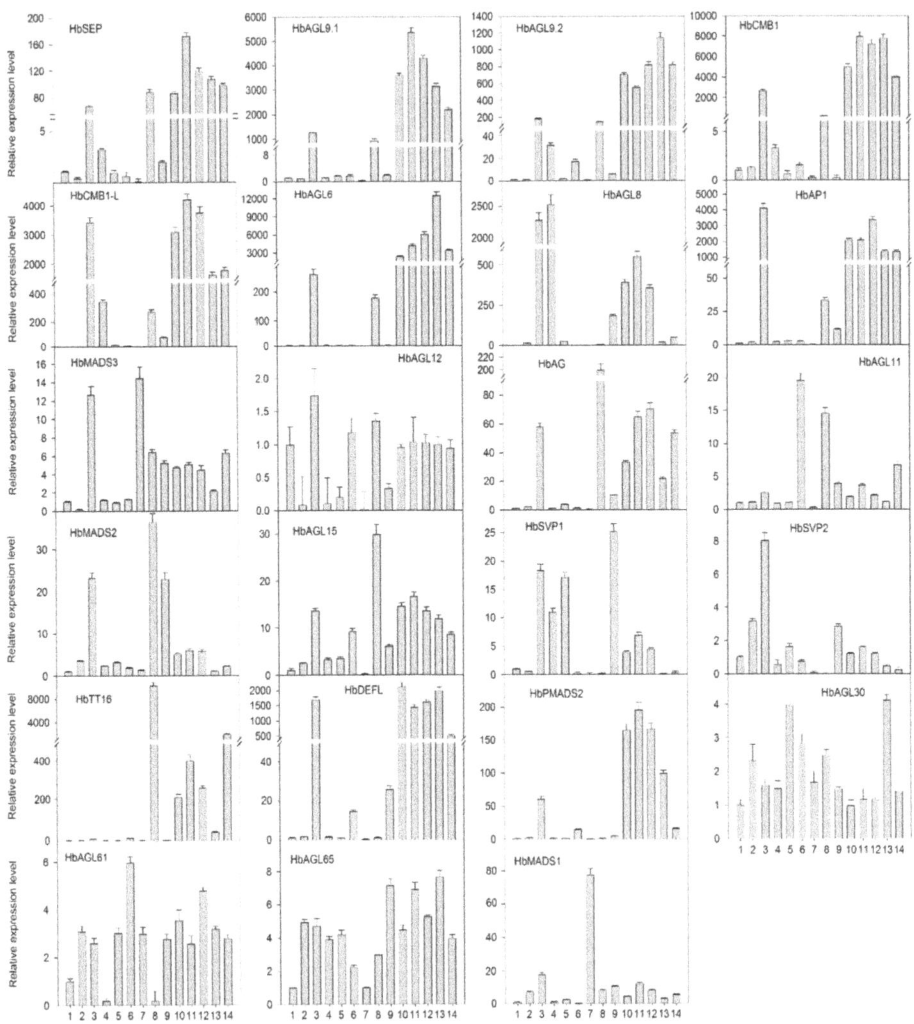


**Figure 3.** Relative expression levels of HbMad-box genes were determined by qRT-PCR and normalized by 18S rRNA gene expression. Relative expression levels of HbMad-box genes were determined by qRT-PCR and normalized by the 18S rRNA gene expression. For each gene, the transcript level in the root was used to normalize the transcript levels in other tissues. Values are means ± SD (standard deviation) of three biological replicates. 1–14 represent roots; stem; stem tips; leaves; labeled bark; xylem; latex; fruits; 3 cm inflorescence; 6 cm inflorescence; 9 cm inflorescence; 12 cm inflorescence; male flowers; and female flowers, respectively.

### 3.6. Expression Patterns of HbMADS-box Genes in Response to Abiotic Stress

Analyzing the expression patterns of 12 floral-enriched HbMADS-box genes under cold, high-temperature, drought, and salt stresses by qRT-PCR, the results indicated that the expressions of 12 HbMADS-box genes specifically related to the floral organ were affected by low-temperature stress (Figure S4). Among the 12 cold-regulated HbMADS-box genes related to floral organ development, most of them (HbSEP, HbAGL9.1, HbAGL9.2, HbCMB1, HbCMB1-L, HbAGL6, HbAG, HbDEFL, and HbMADS2) were rapidly induced by cold stress, and reached the highest levels at 3 or 6 h after treatment, and then declined at 12 or 24 h after treatment. Only three HbMADS-box genes, HbAGL8, HbAP1, and HbTT16, exhibited down-regulated expression at all analyzed time-points in response to cold stress.

Under high-temperature stress, only HbDEFL expression did not significantly change (Figure S5). Of the 11 high temperature-responsive HbMADS-box genes, two genes (HbSEP and HbCMB1) were significantly up-regulated at all treated time points. Seven genes (HbAGL9.1, HbAGL9.2, HbCMB1-L, HbAGL6, HbAP1, HbAG, and HbTT16) were significantly up-regulated at at least one time point (Figure S5). By contrast, HbAGL8 and HbPADS2 were significantly down-regulated at all analyzed time points.

Under polyethylene glycol induced drought stress, the expression patterns of 12 HbMADS-box genes related to floral organ development were markedly affected (Figure S6). Of the 12 drought-responsive HbMADS-box genes, 10 genes (HbSEP, HbAGL9.1, HbAGL9.2, HbCMB1, HbCMB1-L, HbAP1, HbAG, HbTT16, HbDEFL, and HbMADS2) were significantly up-regulated during at least one time point after drought stress (Figure S6). Among them, HbAGL9.1, HbCMB1-L, HbAG, HbTT16, and HbDEFL were strongly affected by drought stress, reaching the highest levels that were 2-, 2.5-, 3.7-, 4-, and 5.8-fold increases compared with the control at 6 h, 12 h, 24 h, 7 day, and 4 day after treatment, respectively. Interestingly, the expression of HbAGL6 and HbAGL8 increased continuously after drought stress, and resulted in a 120- and 11-fold increase, respectively, at 7 day after treatment.

Under salt stress, the expressions of 12 floral-enriched HbMADS-box genes were regulated by salt stress. Eight genes (HbAGL9.1, HbAGL9.2, HbAGL6, HbAGL8, HbAP1, HbAG, HbDEFL, and HbPADS2) were significantly up-regulated at all of the treated time points (Figure S7). Their expression levels showed rapid up-regulation under salt stress, which peaked at 2, 48, 48, 12, 24, 48, 2, and 48 h of treatment, respectively. Four genes (HbSEP, HbCMB1, HbCMB1-L, and HbTT16) were significantly up-regulated at at least one time point under salt stress (Figure S7), and their transcripts reached maximum levels that were 2-, 2.9-, 12-, and 24-fold up-regulated compared with the control at 12, 2, 2, and 2 h after treatment, respectively.

### 3.7. Expression Patterns of HbMADS-box Genes in Response to Phytohormones

After ABA treatment, HbAGL9.1, HbAGL6, HbAP1, and HbPADS2 displayed obvious up-regulation across all time points (Figure S8). Their expression levels exhibited rapid up-regulation under ABA treatment, and peaked at 48, 48, 12, and 48 h after treatment, respectively. HbAGL9.2 and HbCMB1-L showed similar expression patterns; their transcripts exhibited down-regulated expression at the 6-h time point, but they were significantly up-regulated at other time points. HbSEP, HbCMB1, HbAGL8, HbAG, and HbDEFL were down-regulated at the time points of 24, 12, 0.5, 24, and 24 h compared with the control, respectively, but they displayed obvious up-regulation at other time points. In comparison, HbTT16 showed an irregular expression pattern under ABA treatment; it was up-regulated at 48 h, but was down-regulated at 0.5, 2, 12, and 24 h.

After GA3 treatment, HbSEP, HbAGL9.1, HbAGL9.2, HbCMB1, HbAGL6, HbAGL8, HbAP1, HbTT16, and HbDEFL showed obvious up-regulation across all time points (Figure S9). Their expression levels exhibited rapid up-regulation under GA3 treatment, and reached a maximum level at 0.5, 12, 6, 48, 12, 48, 6, and 48 h after treatment, respectively. However, HbAG and HbPADS2 were slightly down-regulated at the 24-h time point, but their expression levels showed obvious up-regulation at other time points. By contrast, HbCMB1-L displayed an irregular expression pattern; it was strongly up-regulated at 0.5, 2, and 48 h, but was down-regulated at 6, 12, and 24 h.

## 4. Discussion

Understanding the flowering regulation of *H. brasiliensis* is important for accelerating the breeding process of this species. Based on studies of Arabidopsis and other plants, extensive efforts have been devoted to clarifying the molecular mechanism of reproductive development in plants [48,49], and techniques such as RT-qPCR and high-throughput RNA sequencing have been used in many transcriptional-level studies to identify the genes that regulate the metabolism of plant reproductive development. However, available information about TFs related to reproductive development in *H. brasiliensis* is still limited. To date, no study has been performed to determine the expression patterns of HbMad-box genes relative to rubber tree reproductive development.

As an important gene family, Mad-box genes are widespread in the plant kingdom. The number of Mad-box genes varies considerably in genomes of different species, from 20 in *Physcomitrella patens* (Hedw.) Bruch & Schimp. to 167 in *Brassica rapa* L. [50,51]. In this study, we performed genome-wide identification and analysis of the MADS-box gene family related to flower development in *H. brasiliensis*. A total of 20 MADS-box genes were newly identified in the *H. brasiliensis* genome, which was lower than the number of MADS-box genes in the woody plant *Malus pumila* Mill. [52]. A domain search using EMBL with the corresponding *H. brasiliensis* candidate protein sequences confirmed that 26 of the sequences contained a 'MADS' domain. We classified all 26 putative *H. brasiliensis* MADS-box proteins into three clades, which were consistent with the classifications of MADS-box gene family members in other flowering plants [53]. Sequence alignment analysis of 26 MADS-box genes revealed that their ORFs ranged from 609 to 1116 bp, and predicted protein lengths that ranged from 202 to 371 amino acids (Table 1). Subsequent gene structures analysis showed that most of the HbMad-box genes usually contained multiple introns, with a maximum of 11 introns; the exception was HbAGL61, which did not have any introns. Interestingly, we observed that only MIKC-type (no M-type) MADS-box genes existed in *H. brasiliensis*. In contrast, the number of M-type MADS-box genes in *Arabidopsis* is more than that of the MIKC$^{C}$-type [54]. Furthermore, we found that the expansion of MIKCC-type and MIKC*-type MADS-box genes was disproportionate, and there were more MIKC$^{C}$-type (twenty-one members) than MIKC*-type (five members) MADS-box genes presented in *H. brasiliensis*, which might be related to the fact that MIKC$^{C}$-type genes conducted as functional genes to perform more complex functions in *H. brasiliensis* flower organogenesis.

Most studies on the functional identification of MADS-box genes were conducted on model plants [55–57], which also substantially contributed to revealing the diverse functions of MADS-box genes. In order to detect the expression of HbMADS-box genes in different tissues, we analyzed the expression patterns of HbMADS-box genes in 11 sample tissues and four stages of inflorescence development by qRT-PCR. The results revealed that for close orthologs of SEP and AGL6 in the rubber tree, HbSEP, HbAGL9.1, HbAGL9.2, HbCMB1, HbCMB1-L, and HbAGL6 showed higher expression during stem tip and floral organ development; for close orthologs of FUL-AP1 and AG in the rubber tree, HbAGL8, HbAP1, and HbAG, also exhibited higher expression during stem tip and floral organ development (Figure 3). The same expression pattern was also observed in subclade AP3-PI; HbDEFL and HbPADS2 were found to be highly expressed during stem tip and floral organ development, which indicates association with the development of reproductive organs. These findings are consistent with those of previous results that revealed the MIKC-type MADS-box gene as the flower homeotic gene that plays a dominant role in floral organ development [58].

Interestingly, HbAGL6 was progressively increased in four stages of inflorescence development, and the expression level of HbAGL6 in the male flowers was almost four-fold more than the female flowers in *H. brasiliensis*, which indicates that HbAGL6 may be involved in the development of inflorescence and floral organs, and especially participates in the development of male flowers organs (Figure 3). In addition, HbDEFL also displayed strong expression in inflorescence and was almost four-fold more abundant in the male flowers than the female flowers in *H. brasiliensis*, which indicates that HbDEFL may play an important role in the development of male flowers. In contrast, HbTT16 showed strong expression in the fruits and female flowers, and the expression level of HbTT16 in the female flowers was almost 40-fold more than the male flowers in *H. brasiliensis*, which indicates association with the development of female organs in the rubber tree. In other plant species, TT16 homologs were also demonstrated to mediate the crosstalk between the endothelium and nucellus for the development of female organs [59,60].

Besides being involved in the essential regulation of the development of floral organs, many studies have shown that MADS-box genes also play an essential role in the response to various stresses and exhibit differential expression patterns under abiotic stress [61,62]. Rubber trees originated in the Amazon Basin of South America, but are now widely planted in the northern margin of tropical areas, such as Southeast Asia countries (e.g., Thailand, Vietnam, and southern China), which are often suffer from cold, drought, typhoon, and other abiotic stresses.

To understand the responses of HbMADS-box genes to various stresses, the expression profiles of 12 floral-enriched HbMADS-box genes were examined in leaves of *Hevea* seedlings after hormone, salt, cold, high-temperature, and drought stress treatments. In the present study, almost all of the HbMADS-box genes specifically related to floral organs exhibited changes in expression patterns when responding to hormone, salt, cold, high-temperature, and drought stresses (Figures S4–S9). Our results show that different HbMADS-box genes responded differently to abiotic stress, which indicates that stress responses of *H. brasiliensis* are regulated by many factors, involving a variety of signaling pathways. In addition, the hormone and salt stress responses of 12 floral-enriched HbMADS-box genes were more intense than those of temperature and drought stress responses. Among them, HbAGL9.2, HbAGL6, and HbAP1 were strongly induced by ABA; HbAGL9.1, HbAGL9.2, HbCMB1-L, HbAGL6, HbAGL8, HbAP1, HbTT16, and HbPADS2 were strongly induced by GA3; and HbAGL9.2, HbCMB1-L, HbAGL6, HbAGL8, HbAP1, HbTT16, and HbPADS2 were strongly induced by salt treatments, whereas their expressions were slightly affected by cold, high-temperature, and drought stresses. The expressions of HbAGL9.2, HbAGL6, and HbAP1 were strongly induced by a variety of hormones, which revealed that these genes might play a crucial role in hormone signaling in *H. brasiliensis*.

Overall, we characterized HbMADS-box genes as multifunctional regulators in *H. brasiliensis* based on tissue specific expression analysis, as well as various stress responses. Our results revealed that HbSEP, HbAGL9.1, HbAGL9.2, HbCMB1, HbCMB1-L, HbAGL6, HbAGL8, HbAP1, HbAG, HbDEFL, and HbPADS2 may mainly regulate the differentiation of flower buds and could help regulate reproduction in *H. brasiliensis*; whereas HbTT16 may mainly regulate the development of fruits and female organs in the rubber tree. Expression profiling revealed that different HbMADS-box genes responded differently to abiotic stresses, which indicates that abiotic stresses in *H. brasiliensis* are regulated by many factors and various signaling pathways. Our findings will promote the development of technology that can control the reproduction of *H. brasiliensis* by genetic engineering. In the near future, we will be able to verify the role of these HbMADS-box genes in the transition of vegetative to reproductive development using transgenic *H. brasiliensis*, as well as transgenic *Arabidopsis*.

## 5. Conclusions

In this study, 20 new HbMad-box genes were identified in the rubber tree. Subsequently, the bioinformatics characteristics, sequence identity, conserved domains, gene structure, and phylogenetic relationship of these genes were systematically analyzed. Expression profiling revealed that HbMad-box genes were differentially expressed in various tissues, which indicated that HbMad-box genes may exert different functions throughout the life cycle. Additionally, HbSEP, HbAGL9.1, HbAGL9.2, HbCMB1, HbCMB1-L, HbAGL6, HbAGL8, HbAP1, HbAG, HbDEFL, HbTT16, and HbPADS2 were found to be associated with the differentiation of flower buds and may be involved in flower development. All of these floral-enriched HbMADS-box genes were regulated by hormone, salt, cold, high-temperature, and drought stresses, which revealed that abiotic stresses in

*H. brasiliensis* are regulated by many factors and various signaling pathways. Our results provide a comprehensive overview of the HbMADS-box gene family related to floral development and lay the foundation for further functional characterization of this gene family in *H. brasiliensis*.

**Supplementary Materials:** The following are available online at www.mdpi.com/1999-4907/9/6/304/s1, Figure S1: Tissue samples of rubber tree flowers, Figure S2: The conserved domain logo and motif composition of HbMADS-box genes, Figure S3: Exon-intron structures of HbMads-box genes, Figure S4: Expression profiles of the floral-enriched HbMad-box genes under cold stress, Figure S5: Expression profiles of the floral-enriched HbMad-box genes under high temperature stress, Figure S6: Expression profiles of the floral-enriched HbMad-box genes under drought stress, Figure S7: Expression profiles of the floral-enriched HbMad-box genes under salt stress, Figure S8: Expression profiles of the floral-enriched HbMad-box genes responding to ABA treatment, Figure S9: Expression profiles of the floral-enriched HbMad-box genes responding to GA3 treatment, Table S1: List of primer sequences used for HbMad-box TFs amplification, Table S2: List of primer sequences used for qRT-PCR analysis, Table S3: Sequence features of AtMADSs in *A. thaliana*. (L.) Heynh. Gene IDs, protein length, PI, and molecular weight of corresponding AtMADSs in *A. thaliana* are shown, Table S4: Percent identity matrix of HbMad-box TFs proteins, Table S5: Statistics of exon and intron number distribution among HbMad-box TFs.

**Author Contributions:** M.W. and W.L. designed the experiments. R.P. performed the tissue and organ collection. M.W. and Y.W. performed the experiments. All authors read and approved the final manuscript.

**Funding:** This research was funded by the Central Public-interest Scientific Institution Basal Research Fund for Chinese Academy of Tropical Agricultural Sciences (No. 1630022017010) and the earmarked Fund for Modern Agro-industry Technology Research System (CARS-34).

**Conflicts of Interest:** The authors declare no conflict of interest.

## References

1. Li, D.; Zeng, R.; Yan, L.; Zhao, M.; Chao, J.; Yu, L. Gene expression analysis and SNP/InDel discovery to investigate yield heterosis of two rubber tree F1 hybrids. *Sci. Rep.* **2016**, *6*, 24984. [CrossRef] [PubMed]
2. Mooibroek, H.; Cornish, K. Alternative sources of natural rubber. *Appl. Microbiol. Biotechnol.* **2000**, *53*, 355–365. [CrossRef] [PubMed]
3. Ahmad, N.; Abnisaa, F.; Daud, W. Potential use of natural rubber to produce liquid fuels using hydrous pyrolysis—A review. *RSC Adv.* **2016**, *6*, 68906–68921. [CrossRef]
4. Pethin, D.; Nakkanong, K.; Nualsri, C. Performance and genetic assessment of rubber tree clones in Southern Thailand. *Sci. Agric.* **2015**, *72*, 306–313. [CrossRef]
5. Ahrends, A.; Hollingsworth, P.M.; Ziegler, A.D.; Fox, J.M.; Chen, H.; Su, Y. Current trends of rubber plantation expansion may threaten biodiversity and livelihoods. *Glob. Environ. Chang.* **2015**, *34*, 48–58. [CrossRef]
6. Dornelas, M.C.; Rodriguez, A.P. The rubber tree (*Hevea brasiliensis* Muell. Arg.) homologue of the *LEAFY/FLORICAULA* gene is preferentially expressed in both male and female floral meristems. *J. Exp. Bot.* **2005**, *56*, 1965–1974. [CrossRef] [PubMed]
7. Cubry, P.; Pujade-Renaud, V.; Garcia, D.; Espeout, S.; Le, G.V.; Granet, F. Development and characterization of a new set of 164 polymorphic EST-SSR markers for diversity and breeding studies in rubber tree (*Hevea brasiliensis* Mull. Arg.). *Plant Breed.* **2014**, *133*, 419–426. [CrossRef]
8. Ó'Maoiléidigh, D.S.; Wuest, S.E.; Rae, L.; Raganelli, A.; Ryan, P.T.; Kwaśniewska, K.; Wellmer, F. Control of reproductive floral organ identity specification in *Arabidopsis* by the C function regulator AGAMOUS. *Plant Cell* **2013**, *25*, 2482–2503. [CrossRef] [PubMed]
9. Yu, H.; Ito, T.; Wellmer, F.; Meyerowitz, E.M. Repression of AGAMOUS-LIKE 24 is a crucial step in promoting flower development. *Nat. Genet.* **2004**, *36*, 157–161. [CrossRef] [PubMed]
10. Patharkar, O.R.; Walker, J.C. Floral organ abscission is regulated by a positive feedback loop. *PNAS* **2015**, *112*, 2906–2911. [CrossRef] [PubMed]
11. Adamczyk, B.J.; Lehtishiu, M.D.; Fernandez, D.E. The MADS domain factors AGL15 and AGL18 act redundantly as repressors of the floral transition in *Arabidopsis*. *Plant J.* **2007**, *50*, 1007–1019. [CrossRef] [PubMed]
12. García-Maroto, F.; Carmona, M.J.; Garrido, J.A.; Vilches-Ferrón, M.; Rodríguez-Ruiz, J.; Alonso, D.L. New roles for MADS-box genes in higher plants. *Biol. Plant.* **2003**, *46*, 321–330. [CrossRef]

13. Nam, J.; Ma, H.; Nei, M. Antiquity and evolution of the MADS-box gene family controlling flower development in plants. *Mol. Biol. Evol.* **2003**, *20*, 1435–1447. [CrossRef] [PubMed]
14. Wong, C.E.; Singh, M.B.; Bhalla, P.L. Novel members of the AGAMOUS LIKE 6 subfamily of MIKCc-type MADS-box genes in soybean. *BMC Plant Biol.* **2013**, *13*, 105. [CrossRef] [PubMed]
15. Leseberg, C.H.; Li, A.; Kang, H.; Duvall, M.; Mao, L. Genome-wide analysis of the MADS-box gene family in populus trichocarpa. *Gene* **2006**, *378*, 84–94. [CrossRef] [PubMed]
16. Kofuji, R.; Sumikawa, N.; Yamasaki, M.; Kondo, K.; Ueda, K.; Ito, M.; Hasebe, M. Evolution and divergence of the MADS-box gene family based on genome-wide expression analyses. *Mol. Biol. Evol.* **2003**, *20*, 1963–1977. [CrossRef] [PubMed]
17. Torti, S.; Fornara, F.; Vincent, C.; Andrés, F.; Nordström, K.; Göbel, U. Analysis of the *Arabidopsis* shoot meristem transcriptome during floral transition identifies distinct regulatory patterns and a leucine-rich repeat protein that promotes flowering. *Plant Cell* **2012**, *24*, 444–462. [CrossRef] [PubMed]
18. Honma, T.; Goto, K. Complexes of MADS-box proteins are sufficient to convert leaves into floral organs. *Nature* **2001**, *409*, 525–529. [CrossRef] [PubMed]
19. Lalusin, A.G.; Nishita, K.; Kim, S.H.; Ohta, M.; Fujimura, T. A new MADS-box gene (IbMADS10) from sweet potato (*Ipomoea batatas* L. Lam) is involved in the accumulation of anthocyanin. *Mol. Genet. Genom.* **2006**, *275*, 44–54. [CrossRef] [PubMed]
20. Zhang, B.; Su, X.; Zhou, X. A MADS-box gene of *Populus deltoides* expressed during flower development and in vegetative organs. *Tree Physiol.* **2008**, *28*, 929–934. [CrossRef] [PubMed]
21. Liu, J.; Xu, B.; Hu, L.; Li, M.; Su, W.; Wu, J. Involvement of a banana MADS-box transcription factor gene in ethylene-induced fruit ripening. *Plant Cell Rep.* **2009**, *28*, 103–111. [CrossRef] [PubMed]
22. Tani, E.; Polidoros, A.N.; Tsaftaris, A.S. Characterization and expression analysis of Fruitfull- and shatterproof-like genes from peach (*Prunus persica*) and their role in split-pit formation. *Tree Physiol.* **2007**, *27*, 649–659. [CrossRef] [PubMed]
23. Ge, J.; Zhao, D.; Han, C.; Wang, J.; Hao, Z.; Tao, J. Cloning and expression of floral organ development-related genes in herbaceous peony (*Paeonia lactiflora*, Pall.). *Mol. Biol. Rep.* **2014**, *41*, 6493–6530. [CrossRef] [PubMed]
24. Bemer, M.; Heijmans, K.; Airoldi, C.; Davies, B.; Angenent, G. An atlas of type I MADS box gene expression during female gametophyte and seed development in *Arabidopsis*. *Plant Physiol.* **2010**, *154*, 287–300. [CrossRef] [PubMed]
25. Thakare, D.; Tang, W.; Hill, K.; Perry, S.E. The MADS-domain transcriptional regulator AGAMOUS-LIKE15 promotes somatic embryo development in *Arabidopsis* and soybean. *Plant Physiol.* **2008**, *146*, 1663–1672. [CrossRef] [PubMed]
26. Colombo, M.; Masiero, S.; Vanzulli, S.; Lardelli, P.; Kater, M. M.; Colombo, L. AGL23, a type I MADS-box gene that controls female gametophyte and embryo development in Arabidopsis. *Plant J.* **2008**, *54*, 1037–1048. [CrossRef] [PubMed]
27. Lee, S.; Woo, Y.; Ryu, S.; Shin, Y.; Kim, W.; Park, K.; Lee, I.; An, G. Further characterization of a rice AGL12 group MADS-Box gene, OsMADS26. *Plant Physiol.* **2008**, *147*, 156–168. [CrossRef] [PubMed]
28. Elitzur, T.; Vrebalov, J.; Giovannoni, J.J.; Goldschmidt, E.E.; Friedman, H. The regulation of MADS-box gene expression during ripening of banana and their regulatory interaction with ethylene. *J. Exp. Bot.* **2010**, *61*, 1523–1535. [CrossRef] [PubMed]
29. Lozano, R.; Angosto, T.; Gomez, P.; Payan, C.; Capel, J.; Huijser, P. Tomato flower abnormalities induced by low temperatures are associated with changes of expression of MADS-Box genes. *Plant Physiol.* **1998**, *117*, 91–100. [CrossRef] [PubMed]
30. Arora, R.; Agarwal, P.; Ray, S.; Singh, A.K.; Singh, V.P.; Tyagi, A.K. MADS-box gene family in rice: Genome-wide identification, organization and expression profiling during reproductive development and stress. *BMC Genom.* **2007**, *8*, 242. [CrossRef] [PubMed]
31. Bonhomme, F.; Kurz, B.; Melzer, S.; Bernier, G.; Jacqmard, A. Cytokinin and gibberellin activate SaMADS A, a gene apparently involved in regulation of the floral transition in *Sinapis alba*. *Plant J.* **2000**, *24*, 103–111. [CrossRef] [PubMed]
32. Ando, S.; Sato, Y.; Kamachi, S.; Sakai, S. Isolation of a MADS-box gene (ERAF17) and correlation of its expression with the induction of formation of female flowers by ethylene in cucumber plants (*Cucumis sativus* L.). *Planta* **2001**, *213*, 943–952. [CrossRef] [PubMed]

33. Zhu, C.; Perry, S.E. Control of expression and auto regulation of AGL15, a member of the MADS-box family. *Plant J.* **2005**, *41*, 583–594. [CrossRef] [PubMed]
34. Li, H.L.; Wei, L.R.; Guo, D.; Wang, Y.; Zhu, J.H.; Chen, X.T. HbMADS4, a MADS-box transcription factor from *Hevea brasiliensis*, negatively regulates HbSRPP. *Front. Plant Sci.* **2016**, *7*, 1709–1712. [CrossRef] [PubMed]
35. Cai, H.B.; Tu, M.; Hua, Y.W.; Huang, H.S. Cloning and expression analysis of HbAGL62 of rubber tree (*Hevea brasiliensis* Arg.). *Chin. J. Trop. Crops.* **2012**, *33*, 1766–1771.
36. Wang, H.; Li, L.; Liang, Z.; Wu, Y.J. Cloning and bioinformatic analysis of the gene HbMADS-27 from *Hevea brasiliensis*. *Chin. J. Trop. Agric.* **2013**, *33*, 19–24.
37. Li, H.L.; Wang, Y.; Guo, D.; Tian, W.M.; Peng, S.Q. Three MADS-box genes of *Hevea brasiliensis* expressed during somatic embryogenesis and in the laticifer cells. *Mol. Biol. Rep.* **2011**, *38*, 4045–4052. [CrossRef] [PubMed]
38. Cheng, H.; Chen, X.; Zhu, J.S.; Huang, H.S. Overexpression of a *Hevea brasiliensis* ErbB-3 Binding protein 1 Gene Increases Drought Tolerance and Organ Size in *Arabidopsis*. *Front. Plant Sci.* **2016**, *7*, 1703. [CrossRef] [PubMed]
39. Xu, J.; Aileni, M.; Abbagani, S.; Zhang, P. A reliable and efficient method for total RNA isolation from various members of spurge family (*Euphorbiaceae*). *Phytochem. Anal.* **2010**, *21*, 395–398. [CrossRef] [PubMed]
40. Solovyev, V. Statistical Approaches in Eukaryotic Gene Prediction. In *Handbook of Statistical Genetics*; John Wiley & Sons, Ltd.: Clifton, NJ, USA, 2004; pp. 7–39.
41. Tang, C.; Huang, D.; Yang, J.; Liu, S.; Sakr, S.; Li, H. The sucrose transporter HbSUT3 plays an active role in sucrose loading to laticifer and rubber productivity in exploited trees of *Hevea brasiliensis* (para rubber tree). *Plant Cell Environ.* **2010**, *33*, 1708–1720. [CrossRef] [PubMed]
42. Long, X.; He, B.; Wang, C.; Fang, Y.; Qi, J.; Tang, C. Molecular identification and characterization of the pyruvate decarboxylase gene family associated with latex regeneration and stress response in rubber tree. *Plant Physiol. Biochem.* **2015**, *87*, 35–44. [CrossRef] [PubMed]
43. Livak, K.J.; Schmittgen, T.D. Analysis of relative gene expression data using real-time quantitative PCR and the $2^{-\Delta\Delta CT}$ method. *Methods* **2001**, *25*, 402–408. [CrossRef] [PubMed]
44. Nakatsuka, T.; Saito, M.; Yamada, E.; Fujita, K.; Yamagishi, N.; Yoshikawa, N. Isolation and characterization of the C-class MADS-box gene involved in the formation of double flowers in Japanese gentian. *BMC Plant Biol.* **2015**, *15*, 182. [CrossRef] [PubMed]
45. Liu, Y.; Zhang, D.; Ping, J.; Li, S.; Chen, Z.; Ma, J. Innovation of a regulatory mechanism modulating semi-determinate stem growth through artificial selection in soybean. *PLoS Genet.* **2016**, *12*, e1005818. [CrossRef] [PubMed]
46. Par Enicová, L.; Folter, S.D.; Kieffer, M.; Horner, D.S.; Favalli, C.; Busscher, J. Molecular and Phylogenetic Analyses of the Complete MADS-Box Transcription Factor Family in *Arabidopsis*: New openings to the MADS world. *Plant Cell* **2003**, *15*, 1538–1551. [CrossRef]
47. Díazriquelme, J.; Lijavetzky, D.; Martínezzapater, J.M.; Carmona, M.J. Genome-wide analysis of MIKCc-type MADS-box genes in grapevine. *Plant Physiol.* **2009**, *149*, 354–369. [CrossRef] [PubMed]
48. Zinta, G.; Khan, A.; Abdelgawad, H.; Verma, V.; Srivastava, A.K. Unveiling the redox control of plant reproductive development during abiotic stress. *Front. Plant Sci.* **2016**, *7*, 700. [CrossRef] [PubMed]
49. Fromm, S.; Braun, H.; Peterhansel, C. Mitochondrial gamma carbonic anhydrases are required for complex I assembly and plant reproductive development. *New Phytol.* **2016**, *211*, 194–207. [CrossRef] [PubMed]
50. Yi, J.M.; Huang, T.; Huang, Y.; Chen, D.H. Phylogenetic analysis of MADS-box gene family in *Physcomitrella patens*. *Plant Physiol. J.* **2015**, *51*, 197–206. [CrossRef]
51. Saha, G.; Park, J.I.; Jung, H.J.; Ahmed, N.U. Genome-wide identification and characterization of MADS-box family genes related to organ development and stress resistance in *Brassica rapa*. *BMC Genom.* **2015**, *16*, 178. [CrossRef] [PubMed]
52. Kumar, G.; Arya, P.; Gupta, K.; Randhawa, V.; Acharya, V.; Singh, A.K. Comparative phylogenetic analysis and transcriptional profiling of MADS-box gene family identified DAM and FLC-like genes in apple (*Malusx domestica*). *Sci. Rep.* **2016**, *6*, 20695. [CrossRef] [PubMed]
53. Grimplet, J.; Martínez-Zapater, J.M.; Carmona, M.J. Structural and functional annotation of the MADS-box transcription factor family in grapevine. *BMC Genom.* **2016**, *17*, 80. [CrossRef] [PubMed]
54. Nam, J.; Kim, J.; Lee, S.; An, G.; Ma, H.; Nei, M. Type I MADS-box genes have experienced faster birth-and-death evolution than type II MADS-box genes in angiosperms. *PNAS* **2004**, *101*, 1910–1915. [CrossRef] [PubMed]

55. Smaczniak, C.; Immink, R.G.; Angenent, G.C.; Kaufmann, K. Developmental and evolutionary diversity of plant MADS-domain factors: Insights from recent studies. *Development* **2012**, *139*, 3081–3098. [CrossRef] [PubMed]
56. Masiero, S.; Li, M.A.; Will, I.; Hartmann, U.; Saedler, H.; Huijser, P. Incomposita: A MADS-box gene controlling prophyll development and floral meristem identity in antirrhinum. *Development* **2004**, *131*, 5981–5990. [CrossRef] [PubMed]
57. Toshitsugu, N.; Junji, K.; Masaki, F.; Mamiko, K.; Nao, I.; Hideo, M. Macrocalyx and jointless interact in the transcriptional regulation of tomato fruit abscission zone development. *Plant Physiol.* **2012**, *158*, 439–450. [CrossRef]
58. Huang, Q.; Li, W.; Fan, R.; Chang, Y. New mads-box gene in fern: Cloning and expression analysis of dfmads1 from *Dryopteris fragrans*. *PLoS ONE* **2014**, *9*, e86349. [CrossRef] [PubMed]
59. Lepiniec, L.; Debeaujon, I.; Routaboul, J.M.; Baudry, A.; Pourcel, L.; Nesi, N.; Caboche, M. Genetics and biochemistry of seed flavonoids. *Annu. Rev. Plant Biol.* **2006**, *57*, 405–430. [CrossRef] [PubMed]
60. Xu, W.; Fiume, E.; Coen, O.; Pechoux, C.; Lepiniec, L.; Magnani, E. Endosperm and nucellus develop antagonistically in *Arabidopsis* seeds. *Plant Cell* **2016**, *28*, 1343–1360. [CrossRef] [PubMed]
61. Chen, G.; Deng, W.; Peng, F.; Truksa, M.; Singer, S.; Snyder, C.L. *Brassica napus* TT16 homologs with different genomic origins and expression levels encode proteins that regulate a broad range of endothelium-associated genes at the transcriptional level. *Plant J.* **2013**, *74*, 663–677. [CrossRef] [PubMed]
62. Zhang, Z.; Li, H.; Zhang, D.; Liu, Y.; Fu, J.; Shi, Y. Characterization and expression analysis of six MADS-box genes in maize (*Zea mays* L.). *J. Plant Physiol.* **2012**, *169*, 797–806. [CrossRef] [PubMed]

*forests*

MDPI

*Article*

# Genome-Wide Analysis and Expression Profiling of the Heat Shock Factor Gene Family in *Phyllostachys edulis* during Development and in Response to Abiotic Stresses

**Lihua Xie, Xiangyu Li, Dan Hou, Zhanchao Cheng, Jun Liu, Juan Li, Shaohua Mu and Jian Gao ***

Key Laboratory of Bamboo and Rattan Science and Technology, International Center for Bamboo and Rattan, State Forestry and Grassland Administration, Beijing 100102, China; xielihua0227@163.com (L.X.); leerduo727@163.com (X.L.); hou120314@foxmail.com (D.H.); chengzhan_chao@126.com (Z.C.); liujun_0325@163.com (J.L.); ljgx2003@126.com (J.L.); mush@icbr.ac.cn (S.M.)

* Correspondence: gaojian@icbr.ac.cn or gaojianicbr@163.com; Tel.: +86-010-8478-9801

Received: 3 December 2018; Accepted: 23 January 2019; Published: 26 January 2019

**Abstract:** Heat shock transcription factors (Hsfs) play crucial roles in regulating plant responses to heat and other stresses, as well as in plant development. As the largest monopodial bamboo species in the world, how to adapt to various stresses under the background of global climate change is very important for the sustainable development of bamboo forest. However, our understanding of the function of Hsfs in moso bamboo (*Phyllostachys edulis*) is limited. In this study, a total of 22 non-redundant Hsf genes were identified in the moso bamboo genome. Structural characteristics and phylogenetic analysis revealed that members of the PheHsf family can be clustered into three classes (A, B and C). Furthermore, PheHsfs promoters contained a number of stress-, hormone- and development-related cis-acting elements. Transcriptome analysis indicated that most *PheHsfs* participate in rapid shoot growth and flower development in moso bamboo. Moreover, the expression patterns of all 12 members of class A were analyzed under various stresses (heat, drought, salt and cold treatment) through Figurereal-time quantitative polymerase chain reaction (qRT-PCR). Within the class A *PheHsf* members, *PheHsfA1a* was expressed mainly during moso bamboo development. Expression of four *PheHsfA4s* and one *PheHsfA2* (*PheHsfA4a-1*, *PheHsfA4a-2*, *PheHsfA4d-1*, *PheHsfA4d-2*, and *PheHsfA2a-2*) was up-regulated in response to various stresses. *PheHsfA2a-2*, *PheHsfA4d-1* and *PheHsfA4d-2* were strongly induced respectively by heat, drought and NaCl stress. Through co-expression analysis we found that two hub genes *PheHsfA4a-2* and *PheHsfA4a-1* were involved in a complex protein interaction network. Based on the prediction of protein interaction networks, five *PheHsfAs* (*PheHsfA4a-1*, *PheHsfA4a-2*, *PheHsfA4d-1*, *PheHsfA4d-2*, and *PheHsfA2a-2*) were predicted to play an important role in flower and shoot development and abiotic stress response of moso bamboo. This study provides an overview of the complexity of the PheHsf gene family and a basis for analyzing the functions of PheHsf genes of interest.

**Keywords:** moso bamboo; heat shock factor gene; abiotic stresses; co-expression

---

## 1. Introduction

Moso bamboo (*Phyllostachys edulis* (Carrière) J. Houzeau, synonym *Phyllostachys heterocycla* (Carrière) is a large woody bamboo of high ecological, economical and cultural value in Asia. Under suitable spring conditions, its shoot can grow from 0 to 20 m in 45–60 days [1]. Moso bamboo forest covers an area of 3.87 million hm$^2$, accounting for up to 70% of the Chinese bamboo forest area [2,3]. Because of its rapid growth and highly lignified culms, the annual economic value of moso bamboo production, including timber and wood production, reaches 184 billion dollars [4]. Moreover, carbon sequestration in moso bamboo is two to four times greater than that of Chinese fir, making it an important global non-timber forest resource [5]. The growth of bamboo is dependent on natural precipitation and is vulnerable to high temperature and drought. Liu et al. [6] has shown that temperatures >40 °C and drought for >10 days during August result in severe losses in moso bamboo forests. Drought during spring can reduce moso bamboo shoot growth, yield and quality. From July to September, high temperatures and drought affect the sprouting phase of bamboo. These stresses affect the yield and quality of winter shoots, as well as new bamboo yield in the following year and the yield of wood during subsequent years [7]. Climate change has also been associated with more frequent high temperatures and drought, which in turn reduce the ecological and economical value of moso bamboo. Therefore, it is essential to elucidate the molecular mechanisms involved in heat and drought stress responses to improve stress tolerance in moso bamboo.

To survive high temperature and other stresses, plants have evolved a series of defense strategies [8,9]. Heat shock proteins (HSPs) act as molecular chaperones that protect cells against heat and other stress damage by preventing protein aggregation [10,11]. As the terminal components of the stress signal transduction chain, heat shock stress transcription factors (Hsfs) bind to the promoter regions of HSP genes to regulate transcription in response to stress [12,13], particularly heat stress [14]. Hsfs also contain a highly conserved DNA-binding domain (DBD) at their N-terminal and an oligomerization domain (OD or HR-A/B region) composed of two hydrophobic heptad repeats. Based on the amino acids of the HR-A/B region, plant Hsfs are grouped into three main classes (HsfA, HsfB and HsfC) [15]. Certain Hsfs contain a nuclear localization signal (NLS), a nuclear export signal (NES) and an activator motif (AHA). The AHA motif located at the C-terminus in class A Hsfs confers transcriptional activation. Moreover, a repressor domain (RD) that contains the tetrapeptide LFGV occurs at the C-terminal of class B Hsfs.

Recent studies have revealed that plant Hsfs plays important roles in generating responses to heat and other stimuli, as well as in organ development [16]. Class A HsfA1a is regarded as a master regulator and has a unique role in eliciting heat stress responses in tomato (*Solanum lycopersicum* L.) [17]. HsfA2 is functionally similar to HsfA1 in regulating thermotolerance, as well as serving as a key regulator in osmotic and oxidative stress responses [18–20]. The expression of *HsfA3* is induced in *Arabidopsis* by heat and drought stress, indicating that *HsfA3* might play a role in drought and heat stress signaling [21,22]. The thermotolerance of *Arabidopsis* with *hsfA3* T-DNA insertion mutants was decreased [23]. Moreover, the ectopic overexpression of *SIHsfA3* increases thermotolerance and salt hypersensitivity in germination in transgenic *Arabidopsis* [24]. The Arabidopsis mutant *athsfa4a* was more sensitive to dehydration. Furthermore, desiccation tolerance was rescued in *athsfa4a/BnHSFA4a* seeds to similar levels compared with those of Col-0 [25]. Transgenic chrysanthemum overexpressing *CmHSFA4* displayed enhanced salinity tolerance partly due to enhanced $Na^+/K^+$ ion and ROS homeostasis [26]. Interestingly, in rice, wheat and *Sedum Alfredii*, HsfA4a possibly confers cadmium tolerance [27,28]. In addition, AtHsfA9 plays a role in embryonic development and seed maturation in *Arabidopsis* [29]. In group B, the majority of HsfBs act as repressors due to the RD region [15]. However, AtHsfB1 act as repressors of the heat shock response under non-heat-stress conditions, but act as positive regulators of heat shock proteins under heat-stress conditions [30]. In group C, *FaHsfC1b* from *Festuca arundinacea* confers heat tolerance in *Arabidopsis* [31].

Although several Hsf family genes in *Arabidopsis* and other plant species have been characterized, functional analysis of those in moso bamboo has been limited. The completion of the draft genome sequence of moso bamboo has greatly facilitated the identification of Hsf family at the whole-genome level [32].

In this study, we describe the genome-wide identification and analysis of the PheHsf family of moso bamboo for the first time. In addition, expression patterns of *PheHsf* genes during development, as well as in response to various abiotic stresses, were also investigated. Our results will provide a foundation and valuable information for future functional analysis of PheHsfs.

## 2. Materials and Methods

### 2.1. Database Searches for Hsf Genes in Moso Bamboo and Analyses of Physicochemical Characteristics

For more accurate identification of Hsf genes in Moso bamboo, multiple database searches were performed according Hou et al. [33]. First, Hsf mRNA sequences of *Oryza sativa* and *Brachypodium distachyon*, obtained from NCBI Nucleotide database (https://www.ncbi.nlm.nih.gov/) as query sequences to blast against the moso bamboo genome database. For the filtration step of the blast process, the Hsf genes were obtained by blast bamboo transcriptome using Hsf mRNA from some other species identified as query with loose *e*-value of <0.00001. Next, the protein sequences of putative genes obtained from the first step were blast against the NCBI non-redundant protein database with *e*-value of <0.0000000001 by Blast2GO to confirm the identification. The putative genes described as Hsf proteins or proteins belonging to Hsf family were kept, and genes described as other family proteins were abandoned. Then the HSF domain (PF00447) of the HSF family were researched in these putative Hsf proteins to reconfirm our data using the Pfam database (http://pfam.sanger.ac.uk/) and Conserved Domains Database (http://www.ncbi.nlm.nih.gov/Structure/cdd/cdd.shtml) with e-values <0.001. Only the gene models containing HSF domain were considered to belong to the Hsf family. Finally, the selected Hsfs were further screened using the full-length non-chimeric (FLNC) reads (http://www.forestrylab.org/db/PhePacBio/) [34]. The moso bamboo Hsf genome sequences, coding sequences, protein sequences, and putative novel or mis-annotated *Hsf* genes were obtained from moso bamboo genome database. The amino acid sequences of *Arabidopsis*, rice (*Oryza sativa*), and *B. distachyon* Hsf proteins were downloaded from Plant Transcription Factor Database v4.0 (PlantTFDB 4.0, http://planttfdb.cbi.pku.edu.cn/) and Heatster (http://www.cibiv.at/services/Hsf/) [35,36]. Molecular weights and theoretical isoelectric point (pI) were determined using ExPASY (http://web.expasy.org/compute_pi/). CELLO v2.5 Server (http://cello.life.nctu.edu.tw/) was used to predict the protein subcellular locations for candidate PheHsfs [37]

### 2.2. Phylogenetic Analysis

Arabidopsis and rice Hsf gene datasets [38] were used to classify the moso bamboo Hsf genes and predict their functional roles. Multiple sequence alignments of full-length Hsf proteins from Arabidopsis, rice, B. distachyon, and moso bamboo were performed using ClustalX 1.83 (http://www.clustal.org/) and two online programs (Clustal Omega and MUSCLE) [27,39–41]. An unrooted neighbor-joining (NJ) phylogenetic tree was constructed using MEGA7.0 with 1000 bootstrap replicates. Another phylogenetic tree with only the Moso bamboo Hsfs was also constructed using the amino acid sequences according to the same method.

### 2.3. Structural and Motif Analyses of PheHsf Genes

To confirm subgroup designation through phylogenetic analysis, the Gene Structure Display Server (http://gsds.cbi.pku.edu.cn/) was used to illustrate exon-intron organization by aligning every cDNA sequence and its corresponding genomic DNA sequence [42]. The conserved motifs in the candidate PheHsf sequence were defined by MEME version 4.12.0 (http://meme-suite.org/tools/meme) [43] using the following parameters: number of repetitions = any, maximum number of motifs = 30, minimum width $\geq$4, maximum width $\leq$200, and only motifs with an e-value <0.01 were retained for further analysis. The DBD (DNA Binding domain) and HR-A/B domains were identified using Heatster (http://www.cibiv.at/services/Hsf/). NES domains in the PheHsfs were predicted with the NetNES 1.1 server software (http://www.cbs.dtu.dk/services/NetNES/). NLS domains were predicted using cNLS Mapper software (http://nls-mapper.iab.keio.ac.jp/cgi-bin/NLS_Mapper_form.cgi). AHA domains were predicted according to the conserved motif sequence FWxxF/L, F/I/L [16].

### 2.4. Cis-Regulatory Element Analysis of PheHsf Genes

The 1500-bp sequence upstream from the initiation codon of each *PheHsf* gene was obtained from the moso bamboo genome database. These sequences were used to identify *cis*-acting regulatory elements with the online program Plant CARE (http://bioinformatics.psb.ugent.be/webtools/plantcare/html/).

### 2.5. Plant Material

After the surface was sterilized with 1% formaldehyde, the moso bamboo seeds were germinated in Petri dishes (12-cm diameter) lined with filter paper and containing 10 mL of sterile water. After 4 days, the germinated seedlings were planted in vermiculite and watered with 1/2 Hoagland's nutrient medium weekly (all plants were grown in 16 h day/8 h night at 22 °C). Two-month-old seedlings were used for abiotic stress treatments. According to Cheng et al. [44], drought and stress was conducted by incubated seedlings with 20% (*m*/*v*) PEG6000 and 200 mM NaCl, respectively. Heat stress and low temperature treatments were respectively created by placing seedlings in a 42 °C and 4 °C lighted growth chamber according to Liu et al. [45]. The control seedlings were grown without any stress treatment. The second or third mature leaves were collected at 0 h, 15 min, 30 min, 1 h, 3 h, 6 h, 12 h, and 24 h after abiotic stress treatments. These materials were immediately frozen and stored in liquid nitrogen until total RNA extraction and real-time quantitative polymerase chain reaction (qRT-PCR).

### 2.6. RNA Isolation and Relative Expression Level Analysis of PheHsfs

For the tissue-specific expression analysis, the RPKM (the reads per kilobase of exon model per million mapped reads) values of *PheHsfs* were retrieved from transcriptome sequencing of developing flowers (four stages of flowering and leaves) and shoots of moso bamboo (winter shoots, six shoot heights, and culms) [1,46]. RPKM values were used to analyze the relative expression levels of the *PheHsf* genes. For the flower development samples, four developmental stages (F1: the floral buds begun to form; F2: the floral organs gradually matured; F3: the flowers were in full blossom; and F4: the embryo formation stage) were defined based on the anatomical structure of floral organs by Gao et al. [46]. For F1 and F2, the buds were collected, respectively. For F3 and F4, spikelets were collected, respectively. Leaves collected from non-flowering moso bamboo were defined as CK1. For the shoot development stage sample, four development stages of shoots were defined based on the continuing measurement of bamboo shoot height and anatomical changes by Li et al. [1]. The eight samples according to the four developmental stages of moso bamboo shoots were as follows: S1 (winter shoots), S2–S5 (0.5 m, 1 m, 3 m, and 6 m, early growth period), S6–S7 (9 m and 12 m, late growth), and CK (clum after leaf expansion, mature period). For S1–S7, the shoot tips of different heights were collected, respectively. For CK2, each top internode was cut from the top to 1/2, then each top internode was divided into basal, middle and top. After that, the samples were cut from the tissue located in the top part of the three internodes above and then mixed.

For qRT-PCR, *PheHsfs* primers were designed using Primer 3.0 (http://primer3.ut.ee/). Primer sequences, amplicon Length, amplification efficiency, and correlation coefficients are listed in Table S1, and their specificity was verified using the online tool Primer-BLAST (https://www.ncbi.nlm.nih.gov/tools/primer-blast/index.cgi) and the melting curves of PCR products. For every primer pair, a standard curve was constructed to calculate the gene-specific PCR efficiency from the 10-fold series dilution of the mix cDNA template. The $R^2$ (correlation coefficients) and slope values can be obtained from the standard curve. The following formula was used to calculate the corresponding PCR amplification efficiencies (E): $E = (10^{-1/\text{slope}} - 1) \times 100$ [47]. Tonoplast intrinsic protein 41 gene (*TIP41*) was used as an internal control [48]. The qRT-PCR reactions were conducted using a SYBR Green I master mix (Roche, Mannheim, Germany). The qRT-PCR conditions were as follows: 45 cycles of 95 °C for 10 s, 60 °C for 10 s, and 72 °C for 20 s. Three replicates were performed for each gene. Gene expression was evaluated using the $2^{-\Delta\Delta Ct}$ method [49].

*2.7. Co-Expression Network and Protein Interactions of PheHsfs*

The expression correlation of the PheHsfs was calculated by Pearson correlation coefficient (PCC; R-value) using gene expression RPKM values from the high-throughput transcriptome data in R. Expression correlation data were used for the correlation network, and co-expressed gene pairs were filtered with a PCC cut-off of 0.85 as previously described [50]. Cytoscape version 3.4.0 were used to analyze and visualize the network [51].

For the protein interaction networks, the homolog Hsf proteins in rice were constructed by STRING (http://stringdb.org/) using an option value >0.7. The homolog proteins of the determined interactive rice proteins were identified in moso bamboo by BLASTP analysis.

## 3. Results

*3.1. The Hsf Family Genes in Moso Bamboo*

To identify Hsf genes in moso bamboo, a total of 41 candidate PheHsf genes were retrieved from the annotation in the Bamboo Genome Database. From these, 19 candidate PheHsf genes with incomplete HSF domains were considered as PheHsf-like genes, which were not selected for subsequent analysis in this study. Twenty-two putative moso bamboo PheHsf genes containing full HSF domains (PF00447) were confirmed by searching the Pfam and the Conserved Domain Databases. However, the CDS sequence for PheHsfA5, PheHsfB4a-1 and PheHsfB4c-1 contained 42 bases (from 268 to 309 base), 102 bases (233 to 336) and 123 base (256 to 377) inserts, respectively, when compared to the cloned cDNA sequence (Fasta S3). For this analysis, the CDS and amino acid sequences of these three genes are based on the cDNA sequences. The CDS and amino acid sequences are listed in supplementary files (Fasta S1 and S2). The identified 22 PheHsf genes were distributed among 22 scaffolds (Table 1). The sequences of 22 PheHsf genes were named according to the corresponding relationship among in rice, *B. distachyon* and moso bamboo. The number of amino acid (aa) sequences of PheHsf proteins ranged from 247 (PheHsfC1b-1) to 679 (PheHsfA1a), the predicted isoelectric points (pI) varied between 4.70 (PheHsfA6a) and 9.81 (PheHsfB4c-1), and the molecular weight (MW) ranged from 26.77 kDa (PheHsfC1b-1) to 75.05 kDa (PheHsfA1a) (Table 1).

*3.2. Phylogenetic Relationships and Multiple Sequence Alignment of PheHsf Genes*

To predict the potential function, an unrooted phylogenetic tree was constructed from an alignment of 94 full length Hsf protein sequences from different species (23 AtHsf, 25 OsHsf, 24 BdHsf, and 22 PheHsf) (Figure 1). Because the alignment results of Hsf proteins in moso bamboo using ClustalX 1.83, MUSCLE, and Clustal Omega were similar, we employed the results of ClustalX 1.83 and MEGA7.0 to illustrate the phylogenetic relationships of the PheHsf family. The PheHsf genes were also grouped into three subgroups, including class A (PheHsfA1a, PheHsfA2a-1, PheHsfA2a-2, PheHsfA4a-1, PheHsfA4a-2, PheHsfA4d-1, PheHsfA4d-2, PheHsfA6b-2, PheHsfA6b-1, PheHsfA6a, PheHsfA5, and PheHsfA7a), class B (PheHsfB1, PheHsfB2a, PheHsfB2c, PheHsfB4c-1, PheHsfB4c-2, PheHsfB4a-2, and PheHsfB4a-1), and class C (PheHsfC1b-1, PheHsfC1b-2, and PheHsfC2b), which is similar to that described in Arabidopsis [52], rice [38], and *B. distachyon* [53]. In moso bamboo, class A was the largest and consisted of 12 members from six subclasses, however, three subclasses (A3, A8, and A9) in this class were not detected. There were seven members in class B from subclasses B1, B2, and B4, but no members in subclass B3. Class C was the smallest, with only three members.

**Table 1.** Overview of *PheHsf* genes in moso bamboo.

| Name | Gene ID | Scaffolds | Number (aa) | MW (kDa) | PI | CELLO Localization (Reliability) |
|---|---|---|---|---|---|---|
| *PheHsfA4a-1* | PH01000018G0140 | PH01000018 | 443 | 49.07 | 5.22 | Nuclear (3.717) |
| *PheHsfB2c* | PH01000314G0470 | PH01000314 | 382 | 40.12 | 4.86 | Nuclear (2.907) |
| *PheHsfC1b-1* | PH01000333G0160 | PH01000333 | 247 | 26.77 | 8.71 | Cytoplasmic (1.219)<br>Mitochondrial (1.205)<br>Chloroplast (1.221) |
| *PheHsfA1a* | PH01000371G0730 | PH01000371 | 679 | 75.05 | 6.06 | Nuclear (2.682) |
| *PheHsfB4a-1* | PH01000700G0690 | PH01000700 | 385 | 41.99 | 8.87 | Cytoplasmic (1.061)<br>Nuclear (1.481) |
| *PheHsfC1b-2* | PH01000701G0030 | PH01000701 | 296 | 32.24 | 8.84 | Cytoplasmic (1.557) |
| *PheHsfA4a-2* | PH01000746G0370 | PH01000746 | 444 | 49.50 | 5.36 | Nuclear (3.999) |
| *PheHsfA4d-1* | PH01000814G0580 | PH01000814 | 554 | 61.26 | 4.95 | Nuclear (2.593) |
| *PheHsfC2b* | PH01000849G0330 | PH01000849 | 276 | 29.22 | 7.46 | Extracellular (1.087) |
| *PheHsfA5* | PH01001018G0530 | PH01001018 | 499 | 55.33 | 5.21 | Nuclear (3.607) |
| *PheHsfA7a* | PH01001028G0180 | PH01001028 | 381 | 42.51 | 5.94 | Cytoplasmic (1.751) |
| *PheHsfB4c-2* | PH01001228G0360 | PH01001228 | 276 | 29.72 | 8.34 | Nuclear (3.293) |
| *PheHsfB4c-1* | PH01001554G0080 | PH01001554 | 420 | 45.70 | 9.81 | Nuclear (1.699) |
| *PheHsfA4d-2* | PH01002437G0340 | PH01002437 | 453 | 51.34 | 5.18 | Nuclear (4.154) |
| *PheHsfB4a-2* | PH01004959G0060 | PH01004959 | 275 | 30.24 | 8.37 | Chloroplast (1.119)<br>Nuclear (1.604) |
| *PheHsfB2a* | PH01000000G3800 | PH01000000 | 292 | 31.22 | 4.98 | Nuclear (3.984) |
| *PheHsfA6b-2* | PH01000081G0140 | PH01000081 | 362 | 41.44 | 4.97 | Nuclear (3.466) |
| *PheHsfA2a-2* | PH01000174G0590 | PH01000174 | 342 | 38.47 | 5.32 | Nuclear (4.565) |
| *PheHsfA6a* | PH01000194G0800 | PH01000194 | 378 | 42.59 | 4.70 | Nuclear (1.785) |
| *PheHsfA6b-1* | PH01000208G0690 | PH01000208 | 348 | 40.04 | 4.89 | Cytoplasmic (2.207)<br>Nuclear (2.182) |
| *PheHsfA2a-1* | PH01003916G0010 | PH01003916 | 358 | 40.31 | 5.49 | Nuclear (3.869) |
| *PheHsfB1* | PH01000149G1320 | PH01000149 | 297 | 32.42 | 8.55 | Nuclear (3.609) |

Gene ID: refer to Bamboo Genome Datebase (http://202.127.18.221/bamboo/index.php); MW: molecular weight represents the predicted weights of PheHsf proteins; PI: represents the predicted isoelectric point of PheHsf proteins.

### 3.3. Structure and Motif Analyses of PheHsf Genes

To better understand the gene structure diversity of PheHsfs, we compared the intron-exon arrangements and the conserved motifs (Figure S1). The number of introns in the Hsf genes of moso bamboo ranged from zero to three. Most of the PheHsfs (19/22) contained one to two introns (Figure S1b). Three introns were found in PheHsf4, whereas none were detected in PheHsfB4a-1 and PheHsfB4c-1.

Based on the known information on functional domains of Hsfs in some model plants [15,54], the sequence and positions of similar domains were identified in the PheHsfs by sequence alignment. As shown in Table 2, five conserved domains (DBD, HR-A/B, NLS, NES, and AHA) were identified. The DBD domain comprised of three α-helices (α1–3) and four β-sheets (β1–4) were found in all PheHsfs (Figure S2a). HR-A/B domain is critical for one Hsf interacting with other Hsfs to form trimer [15]. All class A PheHsfs have a 21 amino-acid (aa) insertion between HR-A and HR-B regions; class C PheHsfs have seven aa insertions; all class B PheHsfs have no insertion (Figure S2b). NLS and NES domain function in the assembly of a nuclear import complex and the receptor-mediated export in complex with the NES receptor [15]. The majority of the PheHsfs showed the presence of a NES and/or NLS domain. In addition, the activation domain AHA was found in all class A members but not in classes B and C (Table 2).

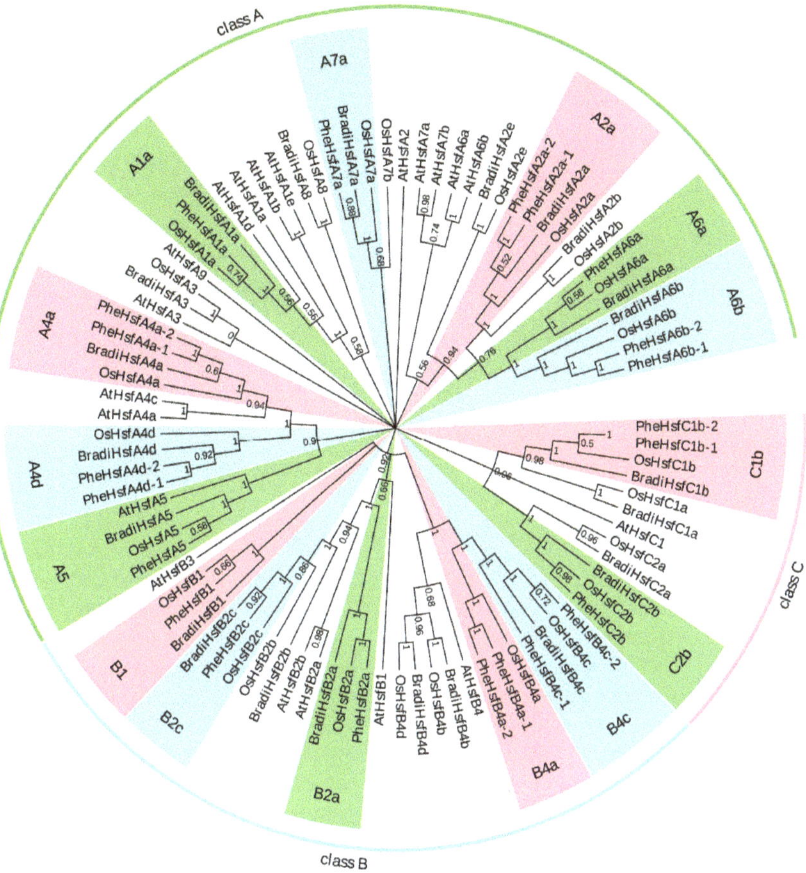


**Figure 1.** Phylogenetic relationship of PheHsf, AtHsf and OsHsf proteins. Neighbor-joining method and MEGA 7.0 software were used for phylogenetic analysis of Hsf proteins from *Phyllostachys edulis* (22 PheHsfs), *Arabidopsis thaliana* (23 AtHsfs), *Oryza sativa* (25 OsHsfs), and *Brachypodium distachyon* (24 BdHsfs). The names of subclass are shown outside of the circle. Branch lines of subclass are colored, indicating different Hsf subclasses.

A MEME motif search revealed a total of 19 motifs in the PheHsfs (Figure S1c). Three motifs (1, 2, and 4) constituting the DBD domain were identified. Motif 3 and 5 for the OD domain were detected in class A and C, and motif 6 for the OD domain was observed in class B. Motif 6 and motif 11, motif 8, and motif 17 (NSL) were observed in class A, class B and C, respectively. In general, the structure of the PheHsf proteins is conserved in moso bamboo. Furthermore, motif 7 represented the AHA domain close to the PheHsfA C-terminus (Figure S1c and Table 2).

**Table 2.** Functional domains of PheHsfs.

| Gene | DBD | HR-A/B | NLS | AHA | RD | NES |
|---|---|---|---|---|---|---|
| PheHsfA1a | 1–84 | 121–171 | (201) ANKKRRLPKQ | (410) SFWEQFLVA | nd | nd |
| PheHsfA2a-1 | 39–128 | 151–202 | (237) ISKKRRRRID | (314) DDFWEDLLHE | nd | nd |
| PheHsfA2a-2 | 40–129 | 156–197 | (238) ISKKRRRRID | (315) DDFWEDLLHE | nd | (165) LLM |
| PheHsfA4a-1 | 14–103 | 124–165 | (204) DHHRKKRRLPKPISF | (379) GFWQQFLTE | nd | nd |
| PheHsfA4a-2 | 14–103 | 124–172 | (204) DHHRKKRRLPKPISF | (380) GFWQQFLTE | nd | nd |
| PheHsfA4d-1 | 13–96 | 112–140 | (199) FSKKRRAPKI | (363) LFWERFLTE | nd | nd |
| PheHsfA4d-2 | 16–105 | 120–149 | (208) FSKKRRVPKI | (382) LFWERFLTE | nd | (252) MELAL |
| PheHsfA5 | 30–134 | 152–179 | (241) FQKKRRLTGL | (446) KFWEQFLTE | nd | nd |
| PheHsfA6a | 46–135 | 165–206 | (247) ISKKRRRPID | (319) DDFWAELLVE | nd | (288) LENLAL |
| PheHsfA6b-1 | 45–134 | 163–204 | (133) LLKMIKRRRLLYY | (318) DDFWEELLNE | nd | nd |
| PheHsfA6b-2 | 45–134 | 163–204 | (135) KMIKRRRPLS | (318) EDFWEELLNE | nd | nd |
| PheHsfA7a | 74–165 | 193–227 | (166) KNIKRRRASK | (329) DDVWEELDAI | nd | nd |
| PheHsfB1 | 30–117 | 176–205 | (253) EDATRKRKRCEEAAARERPFKMIRI | nd | (246) KLFGVLL | nd |
| PheHsfB2a | 7–100 | 160–183 | nd | nd | (203) TLFGVTI | (264) LDVLALSL |
| PheHsfB2c | 29–118 | 199–232 | nd | nd | (304) RLFGVSI | nd |
| PheHsfB4a-1 | 19–112 | 110–139 | (237) RKRLLQEQPPTSPEWKRSMV | nd | (215) KLFGVNL | nd |
| PheHsfB4a-2 | 19–103 | 201–230 | (347) RKRSLQEQPPTSPDWKRSMV | nd | (325) KLFGVDL | nd |
| PheHsfB4c-1 | 26–116 | 256–285 | nd | nd | (323) KLFGVHI | nd |
| PheHsfB4c-2 | 26–116 | 124–143 | nd | nd | (213) KLFGVHL | (251) LESDDLSL |
| PheHsfC1b-1 | 25–114 | 136–165 | (209) PGKKRRIGAE | nd | nd | nd |
| PheHsfC1b-2 | 72–161 | 186–213 | (257) TPGKRRRIG | nd | nd | nd |
| PheHsfC2b | 12–105 | 140–176 | nd | nd | nd | (173) LKV |

DBD: DNA-binding domain; HR–A/B: OD (oligomerisation domain); NLS: Nuclear localization signal; AHA: Activator motifs; RD: Tetrapeptid motif LFGV as core of repressor domain; NES: Nuclear export signal; nd, no motifs detectable by sequence similarity search.

### 3.4. Cis-Regulatory Element Analysis in Promoters of PheHsfs

To predict the biological function of PheHsfs, the 1500 bp upstream sequence from the translation start sites of PheHsf genes were analyzed using the PlantCARE database. The results show that the promoter of each PheHsf has several cis-regulatory elements such as phytohormone- (abscisic acid, jasmonic acid and gibberellic acid), abiotic stress- (low temperature, heat stress, drought, and fungal elicitor), and developmental process-related elements. Figure S3 shows that the ABA-responsive element (ABRE), the MeJA-responsive element (CGTCA-motif), and SA-responsive element (TCA-element) were found in the promoters of 17, 16, and 11 PheHsf genes, respectively. The promoters of 12 and 10 PheHsf genes contained the HSE and LTR, respectively. MYB-binding sites involved in drought inducibility (MBS), fungal elicitor-responsive elements (Box-W1) and defense- and stress-responsive elements (TC-rich) were found in 17, 15 and 8 PheHsf genes, respectively. Additionally, meristem expression (CAT-box), meristem-specific activation (CCGTCC-box), and endosperm expression (Skn-1_motif) motifs were found in the 13, 10 and 18 PheHsf genes, respectively. These findings indicate that PheHsfs might be associated with different transcriptional regulatory mechanisms for developmental, hormone and stress processes.

### 3.5. Expression Pattern of the PheHsf Genes in Shoot and Flower Development

Based on the RNA-Seq data of different flowering developmental stages [43] and the internodes of shoots at different heights [1], a heat map was constructed according to the RPKM of 22 PheHsfs (Figure 2). During four flowering developmental stages of moso bamboo, *PheHsf* genes could be classified into four groups (A, B, C, and D) according to their relative expression levels (Figure 2a). Most of the *PheHsfs* were highly expressed (RPKM > 10) in at least one stage, and only four PheHsf genes were expressed at low levels (RPKM < 2) in at least two stages during floral development (Table S2). Four members (*PheHsfB4a-1*, PheHsf15, *PheHsf12*, and *PheHsfB4c-1*) of group C had higher transcript accumulation during two earlier stages (F1 and F2) and were downregulated at two later stages (F3 and F4 stage). Group D consisted of 10 PheHsf genes (*PheHsfA4a-1*, *PheHsfB2c*, *PheHsfA1a*, *PheHsfA4a-2*, *PheHsfC2b*, *PheHsfA7a*, *PheHsfB2a*, *PheHsfA2a-2*, *PheHsfA6b-1*, and *PheHsfB1*), and their expression levels steadily increased at the F3 and F4 stages. Group B comprised of six *PheHsfs* (*PheHsfA4d-1*, *PheHsfA4d-2*, *PheHsfA5*, *PheHsfA6b-2*, *PheHsfA6a*, and *PheHsfA2a-1*), showing higher transcript accumulation at two later stages (F3 and F4) and in the leaves. Group A only had two genes, *PheHsfC1b-1* and *PheHsfC1b-2*, with expression levels six times and three times higher in leaves than the four stages of floral development, respectively. During shoot growth, Most *PheHsfs* (19/22) showed very low expression levels (RPKM < 8) at all seven stages, and only two genes (*PheHsfA1a* and *PheHsfA6b-2*) had higher expression levels (RPKM > 20) in at least one stage of bamboo shoot growth (Table S3). Based on the expression profiles, the *PheHsfs* were classified into four groups (Figure 2b). Among these, *PheHsfA1a* and *PheHsfA2a-2* were clustered into the same group with continuous down-regulated expression from the S1 to S7 stage. However, *PheHsfA6b-2* and *PheHsfB2c* were clustered in the same group, which showed twin peaks during the S2 and S5 stage, respectively.

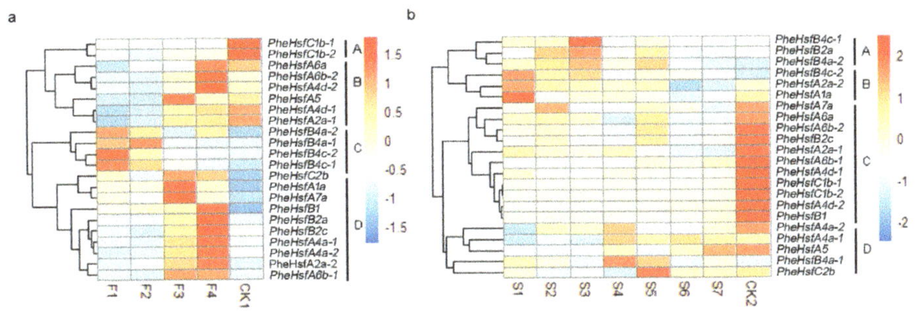


**Figure 2.** Expression pattern of *PheHsfs* in developmental flowers and shoots of moso bamboo. (**a**) The expression profile of *PheHsfs* in different stages of flowering. F1–F4: The four stages of developmental flowers (the floral buds begun to form, the floral organs gradually matured but did not undergo flowering, the flowers were in full blossom, and the embryo formation stage); CK1: leaves. (**b**) The expression profile of *PheHsfs* in different stages of shoots. S1: winter shoot; S2–S7: different heights of shoots (0.5 m, 1 m, 3 m, 6 m, 9 m and 12 m), and CK2 (culms after leaf expansion). All the samples had one repeat. The heatmap was pictured using R based on the RPKM of *Phehsfs* in each sample. Details of the RPKM are shown in Tables S2 and S3.

For different periods of flower and shoot development in moso bamboo, the *PheHsfA1a* and *PheHsfA6b-2* genes showed high transcript accumulation in all 11 stages, whereas four genes (*PheHsfB4a-1*, *PheHsfA4d-1*, *PheHsfA7a*, and *PheHsfB4a-2*) exhibited extremely low transcript accumulation at all stages. Moreover, the other 15 *PheHsfs* depicted significantly higher transcript levels during flower development than during shoot growth. These findings indicate that these genes have different regulatory roles in moso bamboo flower development and shoot growth.

### *3.6. PheHsfAs Expression in Moso Bamboo in Response to Various Stresses*

The expression of 12 PheHsfAs were analyzed by qRT-PCR under four abiotic stresses: high temperature, cold, drought and salt (Figure 3). The expression levels of all members of subclass A were upregulated (>2 folds) after at least one treatment (Table S4, Figure 3b). For heat stress, eight PheHsfAs were upregulated (>2 folds). PheHsfA4d-1, PheHsfA2a-2, and PheHsfA6b-1 rapidly responded to high temperatures, showing upregulation up to ~nine-fold within 0–1 h of 42 °C (Table S4) treatment. Among these, PheHsfA2a-2 (~22-fold higher than the control) was the most strongly induced gene. For cold stress, the expression levels of seven PheHsfAs (PheHsfA4a-1, PheHsfA1a, PheHsfA4a-2, PheHsfA4d-2, PheHsfA4d-2, PheHsfA5 and PheHsfA2a-2) were at least two-fold greater than the control for at least one of the time points. For drought and salt stress, eight PheHsfAs (PheHsfA4a-1, PheHsfA4a-2, PheHsfA4d-1, PheHsfA4d-2, PheHsfA5, PheHsfA7a, and PheHsfA6a) were upregulated. The expression level of PheHsfA4d-2 was ~20-fold higher than control after 3 h of 20% PEG treatment. The expression level of PheHsfA4d-1 was ~70-fold higher than the control after 3 h of 200 mM NaCl treatment (Table S4). Among the PheHsfA genes, five genes (PheHsfA4a-1, PheHsfA4a-2, PheHsfA4d-1, PheHsfA4d-2, and PheHsfA2a-2) were upregulated relative to the control treatment. (Figure 3b). These findings indicate that these genes might play vital roles in different stress response pathways.

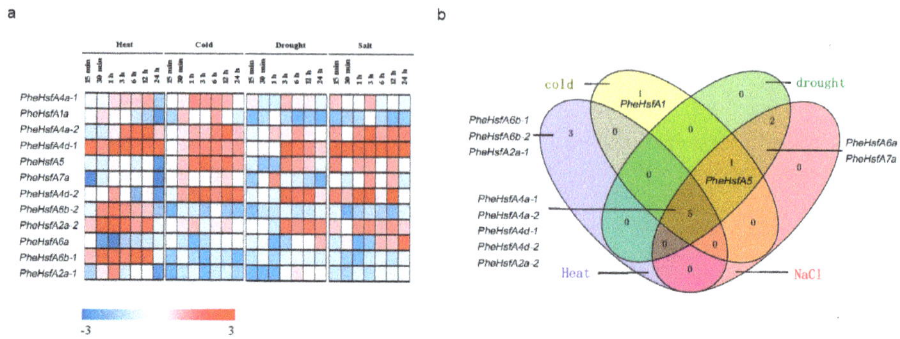


**Figure 3.** Expression analysis of PheHsfAs under different abiotic stress treatments in moso bamboo. (**a**) Heat map representation for the expression patterns of PheHsfAs after 15 min, 30 min, 1 h, 3 h, 6 h, 12 h, and 24 h of heat, cold, drought, and salt stresses: expression levels under stress vs. control; the expression results were obtained by qRT-PCR. The different colors correspond to log2 transformed value. (**b**) Venn diagram showing the number of overlapping PheHsfAs that are up-regulated > two-fold under abiotic stress: heat, cold, salt, and drought. Details of the expression data are shown in Table S4.

### *3.7. Expression Correlation and Interaction Networks*

To further investigate the PheHsf proteins and how they interact with each other, a co-expression network was constructed using expression values of PheHsf genes during shoot and flower development. The connecting gene with PCC magnitude >0.85 was recognized as strongly co-expressed genes [36]. The result showed that 18 PheHsf genes and 38 correlations in a co-expression network with PPC >0.85 cutoffs were obtained (Figure 4a). Genes with stronger correlation might play roles as interacting partners in similar biological pathways. Based on the results, *PheHsfA4a-1* and *PheHsfA4a-2* were recognized as hub genes as nodes with 10 and 9 connectivity in the whole network, respectively, which had more connectivity than other PheHsf genes. Both *PheHsfA4a-1* and *PheHsfA4a-2* had an up-regulated expression in response to different stress treatments (heat stress, cold, drought and salt). These results indicated that the two *PheHsfA4s* with a greater role in shoot and flower development also had an important role in stress responses.

To identify the two PheHsfA4a-associated proteins and protein complexes, prediction networks were built with STRING (http://www.string-db.org/) based on the interaction network of rice orthologous genes (Figure 4b). Because the rice orthologous gene of both *PheHsfA4a-1* and *PheHsfA4a-2* were *OsHsfA4a*, the identified moso bamboo proteins predicted to participate in the interaction network with *PheHsfA4a-1* and *PheHsfA4a-2* were the same. They interacted directly with 10 identified proteins, including five HSP70 proteins, one HSP90 protein, and four MAPK proteins. Because *PheHsfA2a-2* (~22-fold higher than the control), *PheHsfA4d-2* (~20-fold higher than the control) and *PheHsfA4d-1* (~79-fold higher than the control) were strongly induced by heat, drought and NaCl stress, respectively. We also identified the two PheHsfA4d- and *PheHsfA2a-2*- associated proteins and protein complexes, and PheHsfA4ds and *PheHsfA2a-2* were also predicted to interact with HSP70 proteins, HSP90 protein, and MAPK proteins (Figure 4c,d).

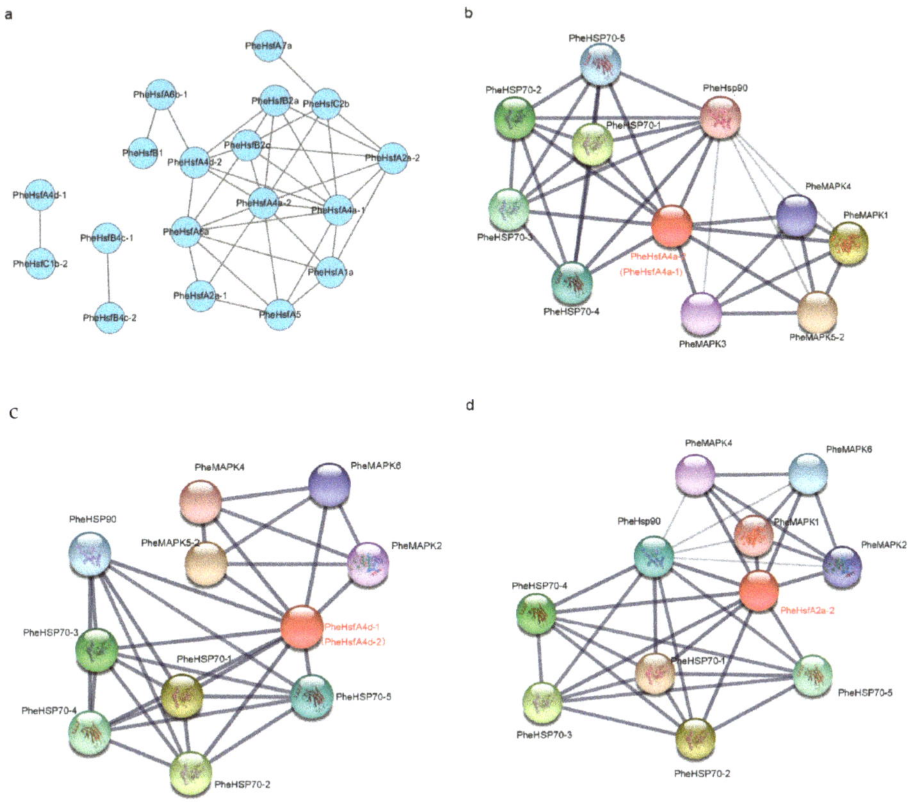


**Figure 4.** Co-expression network and interaction network of selected PheHsf genes in moso bamboo. (**a**) The model was built based on RPKM of RNA-seq of shoot and flower development. (**b–d**) Interaction network of PheHsfA4as, PheHsfA4ds and PheHsfA2a-2 in moso bamboo, respectively. Colored balls (protein nodes) in the network were used as a visual aid to indicate different input proteins and predicted interactors. Protein nodes which are enlarged indicate the availability of 3D protein structure information. Gray lines connect proteins which are associated by recurring textmining evidence. Line thickness indicates the strength of data support.

## 4. Discussion

### *4.1. Characterization of the Moso Bamboo Hsf Genes Family*

Hsf genes play essential roles in plant development and in responding to various stress conditions [55,56]. To explore the functions of *PheHsf*s in moso bamboo for the first time, the present study identified a total of 22 PheHsf genes according to the moso bamboo genome database and FLNC reads database [32,34,57].

We found that the moso bamboo has a similar number of *Hsf*s as rice, *B. distachyon*, maize, and *A. thaliana* (22–25). This partially accounts for the support of Hsf conservation during evolution. Phylogenetic analysis of Hsfs in moso bamboo, *O. sativa*, *B. distachyon*, and *A. thaliana* indicated that PheHsfs have a higher degree of sequence similarity with OsHsfs and BdHsfs than AtHsfs, which coincides with the evolutionary relationships among the four species. All three *Hsf* classes (A, B, and C) were identified in the three monocots and one dicot, implying that the Hsf genes originated prior to the divergence of monocots and dicots.

In the investigation of conserved Hsf domains and intron-exon structures, all 22 of the PheHsfs contain the necessary (DBD and OD) and/or specific protein domains (NLS, NES, and AHA). The hydrophobic core of DBD domain ensures precise positioning and highly selective interaction with heat stress promoter elements [15]. The OD of plant Hsfs confers distinct patterns of specificity for hetero oligomerization [15]. These AHA motifs, which are located in the C-terminal, are characterized by aromatic (W, F, Y), large hydrophobic (L, I, V) and acidic (E, D) amino acid residues [15]. Those domains might be essential for functional conservation [56]. Twenty of twenty-two PheHsf genes have one intron in their DBD domain (Figure S1), which is an evolutionarily conserved intron [13]. However, no intron was identified in *PheHsfB4a-1* and *PheHsfB4c-1* of subclass B4, which is different from subclass B4 of rice and *B. distachyon* [53].

### *4.2. Cis-Regulatory Element Analysis in the Promoters of PheHsfs*

Previous studies have illustrated the key roles of *Hsf*s in developmental processes and stress tolerance through their regulation of target genes [58]. The number and form of *cis*-elements in the promoter region might play an essential function in the regulation of gene expression in relation to metabolic pathways [59]. The in silico survey of putative *cis*-elements using PlantCARE showed that 12 of the 22 *PheHsfs* promoters contained HSEs. This implies that *PheHsfs* might regulate themselves [55]. Additionally, the promoter region of *PheHsfA1a* contained more types of development-related *cis*-elements (Figure S3), which coincides with its constitutive expression during shoot and flower development in moso bamboo (Figure 2).

### *4.3. PheHsfAs Involvement in Development Processes*

The 22 *PheHsfs* exhibited diverse expression patterns during shoot and flower development of moso bamboo under normal conditions. *PheHsfA1a* was upregulated during shoot and flower development and was constitutively expressed in different tissues (Figure 2). These findings are similar to that in *Arabidopsis* [60] and *Salix suchowensis* [61]. Under normal conditions, class A1 *Hsfs* of *Arabidopsis* are involved in housekeeping processes, and *SsuHsf-A1a* of *Salix suchowensis* are constitutively expressed in different tissues [60,61]. In rice, nearly all classA members showed high expression levels in all tissues [45,62]. In this study, nearly all the *PheHsfAs* (except *PheHsfA7a*) were found to show high transcription levels in the leaves, culms, and the four stages of flower development, similar to that in rice. Only two members of *PheHsfAs* (*PheHsfA1a* and *PheHsfA6b-2*) have high transcription levels in the seven stages of shoot development. This indicated that the function of *PheHsfAs* is conserved and/or specific in regulating flower development and shoot growth in moso bamboo.

### *4.4. PheHsfAs are Involved in Stress Responses*

Under heat or other stress conditions, plant Hsfs regulate the transcription of target genes (*Hsps* and other stress-inducible genes) to enhance stress resistance. Recent genome-wide expression profile analyses showed that most of the Hsf genes are upregulated after heat, cold, drought, and salt stress [56,58,60]. An increase in the number of Hsf genes has been shown to improve plant stress tolerance [54,63]. In moso bamboo, Zhao et al. identified seven and two *PheHsfs* that were upregulated after 0.5 h and 8 h of high light stress (1200 μmol · $m^{-2}$ · $s^{-1}$), suggesting that these play vital roles both in response to short-term (0.5 h) and mid-term (8 h) high light [64]. However, the regulatory roles of Hsfs in response to other abiotic stresses are unclear. Classes B and C do not harbor AHA motifs, which are essential for the activity of class A Hsfs [16]. Therefore, class A PheHsf genes were selected to further study their response patterns under stress and hormone treatments.

Under heat treatment, 9 of 12 *OsHsfAs* and 9 of 13 *BdHsfAs* were upregulated in rice and *B. distachyon*, respectively [59,62]. In our study, 8 of 12 *PheHsfAs* were found to be induced by high temperature. Of these, *PheHsfA4d-1*, *PheHsfA2a-2* and *PheHsfA6b* responded more quickly to heat stress than others. *PheHsfA2a-2* showed stronger induced under-heat stress conditions, which is similar to that of its homologous genes in rice, Arabidopsis, and *B. distachyon* [18,45,53]. These findings suggest that *PheHsfA2a-2* plays an important role in response to heat stress of moso bamboo. However, *PheHsfA2a-1*, which is the most similar to *PheHsfA2a-2* and also belongs to A2a-type PheHsfs, was slightly upregulated 1 h after heat stress application. *PheHsfA4d-2*, which is the most similar to *PheHsfA4d-1*, was less upregulated by heat and salt stress compared to *PheHsfA4d-1*, but showed higher relative expression levels than *PheHsfA4d-1* with cold and drought stresses.

HsfA4a of wheat (*Triticum aestivum*) and rice conferred cadmium tolerance in yeast and plants, but other Hsfs with similar structure (OsHsfA4d, AtHsfA4a, and AtHsfA4c) did not [27]. HsfA4a of Arabidopsis and chrysanthemum (chrysanthemum cultivar 'Jinba') confers salt and oxidative stress [26,65]. In this study, *PheHsfA4a-1*, *PheHsfA4a-2*, *PheHsfA4d-1* and *PheHsfA4d-2* were upregulated in response to these four abiotic stress (heat, cold, drought and salt). These findings indicate that the three *PheHsfA4s* could act as potential "nodes" for connecting the above four abiotic stresses.

### *4.5. Expression Correlation and Interaction Networks*

In addition, the expression values of PheHsf genes in shoot and flower developmental stages were used to identify potential underlying co-expression networks using Cytoscape. Based on their degree of connectivity, the hub genes *PheHsfA4a-1* and *PheHsfA4a-2* were identified to play a regulatory role and correlated with other PheHsfs in the complex feedback network. According to the prediction of five PheHsf proteins (four PheHsfA4s and one PheHsfA2a-2) interaction networks, the interactive proteins might include MAPKs and heat shock proteins. MPK5 protein in rice acts as a positive regulator of drought, salt and cold tolerance; is involved in disease resistance and abiotic stress tolerance signaling pathways; and also negatively modulates pathogenesis-related (PR) gene expression and broad-spectrum disease resistance [66]. The MPK1 protein in rice acts downstream of heterotrimeric G protein alpha subunit and small GTPase RAC1 and may regulate the expression of various genes involved in biotic and abiotic stress response [67]. Base on the above information and our results, four PheHsfA4s and one PheHsfA2a-2 may play a very important role in shoot and flower development and stress response.

## 5. Conclusions

In this study, 22 PheHsf genes in moso bamboo were identified for the first time. These genes could be classified into three classes (A, B and C) according to the comparison of the phylogenetic relationships with *O. sativa*, *B. distachyon* and *A. thaliana* Hsf genes. Expression analyses revealed that two hub genes, *PheHsfA4a-1* and *PheHsfA4a-2*, might act as a potential "node" for crosstalk between developmental processes and abiotic stress responses. Furthermore, *PheHsfA2a-2*, *PheHsfA4d-1* and *PheHsfA4d-2* might also act as essential parts in response to stress. These results provide insights into the responses of *PheHsfAs* to abiotic stresses treatments, although their underlying molecular mechanism requires further study.

**Supplementary Materials:** The following are available online at http://www.mdpi.com/1999-4907/10/2/100/s1, Figure S1: Phylogenetic tree and gene structure of PheHsf genes, Figure S2: Multiple sequence alignment of the DBD domains and OD region of the PheHsf proteins, Figure S3: Various cis-acting elements in PheHsf genes promoter regions, Table S1: Primer sequences used in gene expression with qRT-PCR experiments, Table S2: The relative level of gene expression of PheHsf genes during developmental flowers (F1 to F4) and leaves (CK)(RPKM), Table S3: The relative level of gene expression of PheHsf genes during growing shoots (S1–S7) and clum (CK)(RPKM), Table S4: Expression analysis of *PheHsfAs* under different abiotic stress in moso bamboo, Fasta S1: PheHsfs and PheHsfs-like CDS sequence, Fasta S2: PheHsfs and PheHsfs-like protein sequence, Fasta S3: The cloned cDNA sequence of PheHsfA5, PheHsfB4a-1 and PheHsfB4c-1.

**Author Contributions:** X.L. and J.G. designed the experiments; X.L. and J.L. (Juan Li) performed the tissue and organ collection; L.X. and X.L. performed the experiments; Z.C. formal analysis; J.L. (Jun Liu) and S.M. assisted in equipment maintenance and sample collection; L.X. writing—original draft preparation; X.L. and D.H. writing—review and editing; J.G. review and funding acquisition. L.X. and X.L. contributed equally in this work and should be considered first co-authors.

**Funding:** This project was supported by National Keypoint Research and Invention Program in 13th Five-Year (Grant No. 2018YFD0600100).

**Conflicts of Interest:** The authors declare that they have no conflict of interest.

## References

1. Li, L.; Cheng, Z.; Ma, Y.; Bai, Q.; Li, X.; Cao, Z.; Wu, Z.; Gao, J. The association of hormone signaling genes, transcription, and changes in shoot anatomy during moso bamboo growth. *Plant Biotechnol. J.* **2017**, *16*, 72–85. [CrossRef] [PubMed]
2. Lobovikov, M.; Paudel, S.; Piazza, M.; Ren, H.; Wu, J. *World Bamboo Resources: A Thematic Study Prepared in the Framework of the Global Forest Resources Assessment 2005*; Food and Agriculture Organization of the United Nations: Rome, Italy, 2007; pp. 1–73, 1020–3370.
3. Song, X.; Zhou, G.; Jiang, H.; Yu, S.; Fu, J.; Li, W. Carbon sequestration by Chinese bamboo forests and their ecological benefits: Assessment of potential, problems, and future challenges. *Environ. Rev.* **2011**, *19*, 418–428. [CrossRef]
4. Li, L.; Hu, T.; Li, X.; Mu, S.; Cheng, Z.; Ge, W.; Gao, J. Genome-wide analysis of shoot growth-associated alternative splicing in moso bamboo. *Mol. Genet. Genom.* **2016**, *291*, 1695–1714. [CrossRef]
5. Zhou, G.M.; Wu, J.S.; Jiang, P.K. Effects of different management models on carbon storage in *Phyllostachys pubescens* forests. *J. Beijing For. Univ.* **2006**, *28*, 51–55. [CrossRef]
6. Liu, L.; Dong, D.; Li, Y.; Li, X.; Bureau, A.M. Investigation of moso bamboo forest under high temperature and drought disaster. *World Bamboo Rattan* **2014**, *12*, 24–27. [CrossRef]
7. Zhang, P.; Wang, J.; Zhang, H. Measures of water management and increasing drought resistance of moso forests in Anji County, Zhejiang Province. *World Bamboo Rattan* **2008**, *6*, 23–24. [CrossRef]
8. Rizhsky, L.; Liang, H.; Mittler, R. The combined effect of drought stress and heat shock on gene expression in tobacco. *Plant Physiol.* **2002**, *130*, 1143–1151. [CrossRef]
9. Chauhan, H.; Khurana, N.; Agarwal, P.; Khurana, P. Heat shock factors in rice (*Oryza sativa* L.): Genome-wide expression analysis during reproductive development and abiotic stress. *Mol. Genet. Genom.* **2011**, *286*, 171–187. [CrossRef]
10. Hartl, F.U.; Hayerhartl, M. Molecular chaperones in the cytosol: From nascent chain to folded protein. *Science* **2002**, *295*, 1852–1858. [CrossRef]
11. Queitsch, C.; Hong, S.W.; Vierling, E.; Lindquist, S. Heat shock protein 101 plays a crucial role in thermotolerance in Arabidopsis. *Plant Cell* **2000**, *12*, 479–492. [CrossRef]
12. Döring, P.; Treuter, E.; Kistner, C.; Lyck, R.; Chen, A.; Nover, L. The role of AHA motifs in the activator function of tomato heat stress transcription factors *HsfA1* and *HsfA2*. *Plant Cell* **2000**, *12*, 265–278. [CrossRef] [PubMed]
13. Pelham, H.R.; Bienz, M. A synthetic heat-shock promoter element confers heat-inducibility on the herpes simplex virus thymidine kinase gene. *EMBO J.* **1982**, *1*, 1473–1477. [CrossRef] [PubMed]
14. Von Koskull-Döring, P.; Scharf, K.; Nover, L. The diversity of plant heat stress transcription factors. *Trends Plant Sci.* **2007**, *12*, 452–457. [CrossRef] [PubMed]
15. Scharf, K.D.; Berberich, T.; Ebersberger, I.; Nover, L. The plant heat stress transcription factor (Hsf) family: Structure, function and evolution. *Biochim. Biophys. Acta* **2012**, *1819*, 104–119. [CrossRef] [PubMed]
16. Kotak, S.; Port, M.; Ganguli, A.; Bicker, F.; von Koskull-Döring, P. Characterization of C-terminal domains of *Arabidopsis* heat stress transcription factors (Hsfs) and identification of a new signature combination of plant class A Hsfs with AHA and NES motifs essential for activator function and intracellular localization. *Plant J.* **2004**, *39*, 98–112. [CrossRef] [PubMed]
17. Mishra, S.K.; Tripp, J.; Winkelhaus, S.; Tschiersch, B.; Theres, K.; Nover, L.; Scharf, K.D. In the complex family of heat stress transcription factors, HsfA1 has a unique role as master regulator of thermotolerance in tomato. *Genes Dev.* **2002**, *16*, 1555–1567. [CrossRef] [PubMed]
18. Ogawa, D.; Yamaguchi, K.; Nishiuchi, T. High-level overexpression of the *Arabidopsis* HsfA2 gene confers not only increased themotolerance but also salt/osmotic stress tolerance and enhanced callus growth. *J. Exp. Bot.* **2007**, *12*, 3373–3383. [CrossRef]

19. Wang, X.; Huang, W.; Liu, J.; Yang, Z.; Huang, B. Molecular regulation and physiological functions of a novel FaHsfA2c cloned from tall fescue conferring plant tolerance to heat stress. *Plant Biotechnol. J.* **2017**, *15*, 237–248. [CrossRef]
20. Zhang, L.; Li, Y.; Xing, D.; Gao, C. Characterization of mitochondrial dynamics and subcellular localization of ROS reveal that HsfA2 alleviates oxidative damage caused by heat stress in *Arabidopsis*. *J. Exp. Bot.* **2009**, *60*, 2073–2091. [CrossRef]
21. Sakuma, Y.; Maruyama, K.; Qin, F.; Osakabe, Y.; Shinozaki, K. Yamaguchi-Shino-zaki, K. Dual function of an Arabidopsis transcription factor DREB2A in water-stress-responsive and heat-stress-responsive gene expression. *Proc. Natl. Acad. Sci. USA* **2006**, *103*, 18822–18827. [CrossRef]
22. Schramm, F.; Larkindale, J.; Kiehlmann, E.; Ganguli, A.; Englich, G.; Vierling, E.; von Koskull-Döring, P. A cascade of transcription factor DREB2A and heat stress transcription factor HsfA3 regulates the heat stress response of Arabidopsis. *Plant J.* **2008**, *53*, 264–274. [CrossRef] [PubMed]
23. Yoshida, T.; Sakuma, Y.; Todaka, D.; Maruyama, K.; Qin, F.; Mizoi, J.; Kidokoro, S.; Fujita, Y.; Shinozaki, K.; Yamaguchi-Shinozaki, K. Functional analysis of an Arabidopsis heat-shock transcription factor HsfA3 in the transcriptional cascade downstream of the DREB2A stress-regulatory system. *Biochem. Biophys. Res. Commun.* **2008**, *368*, 515–521. [CrossRef] [PubMed]
24. Li, Z.; Zhang, L.; Wang, A.; Xu, X.; Li, J. Ectopic Overexpression of *SlHsfA3*, a Heat Stress Transcription Factor from Tomato, Confers Increased Thermotolerance and Salt Hypersensitivity in Germination in Transgenic Arabidopsis. *PLoS ONE* **2013**, *8*, e54880. [CrossRef] [PubMed]
25. Lang, S.; Liu, X.; Xue, H.; Li, X.; Wang, X. Functional characterization of BnHSFA4a as a heat shock transcription factor in controlling the re-establishment of desiccation tolerance in seeds. *J. Exp. Bot.* **2017**, *68*, 2361–2375. [CrossRef] [PubMed]
26. Li, F.; Zhang, H.; Zhao, H.; Gao, T.; Song, A.; Jiang, J.; Chen, F.; Chen, S. Chrysanthemum CmHSFA4 gene positively regulates salt stress tolerance in transgenic chrysanthemum. *Plant Biotechnol. J.* **2017**, *16*, 1311–1321. [CrossRef] [PubMed]
27. Shim, D.; Hwang, J.U.; Lee, J.; Lee, S.; Choi, Y.; An, G.; Martinoia, E.; Lee, Y. Orthologs of the class A4 heat shock transcription factor HsfA4a confer cadmium tolerance in wheat and rice. *Plant Cell* **2009**, *21*, 4031–4043. [CrossRef] [PubMed]
28. Chen, S.S.; Jiang, J.; Han, X.J.; Zhang, Y.X.; Zhuo, R.Y. Identification, expression analysis of the Hsf family, and characterization of class A4 in *Sedum Alfredii* hance under cadmium stress. *Int. J. Mol. Sci.* **2018**, *19*, 1216. [CrossRef] [PubMed]
29. Kotak, S.; Vierling, E.; Bäumlein, H.; von Koskull-Döring, P. A novel transcriptional cascade regulating expression of heat stress proteins during seed development of *Arabidopsis*. *Plant Cell* **2007**, *19*, 182–195. [CrossRef]
30. Ikeda, M.; Ohme-Takagi, M. Arabidopsis HsfB1 and HsfB2b act as repressors of the expression of heat-inducible *Hsfs* but positively regulate the acquired thermotolerance. *Plant Physiol.* **2011**, *157*, 1243–1254. [CrossRef]
31. Zhuang, L.L.; Cao, W.; Wang, J.; Yu, J.J.; Yang, Z.M.; Huang, B.G. Characterization and functional analysis of *FaHsfC1b* from *Festuca arundinacea* conferring heat tolerance in *Arabidopsis*. *Int. J. Mol. Sci.* **2018**, *19*, 2702. [CrossRef]
32. Peng, Z.; Lu, Y.; Li, L.; Zhao, Q.; Feng, Q.; Gao, Z.; Lu, H.; Hu, T.; Yao, N.; Liu, K.; et al. The draft genome of the fast-growing non-timber forest species moso bamboo (*Phyllostachys heterocycla*). *Nat. Genet.* **2013**, *45*, 456–461. [CrossRef] [PubMed]
33. Hou, D.; Cheng, Z.; Xie, L.; Li, X.; Li, J.; Mu, S.; Gao, J. The R2R3MYB Gene Family in *Phyllostachys edulis*: Genome-Wide Analysis and Identification of Stress or Development-Related R2R3MYBs. *Front. Plant Sci.* **2018**, *9*, 738. [CrossRef] [PubMed]
34. Wang, T.; Wang, H.; Cai, D.; Gao, Y.; Zhang, H.; Wang, Y.; Lin, C.; Ma, L.; Gu, L. Comprehensive profiling of rhizome-associated alternative splicing and alternative polyadenylation in moso bamboo (*Phyllostachys edulis*). *Plant J.* **2017**, *91*, 684–699. [CrossRef]
35. Jin, J.; He, K.; Tang, X.; Li, Z.; Lv, L.; Zhao, Y.; Luo, J.; Gao, G. An *Arabidopsis* transcriptional regulatory map reveals distinct functional and evolutionary features of novel transcription factors. *Mol. Biol. Evol.* **2015**, *32*, 1767–1773. [CrossRef] [PubMed]

36. Jin, J.; Tian, F.; Yang, D.C.; Meng, Y.Q.; Kong, L.; Luo, J.; Gao, G. PlantTFDB 4.0: Toward a central hub for transcription factors and regulatory interactions in plants. *Nucleic Acids Res.* **2017**, *45*, 1040–1045. [CrossRef]
37. Yu, C.S.; Lin, C.J.; Hwang, J.K. Predicting subcellular localization of proteins for Gram-negative bacteria by support vector machines based on n-peptide compositions. *Protein Sci.* **2004**, *13*, 1402–1406. [CrossRef]
38. Guo, J.; Wu, J.; Ji, Q.; Wang, C.; Luo, L.; Yuan, Y.; Wang, Y.; Wang, J. Genome-wide analysis of heat shock transcription factor families in rice and *Arabidopsis*. *J. Genet. Genom.* **2008**, *35*, 105–118. [CrossRef]
39. Edgar, R.C. MUSCLE: Multiple sequence alignment with high accuracy and high throughput. *Nucleic Acids Res.* **2004**, *32*, 1792–1797. [CrossRef]
40. Sievers, F.; Higgins, D.G. Clustal Omega, accurate alignment of very large numbers of sequences. *Methods Mol. Biol.* **2014**, *1079*, 105–116. [CrossRef]
41. Thompson, J.D.; Gibson, T.J.; Plewniak, F.; Jeanmougin, F.; Higgins, D.G. The CLUSTAL_X windows interface: Flexible strategies for multiple sequence alignment aided by quality analysis tools. *Nucleic Acids Res.* **1997**, *25*, 4876–4882. [CrossRef]
42. Hu, B.; Jin, J.; Guo, A.Y.; Zhang, H.; Luo, J.; Gao, G. GSDS 2.0: An upgraded gene feature visualization server. *Bioinformatics* **2015**, *31*, 1296. [CrossRef]
43. Bailey, T.L.; Boden, M.; Buske, F.A.; Frith, M.; Grant, C.E.; Clementi, L.; Ren, J.Y.; Li, W.W.; Noble, W.S. MEME Suite: Tools for motif discovery and searching. *Nucleic Acids Res.* **2009**, *37*, W202–W208. [CrossRef]
44. Cheng, Z.C.; Hou, D.; Liu, J.; Li, X.Y.; Xie, L.H.; Ma, Y.J.; Gao, J. Characterization of moso bamboo (*Phyllostachys edulis*) Dof factors in floral development and abiotic stress responses. *Genome* **2018**, *61*, 151–156. [CrossRef]
45. Liu, A.L.; Zou, J.; Zhang, X.W.; Zhou, X.Y.; Wang, W.F.; Xiong, X.Y.; Chen, L.Y.; Chen, X.B. Expression profiles of class A rice heat shock transcription factor genes under abiotic stresses. *J. Plant Biol.* **2010**, *53*, 142–149. [CrossRef]
46. Gao, J.; Zhang, Y.; Zhang, C.; Qi, F.; Li, X.; Mu, S.; Peng, Z. Characterization of the floral transcriptome of moso bamboo (*Phyllostachys edulis*) at different flowering developmental stages by transcriptome sequencing and RNA-Seq analysis. *PLoS ONE* **2014**, *9*, e98910. [CrossRef]
47. Bustin, S.A.; Benes, V.; Garson, J.A.; Hellemans, J.; Huggett, J.; Kubista, M.; Mueller, R.; Nolan, T.; Pfaffl, M.W.; Shipley, G.L.; et al. The MIQE guidelines: Minimum information for publication of quantitative real-time PCR experiments. *Clin. Chem.* **2009**, *55*, 611–622. [CrossRef]
48. Fan, C.; Ma, J.; Guo, Q.; Li, X.; Wang, H.; Lu, M. Selection of reference genes for quantitative real-time PCR in bamboo (*Phyllostachys edulis*). *PLoS ONE* **2013**, *8*, e56573. [CrossRef]
49. Livak, K.; Schmittgen, T. Analysis of relative gene expression data using real-time quantitative PCR and the $2^{-\Delta\Delta CT}$ Method. *Methods* **2001**, *25*, 402–408. [CrossRef]
50. Smita, S.; Katiyar, A.; Chinnusamy, V.; Pandey, D.M.; Bansal, K.C. Transcriptional regulatory network analysis of MYB transcription factor family genes in rice. *Front. Plant Sci.* **2015**, *6*, 1157. [CrossRef]
51. Shannon, P.; Markiel, A.; Ozier, O.; Baliga, N.S.; Wang, J.T.; Ramage, D.; Amin, N.; Schwikowski, B.; Ideker, T. Cytoscape: A software environment for integrated models of biomolecular interaction networks. *Genome Res.* **2003**, *13*, 2498–2504. [CrossRef]
52. Sultan, S.; Ali, M.; Nawaz, S.; Ali, M.A.; Shahzad, A. Genome wide analysis of heat shock factors (HSF) gene family of *Arabidopsis Thaliana*. *J. Biol. Agric. Healthc.* **2016**, *6*, 69–77.
53. Li, P.S.; Zheng, W.J.; Zhou, Y.B.; Chen, M.; Chai, S.C.; Ma, L.J.; Xu, Z.S. Genome-wide identification, classification and high temperature responses of Hsf family in *Brachypodium distachyon*. *J. China Agric. Univ.* **2015**, *20*, 8–18. [CrossRef]
54. Nover, L.; Bharti, K.; Döring, P.; Mishra, S.K.; Ganguli, A.; Scharf, K. *Arabidopsis* and the heat stress transcription factor world: How many heat stress transcription factors do we need? *Cell Stress Chaperones* **2001**, *6*, 177–189. [CrossRef]
55. Guo, M.; Liu, J.H.; Ma, X.; Luo, D.X.; Gong, Z.H.; Lu, M.H. The plant heat stress transcription factors (HSFs): Structure, regulation, and function in response to abiotic stresses. *Front. Plant Sci.* **2016**, *7*, 114. [CrossRef]
56. Xue, G.; Sadat, S.; Drenth, J.; McIntyre, C.L. The heat shock factor family from *Triticum aestivum* in response to heat and other major abiotic stresses and their role in regulation of heat shock protein genes. *J. Exp. Bot.* **2014**, *65*, 539–557. [CrossRef]
57. Zhao, H.; Peng, Z.; Fei, B.; Li, L.; Hu, T.; Gao, Z.; Jiang, Z. BambooGDB: A bamboo genome database with functional annotation and an analysis platform. *Database* **2014**, *2014*, bau006. [CrossRef]

58. Giorno, F.; Guerriero, G.; Baric, S.; Mariani, C. Heat shock transcriptional factors in *Malus domestica*: Identification, classification and expression analysis. *BMC Genom.* **2012**, *13*, 639. [CrossRef]
59. Lescot, M.; Déhais, P.; Thijs, G.; Marchal, K.; Moreau, Y.; Van de Peer, Y.; Rouzé, P.; Rombauts, S. PlantCARE, a database of plant cis-acting regulatory elements and a portal to tools for in silico analysis of promoter sequences. *Nucleic Acids Res.* **2002**, *30*, 325–327. [CrossRef]
60. Busch, W.; Wunderlich, M.; Schöffl, F. Identification of novel heat shock factor-dependent genes and biochemical pathways in *Arabidopsis thaliana*. *Plant J.* **2010**, *41*, 1–14. [CrossRef]
61. Zhang, J.; Li, Y.; Jia, H.X.; Li, J.B.; Huang, J.; Lu, M.Z.; Hu, J.J. The heat shock factor gene family in *Salix suchowensis*: A genome-wide survey and expression profiling during development and abiotic stresses. *Front. Plant Sci.* **2015**, *6*, 748. [CrossRef]
62. Mittal, D.; Chakrabarti, S.; Sarkar, A.; Singh, A.; Grover, A. Heat shock factor gene family in rice: Genomic organization and transcript expression profiling in response to high temperature, low temperature and oxidative stresses. *Plant Physiol. Biochem.* **2009**, *47*, 785–795. [CrossRef]
63. Liu, A.L.; Zou, J.; Liu, C.F.; Zhou, X.Y.; Zhang, X.W.; Luo, G.Y.; Chen, X.B. Over-expression of *OsHsfA7* enhanced salt and drought tolerance in transgenic rice. *BMB Rep.* **2013**, *46*, 31–36. [CrossRef]
64. Zhao, H.; Lou, Y.; Sun, H.; Li, L.; Wang, L.; Dong, L.; Gao, Z. Transcriptome and comparative gene expression analysis of *Phyllostachys edulis* in response to high light. *BMC Plant Biol.* **2016**, *16*, 34. [CrossRef]
65. Pérez-Salamó, I.; Szabados, L. The heat shock factor A4A confers salt tolerance and is regulated by oxidative stress and the mitogen-activated protein kinases MPK3 and MPK6. *Plant Physiol.* **2014**, *165*, 319–334. [CrossRef]
66. Xie, K.; Chen, J.; Wang, Q.; Yang, Y. Direct phosphorylation and activation of a mitogen-activated protein kinase by a calcium-dependent protein kinase in rice. *Plant Cell* **2014**, *26*, 3077–3089. [CrossRef]
67. Shi, B.; Ni, L.; Liu, Y.; Zhang, A.; Tan, M.; Jiang, M. OsDMI3-mediated activation of OsMPK1 regulates the activities of antioxidant enzymes in abscisic acid signalling in rice. *Plant Cell Environ.* **2014**, *37*, 341–352. [CrossRef]

*forests*

MDPI

*Article*

# DNA Barcoding Analysis and Phylogenetic Relation of Mangroves in Guangdong Province, China

**Feng Wu [1,2], Mei Li [1], Baowen Liao [1,*], Xin Shi [1] and Yong Xu [3,4]**

1 Key Laboratory of State Forestry Administration on Tropical Forestry Research, Research Institute of Tropical Forestry, Chinese Academy of Forestry, Guangzhou 510520, China; wufengcaf@163.com (F.W.); mangrovelimei@163.com (M.L.); hb-shixin@126.com (X.S.)

2 Zhaoqing Xinghu National Wetland Park Management Center, Zhaoqing 526060, China

3 Key Laboratory of Plant Resources Conservation and Sustainable Utilization, South China Botanical Garden, Chinese Academy of Sciences, Guangzhou 510650, China; xuyong@scbg.ac.cn

4 University of Chinese Academy of Sciences, Beijing 100049, China

* Correspondence: baowenliao@caf.ac.cn; Tel.: +86-020-8702-8494

Received: 9 December 2018; Accepted: 4 January 2019; Published: 12 January 2019

**Abstract:** Mangroves are distributed in the transition zone between sea and land, mostly in tropical and subtropical areas. They provide important ecosystem services and are therefore economically valuable. DNA barcoding is a useful tool for species identification and phylogenetic reconstruction. To evaluate the effectiveness of DNA barcoding in identifying mangrove species, we sampled 135 individuals representing 23 species, 22 genera, and 17 families from Zhanjiang, Shenzhen, Huizhou, and Shantou in the Guangdong province, China. We tested the universality of four DNA barcodes, namely *rbcL*, *matK*, *trnH-psbA*, and the internal transcribed spacer of nuclear ribosomal DNA (ITS), and examined their efficacy for species identification and the phylogenetic reconstruction of mangroves. The success rates for PCR amplification of *rbcL*, *matK*, *trnH-psbA*, and ITS were 100%, 80.29% ± 8.48%, 99.38% ± 1.25%, and 97.18% ± 3.25%, respectively, and the rates of DNA sequencing were 100%, 75.04% ± 6.26%, 94.57% ± 5.06%, and 83.35% ± 4.05%, respectively. These results suggest that both *rbcL* and *trnH–psbA* are universal in mangrove species from the Guangdong province. The highest success rate for species identification was 84.48% ± 12.09% with *trnH-psbA*, followed by *rbcL* (82.16% ± 9.68%), ITS (66.48% ± 5.97%), and *matK* (65.09% ± 6.00%), which increased to 91.25% ± 9.78% with the addition of *rbcL*. Additionally, the identification rate of mangroves was not significantly different between *rbcL* + *trnH-psbA* and other random fragment combinations. In conclusion, *rbcL and trnH-psbA* were the most suitable DNA barcode fragments for species identification in mangrove plants. When the phylogenetic relationships were constructed with random fragment combinations, the optimal evolutionary tree with high supporting values (86.33% ± 4.16%) was established using the combination of *matK* + *rbcL* + *trnH-psbA* + ITS in mangroves. In total, the 476 newly acquired sequences in this study lay the foundation for a DNA barcode database of mangroves.

**Keywords:** mangroves; DNA barcoding; species identification; phylogenetic relation

## 1. Introduction

In 2003, Hebert [1] proposed a novel DNA barcoding technology to expedite the process of species identification. Around 2005, the concept of DNA barcoding was introduced into botanical research [2,3]. In 2009, the CBOL(Consortium for the Barcode of Life) Plant Working Group initially identified and recommended the use of the chloroplast-derived DNA barcode fragments *rbcL* and *matK* [4]. In addition, *trnH-psbA*, the internal spacer region of the chloroplast gene, and the ITS region of the nuclear genome were also investigated [5–7]. Further research is required to compare DNA barcode fragments to test their efficacy for species identification [8–11]. Previous studies on the DNA barcoding of plants have mainly focused on tropical and subtropical forests [12–15]. In addition, sequences obtained from DNA barcode fragments can also be used to reconstruct the phylogenetic relationships of specific biological groups, and this has become a new research hotspot in recent years [16–18]. This research promotes the integration of phylogenetic analysis, ecology, and barcoding technology and develops our understanding of evolutionary biology and other related disciplines [19–22].

Mangroves, in the broad sense, are woody plant communities growing in tropical and subtropical intertidal zones, which play an important role in maintaining the ecological balance of coastal zones [23,24]. There are 84 species of 24 genera and 16 families of mangrove plants in the world, including 70 species of 16 genera and 11 families of true mangrove plants and 14 species of eight genera and five families of semi-mangrove plants [25]. There are 25 species of mangrove plants from Guangdong, including 16 true mangroves and 9 semi-mangroves [26]. There are many similar species of mangroves, and their distribution areas are overlapped. It is difficult to identify similar mangrove species using the external morphology of plants, such as *Sonneratia alba* J. Smith, *Sonneratia caseolaris* (L.) Engl., *Sonneratia Hainanesis* Ko, E. Y. Chen et W. Y. Chen, *Sonneratia paracaseolaris* W. C. Ko et al. or *Bruguiera sexangula* (Lour.) Poir and *Bruguiera sexangula* (Lour.) Poir. var. rhynchopetala Ko. Moreover, it is hard to understand the evolutionary relationship between mangrove species with traditional classification methods. DNA barcoding in animals [27,28], insects [29,30], tropical and subtropical plants [19,31], and microorganisms [32,33] have achieved reliable reconstructions of evolutionary relationships, successfully identified species of the same genus, and discovered new species or cryptic species. In this study, we aimed to investigate the universality of DNA barcoding in the mangrove flora, which is in the transition zone between land and water, and to construct the phylogenetic trees of mangrove flora, to provide a scientific basis for the conservation of mangrove biodiversity.

## 2. Materials and Methods

### 2.1. Plant Material

In this study, the main distribution areas of mangroves in Guangdong were selected, namely the mangrove protection areas in Shenzhen, Huizhou, and Shantou in Eastern Guangdong and Zhanjiang in Western Guangdong. Specific sampling location information is shown in Table 1. According to the DNA barcode sample collection specifications of Gao [34], two to three individuals of each mangrove species were sampled. This involved taking fresh leaves and buds to facilitate the extraction of DNA molecular materials. Each individual sampled was more than 20 m apart. The test material was dried with silica gel after collection. A total of 135 individuals of mangrove plants were collected. Based on expert (LIAO and LI) identification, the individuals were classified into 23 species (including two mangrove companion species) of 22 genera belonging to 17 families. The number of true mangrove and semi-mangrove species sampled in this study accounted for 84% of the 25 species of mangrove plants in Guangdong province [26]. Among them, *B. sexangula* (Lour.) Poir and *B. sexangula* (Lour.) Poir. var. rhynchopetala Ko, which were introduced by humans, are almost extinct, so they were not sampled.

**Table 1.** The information of mangrove samples collected in the Guangdong province.

| Site | Collection Places | Longitude | Latitude | Numbers of Tree Species |
|---|---|---|---|---|
| Zhanjiang | Baitoupo<br>Fengdi | 109°45′38″<br>109°46′21″ | 21°34′03″<br>21°32′36″ | 16 |
| Shenzhen | Mangrove reserve | 114°00′33″ | 22°31′33″ | 12 |
| Huizhou | Yanzhou island<br>Xiezhouwan<br>Haozhaolou | 114°56′17″<br>114°47′00″<br>114°52′02″ | 22°43′47″<br>22°49′40″<br>22°45′58″ | 14 |
| Shantou | Su aiwan<br>Waishahe<br>Yifengxi | 116°43′36″<br>116°48′11″<br>116°53′11″ | 23°19′16″<br>23°23′09″<br>23°32′45″ | 9 |

### 2.2. DNA Extraction and Sequence Analysis

The DNA of mangrove plants was extracted from silica gel-dried leaf material following a modified version of the cetyltrimethyl ammonium bromide (CTAB) protocol of DoyleandDoyle [35]. According to the recommendation of the international union for biological barcoding [4] and previous studies [13,36,37] on regional plant DNA barcoding, a total of four molecular sequences including chloroplast *rbcL*, *matK*, *trnH–psbA*, and nuclear genome ITS were selected for use as amplification fragments. Referrals to the PCR system recommended by the plant working group of the international DNA barcode alliance for life, optimization, and adjustment were made. Primer information and amplification procedures are provided in Table 2. All amplification products were sent to Guangzhou after gel electrophoresis detection for complete sequencing; BLAST (Basic Local Alignment Search Tool) searches were performed using GenBank for the sequences obtained after bidirectional sequencing of the four fragments. If significant inconsistencies were found between the sequences and the original species, reasons were found and reextracted or reconfirmed by consulting experts, until the BLAST results of the sequences and the original species were of the same genus or family. SeqMan 5.00 (DNASTAR package, Madison, WI, USA) was used to splice and proofread the obtained sequences. The sequences were aligned in Geneious 11.1.3 (Biomatters Ltd, Auckland, New Zealand) using the MAFFT (Multiple Alignment using Fast Fourier Transform) algorithm with the default parameters.

**Table 2.** The primers used to amplify DNA barcodes and the amplification protocol.

| DNA Fragment | Primers | Sequences (5′–3′) | References | Amplification Protocol |
|---|---|---|---|---|
| *rbcL* | *rbcLa_F* | ATGTCACCACAAACAGAGACTAAAGC | Kress [19] | 94 °C 3 min; 94 °C 30 s, 55 °C 1 min, 72 °C 1 min, 35 cycles; 72 °C 7 min |
| | *rbcLa_R* | GTAAAATCAAGTCCACCRCG | | |
| *matK* | Kim_3F | CGTACAGTACTTTTGTGTTTACGAG | Kim, unpublished | 94 °C 3 min; 94 °C 45 s, 51 °C 45 s, 72 °C 1 min, 35 cycles; 72 °C 7 min |
| | Kim_1R | ACCCAGTCCATCTGGAAATCTTGGTTC | | |
| *trnH-psbA* | psbA3 | GTTATGCATGAACGTAATGCTC | Sang [38] | 94 °C 3 min; 94 °C 30 s, 55 °C 1 min, 72 °C 1 min,35 cycles; 72 °C 7 min |
| | trnH05 | CGCGCATGGTGGATTCACAATCC | Tate [39] | |
| ITS | ITS-leu1 | GTCCACTGAACCTTATCATTTAG | Urbatsh | 94 °C 3 min; 94 °C 30 s, 55 °C 1 min, 72 °C 1 min, 35 cycles; 72 °C 7 min |
| | ITS4 | TCCTCCGCTTATTGATATGC | White [40] | |

ITS: The internal transcribed spacer of nuclear ribosomal DNA.

### 2.3. Statistical Analysis

PCR amplification success rates and sequencing success rates were calculated following Kress [19]. The success rate of PCR amplification refers to the percentage of successful individuals of a segment in all individuals of the segment. BLAST was used to evaluate the efficacy of the species identification method. Firstly, a local database was established for the four DNA fragments in Geneious 11.1.3 [41], and all sequence comparisons were saved as *. fasta files to adjust the sequence direction and clear the gap between sequences. BLAST-2.7.1+ (https://www.ncbi.nlm.nih.gov/ package, National Center for Biotechnology Information, USA) was used to compare each sequence with all sequences in the database, and the percentage of identical sites was used as the quantification standard. If the minimum value of the identical sites of the same species was greater than the value between individuals of all other species, then we considered that the sequence of this species had been accurately identified. The success rate of identifications was determined by multiplying the percentage of species successfully identified by the success rate of sequencing the segment. Joint fragment identification is the result of accumulation on a single fragment [13,41].

## 3. Results

### 3.1. Sequence Universality

Sequence statistics were calculated for 135 individuals of 23 species of mangrove plants. The results (Table 3) showed a total of 496 DNA barcode fragments were obtained, with a sequence acquisition rate of 88.15% (476 divided by 540). Among them, a total of 118 sequences of mangrove plants from the Zhanjiang mangrove reserve were obtained. The highest success rate for PCR amplification were with *rbcL* and *trnH-psbA* fragments, followed by ITS and *matK*. The highest sequencing success rate was 100% with the *rbcL* fragment, followed by *trnH-psbA*, ITS, and *matK*.

**Table 3.** The success rates of PCR amplification and sequencing of the four barcoding fragments in the four mangrove forests.

| Plot | DNA Fragment | PCR Amplification Success Rate (%) | Sequencing Success Rate (%) | Individuals Resolution Rate |
|---|---|---|---|---|
| Zhanjiang | *rbcL* | 100 | 100 | 35 |
| | *matK* | 71.43 | 65.71 | 23 |
| | *trnH-psbA* | 100 | 88.57 | 31 |
| | ITS | 94.29 | 82.86 | 29 |
| Shenzhen | *rbcL* | 100 | 100 | 36 |
| | *matK* | 80.56 | 77.78 | 28 |
| | *trnH-psbA* | 100 | 97.22 | 35 |
| | ITS | 94.44 | 88.89 | 32 |
| Huizhou | *rbcL* | 100 | 100 | 40 |
| | *matK* | 77.5 | 77.5 | 31 |
| | *trnH-psbA* | 97.5 | 92.5 | 37 |
| | ITS | 100 | 82.5 | 33 |
| Shantou | *rbcL* | 100 | 100 | 24 |
| | *matK* | 91.67 | 79.17 | 19 |
| | *trnH-psbA* | 100 | 100 | 24 |
| | ITS | 100 | 79.17 | 19 |
| Total | | | | 476 |

A total of 131 sequences of mangrove plants from the mangrove reserve of Shenzhen were obtained. The highest amplification success rates were with both *rbcL* and *trnH-psbA*, followed by ITS and *matK*. The highest sequencing success rate was 100% with *rbcL*, followed by *trnH-psbA*, ITS, and *matK*. A total of 141 sequences of mangrove plants from the Huizhou mangrove reserve were obtained. The highest amplification success rates were with *rbcL*, followed by ITS, *trnH-psbA*, and *matK*. The highest sequencing success rate was 100% with *rbcL*, followed by *trnH-psbA*, ITS, and *matK*. A total of 86 sequences of mangrove plants from the Shantou mangrove reserve were obtained. The amplification success rates were 100% with *rbcL*, *trnH-psbA* or ITS, and 91.67% with *matK*. The highest sequencing success rate was 100% for *rbcL*, followed by *trnH-psbA*, ITS, and *matK*.

### 3.2. Species Delimitation

The true mangroves and semi-mangroves distributed in Guangdong are all single-genus and single-species, except *Sonneratia* and *Bruguiera*. Therefore, only the success rates of species-level identification are discussed here. BLAST results showed that the highest success rate of species identification was 84.48% ± 12.09% with *trnH-psbA*, followed by 82.16% ± 9.68% with *rbcL*, 66.48% ± 5.97% with ITS, and 65.09% ± 6.00% with *matK* (Figure 1). However, more species were successfully identified with *rbcL* fragments than with *trnH-psbA* fragments in any region, except for Shantou. The main difference was that *trnH-psbA* could accurately distinguish *S. caseolaris* (L.) Engl. and *Sonneratia apetala* Buch.-Ham, but *rbcL* could not.

According to the statistical analysis of different combinations of multiple fragments, the identification success rate of *rbcL* + *trnH-psbA* was 91.25% ± 9.78%, which increased to 91.88% ± 8.62% with the addition of ITS. The success rate of species identification with all fragments was consistent with that of *rbcL* + *trnH-psbA* + *matK* (94.38% ± 4.48%). These results suggest that the combination of certain fragments can increase the success rate of barcoding for species identification.

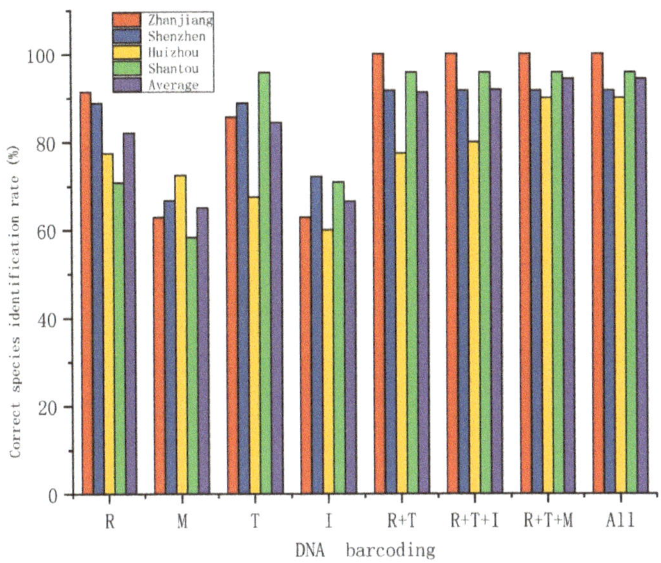


**Figure 1.** Mangrove species discrimination rate of all single and multi-DNA barcoding fragments. R, *rbcL*; M, *matK*; T, *trnH–psbA*; I, ITS; All, the database consisted of all species.

### 3.3. Phylogenetic Trees

Phylogenetic trees were constructed in MEGA6.0 (Tamura, Stecher, Peterson, Filipski, and Kumar) using the neighbor-joining (NJ) method based on the Kimura'2-parameter model. Phylogenetic trees were constructed with the individual or random combined fragments and the average node support rate was calculated. The results showed that the highest average node support rate of phylogenetic trees using the combination *rbcL* + *matK* + *trnH-psbA* + ITS was 86.33% ± 4.16% (as shown in Figures 2–5). Phylogenetic trees were fan-shaped, with one branch of the same or similar species. The average node support rate for mangrove phylogenetic trees in the four regions was 89.66% ± 18.50% in Zhanjiang, 88.49% ± 17.25% in Huizhou, 86.85% ± 15.60% in Shenzhen, and 80.33% ± 19.89% in Shantou.

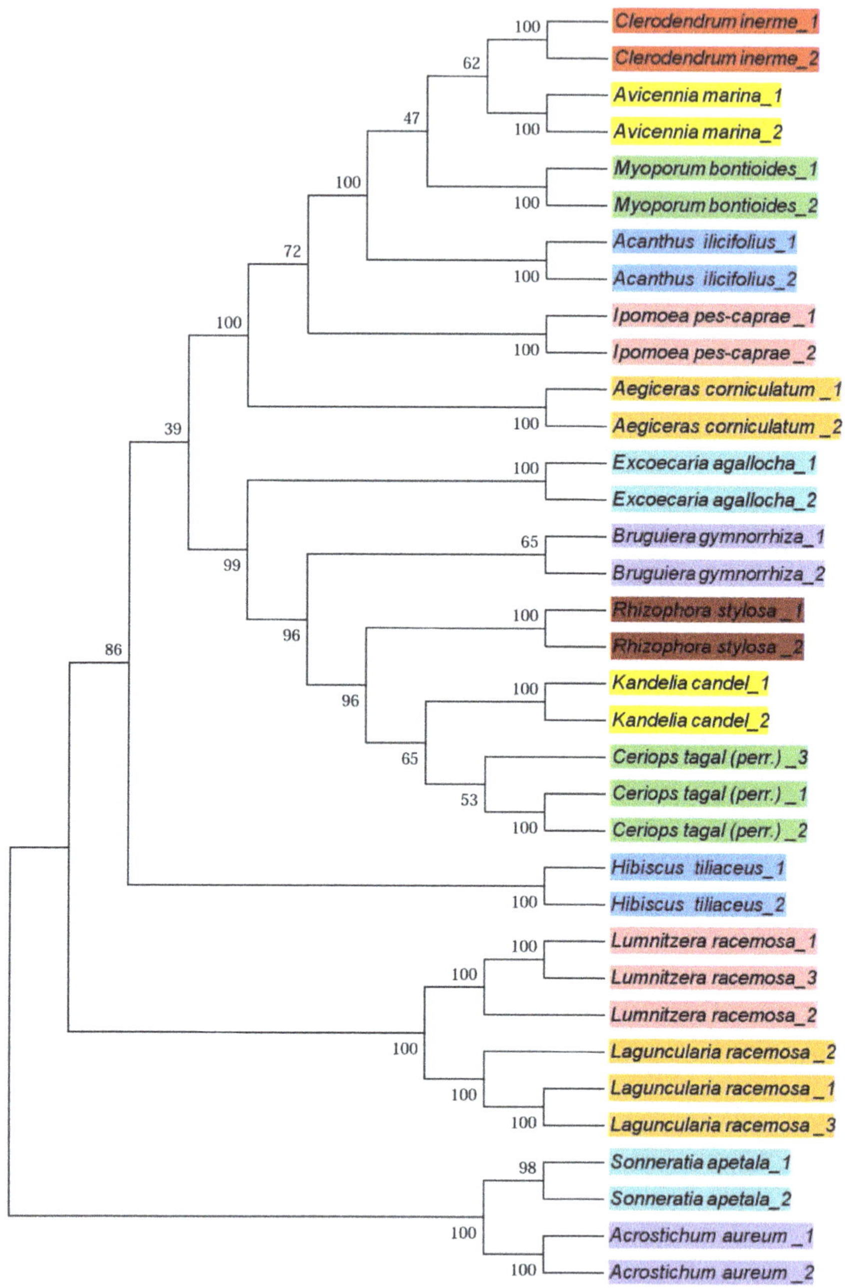


**Figure 2.** The phylogenetic tree of mangroves in Zhanjiang using fragment of *matK* + *rbcL* + *trnH-psbA* + ITS.

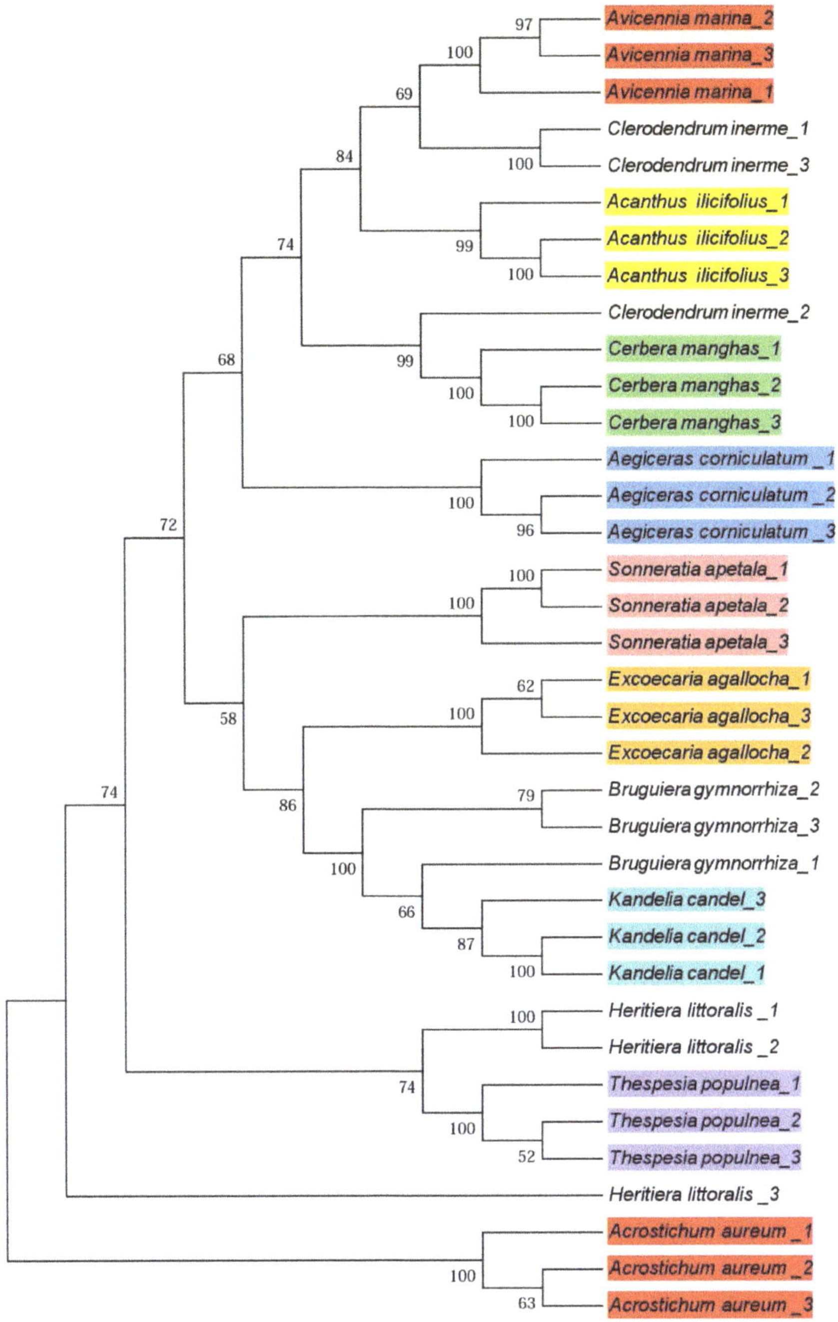


**Figure 3.** The phylogenetic tree of mangroves in Shenzhen using fragment of *matK* + *rbcL* + *trnH-psbA* + ITS.

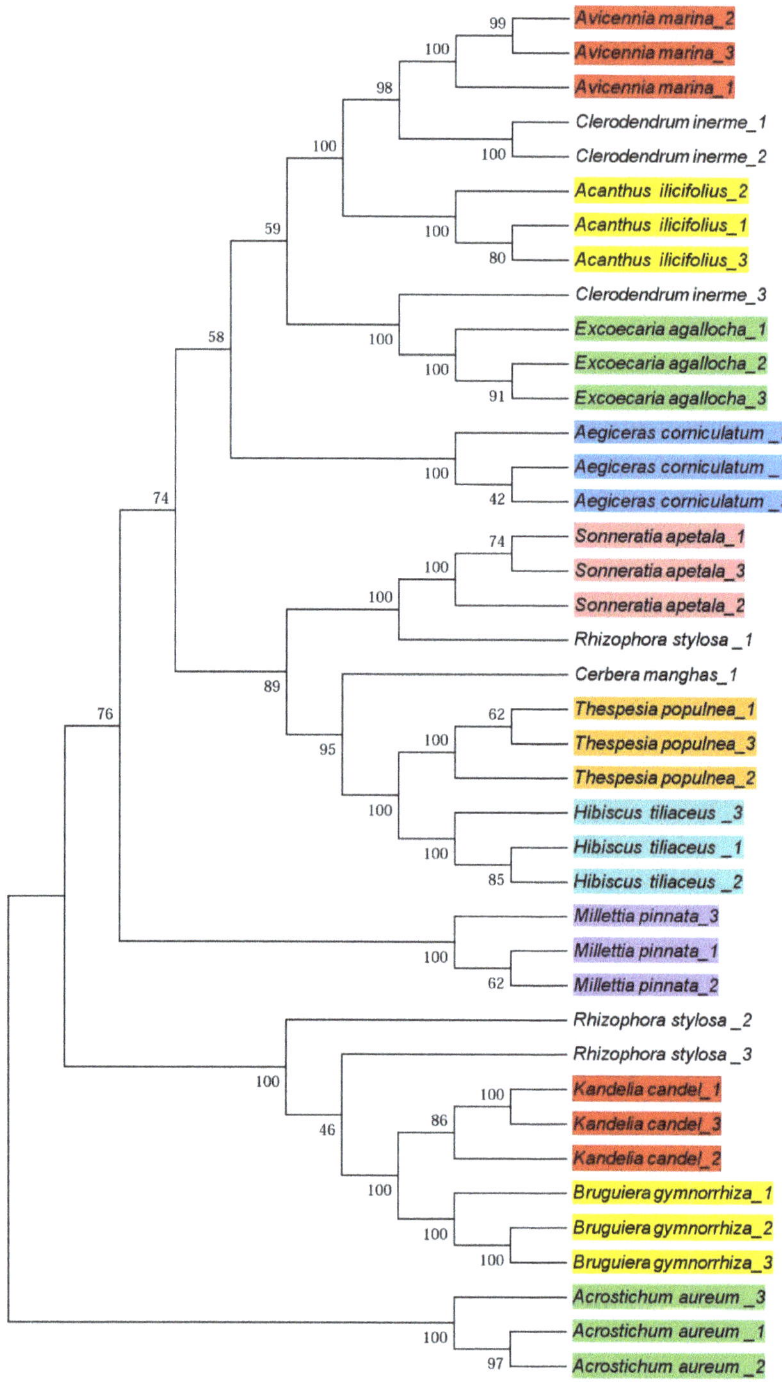


**Figure 4.** The phylogenetic tree of mangroves in Huizhou using fragment of *matK* + *rbcL* + *trnH-psbA* + ITS.

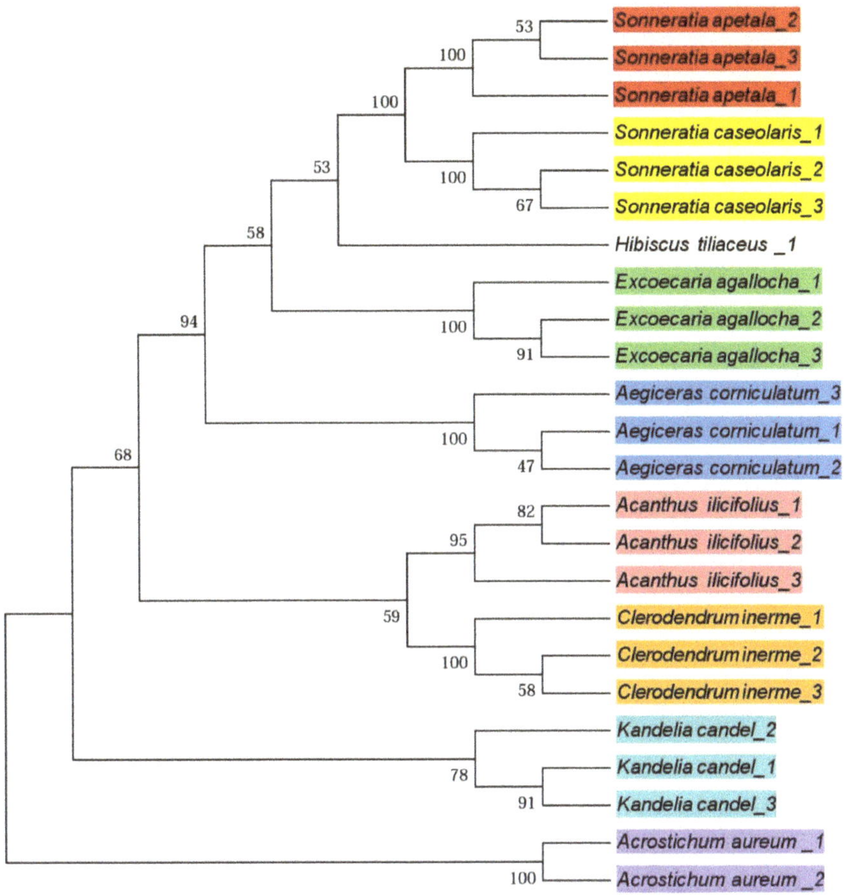


**Figure 5.** The phylogenetic tree of mangroves in Shantou using fragment of *matK* + *rbcL* + *trnH-psbA* + ITS.

## 4. Discussion

Our study investigated mangrove plants in the Guangdong province using DNA barcoding technology. The purpose of this study was to evaluate the performance of DNA barcoding in terms of primer universality, successful identification rate, and phylogenetic tree construction.

### *4.1. The Universality of DNA Barcoding in Mangrove Communities*

The success rate of PCR amplification and the sequencing of *rbcL* fragments in core barcodes of mangrove DNA samples reached 100%. Compared with Kress [19], our results showed higher universality and success rates, and were the same as those of Pei [42], with 90%–100% in the forest plant communities in tropical and subtropical regions. This indicates that *rbcL* has the fewest numbers of variable sites and that the selected primer sequence has strong universality. Therefore, *rbcL* is recommended as an effective fragment for DNA barcoding in mangroves. The success rate of another core fragment, *matK*, was the lowest of the four fragments. With this fragment, the amplification or sequencing of *Ceriops tagal* (perr.) C. B. Rob. and *Kandelia candel* (L.) Druce were not successful, and only a small number of individuals were successfully sequenced in, e.g., *Bruguiera gymnorrhiza* (L.) Poir., *Acrostichum aureum* L., and *Rhizophora stylosa* Griff., indicating a lack of universality for this fragment compared to *rbcL*. This may be due to the existence of single nucleotide repeat sequences. The amplification rate of *matK* fragments in previous studies was lower, e.g., 68.18% [43], as was the success rate, e.g., 64% [14]. Similarly low results were reported in other studies of Liu [13], Lu [44], and Wei [45]. Generally, in this situation, the number of primers increases, and the procedure is repeated multiple times.

The barcode *trnH-psbA* has high universality, and the amplification rate and sequencing success rate can be 99.38% ± 1.25% and 94.57% ± 5.06%, respectively, with only one pair of primers, which is second only to the *rbcL* fragment. This fragment can also successfully distinguish *Sonneratia*. The success rate of the nuclear gene ITS amplification was 97.18% ± 3.25%; the amplifications of *A. aureum*, *B. gymnorrhiza*, *C. tagal*, and *R. stylosa* were not successful. The sequencing success rate of the nuclear gene ITS was 83.35% ± 4.05%, and only a few individuals were successfully sequenced in *K. candel* and *Acanthus ilicifolius* L. Previous studies, such as that of Tripathi [46], showed that the sequencing success rate of ITS sequences in tropical forest species in India was 62.0%. Kang [47] showed that the success rate of ITS sequencing in Hainan tropical cloud forests was 47.20% ± 5.76%. In conclusion, the *rbcL* fragment and *trnH-psbA* fragment are recommended for the amplification and sequencing of mangrove plants.

### 4.2. Species Identification Ability

CBOL Plant Working Group recommend *rbcL* and *matK* for the core DNA barcoding for plants. Burgess [41] found that the core barcode can successfully identify 93% of species in the temperate flora of Canada, and de Vere [48] found that *rbcL* + *matK* can identify 69.4%–74.1% of flowering plants in Wales, UK. Kress [19] studied 296 woody species in Panama and found that the species identification rate of *matK* + *rbcL* was as high as 98%. In this study, the species identification rate of the core barcode *rbcL* + *matK* combination fragment can reach 90.56% ± 6.94%. This indicates that the core barcode is effective for species identification and is consistent with previous research results. However, the identification rate of mangroves decreased when only one of them was used. Either *rbcL* or *trnH-psbA* had a higher identification rate of mangroves than that of *matK*. Additionally, the rate for *rbcL* + *trnH-psbA* was higher than that for *rbcL* + *matK*. Furthermore, the identification rate of mangroves had no significant differences between *rbcL* + *trnH-psbA* and other random fragment combinations.

Although the number of bases of *trnH-psbA* varies greatly among different plant groups, a large number of insertions and deletions make sequencing difficult, as well as the existence of single-nucleotide repeats and other special structures in some groups. In this study, the success rate of *trnH-psbA* sequencing was second only to *rbcL*. At the same time, this fragment was able to successfully identify *S. caseolares* and *S. apetala*. The species identification rate of *trnH-psbA* was the highest among the four single fragments, Gonzalez [12] and Tripathi [46] also showed that *trnH-psbA* is one of the most promising barcodes for species identification. This suggests that *trnH-psbA* may act as a complementary fragment to *rbcL*.

Although studies by Kress [19], Sass [49], and Li [7] support the incorporation of ITS into the core barcode of plant DNA, the supplementary ITS fragment was found to be of low universality in the present study, and the species identification rate, when used singly, was 66.48% ± 5.97%. However, the ITS fragment could identify *Laguncularia racemosa* C.F.Gaertn but not *rbcL* and *trnH-psbA*. In addition, the ITS region of the nuclear genome can provide more genetic information from parents than the chloroplast genes. Given the above, *rbcL* and *trnH-psbA* were effective fragments for mangrove identification, and ITS fragment could be used for specific mangroves.

### 4.3. Phylogenetic Trees

*rbcL* + *matK* + *trnH-psbA* fragments had been used to construct phylogenetic trees in different localities, e.g., Barro Colorado Island [19], the Dinghu mountain forest [50], the Ailao mountain forest [44], and tropical cloud forests in Hainan [47]. However, the average node support for phylogenetic trees of mangrove species in these four subtropical regions constructed with *rbcL* +

*matK* + *trnH-psbA* + ITS was 5.59% higher than that with *rbcL* + *matK* + *trnH-psbA*, which saw no significant difference. In addition, in Figures 2–5, most of the same species were clustered into one branch, also indicating that DNA barcoding can be used to identify species. Due to the different geographical locations of the mangroves, environmental conditions are also different (Table 1), which leads to differences in selective pressure. This could lead to differences in the evolutionary trajectory of mangrove species in the four regions. For example, community composition was observed to be different at different locations, whereby true mangrove species gradually decreased from south to north.

## 5. Conclusions

The results of the present study suggest that *rbcL* and *trnH-psbA* have high success rates for amplification and sequencing, indicating that these two barcodes are common in mangrove species. In terms of species identification, these fragments were relatively successful compared to the other fragments tested. The phylogenetic trees of mangrove plants constructed with a combination of *rbcL* + *matK* + *trnH-psbA* + ITS had the highest node support rate. Due to the low efficiency of *matK* fragment amplification and identification in mangrove plants. And the identification success rate of *rbcL* was higher than that of *trnH-psbA*, except for Shantou region, where the fragment of *trnH-psbA* can be used to identify specific species which cannot be identified by *rbcL*. Thus, this study concluded that *rbcL* and *trnH-psbA* were the most suitable DNA barcode fragments for species identification in mangrove plants.

Data collection for mangrove DNA barcoding in the Guangdong province is ongoing. A total of 476 sequences were obtained, from 135 individuals of 23 species of mangrove plants, accounting for 55.26% of the mangrove plants in China (21/38). In future research, the sampling range can be further expanded to include the DNA barcoding of other mangrove tree species and mangrove companion plants, to build a complete and high-coverage mangrove plant DNA barcoding database.

**Author Contributions:** Conceptualization, B.L. and M.L.; methodology, F.W.; software, F.W., Y.X.; formal analysis, F.W., Y.X.; investigation, F.W., B.L., M.L., X.S.; supervision, B.L., M.L.; writing, F.W.

**Funding:** This research was funded by the Ministry of Science and Technology of China (No. 2017FY100705 and No. 2017FY100700), and the National Natural Science Foundation of China (No. 31570594).

**Acknowledgments:** The authors would like to thank Jiang, Z.M. and Xu, Y.W. for collecting material, and additional help. Our sincere thanks are extended to Yan, H.F. for providing a scientific research platform.

**Conflicts of Interest:** The authors declare no conflict of interest.

## References

1. Hebert, P.D.; Cywinska, A.; Ball, S.L.; de Waard, J.R. Biological identifications through DNA barcodes. *Proc. Biol. Sci.* **2003**, *270*, 313–321. [CrossRef] [PubMed]
2. Chase, M.W.; Salamin, N.; Wilkinson, M.; Dunwell, J.M.; Kesanakurthi, R.P.; Haider, N.; Savolainen, V. Land plants and DNA barcodes: Short-term and long-term goals. *Philos. Trans. R. Soc. Lond. B. Biol. Sci.* **2005**, *360*, 1889–1895. [CrossRef] [PubMed]
3. Kress, W.J.; Wurdack, K.J.; Zimmer, E.A.; Weigt, L.A.; Janzen, D.H. Use of DNA barcodes to identify flowering plants. *Proc. Natl. Acad. Sci. USA* **2005**, *102*, 8369–8374. [CrossRef] [PubMed]
4. CBOL Plant Working Group. A DNA barcode for land plants. *Proc. Natl. Acad. Sci. USA* **2009**, *106*, 12794–12797. [CrossRef] [PubMed]
5. Kress, W.J.; Erickson, D.L. A two-locus global DNA barcode for land plants: The coding *rbcL* gene complements the non-coding *trnH-psbA* spacer region. *PLoS ONE* **2007**, *2*, e508. [CrossRef] [PubMed]
6. Chen, S.; Yao, H.; Han, J.; Liu, C.; Song, J.; Shi, L.; Zhu, Y.; Ma, X.; Gao, T.; Pang, X.; et al. Validation of the ITS2 region as a novel DNA barcode for identifying medicinal plant species. *PLoS ONE* **2010**, *5*, e8613. [CrossRef]
7. Li, D.Z.; Gao, L.M.; Li, H.T.; Wang, H.; Ge, X.J.; Liu, J.Q.; Chen, Z.D.; Zhou, S.L.; Chen, S.L.; Yang, J.B.; et al. Comparative analysis of a large dataset indicates that internal transcribed spacer (its) should be incorporated into the core barcode for seed plants. *Proc. Natl. Acad. Sci. USA* **2011**, *108*, 19641–19646.

8. Kress, W.J.; Erickson, D.L.; Swenson, N.G.; Thompson, J.; Uriarte, M.; Zimmerman, J.K. Advances in the use of DNA barcodes to build a community phylogeny for tropical trees in a Puerto Rican forest dynamics plot. *PLoS ONE* **2010**, *5*, e15409. [CrossRef]
9. Hollingsworth, P.M. Refining the DNA barcode for land plants. *Proc. Natl. Acad. Sci. USA* **2011**, *108*, 19451–19452. [CrossRef]
10. Li, H.Q.; Chen, J.Y.; Wang, S.; Xiong, S.Z. Evaluation of six candidate DNA barcoding loci in *Ficus* (Moraceae) of China. *Mol. Ecol. Resour.* **2012**, *12*, 783–790. [CrossRef]
11. Braukmann, T.W.; Kuzmina, M.L.; Sills, J.; Zakharov, E.V.; Hebert, P.D. Testing the efficacy of DNA barcodes for identifying the vascular plants of Canada. *PLoS ONE* **2017**, *12*, e169515. [CrossRef] [PubMed]
12. Gonzalez, M.A.; Baraloto, C.; Engel, J.; Mori, S.A.; Petronelli, P.; Riera, B.; Roger, A.; Thebaud, C.; Chave, J. Identification of Amazonian trees with DNA barcodes. *PLoS ONE* **2009**, *4*, e7483. [CrossRef] [PubMed]
13. Liu, J.; Yan, H.F.; Newmaster, S.G.; Pei, N.; Ragupathy, S.; Ge, X.J.; Lowe, A. The use of DNA barcoding as a tool for the conservation biogeography of subtropical forests in China. *Divers. Distrib.* **2015**, *21*, 188–199. [CrossRef]
14. Parmentier, I.; Duminil, J.; Kuzmina, M.; Philippe, M.; Thomas, D.W.; Kenfack, D.; Chuyong, G.B.; Cruaud, C.; Hardy, O.J. How effective are DNA Barcodes in the identification of African rainforest trees. *PLoS ONE* **2013**, *8*, e54921. [CrossRef] [PubMed]
15. Kress, W.J.; Erickson, D.L. DNA barcodes: Methods and protocols. *Methods Mol. Biol.* **2012**, *858*, 3–8. [PubMed]
16. Li, D.Z.; Zeng, C.X. Prospects for plant DNA barcoding. *Biodivers. Sci.* **2015**, *23*, 297–298. [CrossRef]
17. Kress, W.J.; Garcia-Robledo, C.; Uriarte, M.; Erickson, D.L. DNA barcodes for ecology, evolution, and conservation. *Trends Ecol. Evol.* **2015**, *30*, 25–35. [CrossRef]
18. Pei, N.C.; Chen, B.F.; Kress, W.J. Advances of community-level plant DNA barcoding in China. *Front. Plant Sci.* **2017**, *8*, 225. [CrossRef]
19. Kress, W.J.; Erickson, D.L.; Jones, F.A.; Swenson, N.G.; Perez, R.; Sanjur, O.; Bermingham, E. Plant DNA barcodes and a community phylogeny of a tropical forest dynamics plot in Panama. *Proc. Natl. Acad. Sci. USA* **2009**, *106*, 18621–18626. [CrossRef]
20. Pei, N.C.; Zhang, J.L.; Mi, X.C.; Ge, X.J. Plant DNA barcodes promote the development of phylogenetic community ecology. *Biodivers. Sci.* **2011**, *19*, 284–294. (In Chinese)
21. Valentini, A.; Pompanon, F.; Taberlet, P. DNA barcoding for ecologists. *Trends Ecol. Evol.* **2009**, *24*, 110–117. [CrossRef] [PubMed]
22. Gonzalez, M.A.; Roger, A.; Courtois, E.A.; Jabot, F.; Norden, N.; Paine, C.E.T.; Baraloto, C.; Thébaud, C.; Chave, J.M. Shifts in species and phylogenetic diversity between sapling and tree communities indicate negative density dependence in a lowland rain forest. *J. Ecol.* **2010**, *98*, 137–146. [CrossRef]
23. Lin, P. *Chinese Mangrove Ecosystem*; Science Press: Beijing, China, 1997; pp. 124–156. ISBN 9787030056924. (In Chinese)
24. Dan, X.Q.; Liao, B.W.; Wu, Z.B.; Wu, H.J.; Bao, D.M. resources, conservation status and main threats of mangrove wetlands in China. *Ecol. Environ. Sci.* **2016**, *25*, 1237–1243. (In Chinese)
25. Wang, B.J.; Liang, S.C.; Zhang, W.Y.; Zan, Q.J. Mangrove flora of the world. *Acta Bot. Sin.* **2003**, *45*, 644–653. (In Chinese)
26. Liao, B.W.; Zhang, Q.M. Area, distribution and species composition of mangroves in China. *Wetl. Sci.* **2014**, *12*, 435–440. (In Chinese)
27. Murphy, R.W.; Crawford, A.J.; Bauer, A.M.; Che, J.; Donnellan, S.C.; Fritz, U.; Haddad, C.F.B.; Nagy, Z.T.; Poyarkov, N.A.; Vences, M.; et al. Cold Code: The global initiative to DNA barcode amphibians and nonavian reptiles. *Mol. Ecol. Resour.* **2013**, *13*, 161–167. [CrossRef]
28. Shen, Y.Y.; Chen, X.; Murphy, R.W. Assessing DNA barcoding as a tool for species identification and data quality control. *PLoS ONE* **2013**, *8*, e57125. [CrossRef]
29. Martoni, F.; Bulman, S.; Pitman, A.; Taylor, G.; Armstrong, K. DNA barcoding highlights cryptic diversity in the New Zealand Psylloidea (Hemiptera: Sternorrhyncha). *Diversity* **2018**, *10*, 50. [CrossRef]
30. Pfeiler, E. DNA barcoding and taxonomic challenges in describing new putative species: Examples from sootywing and cloudywing butterflies (Lepidoptera: Hesperiidae). *Diversity* **2018**, *10*, 111. [CrossRef]
31. Pei, N.C.; Lian, J.Y.; Erickson, D.L.; Swenson, N.G.; Kress, W.J.; Ye, W.H.; Ge, X.J. Exploring tree-habitat associations in a Chinese subtropical forest plot using a molecular phylogeny generated from DNA barcode Loci. *PLoS ONE* **2011**, *6*, e21273. [CrossRef]

32. Pawlowski, J.; Audic, S.; Adl, S.; Bass, D.; Belbahri, L.; Berney, C.; Bowser, S.; Cepicka, I.; Decelle, J.; Dunthorn, M.; et al. CBOL protist working group: Barcoding eukaryotic richness beyond the animal, plant, and fungal kingdoms. *PLoS Biol.* **2012**, *10*, e1001419. [CrossRef] [PubMed]
33. Galimberti, A.; de Mattia, F.; Losa, A.; Bruni, I.; Federici, S.; Casiraghi, M.; Martellos, S.; Labra, M. DNA barcoding as a new tool for food traceability. *Food Res. Int.* **2013**, *50*, 55–63. [CrossRef]
34. Gao, L.M.; Liu, J.; Cai, J.; Yang, J.B.; Zhang, T. A synopsis of technical notes on the standards for plant DNA barcoding. *Plant Divers. Resour.* **2012**, *34*, 592–606. (In Chinese) [CrossRef]
35. Doyle, J.; Doyle, J. A rapid DNA isolation procedure for small quantities of fresh leaf tissue. *Phytochem. Bull.* **1987**, *19*, 11–15.
36. Yan, H.F.; Liu, Y.J.; Xie, X.F.; Zhang, C.Y.; Hu, C.M.; Hao, G.; Ge, X.J. DNA barcoding evaluation and its taxonomic implications in the species-rich genus *Primula* L. in China. *PLoS ONE* **2015**, *10*, e122903. [CrossRef] [PubMed]
37. Liu, J. DNA Barcoding Evaluation of the Dinghusahn Woody Species and the Genus *Gentiana* L. (Gentianaceae). Ph.D. Thesis, South China Botanical Garden, The Chinese Academy of Sciences, Guangzhou, China, 2014; pp. 27–28. (In Chinese)
38. Sang, T.; Crawford, D.; Stuessy, T. Chloroplast DNA phylogeny, reticulate evolution, and biogeography of *Paeonia* (Paeoniaceae). *Am. J. Bot.* **1997**, *84*, 1120. [CrossRef] [PubMed]
39. Tate, J.A.; Simpson, B.B. Paraphyly of *Tarasa* (Malvaceae) and diverse origins of the polyploid species. *Syst. Bot.* **2003**, *28*, 723–737.
40. White, T.J.; Bruns, T.; Lee, S.; Taylor, J. Amplification and direct sequencing of fungal ribosomal RNA genes for phylogenetics. In *PCR Protocols: A Guide to Methods and Applications*; Academic: San Diego, CA, USA, 1990; pp. 315–322.
41. Burgess, K.S.; Fazekas, A.J.; Kesanakurti, P.R.; Graham, S.W.; Husband, B.C.; Newmaster, S.G.; Percy, D.M.; Hajibabaei, M.; Barrett, S.C.H. Discriminating plant species in a local temperate flora using the *rbcL+matK* DNA barcode. *Methods Appl. Fluores* **2011**, *2*, 333–340. [CrossRef]
42. Pei, N.; Erickson, D.L.; Chen, B.; Ge, X.; Mi, X.; Swenson, N.G.; Zhang, J.; Jones, F.A.; Huang, C.; Ye, W.; et al. Closely-related taxa influence woody species discrimination via DNA barcoding: Evidence from global forest dynamics plots. *Sci. Rep.* **2015**, *5*, 15127. [CrossRef]
43. Costion, C.; Ford, A.; Cross, H.; Crayn, D.; Harrington, M.; Lowe, A. Plant DNA barcodes can accurately estimate species richness in poorly known floras. *PLoS ONE* **2011**, *6*, e26841. [CrossRef]
44. Lu, M.M.; Ci, X.Q.; Yang, G.P.; Li, J. DNA barcoding of subtropical forest trees—A study from Ailao mountains nature reserve, Yunnan, China. *Plant Divers. Resour.* **2013**, *35*, 733–741. (In Chinese)
45. Wei, Y.N.; Wang, X.M.; Yao, P.C.; Chen, X.Y.; Li, H.Q. Comparison of species resolution rates of DNA barcoding for Chinese coastal halo-tolerant plants. *Biodivers. Sci.* **2017**, *25*, 1095–1104. (In Chinese) [CrossRef]
46. Tripathi, A.M.; Tyagi, A.; Kumar, A.; Singh, A.; Singh, S.; Chaudhary, L.B.; Roy, S. The internal transcribed spacer (its) region and *trnH-psbA* [corrected] are suitable candidate loci for DNA barcoding of tropical tree species of India. *PLoS ONE* **2013**, *8*, e57934. [CrossRef] [PubMed]
47. Kang, Y.; Deng, Z.; Zang, R.; Long, W. DNA barcoding analysis and phylogenetic relationships of tree species in tropical cloud forests. *Sci. Rep.* **2017**, *7*, 12564. [CrossRef] [PubMed]
48. De Vere, N.; Rich, T.C.; Ford, C.R.; Trinder, S.A.; Long, C.; Moore, C.W.; Satterthwaite, D.; Davies, H.; Allainguillaume, J.; Ronca, S.; et al. DNA barcoding the native flowering plants and conifers of Wales. *PLoS ONE* **2012**, *7*, e37945. [CrossRef] [PubMed]
49. Sass, C.; Little, D.P.; Stevenson, D.W.; Specht, C.D. DNA barcoding in the cycadales: Testing the potential of proposed barcoding markers for species identification of cycads. *PLoS ONE* **2007**, *2*, e1154. [CrossRef] [PubMed]
50. Pei, N.C. Building a subtropical forest community phylogeny based on plant DNA barcodes from Dinghushan plot. *Plant Divers. Resour.* **2012**, *34*, 263–270. (In Chinese) [CrossRef]

forests

MDPI

Article

# Diversity and Utilization of Edible Plants and Macro-Fungi in Subtropical Guangdong Province, Southern China

**Juyang Liao [1,2], Linping Zhang [3], Yan Liu [2], Qiaoyun Li [2], Danxia Chen [2], Qiang Zhang [4] and Jianrong Li [5,*]**

1 College of Forestry, Beijing Forestry University, Beijing10083, China; liaojuyang@163.com
2 Hunan Forest Botanical Garden, Changsha 410116, China; ly446864594@163.com (Y.L.); liqiaoyun99@163.com (Q.L.); cdx873898657@163.com (D.C.)
3 Key Laboratory of State Forestry Administration on Forest Ecosystem Protection and Restoration of Poyang Lake Watershed (JXAU), Nanchang 330045, China; zlping619@163.com
4 Guangdong Institute of Applied Biological Resources, Guangzhou 510260, China; zhangqiang06@giabr.gd.cn
5 South China Botanical Garden, Chinese Academy of Sciences, Guangzhou 510650, China
* Correspondence: lijianrong@scbg.ac.cn; Tel.: +86-20-3725-2692

Received: 17 September 2018; Accepted: 22 October 2018; Published: 25 October 2018

**Abstract:** Food supply from forests is a fundamental component of forest ecosystem services, but information relating to suitability for human consumption and sustainable utilization of non-timber forest products (NTFPs) in developing countries is lacking. To address this gap in knowledge, diverse datasets of edible plants and macro-fungi were obtained from field collections, historical publications, and community surveys across seven cities in Guangdong Province (GP), southern China. Seven edible parts and five food categories of plant species were classified according to usage and specific nutrient components. Edible plant species were also categorized into different seasons and life forms. Our results show that at least 100 plant species (with 64 plant species producing fruit) and 20 macro-fungi were commonly used as edible forest products in subtropical GP. There were 55 and 57 species providing edible parts in summer and autumn, respectively. Many edible plants had multiple uses. Tree and herbaceous species made up the majority of the total. Our study provides evidence that both edible plants and macro-fungi can enrich the food supply for residents in rural and urban areas by acting as supplemental resources. We therefore suggest that, in spite of the prevalence of imported foods due to modern infrastructure, edible NTFPs from subtropical forests might be leveraged to support the increasing demand for food in an era of rapid urbanization and global change.

**Keywords:** edible forest product; forest biology; macro-fungi; non-timber forest products (NTFPs); Pan-Pearl River Delta

## 1. Introduction

Forests and trees are principal components of terrestrial ecosystems, providing the earth with a vast array of socio-ecological benefits [1–3]. Among these benefits, forest biodiversity is a crucial dimension, important for valuing ecosystem services, and has been attracting growing attention from municipal authorities, research agencies, and the public. The diversity of edible plants and macro-fungi is a special category of forest biodiversity. From a global perspective, food security is a major concern and will remain a great challenge for the rest of the 21st century, since crop yields have fallen in many countries and regions due to insufficient investments, irregular climate, and intense disturbance from agroforestry [4–7]. Forests, including natural and urban forests, are highly valuable to world food security, annually providing livelihoods and food for over 300 million people [8–11]. Moreover, urban greenspaces contribute to and preserve a considerable proportion of terrestrial biodiversity [12]. Investigation and evaluation of edible species diversity from natural forests and urban green spaces can increase our understanding of the food supply required to meet the demands of a growing population. Furthermore, efforts to describe edible species diversity are of theoretical significance to understand diversity patterns of local food species, in addition to increasing citizens' awareness of biodiversity protection. Currently, edible fungi, wheat, rice, corn, sorghum, cassava, and potato are the top seven major food crops in the world [9,13]. Edible forest products from diverse ecosystems are important supplements for the global food supply. As the largest developing country in the world and a major economic entity, China has the largest population in Asia—more than 1.41 billion as of 2016—and has diverse forest types, including tropical, subtropical, and temperate zones. Therefore, enriching the food supply and achieving food security are important tasks for the international community, regional governments, and local authorities.

Food diversity from forests (i.e., edible plants and macro-fungi) is described in order to reveal a vital component of forest biodiversity in subtropical Guangdong Province (GP), southern China. The objectives of the present study were three fold: (1) to obtain a comprehensive species list of edible plants and macro-fungi commonly used in the region; (2) to explore general patterns of edible plants among different edible parts, categories, seasons, and life forms; and (3) to evaluate the utilization and conservation of edible forest products. Overall, exploring the diversity of edible forest products adds to our understanding of the socio-ecological benefits of forests. These efforts assess the suitability of non-timber forest products (NTFPs) for human consumption, as well as the necessity and feasibility of green infrastructure construction in urbanizing regions.

## 2. Materials and Methods

### 2.1. Research Area

Guangdong Province (20°09′–25°31′ N, 109°45′–117°20′ E) is located in southern China (Figure 1), geographically adjacent to Hong Kong SAR, Macao SAR, and 4 other provinces. GP consists of 21 prefecture-level cities and covers an area of 179,700 km$^2$, with urbanized areas in the southeast and relatively rural areas in northwest. The typical vegetation is subtropical evergreen broadleaved forest, with natural habitats mainly distributed in the northwest, while there is a large proportion of urban forests and green spaces in the southeast. GP has a massive pool of forest resources, estimated to exceed 7400 vascular plant species and 1100 known macro-fungi [14,15]. Several major socioeconomic indicators were retrieved from the Guangdong Statistical Yearbook (2003–2016) (http://www.gdstats.gov.cn/) (Table 1). As a highly urbanized region and major municipality in southern China, GP is representative of China's modernization process and forest biodiversity patterns. According to the official statistics, as of 2016 GP hosted ~110 million residents and thus requires a large amount of imported food.

**Figure 1.** Location of Guangdong Province, southern China (highlighted in black).

*2.2. Edible Species Diversity and Taxonomic Information*

Records of edible species diversity (plants and fungi) and their illustrations were obtained from field observations and historical publications focusing on provincial forests (mostly natural reserves and forest parks), accompanied by community surveys at fruit stores and supermarkets. Sampling sites occurred in 4 cities (Foshan, Guangzhou, Shenzhen, and Zhongshan) located in highly urbanized areas of GP, and another 3 cities (Shaoguan, Zhanjiang, and Zhaoqing) located in relatively rural areas. Taxonomic information on all plant species (Supplementary Materials) refer to the English revision of Flora of China (http://foc.eflora.cn/). Specifically, species name, family name, lifestyle, life form, and harvest time were collected. The names of fungi and their taxonomy are from Mycobank (http://www.mycobank.org/).

Each edible plant species was classified into 1 of the following 7 categories: root (tuber), stalk (bark, and/or shoot), leaf, flower, fruit, seed, and/or whole plant. For fungi, the whole sporocarp was considered an edible part; these are usually used as vegetables and medicinal supplements. Four seasons of food harvest were classified according the climatic conditions in Guangdong Province as follows: March to May, spring; June to August, summer; September to November, autumn; and December to February of the second year, winter. Five food types were included in this study according to their specific nutrient components: cereals and saccharides (providing carbohydrate and starch), fruits (including all fruit types, providing vitamins and protein), vegetables (providing vitamins), oil plants (including oil crops and products, providing vegetable fat), medicinal plants (providing primary healthcare), and spice plants (providing flavoring agents). Field surveys and sampling methods were conducted according to the "Observation Methodology for Long-term Forest Ecosystem Research" of the National Standards of the People's Republic of China (GB/T 33027-2016). All figures were produced using Adobe Photoshop CC 2015 (Adobe Systems Inc, San Jose, CA, USA). Statistical difference (one-sample test) was performed using SPSS (Standard version 13.0; SPSS, Chicago, IL, USA) among the 6 edible parts, 4 seasons, 5 categories, and 4 life forms.

**Table 1.** General statistics of edible plants in Guangdong Province (2003–2016).

| Year | Provincial GDP (Billion CYN) | Primary Industry (Billion CYN) | Proportion of Primary Industry (%) | Components of Edible Plants | | | | | | Population (Million) | Proportion of Urban Population (%) | Forest Coverage (%) |
|---|---|---|---|---|---|---|---|---|---|---|---|---|
| | | | | Cereals ($10^9$ kg) | Saccharides ($10^9$ kg) | Oils ($10^9$ kg) | Vegetables ($10^9$ kg) | Fruits ($10^9$ kg) | Tea ($10^6$ kg) | | | |
| 2003 | 1344.99 | 105.16 | 7.82 | 14.88 | 9.53 | 0.82 | 25.84 | 7.19 | 41.40 | 79.54 | NA | 57.3 |
| 2004 | 1603.95 | 124.54 | 7.76 | 13.90 | 9.41 | 0.78 | 25.58 | 7.88 | 40.40 | 83.04 | NA | 57.40 |
| 2005 | 2170.13 | 137.46 | 6.33 | 13.95 | 9.46 | 0.77 | 25.96 | 8.32 | 44.50 | 91.94 | NA | 57.50 |
| 2006 | 2596.86 | 157.14 | 6.05 | 13.88 | 11.24 | 0.78 | 26.44 | 8.93 | 47.40 | 93.04 | 63.00 | NA |
| 2007 | 3067.37 | 174.62 | 5.69 | 12.85 | 10.56 | 0.79 | 26.54 | 9.51 | 48.90 | 94.49 | 63.10 | 56.00 |
| 2008 | 3569.65 | 197.02 | 5.52 | 12.43 | 10.79 | 0.82 | 24.31 | 9.83 | 48.40 | 95.44 | 63.40 | 56.30 |
| 2009 | 3908.16 | 200.60 | 5.13 | 13.14 | 11.16 | 0.85 | 25.67 | 10.62 | 51.40 | 96.38 | 63.40 | 56.70 |
| 2010 | 4547.28 | 228.69 | 5.03 | 13.17 | 11.34 | 0.88 | 27.21 | 11.29 | 52.60 | NA | NA | 57.00 |
| 2011 | 5267.36 | 265.98 | 5.05 | 13.61 | 12.03 | 0.92 | 28.60 | 12.10 | 58.10 | 105.05 | 66.50 | 57.30 |
| 2012 | 5706.79 | 284.89 | 4.99 | 13.96 | 12.79 | 0.97 | 29.83 | 12.79 | 63.10 | 105.94 | 67.40 | 57.70 |
| 2013 | 6216.40 | 304.75 | 4.90 | 13.16 | 13.59 | 1.01 | 31.44 | 13.69 | 69.80 | 106.44 | 67.76 | 58.20 |
| 2014 | 6779.22 | 316.67 | 4.67 | 13.57 | 13.08 | 1.05 | 32.75 | 14.36 | 73.90 | 107.24 | 68.00 | 58.70 |
| 2015 | 7281.26 | 334.48 | 4.59 | 13.58 | 12.51 | 1.10 | 34.39 | 15.21 | 79.30 | 108.49 | 68.71 | 58.88 |
| 2016 | 7951.20 | 369.36 | 4.64 | 13.60 | 12.70 | 1.13 | 35.80 | 15.80 | 85.00 | 109.99 | 69.20 | 58.98 |

## 3. Results

### 3.1. Major Crops in Guangdong Province

In recent years, social and economic growth has increased in Guangdong Province (Table 1). Provincial gross domestic product (GDP) and primary industry increased by 490% and 250% from 2003 to 2016, respectively. Primary industry accounted for a relatively low and stable proportion of GDP in GP, ranging from 4.59% in 2015 to 7.82% in 2003. The permanent resident population increased by 38% (30.45 million), whereas the proportion of urban population increased by 6%, to nearly 70%, from 2006 to 2016. Cereals decreased by 8% from 2003 to 2016, although yield remained relatively stable. Saccharides, oils, and vegetables all increased by one-third. Fruits increased by 120%; interestingly, the famous local lychee fruit tree (*Litchi chinensis* Sonn.) contributed slightly over 10% of the total fruit yield.

### 3.2. Patterns of Species Diversity and Edible Parts

Species diversity of edible plants was found to be relatively high in the study region. As many as 100 plant species (including var. and cv.) belonging to 88 genera in 51 families were identified (Supplementary Materials). Interestingly, 37% of edible plants were from several common families: nine species from Rosaceae; six species from Fabaceae, Myrtaceae, and Rutaceae; and five species from Poaceae. Species producing fruit parts contributed >60% of the total species, and some plants had multiple edible parts. Specifically, 64 species were identified with edible fresh or dried fruits, 11 species with edible roots or tubers, 10 species with edible seeds, 7 species with edible leaves, 6 species with edible flowers or whole plant, and 3 species with edible bark, stalks, or shoots. Statistical difference was not significant ($t = 1.771$, df (degree of freedom) = 5, $p = 0.137$; $t = 2.402$, $p = 0.061$ when six species with edible whole plant were accumulated to each part accordingly) among the six edible parts (Figure 2A).

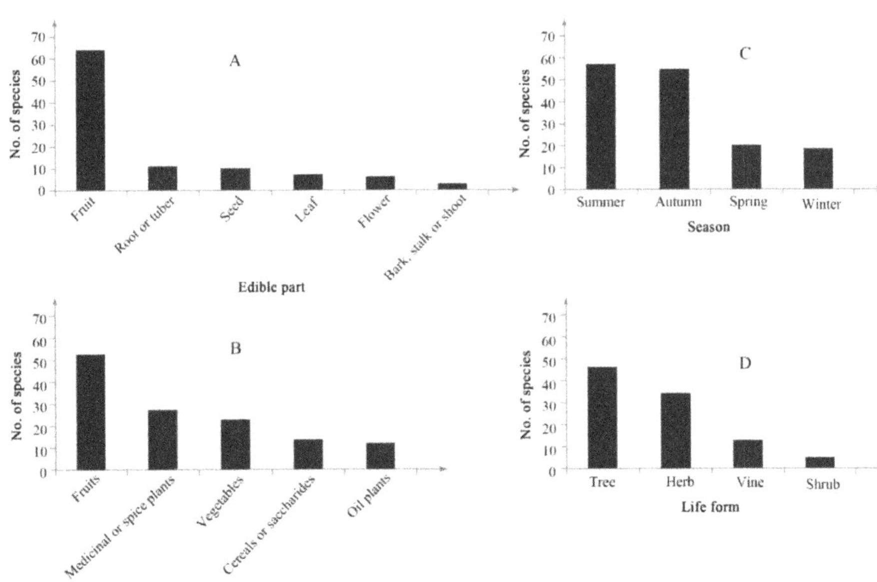


**Figure 2.** Categories of edible plant species commonly found in Guangdong Province, southern China: (**A**) edible parts; (**B**) food categories; (**C**) seasonal patterns; and (**D**) life forms.

### 3.3. Seasonal Patterns and Food Categories of Edible Forest Products

Over half of the total plant species provided edible parts in the summer and autumn, with some plants spanning multiple seasons (Supplementary Materials). Specifically, 57 species were available in autumn, 55 species in summer, 20 species in spring, and 18 species in winter. Furthermore, eight species were available across all four seasons. Statistical difference was significant (t = 3.506, df = 3, $p$ = 0.039; t = 4.254, $p$ = 0.024 when eight species available during the full year were accumulated to each season accordingly) among the four seasons (Figure 2C). Many edible plants had multiple uses. Specifically, 53 species were used as fruits, 27 species as medicinal or spice plants, 23 species as vegetables, 14 species as cereals or saccharides, and 12 species as oil plants. Statistical difference was significant (t = 3.513, df = 4, $p$ = 0.025) among the five categories (Figure 2B). Tree and herbaceous species made up approximately half and one-third of the total, respectively, while shrub species accounted for only 5% of the total. Statistical difference was not significant (t = 2.579, df = 3, $p$ = 0.082) among the four life forms (Figure 2D).

Furthermore, 20 common edible macro-fungi were recorded, belonging to 16 families (Supplementary Materials). Three out the 20 species were from the family Pleurotaceae. The majority of fungus species were cultivated, accompanied by some wild species found in forests. For instance, wild species *Boletus speciosus* Frost and *Russula vinosa* Lindblad usually appeared in the summer and autumn. The whole sporocarp was edible, mostly used as a vegetable. Moreover, about two-thirds of the species pool could be used as medicinal supplements.

## 4. Discussion

The benefits of forest ecosystems are extremely diverse. The composition of food consumption, from the global perspective, is trending toward better, healthier, and more diverse foods [4,7,16,17]. Edible plants and macro-fungi are important because they provide livelihoods for people [11,18,19]. A growing number of Chinese people pursue a high-quality lifestyle (e.g., realizing the importance of a vegetarian diet) [16,20]. The reported 100 edible plants and 20 macro-fungi were identified based on a preliminary investigation within Guangdong Province, southern China. Particularly, biodiversity conservation and food supply in the region were investigated in order to understand how they might address poverty alleviation and sustainable utilization, which are increasingly important due to global changes and rapid urbanization. Our findings might also be utilized by local authorities and stakeholders adjacent to GP (such as Hunan and Jiangxi Provinces and the Pan-Pearl River Delta).

### 4.1. Biodiversity Conservation of Forest Resources

Species diversity is the basis of biodiversity conservation. High species diversity can help counteract homogeneous food resources in the region [6,13,21]. In light of the more than 7400 woody species and 300 macro-fungi already described in GP, additional species are likely to be edible [15,22]. In particular, the presence of endemic species (e.g., fruit species *Dimocarpus longan* Lour., *Litchi chinensis* Sonn., and *Mangifera indica* L.; vegetable species *Hylocereus undatus* Haw., *Lycium chinense* Mill. *var. chinense*, and *Sechium edule* (Jacq.) Swartz; and medicine species *Ficus hirta* Vahl., *Archidendron clypearia* Jack., and *Plumeria rubra* L.) demonstrates the value of preserving regional biodiversity. Genetic diversity serves as another important component of biodiversity. High genetic diversity is important to maintain phylogenetic community structure [23–26]. Interestingly, numerous studies have reported that hidden/cryptic genetic biodiversity can be revealed through modern taxonomic methods such as DNA barcoding [27,28]. Our results show that there is a moderate proportion of wild plants (14% wild, i.e., the species seldom exists in the form of a cultivar; and another 23% partially wild, i.e., the species has the potential to be domesticated) (Supplementary Materials), which suggests that the genetic diversity of edible plants needs to be improved, and may function as a reservoir of biodiversity in the region. In terms of landscape diversity, the third component of biodiversity, the present study provides very little information. However, we agree with previous findings that actions and policies that protect living habitats and landscape components for wild plants, animals, and fungi are vital to regional biodiversity conservation in natural forests and urban green spaces [29,30].

Forests provide humans with multiple benefits, from subsistence to safety nets and cash income. In most cases, however, trade-offs may exist between exploitation and protection of forest resources. People could largely depend on local forest products, but deforestation and overexploitation might prevail in the absence of adequate protection, especially in face of strong global demand [31,32]. We suggest that one way to protect forest-based livelihoods while avoiding overexploitation would be to use an alternative solution that synergizes social and ecological benefits [33–35], with goals for sustainable forestry practices and win-win solutions. Future efforts might include the regions adjacent to GP, such as Hunan and Jiangxi Provinces, which geographically share the Nanling Mountains, a global biodiversity hotspot [36,37].

### 4.2. Food Supply in Different Seasons

Urban landscapes (e.g., ground-level and green rooftop gardens) have the potential to produce a large proportion of food crops (e.g., fresh fruits and vegetables) for a dense population, developing a local urban food system and providing eco-environmental benefits [38–40]. Due to the rapid socioeconomic development in China, attributed to the reform and opening-up policy (since 1978) and entry into the World Trade Organization (WTO) (since 2001), the dietary structure of residents has dramatically changed in GP with the availability of a diverse food supply throughout the year. It is now easy to obtain fresh fruits and vegetables via novel technologies like soilless culture, greenhouse cropping, and convenient transportation. However, we should also respect the natural life cycle of plants. Our results show that only a small proportion (~20%) of edible plant species were available in GP in the winter (such as *Chrysanthemum morifolium* cv. Hangju, *Garcinia mangostana* L., *Manihot esculenta* Crantz, *Trachycarpus fortune* (Hook.) H. Wendl, *Puerariamontana* var. *thomsonii* (Willd.), etc.) or spring (such as *Ananascomosus* L., *Bombax ceiba* L., *Houttuynia cordata* Thunb., etc.), though without specific everyday consumption, similar to previous studies calling for sustainable agroforestry and uninterrupted supplies of edible crops [9,19,41,42].

Edible forest products could be treated as an alternative defense against crises that affect food supply. Cereals are the most important food type supporting the current and projected global human population [43]. However, the yield of cereals largely depends on nutrient investments, climate change, and human activities, which might result in an imbalance of food supply at a particular time or place [44]. Our results show that there is tremendous potential for fruit trees, garden vegetables, and edible macro-fungi from nature to supplement the human food supply in the future. These edible forest products, though they are inflexible and sometimes have low returns, may still ensure sustainable food security for communities and generate as much income as cultivating crops in the long term [16,23,45].

### 4.3. Estimation of Livelihoods from Forests

Foods from forests and urban green spaces contribute a small proportion of everyday food consumption. In Guangdong Province, *Flammulina velutipes* (Curtis) Singer and *Pleurotus ostreatus* (Jacq.) P. Kumm. were the top two edible fungi, with a yield of $1.71 \times 10^8$ kg and $1.36 \times 10^8$ kg, respectively, according to a public report released by the Edible Fungus Index (http://www.mushroommarket.net/datas/). Among the species pool of 100 edible plants in this study, at least one-third were wild or partially wild species, which are not yet domesticated and exploited as are other major cultivars (e.g., lychee). Compared to the high regional species diversity, the proportions of edible plant diversity (~1.35%; 100/7400) and known macro-fungi (~1.82%; 20/1100) were relatively low [46], which indicates that wild plant and fungus resources were not overexploited and in situ conservation generally works well in GP. Meanwhile, many more wild species are likely to be edible and are expected to enrich the species diversity of forest food in this region. However, it should be noted that many wild plants and fungi are poisonous (sometimes fatal) if eaten by mistake, which calls for the involvement of knowledgeable professionals.

Currently, poverty remains an important problem for over 1.2 billion people, mostly in developing countries and especially in rural areas [9,11,18,47]. Three principal ideas may achieve forest-based poverty alleviation: prevent forest resources from shrinking, if they are necessary to maintain well-being (protect the pie); make forests accessible and redistribute resources and rents (divide the pie differently); and increase the value of forest production (enlarge the pie) [48]. It is estimated that the livelihoods of ~20% of the global population are supported by forest products, serving as subsistence, safety nets, and pathways to prosperity [11]. Environmental income (e.g., transfer of ownership of natural forests and other urban green spaces to local communities, coupled with payments for improved ecosystem services) accounts for one-third of total household income [18,49]. In addition to plants and macro-fungi, animal proteins provided by wildlife and insects could benefit human nutrition and provide special livelihoods in some rural areas [17,50,51].

## 5. Conclusions

This study reports state-of-the-art knowledge on common forest products (i.e., edible plants and macro-fungi) in lower subtropical forests in southern China. Efforts to protect natural forests and urban green spaces should continue in order to prevent biodiversity loss. Domestication of edible wild plants could be reinforced to enrich livelihoods for highly urbanized regions. Exploration of more edible plants, macro-fungi, and insects might be expected to add to the food supply. We show that endeavors to explore the diversity of edible forest products could strengthen our understanding of the socio-ecological benefits of subtropical forests.

**Supplementary Materials:** The following are available online at http://www.mdpi.com/1999-4907/9/11/666/s1.

**Author Contributions:** J.L. (Jianrong Li), J.L. (Juyang Liao) and L.Z. conceived the idea, compiled the datasets, and conducted analyses. J.L. (Juyang Liao), L.Z., Y.L., Q.L., Q.Z., and J.L. (Jianrong Li) established the direction. J.L. (Juyang Liao) and J.L. (Jianrong Li) wrote the first draft of the manuscript. J.L. (Juyang Liao), L.Z., Y.L., Q.L., D.C., Q.Z., and J.L. (Jianrong Li) contributed with suggestions and corrections, and approved the final manuscript.

**Funding:** This study was funded by the Fundamental Research Funds of CAF (CAFYBB2017QB002), Science and Technology Key R & D Program of Hunan Province (2017SK2332), National Natural Science Foundation of China (31660189), Pearl River S&T Nova Program of Guangzhou (201610010001), and CFERN & BEIJING TECHNO SOLUTIONS Award Funds on excellent academic achievements.

**Acknowledgments:** We are grateful to three anonymous reviewers for valuable ideas and/or comments on the manuscript.

**Conflicts of Interest:** The authors declare no conflict of interest.

## References

1. Boyd, I.L.; Freer-Smith, P.H.; Gilligan, C.A.; Godfray, H.C.J. The Consequence of Tree Pests and Diseases for Ecosystem Services. *Science* **2013**, *342*, 1235773. [CrossRef] [PubMed]
2. Nowak, D.J.; Crane, D.E.; Stevens, J.C.; Hoehn, R.E.; Walton, J.T.; Bond, J. A ground-based method of assessing urban forest structure and ecosystem services. *Arboricult. Urban For.* **2008**, *34*, 347–358.
3. Costanza, R.; de Groot, R.; Sutton, P.; van der Ploeg, S.; Anderson, S.J.; Kubiszewski, I.; Farber, S.; Turner, R.K. Changes in the global value of ecosystem services. *Glob. Environ. Chang.* **2014**, *26*, 152–158. [CrossRef]
4. Poleman, T.T. World Food: A Perspective. *Science* **1975**, *188*, 510–518. [CrossRef] [PubMed]
5. Schmidhuber, J.; Tubiello, F.N. Global food security under climate change. *Proc. Natl. Acad. Sci. USA* **2007**, *104*, 19703–19708. [CrossRef] [PubMed]
6. Campbell, B.M.; Vermeulen, S.J.; Aggarwal, P.K.; Corner-Dolloff, C.; Girvetz, E.; Loboguerrero, A.M.; Ramirez-Villegas, J.; Rosenstock, T.; Sebastian, L.; Thornton, P.K.; et al. Reducing risks to food security from climate change. *Glob. Food Secur.* **2016**, *11*, 34–43. [CrossRef]
7. Vermeulen, S.J.; Campbell, B.M.; Ingram, J.S.I. Climate Change and Food Systems. *Annu. Rev. Environ. Resour.* **2012**, *37*, 195–222. [CrossRef]

8. Pimentel, D.; McNair, M.; Buck, L.; Pimentel, M.; Kamil, J. The Value of Forests to World Food Security. *Hum. Ecol.* **1997**, *25*, 91–120. [CrossRef]
9. Alexandratos, N.; Bruinsma, J. *World Agriculture towards 2030/2050: The 2012 Revision*; Food and Agriculture Organization of the United Nations: Rome, Italy, 2012.
10. Rosegrant, M.W.; Cline, S.A. Global Food Security: Challenges and Policies. *Science* **2003**, *302*, 1917–1919. [CrossRef] [PubMed]
11. Cheng, S.H.; Ahlroth, S.; Onder, S.; Shyamsundar, P.; Garside, R.; Kristjanson, P.; McKinnon, M.C.; Miller, D.C. What is the evidence for the contribution of forests to poverty alleviation? A systematic map protocol. *Environ. Evid.* **2017**, *6*, 10. [CrossRef]
12. MacGregor-Fors, I.; Escobar, F.; Rueda-Hernández, R.; Avendaño-Reyes, S.; Baena, M.; Bandala, V.; Chacón-Zapata, S.; Guillén-Servent, A.; González-García, F.; Lorea-Hernández, F.; et al. City "Green" Contributions: The Role of Urban Greenspaces as Reservoirs for Biodiversity. *Forests* **2016**, *7*, 146. [CrossRef]
13. Pimentel, D.; Wilson, C.; McCullum, C.; Huang, R.; Dwen, P.; Flack, J.; Tran, Q.; Saltman, T.; Cliff, B. Economic and Environmental Benefits of Biodiversity. *BioScience* **1997**, *47*, 747–757. [CrossRef]
14. Song, B.; Zhong, Y.; Deng, W.; Shen, Y.; Li, J.; Lin, M. Resouces of wild macrofungi and prospects of their exploitation in Guangdong China. *J. Microbiol.* **2007**, *27*, 59–63. (In Chinese)
15. Ye, H.; Xing, F. *Checklist of Guangdong Plants*; World Publishing Guangdong Corporation: Guangzhou, China, 2005.
16. Fan, S.; Brzeska, J. Sustainable food security and nutrition: Demystifying conventional beliefs. *Glob. Food Secur.* **2016**, *11*, 11–16. [CrossRef]
17. Golden, C.D.; Fernald, L.C.H.; Brashares, J.S.; Rasolofoniaina, B.J.R.; Kremen, C. Benefits of wildlife consumption to child nutrition in a biodiversity hotspot. *Proc. Natl. Acad. Sci. USA* **2011**, *108*, 19653–19656. [CrossRef] [PubMed]
18. Angelsen, A.; Jagger, P.; Babigumira, R.; Belcher, B.; Hogarth, N.J.; Bauch, S.; Börner, J.; Smith-Hall, C.; Wunder, S. Environmental Income and Rural Livelihoods: A Global-Comparative Analysis. *World Dev.* **2014**, *64*, S12–S28. [CrossRef]
19. Russo, A.; Escobedo, F.J.; Cirella, G.T.; Zerbe, S. Edible green infrastructure: An approach and review of provisioning ecosystem services and disservices in urban environments. *Agric. Ecosyst. Environ.* **2017**, *242*, 53–66. [CrossRef]
20. Wang, Z.H.; Zhai, F.Y.; Du, S.F.; Popkin, B. Dynamic shifts in Chinese eating behaviors. *Asia Pac. J. Clin. Nutr.* **2008**, *17*, 123–130. [PubMed]
21. Kirkpatrick, J.B.; Davison, A. Home-grown: Gardens, practices and motivations in urban domestic vegetable production. *Landsc. Urban Plan.* **2018**, *170*, 24–33. [CrossRef]
22. Zhou, L.; Yang, Z.; Wen, H.; Bau, T.; Li, T. A revised checklist of edible fungi in China. *Mycosystema* **2010**, *29*, 1–21. (In Chinese)
23. Charnley, S.; Poe, M.R. Community Forestry in Theory and Practice: Where Are We Now? *Annu. Rev. Anthropol.* **2007**, *36*, 301–336. [CrossRef]
24. Pereira, H.M. A latitudinal gradient for genetic diversity. *Science* **2016**, *353*, 1494–1495. [CrossRef] [PubMed]
25. Harlan, J.R. Our Vanishing Genetic Resources. *Science* **1975**, *188*, 618–621. [CrossRef] [PubMed]
26. Swenson, N.G.; Erickson, D.L.; Mi, X.; Bourg, N.A.; Forero-Montana, J.; Ge, X.; Howe, R.; Lake, J.K.; Liu, X.; Ma, K.; et al. Phylogenetic and Functional Alpha and Beta Diversity in Temperate and Tropical Tree Communities. *Ecology* **2012**, *93*, S112–S125. [CrossRef]
27. Lahaye, R.; van der Bank, M.; Bogarin, D.; Warner, J.; Pupulin, F.; Gigot, G.; Maurin, O.; Duthoit, S.; Barraclough, T.G.; Savolainen, V. DNA barcoding the floras of biodiversity hotspots. *Proc. Natl. Acad. Sci. USA* **2008**, *105*, 2923–2928. [CrossRef] [PubMed]
28. Janzen, D.H.; Burns, J.M.; Cong, Q.; Hallwachs, W.; Dapkey, T.; Manjunath, R.; Hajibabaei, M.; Hebert, P.D.N.; Grishin, N.V. Nuclear genomes distinguish cryptic species suggested by their DNA barcodes and ecology. *Proc. Natl. Acad. Sci. USA* **2017**, *114*, 8313–8318. [CrossRef] [PubMed]
29. Ashley, R.; Russell, D.; Swallow, B. The Policy Terrain in Protected Area Landscapes: Challenges for Agroforestry in Integrated Landscape Conservation. *Biodivers. Conserv.* **2006**, *15*, 663–689. [CrossRef]
30. Livesley, S.J.; McPherson, E.G.; Calfapietra, C. The Urban Forest and Ecosystem Services: Impacts on Urban Water, Heat, and Pollution Cycles at the Tree, Street, and City Scale. *J. Environ. Qual.* **2016**, *45*, 119–124. [CrossRef] [PubMed]

31. DeFries, R.S.; Rudel, T.; Uriarte, M.; Hansen, M. Deforestation driven by urban population growth and agricultural trade in the twenty-first century. *Nat. Geosci.* **2010**, *3*, 178. [CrossRef]
32. Edwards, D.P.; Magrach, A.; Woodcock, P.; Ji, Y.; Lim, N.T.L.; Edwards, F.A.; Larsen, T.H.; Hsu, W.W.; Benedick, S.; Khen, C.V.; et al. Selective-logging and oil palm: Multitaxon impacts, biodiversity indicators, and trade-offs for conservation planning. *Ecol. Appl.* **2014**, *24*, 2029–2049. [CrossRef] [PubMed]
33. Laurance, W.F.; Carolina Useche, D.; Rendeiro, J.; Kalka, M.; Bradshaw, C.J.A.; Sloan, S.P.; Laurance, S.G.; Campbell, M.; Abernethy, K.; Alvarez, P.; et al. Averting biodiversity collapse in tropical forest protected areas. *Nature* **2012**, *489*, 290–294. [CrossRef] [PubMed]
34. Lindenmayer, D.B.; Laurance, W.F. The ecology, distribution, conservation and management of large old trees. *Biol. Rev.* **2017**, *92*, 1434–1458. [CrossRef] [PubMed]
35. Persha, L.; Agrawal, A.; Chhatre, A. Social and Ecological Synergy: Local Rulemaking, Forest Livelihoods, and Biodiversity Conservation. *Science* **2011**, *331*, 1606–1608. [CrossRef] [PubMed]
36. Chen, Y.-S.; Deng, T.; Zhou, Z.; Sun, H. Is the East Asian flora ancient or not? *Natl. Sci. Rev.* **2018**, in press. [CrossRef]
37. Myers, N.; Mittermeier, R.A.; Mittermeier, C.G.; Fonseca, G.A.B.D.; Kent, J. Biodiversity hotspots for conservation priorities. *Nature* **2000**, *403*, 853–858. [CrossRef] [PubMed]
38. Saha, M.; Eckelman, M.J. Growing fresh fruits and vegetables in an urban landscape: A geospatial assessment of ground level and rooftop urban agriculture potential in Boston, USA. *Landsc. Urban Plan.* **2017**, *165*, 130–141. [CrossRef]
39. Dieleman, H. Urban agriculture in Mexico City; balancing between ecological, economic, social and symbolic value. *J. Clean. Prod.* **2017**, *163*, S156–S163. [CrossRef]
40. Camps-Calvet, M.; Langemeyer, J.; Calvet-Mir, L.; Gómez-Baggethun, E. Ecosystem services provided by urban gardens in Barcelona, Spain: Insights for policy and planning. *Environ. Sci. Policy* **2016**, *62*, 14–23. [CrossRef]
41. Guarrera, P.M.; Savo, V. Wild food plants used in traditional vegetable mixtures in Italy. *J. Ethnopharmacol.* **2016**, *185*, 202–234. [CrossRef] [PubMed]
42. Lindenmayer, D.B.; Franklin, J.F.; Lõhmus, A.; Baker, S.C.; Bauhus, J.; Beese, W.; Brodie, A.; Kiehl, B.; Kouki, J.; Pastur, G.M.; et al. A major shift to the retention approach for forestry can help resolve some global forest sustainability issues. *Conserv. Lett.* **2012**, *5*, 421–431. [CrossRef]
43. Ladha, J.K.; Tirol-Padre, A.; Reddy, C.K.; Cassman, K.G.; Verma, S.; Powlson, D.S.; van Kessel, C.; de B. Richter, D.; Chakraborty, D.; Pathak, H. Global nitrogen budgets in cereals: A 50-year assessment for maize, rice, and wheat production systems. *Sci. Rep.* **2016**, *6*, 19355. [CrossRef] [PubMed]
44. Lyu, Y.; Tang, H.; Li, H.; Zhang, F.; Rengel, Z.; Whalley, W.R.; Shen, J. Major Crop Species Show Differential Balance between Root Morphological and Physiological Responses to Variable Phosphorus Supply. *Front. Plant Sci.* **2016**, *7*, 1939. [CrossRef] [PubMed]
45. Wunder, S.; Angelsen, A.; Belcher, B. Forests, Livelihoods, and Conservation: Broadening the Empirical Base. *World Dev.* **2014**, *64*, S1–S11. [CrossRef]
46. Zhang, W.; Li, T.; Song, B. A survey on the macrofungi in Guangdong province. *Ecol. Sci.* **2001**, *20*, 48–58.
47. Zhuo, L.; Wang, L.; Chen, Q.; Liu, W. Wild fruit resources and exploitation in Xiaoxing'an Mountains. *J. For. Res.* **1999**, *10*, 31–33.
48. Sunderlin, W.D.; Angelsen, A.; Wunder, S. Forests and poverty alleviation. In *State of the World's Forests*; Center for International Forestry Research: Bogor Barat, Indonesia, 2003; pp. 61–73.
49. Chhatre, A.; Agrawal, A. Trade-offs and synergies between carbon storage and livelihood benefits from forest commons. *Proc. Natl. Acad. Sci. USA* **2009**, *106*, 17667–17670. [CrossRef] [PubMed]
50. Yen, A.L. Edible insects: Traditional knowledge or western phobia? *Entomol. Res.* **2009**, *39*, 289–298. [CrossRef]
51. Lee, T.M.; Sigouin, A.; Pinedo-Vasquez, M.; Nasi, R. *The Harvest of Wildlife for Bushmeat and Traditional Medicine in East, South and Southeast Asia: Current Knowledge Base, Challenges, Opportunities and Areas for Future Research*; CIFOR: Bogor, Indonesia, 2014; Volume 115.

*forests*

MDPI

*Article*

# Heterodichogamy, Pollen Viability, and Seed Set in a Population of Polyploidy *Cyclocarya Paliurus* (Batal) Iljinskaja (Juglandaceae)

**Xia Mao, Xiang-Xiang Fu *, Peng Huang, Xiao-Ling Chen and Yin-Quan Qu**

Co-Innovation Centre for Sustainable Forestry in Southern China, College of Forestry, Nanjing Forestry University, Nanjing 210037, China; 15380924779@163.com (X.M.); nanjinghp111@163.com (P.H.); chenxl0811@163.com (X.-L.C.); 18761808762@163.com (Y.-Q.Q.)

* Correspondence: xxfu@njfu.edu.cn; Tel.: +86-025-85427403; Fax: +86-025-85427402

Received: 21 February 2019; Accepted: 17 April 2019; Published: 19 April 2019

**Abstract:** *Research Highlights: Cyclocarya paliurus*, native to the subtropical region of China, is a monoecious species with a heterodichogamous mating system. Its flowering phenology and low seed success characteristics differ from other typical heterodichogamous Juglandaceae species. This could be caused by the existence of polyploidy in the population. *Background and Objectives: C. paliurus* has been attracting more attention as a result of its medicinal value. To meet the needs for leaf harvest, cultivation expansion is required, but this is limited by a shortage of seeds. This study aims to profile the flowering phenology and the efficacy of pollen dispersal as well as elucidate on the mechanism of low seed success in the population. *Materials and Methods:* The flowering phenology pattern of *C. paliurus* was observed in a juvenile plantation containing 835 individuals of 53 families from 8 provenances at the individual (protandry, PA and protogyny, PG) and population levels for 5 consecutive years (2014–2018). Slides with a culture medium of 10% sucrose and 0.01% boric acid were used to estimate pollen density and viability in the population, and seeds were collected from 20 randomly selected PA and PG individuals to assess seed success during 2017–2018. *Results:* Four flowering phenotypes and strongly skewed ratios of PA/PG and male/female occurred in the juvenile population. Sexual type and ratio changed significantly with the growth of the population over the years, showing an increasing monoecious group (11.1% to 57.2%) and a decreasing unisexual group (33.6% to 16.3%), as well as a tendency for the sexual ratio to move towards equilibrium (5.42:1 to 1.39:1 for PG:PA). Two flowering phases and bimodality in gender were displayed, as in other heterodichogamous species. However, the high overlap of inter-phases and within individuals was quite different from many previous reports. Owing to the low pollen viability of *C. paliurus* (~30%), low seed success was monitored in the plantation, as well as in the investigated natural populations. *Conclusions:* Female-bias (PG and F) and a skewed ratio of mating types corresponded to nutrient accumulation in the juvenile population. Heterodichogamy in *C. paliurus* was verified, but was shown to be different from other documented species in Juglandaceae. The latest finding of major tetraploidy in a natural population could explain the characteristics of the flowering phenology and seed set of *C. paliurus* and also give rise to more questions to be answered.

**Keywords:** protogyny (PG); protandry (PA); pollen viability; seed success; polyploidy

## 1. Introduction

Heterodichogamy, a transitional type from dichogamy to dioecy, is a polymorphic phonologic sexual system [1,2]. It is defined as two complementary morphs, protogyny (PG, female function before male) and protandry (PA, male function before male), that function synchronously and reciprocally to one another at a population level [3]. Generally, disassortative mating between two morphs, regarded as the main pattern for heterodichogamous species, promotes proficient inter-morph cross-pollination, avoids selfing, and reduces inbreeding of intra-morph through temporal separation of the male and female functions in flowers [1,4]. To avoid frequency-dependent selection, the population has evolved to have a 1:1 morph ratio [5]. Based on this equilibrium ratio, genetic theory suggests that heterodichogamy was simply inherited with a single diallelic Mendelian locus [1], which has been confirmed in both *Juglans* and *Carya* genera from the Juglandaceae family [6,7].

Hitherto, plants with heterodichogamy have been documented in 13 families and 21 genera of flowering plants [8,9]. Heterodichogamy is particular found in Juglandaceae genera, including *Juglans, Carya, Peterocarya, Platycarya*, and *Cyclocarya* [1,9,10]. Renner [1] reported that about half of the heterodichogamous taxa are self-incompatible, but the Juglandaceae family is completely self-compatible [6,7]. However, Bai [11] demonstrated that the mating system of *J. mandshurica* is completely out-crossing.

Walnut and pecan species are regarded as a group of economically useful plants whose heterodichogamous features have gained extensive attention. This includes *J. regia*, *J. mandshurica*, *J. ailanthifolia*, *J.nigra*, *J. hindsii*, *J. cinerea*, and *J. cordiformis* in *Juglans*; *C. illinoensis*, *C. ovata*, *C. tomentosa*, and *C. laciniosa* in *Carya*; and some inter-specific hybrids in this family [7,8]. Different from nut- and/or timber-use juglandaceous plants, *C. paliurus* is important for its leaf-use value. In the past two decades, an extraordinary amount of progress has been made in determining the medicinal functions of this species. Publications have mostly stressed the actions of bioactive components, including cyclocariosides, cyclocaric acids, flavonoids, and steroids. These leaf metabolites have been verified to exert important medicinal and health functions on conditions such as hypoglycemia, hypolipidemia, and diabetes mellitus, as well as carrying out antioxidant activities [12–16]. However, leaf resources only from natural stands limited its wide applications. Therefore, the cultivation of *C. paliurus* with improved accumulation of bioactive metabolites for leaf harvest is concerned [17,18]. To date, sexual rather than vegetative propagation has been successful for extensive cultivation. Importantly, the release of deep seed dormancy has greatly driven the development of plantations [19]. However, seed shortages resulting from low plumpness is now the main issue for large-scale plantations. Therefore, the following questions need to be answered: Why does such a low seed plumpness occur in populations of *C. paliurus*? Is this affected by genetic heterodichogamy or yearly changing climatic factors? Is the flowering phenology in *C. paliurus* similar to that of other recorded heterodichogamous Juglandaceae species? Therefore, the aims of this study are the following: (1) To profile the flowering patterns of the two mating types of *C. paliurus* (PA and PG) and the population as a whole using consecutive 5-year observations on the juvenile plantation; (2) to monitor the density and viability of pollen at the population level during the flowering season; (3) to analyze and hypothesize the reasons for the low seed success in the population using the chromosomal ploidy level of *C. paliurus*.

## 2. Materials and Methods

### 2.1. Study Species

*Cyclocarya paliurus*, a species of monotypic genus in Juglandaceae, is endemic in subtropical region of China (from 24°16′12″ N to 33°22′12″ N and from 103°28′12″ E to 119°22′48″ E). Often, small populations (<25 individuals) grow in moist valleys in mountain regions at an altitude of 390–1836 m, and mature trees reach a height of 10–30 (40) m because of their heliophilous characteristic [20].

A single female inflorescence of *C. paliurus* is found at the apex of the growing shoot (Figure 1A); rarely, a single male instead of a female inflorescence is observed at this position (Figure 1C). A cluster containing 2–4 catkins is found at the lateral short branch (Figure 1A). Besides unisexual inflorescences, a small number of female inflorescences mixed with male flower(s) are observed in some individuals at certain years (Figure 1D). However, similar to maples, the male flowers in female inflorescence are abortive [21]. Previous studies revealed that *C. paliurus* is a typically heterodichogamous species that undergoes wind-pollination [10,22].

**Figure 1.** Expressions of male and female inflorescences in *Cyclocarya paliurus*. (**A**) Male and female inflorescences of the protogyny (PG) type. The female flowers are mature, while the male flowers are still enclosed. (**B**) Protandry (PA) inflorescences at the elongation stage, showing the developmental differences between the two sexual inflorescences. Abnormal inflorescences include (**C**) a single male instead of a female inflorescence at the apex of growing shoot and (**D**) a female inflorescence mixed with a male flower.

### 2.2. Plantation of C. paliurus

A plantation of *C. paliurus* used in this study is located on Hongya Mountain and in Chuzhou City and Anhui Province, China (32°21′ N, 117°58′ E), where the climate is classified as northern subtropical humid monsoon, with an annual mean temperature of 15.5 °C, an annual rainfall of 1038 mm, and a frost-free period of 210 days. The site is dominated by a gravel mountain with a slight slope (5°–6°).

The plantation was established in 2008, containing 835 individuals of 53 families from 8 provenances (Lushan, Jiangxi; Hefeng Hubei; Shucheng, Anhui; Jianhe, Guizhou). The plant spacing was 3 m × 4 m, and the stand was not closed until 2018. A small number of flowering plants (mostly only one sexual inflorescence occurred) were observed in 2012; subsequently, an increasing number of plants exhibited sexual polymorphism.

### 2.3. Investigation of Flowering Phenology

Phenological monitoring was performed for a 5-year consecutive period (2014–2018), and observations were recorded from 2015, when at least one sexual flower (male/female) occurred in 3/8 individuals in the population. Characteristics recorded included flowering phenotype, flowering progress, and sex expression for all individuals. We also observed whether the flowering phenotype involved flowering or not, as well as the dates of onset, duration, and termination for male and female flowering each year.

The female flowering period was determined by two feathery stigmas with an angle of 120° and the change of the stigma color from green to brown (withering); the period of male flowering was from pollen shedding to shedding completion. The full-bloom stage for the population was defined as 50%–75% of flowers being in each flowering phase for monoecious individuals. Observations were made on male and female inflorescences tagged in the middle section of each plant.

### 2.4. Pollen Dispersal

To monitor the density and viability of pollen within the flowering season, pollen was collected from 5 trapping sites on stocks at a height of 1.5 m along a line of "S" in the stand. Six slides covered with a culture medium of 10% sucrose and 0.01% boric acid were placed on each stock. Three slides collected every two days, each for 24 h, were used for pollen density counting, while the remaining 3 slides were collected from 9:00–11:00 (when the highest pollen viability is observed) each day (except for rainy days) during flowering duration and used for pollen viability testing.

The slides were examined by (1) counting the number of pollen grains adhering to the surface of the culture medium under a microscope at ×50 magnification, and (2) culturing pollen in an incubator at 25 °C for 8 h. The pollen grain was considered viable when the pollen tube grew to the length of itself. The mean and standard error of the pollen density and viability were calculated over ten visual fields for each side.

### 2.5. Seed Collection

Respectively, 20 PA and 20 PG plants were randomly selected in 2017 and 2018, a total of 300–500 seeds were collected from the middle of each selected tree. Seed plumpness was judged by cutting seed along the hilum linear direction.

Values of pollen density and viability and seed plumpness are presented as mean ± SE, and an ANOVA analysis was conducted to determine differences among years and between mating types using SAS 18.1.

## 3. Results

### 3.1. Sex Expression of the Juvenile Population

As reported in the previous document, the flowering season of *C. paliurus* lasted about one month, from mid-April to the end of May. Two separated flowering phases occurred in the population: one during late-April to mid-May, while the other happened across all of May.

#### 3.1.1. Sexual Diversity

A total of five phenotypes from three groups were recognized for individuals in the population: The monoecious group (MO) containing individuals with both male and female inflorescences, including protogyny (PG) and protandry (PA); the unisexual group (UN) including individuals with only one sexual inflorescence, either male (M) or female (F); and unflowering trees (UF). MO and UN were both classified as flowering plants (FP).

#### 3.1.2. Expression Features of Various Flowering Phenotypes

In a natural population, initial flowering usually occurs at the age of 10–15 years. However, the onset of flowering in a plantation is significantly earlier than in a natural population. In the 5-year-old plantation (in 2013), the F type was predominant among FP plants, up to 75.7% and 53.8% in 2014 and 2015, respectively; the second most dominant type was the PG type. Comparatively, few plants were of the PA type, and this minority was assigned to the M type (Table 1).

**Table 1.** Annual expression of various flowering phenotypes in the juvenile plantation of *C. paliurus*.

| Flowering Group (FP) | Mating Type | Observed Year 2014 | 2015 | 2016 | 2017 | 2018 |
|---|---|---|---|---|---|---|
| Monoecious group (MO) | number of PA | 12 | 40 | 127 | 151 | 185 |
| | number of PG | 65 | 86 | 173 | 200 | 257 |
| | PA:PG | 5.42:1 | 2.15:1 | 1.36:1 | 1.32:1 | 1.39:1 |
| | PG/MO (%) | 84.4 | 68.3 | 57.7 | 57.0 | 58.1 |
| | PG/FP (%) | 18.8 | 28.8 | 37.5 | 37.0 | 45.2 |
| | MO/T (%) | 11.1 | 15.8 | 38.4 | 45.1 | 57.2 |
| Unisexual group (UN) | number of F | 262 | 161 | 147 | 169 | 98 |
| | number of M | 7 | 12 | 14 | 20 | 28 |
| | F:M | 37.4:1 | 13.4:1 | 10.5:1 | 8.5:1 | 3.5:1 |
| | F/UN (%) | 97.4 | 93.1 | 91.3 | 89.4 | 77.8 |
| | F/FP (%) | 75.7 | 53.8 | 39.0 | 31.3 | 17.3 |
| | UN/T (%) | 33.6 | 21.6 | 20.6 | 24.3 | 16.3 |
| FP/T (%) | | 43.2 | 37.4 | 59.0 | 69.4 | 73.5 |
| Total plants in the population (T) * | | 801 | 800 | 781 | 778 | 773 |

* Data changed annually because of the death of individuals. PA: protandry, PG: protogyny; MO: monoecious group, FP: flowering plants, UN: Unisexual group; F: females, M: males; T: total individuals in population.

With the growth of the population, more individuals flowered. The FP/T showed a tendency to increase, from 43.2% in 2014 to 73.5% in 2018 (Table 1). More significantly, the promotion of MO/T increased rapidly from 11.1% (in 2014) to 57.2% (in 2018), whereas there was a descent in UN/T, from 33.6% (2014) to 16.3% (2018).

This data suggests that the ages of 6–10 years (defined as the juvenile population) could be regarded as the transitional stage towards maturation for *C. paliurus*.

### 3.1.3. Changing Patterns of Mating Types across Years

Although there was a tendency for the number of flowering individuals to increase (Table 1), the changing trends were divergent for various flowering phenotypes (Figure 2). In 2016, there were rapid increments of PA and PG morphs, accompanied by a rapid decrease in individuals of the F type. Of the four flowering phenotypes, similar increasing tendencies were observed in PA and PG types, but the proportion of PA was always less than that of PG. In addition, only a slight increment was observed in the M type, rising from 0.9% to 3.2%. On the contrary, the proportion of F type individuals dropped significantly from 32.71% to 12.68% across the five-year period (Figure 2).

Meanwhile, strongly skewed ratios of mating types, namely, PG/PA or F/M, always existed in the juvenile population (Table 1). Significantly, the ratio of PA/PG fell obviously from 5.42:1 in 2014 to 1.39:1 in 2018. Further the F/M ratio fell from 37.43:1 in 2014 to 3.5:1 in 2018. An $X^2$ test indicated the protogynous-biased ratio in the population during 2014 to 2018 (PG/PA: 65/12, $x^2 = 36.5$, d.f = 1, $p < 0.01$ in 2014; 86/40, $x^2 = 16.8$, d.f = 1, $p < 0.01$ in 2015; 173/127, $x^2 = 7.2$, d.f = 1, $p < 0.01$ in 2016; 200/151, $x^2 = 7.7$, d.f = 1, $p < 0.01$ in 2017; and 257/185, $x^2 = 11.7$, d.f = 1, $p < 0.01$ in 2018) significantly deviated from the equilibrium ratio of 1:1, but showed a tendency towards relative equilibrium.

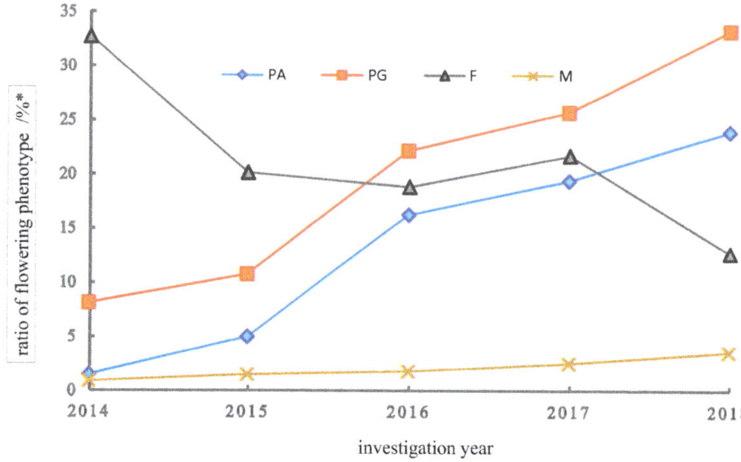


**Figure 2.** Changing trends of flowering phenotypes with the growth of the *C. paliurus* population.* Number of each flowering phenotypes/total plants in the population. F: females, M: males.

### 3.1.4. Reciprocal Transitions between Mating Types

Owing to the immaturity of the juvenile population, plants were sensitive to nutrient shortage and unstable environmental factors. Therefore, reciprocal transitions among five phenotypes happened frequently between adjacent years (Table 2). As a whole, with the growth of population, the ratio of transition from UF to FP showed a decreasing trend from 68.8% to 51.2%, whereas there was an opposite tendency for the transition from UN to MO (Table 2).

It seems that F and PG types played the core roles for all transitions. As shown in Table 2, reciprocal transitions happened more often between UF↔F, UF↔/PG, and F↔PG/PA, and scarcely between M↔F and PA↔PG. Remarkably, observed transitions between PA and PG were usually ambiguous and mainly happened in individuals with a high overlap of male and female flowering (Figure 3).

**Table 2.** Transitions between flowering phenotypes in the juvenile plantation of *C. paliurus*.

| Transition Pattern | | Number of Transitions between Phenotypes/Total Number of Transitions (%) 2014–2015 | 2015–2016 | 2016–2017 | 2017–2018 |
|---|---|---|---|---|---|
| UF↔FP | UF→F, F→UF | 53/20.2, 101/38.4 | 91/30.3, 3/1.0 | 85/31.7, 30/11.2 | 38/13.3, 32/11.2 |
| | UF→M, M→UF | 5/1.9, 4/1.5 | 9/3.0, – | 3/1.1, 4/1.5 | 5/1.8, 2/0.7 |
| | UF→PG, PG→UF | 6/2.3, 8/3.0 | 42/14.0, – | 15/5.6, 3/1.1 | 24/8.4, 10/3.5 |
| | UF→PA, PA→UF | 3/1.1, 1/0.4 | 38/12.7, – | 11/4.1, 4/1.5 | 23/8.1, 12/4.2 |
| | Subtotal of UF | 181/68.8 | 183/61.0 | 155/57.8 | 146/51.2 |
| UN↔MO | F→PG, PG→F | 36/13.7, 8/3.0 | 46/15.3, 5/1.7 | 36/13.4, 14/5.2 | 48/16.8, 2/0.7 |
| | F→PA, PA→F | 28/10.6, 4/1.5 | 43/14.3, 5/1.7 | 18/6.7, 18/6.7 | 39/13.7, 10/3.5 |
| | M→PG, PG→M | 2/0.8, 2/0.8 | 4/1.3, 2/0.7 | 9/3.4, 3/1.1 | 3/1.1, 11/3.9 |
| | M→PA, PA→M | –, – | 8/2.7, 1/0.3 | 7/2.6, 1/0.4 | 6/2.1, 2/0.7 |
| | subtotal | 80/30.1 | 114/38.0 | 106/39.6 | 121/42.5 |
| MO↔MO | PG→PA, PA→PG | –, – | 1/0.3 | –, – | 5/1.8, 3/1.1 |
| UN↔UN | F→M, M→F | 1/0.4, 1/0.4 | 2/0.7, – | 5/1.7, 2/0.7 | 9/3.2, 1/0.4 |
| total | | 263 | 300 | 268 | 285 |

### 3.2. Flowering Phenology of PG and PA Types across Years

Over the five-year period, 21 PA and 30 PG plants were screened via consecutive observations. As shown in Figure 3, for all marked individuals, a longer flowering duration was observed in females than in males; phenology also coincided with the sequence of flowering expression, but differed in terms of the separation of male and female functions across years. Such labile phenological characters could be mainly affected by environmental factors (e.g., temperature, rain, and wind).

Within each PA individual, the separation between two sexual functions gradually shortened and tended to overlap over the years. In 2015, separation occurred in 17 PA trees; however, it only occurred in four trees in 2018. Although overlap was common within PG individuals, it varied across years. This is in accordance with the statistical data for all monoecious plants, which displayed more overlap in the PG type (50%–90%) than in the PA type (14.3%–47.6%). Therefore, the potential mating probability within PA/PG individuals seems to be dependent on the overlap degree of two sexual functions.

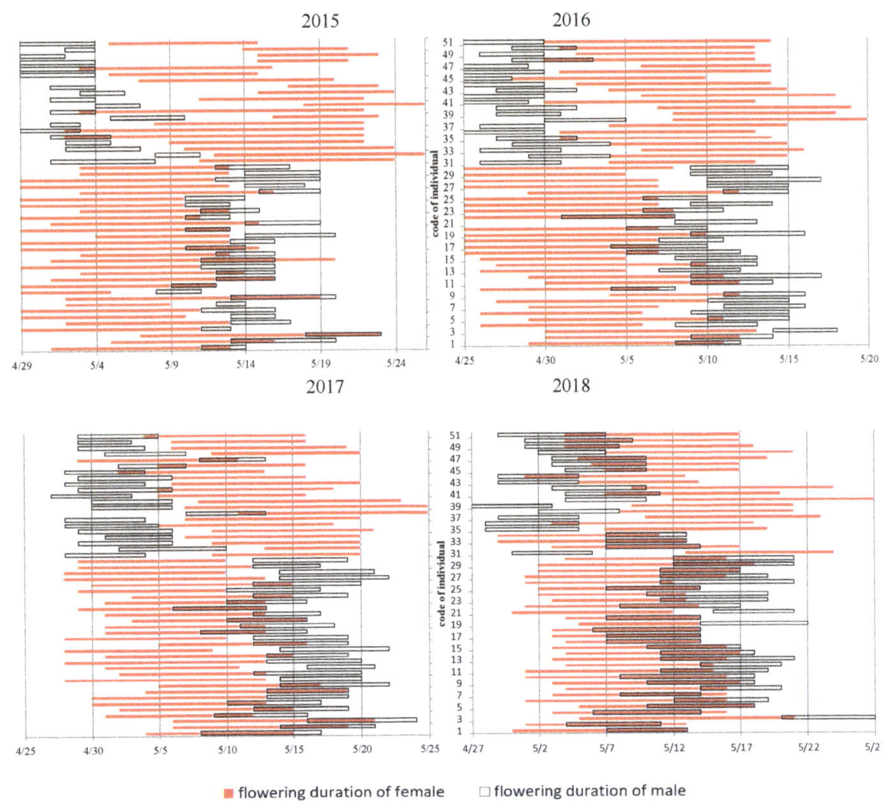


**Figure 3.** Flowering phenology of males and females within individuals based on 21PA and 30PG trees of *C. paliurus* during 2015–2018.

### 3.3. Phenological Characteristics of the Population

PA and PG morphs are the two main types of heterodichogamous species. However, in the juvenile population of *C. paliurus*, monoecious plants were not more than half of the total (57.2%) until 2018 (Table 1). Moreover, some monoecious plants could transfer to other phenotypes, such as F or unflowering ones (Table 2). Here, all individuals of PA or PG morphs were clustered together to describe the flowering characteristics of the population.

Figure 4 illustrates that two incomplete separation flowering phases existed in the population. In the first phase, from 22–26 April to 12–18 May, PG females and PA males were in bloom, while during the second phase, from 28 April 28–1 May to 20–26 May, PG males and PA females were in bloom. Each phase lasted 20–25 days and the obvious overlap (7–10 days) of inter-phases varied across years. Generally, the female flowering duration (>25% individuals in the population in bloom) for two morphs was significantly longer than that of the males (one week).

Comparatively, the male flowering duration was shorter by 4–5 days in the first phase than in the second one. In the first phase, the peak of male flowering was reached earlier than that of female flowering, which lowered the opportunity for inter-morph pollination. In contrast, the flowering peaks of the complementary sexes were rather well synchronized in the second phase (Figure 4). This difference in inter-morph synchronism for each phase could result in higher seed success in PA (pollinated in the second phase) than in PG (pollinated in the first phase) and a lower chance for pollen from PA males than PG males as contributors to pollinate the PG females. In addition, the overlap of inter-phases mainly happened between PG males and PG females, increasing the intra-morph inbreeding probability (Figure 4).

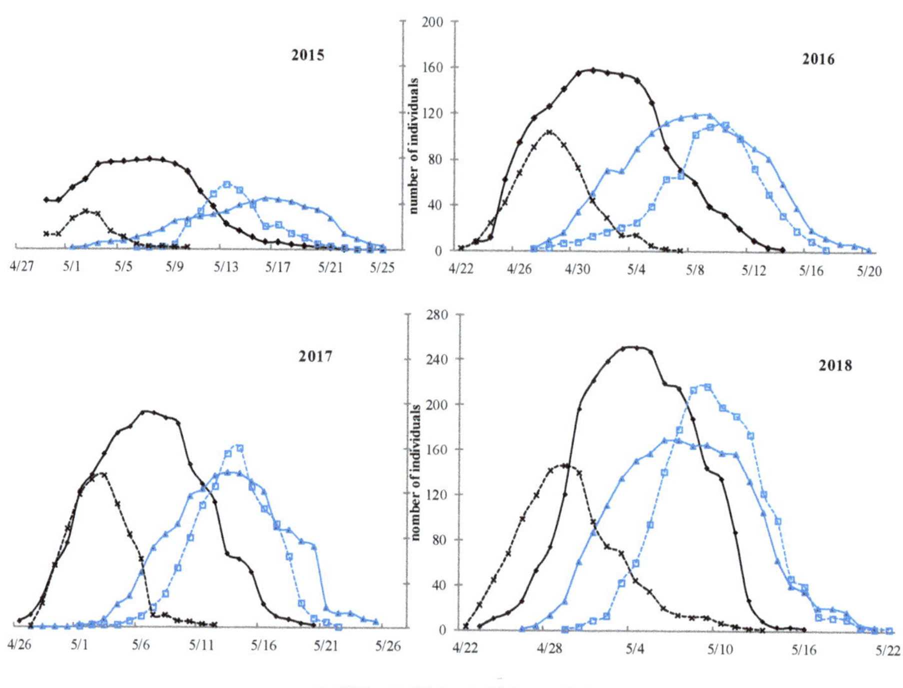


**Figure 4.** Number dynamic patterns of PA and PG flowering individuals in the population of *C. paliurus* during 2015–2018. The back lines describe the first flowering phase, while the blue lines describe the second phase. High overlap was obvious between the two phases and the two sexual functions within each phase.

We found that the flowering expression of population was more complicated than the assembly of PA/PG individuals. Separation of the two sexual functions mainly existed within individuals rather than in the population (Figures 3 and 4). Rather than selfing within individuals, inter-morph and intra-morph pollination were dominant at the population level.

### 3.4. Pollen Dispersal Characteristics of the Population

A sufficient quantity and high viability of pollen guarantee for seed success. To illustrate the effect of pollen on seed bearing, pollen density and viability during flowering in the population were surveyed for three consecutive years.

Generally, pollen density displayed an obviously rising trend with the maturation of the population. In the first phase, the maximum pollen density in 2018 was up to 305 ± 56 grain/cm$^2$, far above that of 2016 (113 ± 19.7 grain/cm$^2$) and 2017, in the second phase, a similar maximum was observed over three years. The pollen density among inter-phases showed significant differences in 2016 and 2017 but was similar in 2018. This indicates that pollen dispersal is affected not only by nutrient conditions, but also by annual climate change.

In addition, the pollen viability kept rising with the growth of the population, showing maximums of 27.6% and 31.1% in 2018, higher than that of 16.4% and 28.6% in 2017 for the two flowering phases, respectively. Overall, the low pollen viability (30% or so) did not show a positive correlation with pollen density; moreover, irregular dynamics were monitored in the population over the three years (Table 3).

**Table 3.** Pollen density and viability during the flowering period in 2017 and 2018.

| Flowering Stage | Observation Date (Month/Date) | Pollen Density (Grain/cm$^2$) 2017 | 2018 | Pollen Viability (%) 2017 | 2018 |
|---|---|---|---|---|---|
| First stage | 4/30 | 136 ± 25 | 305 ± 56 | 20.5 ± 3.9 | 18.3 ± 3.9 |
| | 5/01 | / | 262 ± 31 | 16.4 ± 4.2 | 27.6 ± 6.5 |
| | 5/03 | 178 ± 34 | / | 6.0 ± 3.2 | 11.0 ± 5.0 |
| | 5/04 | / | 47 ± 12 | / | 6.5 ± 2.6 |
| | 5/06 | 105 ± 12 | / | 13.9 ± 2.9 | / |
| Second stage | 5/08 | 151 ± 16 | / | 20.2 ± 6.8 | 26.4 ± 7.1 |
| | 5/10 | 247 ± 33 | 281 ± 29 | 28.6 ± 4.7 | 31.1 ± 5.9 |
| | 5/12 | 216 ± 26 | / | 20.0 ± 4.4 | 28.8 ± 3.4 |
| | 5/14 | 128 ± 23 | 33 ± 7 | 20.3 ± 5.7 | 13.9 ± 4.3 |
| | 5/16 | 23 ± 5 | 9 ± 4 | 5.6 ± 5.6 | 16.6 ± 12.6 |
| | 5/18 | 12 ± 5 | 0 | 0 | 0 |

Note: data was listed as average ± SE; '/' means no data offered because of rainfall; data in red was not from the same day but instead was from the previous/next day to show the changing tendency.

## 4. Discussion

Usually, mature trees (flowering individuals) of *C. paliurus* are up to 10–30 (40) m tall, and both male and female inflorescences cover the outer crown above the canopy of the forest. It is difficult to find natural populations of a sufficient size to monitor flowering phenology; thus, plantation could be a better option for phenological observation [20]. Plants in our plantation included 53 families from eight provenances, representing the majority of the natural germplasm.

### 4.1. Sexual Polymorphism and Bias Associated with Nutrient Conditions and Climatic Factors

In natural populations, several heterodichogamous species (e.g., *A. japonicum*) were reported to include not only monoecious trees (PA/PG), but also unisexual trees (M/F) [23–25]. Similar findings were also observed in the *C. paliurus* population. For the first time, we showed that the F type is overwhelmingly superior in juvenile stands (2014), followed by the PG type. Opposite changing trends—an increment in PG and a reduction in F type-were then monitored over 5 years. Meanwhile, the number of PA trees tended to increase, though this was at a slower rate than the PG trees (Figure 2).

Levy and Dean [26] pointed out that most woody plants can flower only when they reach a certain age or size; flowering expression can be linked to factors such as nutrient supply and adversity. A low nutrient supply may support the formation of female flowers for *C. paliurus*. Therefore, the female superiority (F/PG) could be associated with the nutrient conditions in the juvenile population. This hypothesis is supported by our previous findings that flowering expressions are significantly related to the plant size of *C. paliurus* [10]. Alternatively, adaptive interpretation considers that female-bias favors the production of more seeds for species survival under adverse conditions, like nutrient shortage and changing climatic factors (dry, cold, etc.). However, more extensive studies for longer periods need to be conducted to confirm this.

### 4.2. Morph Ratio and Transition

Generally, an equal morph ratio (PA:PG = 1:1) should occur in a heterodichogamous population, as shown in *Juglans* species, such as *J. regia*, *J. ailanthifolia*, and *J. mandshurica*, as well as in some *Acer* species (e.g., *A. opalus* and *A. japonicum*) (reviewed by Liu [8]). However, derivations from this equilibrium have been found in some species (e.g., *A. pseudoplatanus*) [24] and under certain situations (e.g., small population or human disturbance, like intentional selection) [27]. Similarly, a distinctive bias was found in our study. We speculate that the "juvenile" population is the main reason for this deviation, and the change in the morph ratio (PG/PA) from 5.42:1 in 2014 to 1.39:1 in 2018 predicts a state of equilibrium in the mature population.

Labile sex expression was documented in some heterodichogamous trees across years, like in *Acer pseudoplatanus* [24,28]. It is reported that the sexual phenotype changed mainly from unisexual to the monoecious type, with few or no changes between PA and PG [8,9]. We found that unisexuality is a temporary trait based on the reduction of the ratio of M and F types with the maturation of the stand (Table 2). Similar findings were reported in *J. ailanthifolia* and J. *mandshurica* [9,29]. Researchers believe that the unisexual trait (F and/or M) is an environmental and/or nutrient (young age) response phenotype rather than a separate morph [9,24,25].

Morphs of PA and PG are expected to be stable according to the theory that sex morphs are genetically determined [6,7], which has been demonstrated by some observations in *J. regia* and *P. strobilacea*. [9,29]. However, documents involving transitions between two morphs in *C. ovata*, *C. tomentosa*, and *A. japonica* have been published [24,25,30]. Such a transition in *C. paliurus* was also recorded in several individuals. On these plants, highly overlapping flowering durations of two sexual flowers led to the ambiguous judgment of mating type in 2018 (Figure 3). Thus, further observation needs to be made to confirm such transitions.

### 4.3. Flowering Phenology of the Individual and Population

Heterodichogamous populations showed two flowering phases and bimodality in gender for two distinct sexual morphs. This has been described in *J. regia*, *J. hindii*, *J. ailanthifolia*, and *J. mandshurica*, and *Platycarya strobilacea* [4,6,9,29,31]. Moreover, sufficiently long separation between individuals can avoid selfing but does not prevent intra-morph inbreeding [4]. Therefore, flowering separation and the overlap of two sex inter-phases as well as within-phase and within-individual are the key factors affecting mating patterns. As described by numerous publications, both features are divergent among heterodichogamous species, varying from complete separation to high overlap. In contrast to the pattern of no overlap within individuals and inter-phases in *J. mandshurica* [4] and a short period of overlap in the PG type of *J. ailanthifolia* [29], high overlap occurs extensively, not only inter-phase but also intra-morph and within individuals in *C. paliurus*. High overlap within individuals and intra-morphs seems to favor selfing and inbreeding but is contrary to the mechanism of avoiding assortative mating for heterodichogamous plants [4,32].

Mating fitness also depends on the synchronism of each sexual flower-very early/or late individuals would suffer low reproductive success. Bai et al. [4] observed that female flowering peaks were earlier than male peaks for two periods in *J. mandshurica*, and the PG morphs had a higher level of assortative mating than PA ones. We found that the flourish flowering period of the male preceded that of the female in the first stage, but the female flowering period covered the complete duration in the second stage for four consecutive years (Figure 5), suggesting that a higher seed set could be more successful in PA than in PG. This prediction was further supported by the result that seed plumpness in PA (22% ± 8.2%) was significantly higher than in PG (12% ± 6.8%) ($p < 0.05$).

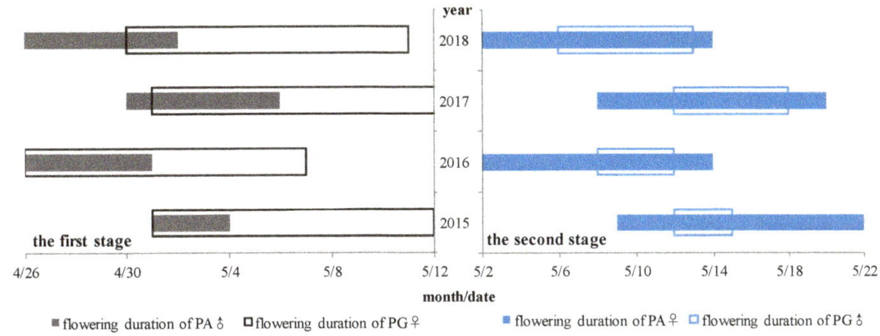


**Figure 5.** Flourish flowering duration (>50% individuals of PA and PG in bloom) of two stages in the population of *C. paliurus*.

### *4.4. Reasons for Low Seed Success in* C. paliurus

Heterodichogamy is recognized as a key factor affecting seed production, and high fruit production has been reported in many species, such as *J. regia* and *C. illinoinesis* [33,34]. Strangely, seeds with low plumpness of 10%–30%, and rarely 50%, were collected from natural populations of *C. paliurus* for nearly ten years. The low seed success (0–30%) in our plantation during 2016–2018 was also in accordance with the natural situation.

Low seed success in a small population can result from the bias of two mating types, like in *A. pseudoplatanus* [28]. Limited by the habitat of *C. paliurus* or by human disturbance, the size of more than two-thirds of the natural populations investigated was less than 25 individuals [20]. This might be one of the reasons for the low seed set in the natural population; however, it cannot be the explanation for our plantation. Furthermore, the overlap of the flowering phenology in *C. paliurus* seems to provide a better chance of pollination, including crossing, inbreeding, and even selfing, and a high seed success should be expected, according to the 100% rate of self-compatibility in *Juglandaceae* species [1]. However, the investigated data did not support these speculations.

### *4.5. Polyploid C. Paliurus Relating to Characters of Flowering and Seed Success*

Different from documented heterodichogamous species in *Juglandaceae*, observations over five years demonstrated a high overlap between the two flowering phases and within individuals, low pollen viability rather than pollen density in the population (Table 3), and low seed plumpness not only in the natural stands and plantation, but also in the control pollination (12%–38%, data unpublished). This infers that unknown reason(s) could be responsible for the previously mentioned phenomenon.

Unexpectedly, data based on a genomic survey found the existence of a tetraploid individual. Subsequently, by screening 1087 individuals collected in our germplasm from 42 provenances from 13 provinces using a cell flow meter, we discovered that tetraploid plants occupied about 95%, and diploid and triploid (uncertain) ones accounted for about 5%. In a word, diploid-tetraploid plants coexist in a natural population. *C. paliurus* is the first species found to have both polyploidy and heterodichogamy characteristics.

Based on these findings, many differences observed in *C. paliurus* from other heterodichogamous species can be elucidated. The dominance of multi-allele rather than two-allele in polyploid *C. paliurus* controlled heterodichogamy might trigger the inter-phase overlap of flowering characteristics as well as within individuals. As we know, low pollen viability from polyploidy [35] could result in low seed success. However, based on the theory that heterodichogamy is simply inherited with a single diallelic Mendelian locus in diploid species [1], it is unclear how heterodichogamy is determined in polyploid *C. paliurus*. Further, it is unknown how *C. paliurus* with its low seed success rate ensures its succession. Thus, further research should be done in *C. paliurus* to disclose these secrets.

## 5. Conclusions

The female-bias (PG and F) and skewed ratio of mating types that occur in *C. paliurus* correspond to the nutrient accumulation in the juvenile population. Heterodichogamy in *C. paliurus* was demonstrated, but with a substantial difference from other recorded species. Low seed success in the population mainly results from low pollen viability. However, the latest finding of a polyploidy majority in the natural population could explain the flowering phenology and seed set characteristics of *C. paliurus*, but also raises more questions that need to be answered.

**Author Contributions:** This work was designed, directed, and coordinated by X.-X.F., who provided technical guidance for all aspects of the project and wrote the manuscript. X.M. was the principal investigator, contributed to the fieldwork and data analysis, performed the literature search, and helped with the writing of the manuscript. P.H. monitored the density and viability of pollens. X.-L.C. and Y.-Q.Q. assisted with the fieldwork and analysis of polyploidy for all individuals.

**Funding:** This research was funded by the National Natural Science Foundation of China (No. 31470637) and Priority Academic Programme Development of Jiangsu Higher Education Institutions, PAPD.

**Acknowledgments:** We also thank Jiaqiu Yuan, JingJing Liu, Qiang Lu, and Biqing Chen for help with fieldwork, and anonymous referees and the editor for valuable comments on the manuscript.

**Conflicts of Interest:** The authors declare no conflicts of interest.

## References

1. Renner, S.S. How common is heterodichogamy? *Trends Ecol. Evol* **2002**, *16*, 595–597. [CrossRef]
2. Barrett, S.C. The evolution of plant sexual diversity. *Nat. Rev. Genet.* **2002**, *3*, 274–284. [CrossRef] [PubMed]
3. Webb, D.J.; Lloyd, D.G. The avoidance of interference between the presentation of pollen and stigmas in angiosperms II. Herkogamy. *N. Z. J. Bot.* **1986**, *24*, 16. [CrossRef]
4. Bai, W.N.; Zeng, Y.F.; Liao, W.J.; Zhang, D.Y. Flowering phenology and wind-pollination efficacy of heterodichogamous *Juglans mandshurica* (Juglandaceae). *Ann. Bot.* **2006**, *98*, 397–402. [CrossRef] [PubMed]
5. Clark, B.C.; Partridge, L.; Robertson, A. *Frequency-Dependent Selection*; Cambridge University Press: New York, NY, USA, 1988.
6. Gleeson, S.K. Heterodichogamy in walnuts: Inheritance and s ratios. *Evolution* **1982**, *36*, 892–902. [CrossRef] [PubMed]
7. Thompson, T.E.; Romberg, L.D. Inheritance of heterodichogamy in pecan. *Heredity* **1985**, *76*, 456–458. [CrossRef]
8. Liu, J.J.; Mao, X.; Li, X.C. A review on flowering mechanism in heterodichogamous plants. *J. Nanjing For. Univ.* **2016**, *40*, 147–154. (In Chinese) [CrossRef]
9. Fukuhara, T.; Tokumaru, S. Inflorescence dimorphism, heterodichogamy and thrips pollination in *Platycarya strobilacea* (Juglandaceae). *Ann. Bot.* **2014**, *113*, 467–476. [CrossRef]
10. Mao, X.; Liu, J.J.; Li, X.C.; Qin, J.; Fu, X.X. Flowering biological characteristics and mating system in immature plantations of heterodichogamous *Cyclocarya paliurus*. *J. Nanjing For. Univ.* **2016**, *40*, 47–55. (In Chinese) [CrossRef]
11. Bai, W.N.; Zeng, Y.F.; Zhang, D.Y. Mating patterns and pollen dispersal in a heterodichogamous tree, *Juglans mandshurica* (Juglandaceae). *New Phytol.* **2007**, *176*, 699–707. [CrossRef] [PubMed]
12. Shu, R.G.; Xu, C.R.; Li, L.N.; Zhi, L.Y. Cyclocariosides II and III: Two secodammarane triterpenoid saponins from *Cyclocarya paliurus*. *Planta Med.* **1995**, *61*, 551–553. [CrossRef] [PubMed]
13. Xie, M.Y.; Wang, Y.X.; Wen, H.L.; Yi, X. Determination of flavonoid compounds and vitamins in the leaves of *Cyclocarya paliurus* (Batal.) Iljinsk. *Food Sci.* **2001**, *22*, 66–68. (In Chinese) [CrossRef]
14. Wu, Z.F.; Meng, F.C.; Jiang, C.H.; Zhao, M.G.; Shang, X.L.; Fang, S.Z.; Ye, W.C.; Zhang, Q.W.; Zhang, J.; Yin, Z.Q. Triterpenoids from *Cyclocarya paliurus* and their inhibitory effect on the secretion of apoliprotein B48 in Caco-2 cells. *Phytochemistry* **2017**, *142*, 76–84. [CrossRef]
15. Yang, H.M.; Yin, Z.Q.; Zhao, M.G.; Jiang, C.H.; Zhang, J.; Pan, K. Pentacyclic triterpenoids from *Cyclocarya paliurus* and their antioxidant activities in FFA-induced HepG2 steatosis cells. *Phytochemistry* **2018**, *151*, 119–127. [CrossRef] [PubMed]

16. Liu, Y.; Wang, T.L.; Fang, S.Z.; Zhou, M.M.; Qin, J. Responses of morphology, gas Exchange, photochemical activity of photosystem II, and antioxidant balance in *Cyclocarya paliurus* to light spectra. *Front. Plant Sci.* **2018**, *23*, 1042. [CrossRef] [PubMed]
17. Liu, Y.; Qian, C.Y.; Ding, S.H.; Shang, X.L.; Yang, W.X.; Fang, S.Z. Effect of light regime and provenance on leaf characteristics, growth and flavonoid accumulation in *Cyclocarya paliurus* (Batal) Iljinskaja coppices. *Bot. Stud.* **2016**, *57*, 28–41. [CrossRef]
18. Liu, Y.; Liu, Q.L.; Wang, T.L.; Fang, S.Z. Leaf nitrogen and phosphorus stoichiometry of *Cyclocarya paliurus* across China. *Forests* **2018**, *9*, 771. [CrossRef]
19. Fang, S.Z.; Wang, J.Y.; Wei, Z.Y.; Zhu, Z.X. Methods to break seed dormancy in *Cyclocarya paliurus* (Batal). Iljinskaja. *Sci. Hortic.* **2006**, *110*, 305–309. [CrossRef]
20. Li, X.C.; Fu, X.X.; Shang, X.L.; Yang, W.X.; Fang, S.Z. Natural population structure and genetic differentiation for heterodicogamous plant: *Cyclocarya paliurus* (Batal.) Iljinskaja (Juglandaceae). *Tree Genet. Genomes* **2017**, *13*, 80. [CrossRef]
21. De Jong, P.C. *Flowering and Sex Expression in Acer L.: A Biosystematic Study*; Mededelingen Land Bouwboge School: Wageningen, The Netherlands, 1976; Volume 76, pp. 1094–1098.
22. Mao, X.; Li, X.C.; Liu, J.J.; Fu, X.X. Scanning electron microscope observation on pollen morphology of six tree species in Juglandaceae. *J. Plant Resour. Environ.* **2016**, *25*, 104–106. (In Chinese) [CrossRef]
23. Dommée, B.; Bompar, J.L.; Denelle, N. Sexual tetra-morphism in *Thymelaea hirsuta* (Thymelaeaceae): Evidence of the pathway from heterodichogamy to dioecy at the infra-specific level. *Am. J. Bot.* **1990**, *77*, 1449–1462. [CrossRef]
24. Asai, T. *Dichogamy in Full Moon Maple (Acer japonicum Thunb.)*; Bulletin of the Hokkaido Forestry Research Institute: Tokyo, Japan, 2000; Volume 37, pp. 27–40.
25. Sato, T. Phenology of sex expression and gender variation in a heterodichogamous maple, *Acer japonicum*. *Ecology* **2002**, *83*, 1226–1238. [CrossRef]
26. Levy, Y.Y.; Dean, C. The transition to flowering. *Plant Cell* **1998**, *10*, 1973–1990. [CrossRef] [PubMed]
27. Pollegioni, P.; Olimpieri, I.; De Simoni, G.; Gras, M.; Malvolti, M. Barriers to interspecific hybridization between *Juglans nigra* L. and *J. regia* L. species. *Tree Genet. Genomes* **2013**, *9*, 291–305. [CrossRef]
28. Tal, O. *Acer pseudoplatanus* (Sapindaceae): Heterodichogamy and thrips pollination. *Plant Syst. Evol.* **2009**, *278*, 211–221. [CrossRef]
29. Kimura, M.; Seiwa, K.; Suyama, Y.; Ueno, N. Flowering system of heterodichogamous *Juglans ailanthifolia*. *Plant Species Biol.* **2003**, *18*, 75–84. [CrossRef]
30. McCarthy, B.C.; Quinn, J.A. Reproductive ecology of system of two sympatric tree species. *Am. J. Bot.* **1990**, *77*, 261–273. [CrossRef]
31. Knuth, P. *Handbook of Flower Pollination*; Clarendon Press: Oxford, UK, 1906.
32. Darwin, C. *The Different Forms of Flowers on Plants of the Same Species*; John Murray: London, UK, 1877.
33. Polito, V.S.; Pinney, K. The relationship between phenology of pistillate flower organogenesis and mode of heterodichogamy in *Juglans regia* L. (Juglandaceae). *Sex. Plant Reprod.* **1997**, *10*, 36–39. [CrossRef]
34. Zhang, R.; Peng, F.R.; Li, Y. Pecan production in China. *Sci. Hortic.* **2015**, *197*, 719–727. [CrossRef]
35. Solis Neffa, V.G.; Fernández, A. Chromosome studies in *Turnera* (Turneraceae). *Genet. Mol. Biol.* **2000**, *23*, 925–930. [CrossRef]

forests

Article

# *Pinus massoniana* Introgression Hybrids Display Differential Expression of Reproductive Genes

**Jiaxing Mo [1,2], Jin Xu [1,2,*], Yuting Cao [1,2], Liwei Yang [1,2], Tongming Yin [1,2], Hui Hua [1,2], Hui Zhao [1,2], Zhenhao Guo [1,2], Junjie Yang [1,2] and Jisen Shi [1,2]**

1 Co-Innovation Center for Sustainable Forestry in Southern China, Nanjing Forestry University, Nanjing 210037, China; jx_mo@njfu.edu.cn (J.M.); 18205190992@163.com (Y.C.); yanglw@njfu.edu.cn (L.Y.); tmyin@njfu.edu.cn (T.Y.); 13218098608@163.com (H.H.); luoyusheng6@163.com (H.Z.); 18915019168@163.com (Z.G.); 18260077751@163.com (J.Y.); jshi@njfu.edu.cn (J.S.)

2 College of Forestry, Nanjing Forestry University, Nanjing 210037, China

* Correspondence: xjinhsh@njfu.edu.cn; Tel.: +86-25-8542-7319

Received: 23 January 2019; Accepted: 28 February 2019; Published: 5 March 2019

**Abstract:** *Pinus massoniana* and *P. hwangshanensis* are two conifer species located in southern China, which are of both economic and ornamental value. Around the middle and lower reaches of the Yangtze River, *P. massoniana* occurs mainly at altitudes below 700 m, while *P. hwangshanensis* can be found above 900 m. At altitudes where the distribution of both pines overlaps, a natural introgression hybrid exists, which we will further refer to as the Z pine. This pine has a morphological character that shares attributes of both *P. massoniana* and *P. hwangshanensis*. However, compared to the other two pines, its reproductive structure, the pinecone, has an ultra-low ripening rate with seeds that germinate poorly. In this study, we aimed to find the reason for the impaired cone maturation by comparing transcriptome libraries of *P. massoniana* and Z pine cones at seven successive growth stages. After sequencing and assembly, we obtained unigenes and then annotated them against NCBI's non-redundant nucleotide and protein sequences, Swiss-Prot, Clusters of Orthologous Groups, Gene Ontology and KEGG Orthology databases. Gene expression levels were estimated and differentially expressed genes (DEGs) of the two pines were mined and analyzed. We found that several of them indeed relate to reproductive process. At every growth stage, these genes are expressed at a higher level in *P. massoniana* than in the Z pine. These data provide insight into understanding which molecular mechanisms are altered between *P. massoniana* and the Z pine that might cause changes in the reproductive process.

**Keywords:** *Pinus massoniana*; introgression hybrid; RNA sequencing; DEGs; reproduction

---

## 1. Introduction

Gymnosperms have their own unique way of reproduction. Microspores grow into pollen carrying sperm cells, while megaspores develop into the megagametophyte. The archegonium forms during development of the megagametophyte. Then, pollen enter ovule through micropyle, they move towards the egg by way of extending a pollen tube, after which fertilization occurs. Finally, the embryo is formed and develops into a gymnosperm seed.

Conifers possess a series of properties that makes exploring their molecular biology through a genomics approach challenging, such as a long life cycle, a reproductive process lasting months or even years, a gigantic genome size and so forth [1,2]. We aimed to explore the molecular mechanism of conifer reproduction by generating transcriptome data through RNA-seq, of successive stages of the developing pinecone. RNA-seq technology has advanced to the stage where it is highly efficient, sensitive and accurate.

By studying the expression dynamics of differentially expressed genes (DEGs) in pinecones, we sought to gain insight into the relevant genes that control the reproductive process of conifers. Previous empirical studies have suggested several genes linked to reproduction, such as *DAL* [3], *MADS-box* [4,5], *MYB* [6] and *MSI* [7] and so forth. However, studies of the determinants of development and regulation of reproduction have concentrated on model angiosperm species so far, while gymnosperms remain largely understudied.

We centered our studies around two types of conifers, naturally occurring in China. First, *Pinus massoniana*, also known as the masson pine, is an economically important conifer species. It is mainly distributed in various southern Chinese provinces, at altitudes below 700 m around the middle and lower reaches of the Yangtze River. It offers wood and pulp for manufacturing furniture and paper and also supplies natural resin, which can be further manufactured into 'resin,' a crucial product used in the maintenance of instrument strings and as an ingredient in medicine. Furthermore, *P. massoniana* fulfills a significant ecological role by replacing or compensating natural forest destruction due to its fast growth and abundant biomass.

The second, *Pinus hwangshanensis*, primarily grows in southeastern China. It grows most abundantly at an altitude above 900 m around the middle and lower reaches of the Yangtze River. Growing at a higher altitude limits its speed of growth, as well as accumulation of biomass and the resin compared to *P. massoniana*. *P. hwangshanensis* is often viewed as luxurious and graceful, making it an ideal ornamental tree.

Mountain Lushan (Figure 1a) is located within the distribution area of both conifer species. Its peak has an altitude of 1474 m and it supports both *P. massoniana* and *P. hwangshanensis* vegetation at the afore mentioned altitudes. Where *P. massoniana* and *P. hwangshanensis* distribution overlaps, a natural introgression hybrid of both species occurs, sharing phenotypic characters of both its parents (Figure 1b) [8,9]. Due to the hybrid not being named yet, we refer to it as the 'Z pine' in this article. The Z pine has an extremely low germination and ripening rate compared to both *P. massoniana* and *P. hwangshanensis* [10]. These characters could indicate that the Z pine displays genetic incompatibilities during fertilization and/or even embryonic development. What causes this phenomenon is still unknown.

In this study, we collected seven successive development stages of open-pollinated cones *P. massoniana* and the Z pine, respectively. Then we characterized the transcriptome of these cones using Illumina high-throughput sequencing technology and forty two cDNA libraries were constructed. A series of experiments was performed to mine candidate genes, focusing on differential expression patterns between these two species. Moreover, differentially expressed genes related to fertilization and embryonic development were determined and analyzed in both taxa. This study could help explaining the defect of the Z pine of its unusual low ripening rate and germination rate, comparing to *P. massoniana* and may provide an approach to understanding difference between species and its introgressive hybrid at the transcriptome level.

(**a**)

(**b**)

**Figure 1.** (**a**) Location of Mountain Lushan. Mt. Lushan lies in Jiujiang City, Jiangxi Province, China. The Great Han Yang Peak, the highest point of the mountain, is 1474 m high. The local climate is humid, subtropical; (**b**) Schematic diagram of main distribution area of *P. hwangshanensis*, the Z pine and *P. massoniana* on Mt. Lushan.

## 2. Materials and Methods

### *2.1. Sample Collection*

Differently staged, openly pollinated cones of *P. massoniana* and the Z pine were collected on Mt. Lushan, Jiujiang, China (Table 1, Figure 2). Due to complex environment of forest land and wind-pollinated way of pine, sampled individuals (especially the Z pine) may possess different level of introgressive background. In order to minimize the different between them, we assigned five maternal trees of each taxa (*P. massoniana* or the Z pine) in its sample plot. Experience tells us that *P. massoniana* carries out its pollination at around 10 April while the Z pine at around 20 April at sample plots. Aware of this condition, we conducted our first sample collection on 27 April (Table 1), when both *P. massoniana* and the Z pine are already pollinated. The cones were packed with aluminum-foil shortly after collection and then immediately submerged in liquid nitrogen, after which they were stored in a −80 °C freezer until RNA extraction.

**Table 1.** Information on the geographical sites of sample collection of *P. massoniana* and the Z pine.

| Species | Sample Code | | | | | | | Location | | |
|---|---|---|---|---|---|---|---|---|---|---|
| | Open Pollinated Cones of the 1st Year | | | | Open Pollinated Cones of the 2nd Year | | | Longitude (E) | Latitude (N) | Altitude (m) |
| | APR. 27 | MAY 25 | JUL. 27 | SEPT. 25 | APR. 27 | JUL. 27 | SEPT. 25 | | | |
| *P. massoniana* | MA | MB | MC | MD | ME | MF | MG | 116.04 | 29.50 | 78 |
| the Z pine | ZA | ZB | ZC | ZD | ZE | ZF | ZG | 115.98 | 29.54 | 730 |

**Figure 2.** Successively staged pine cones of *P. massoniana* (**a**) and the Z pine (**c**). Details of mature cones of MG (**b**) and ZG (**d**) stages are shown. The scale shields of *P. massoniana* are flat or slightly bulged, the transverse ridge is not very obvious, and the scale umbilical has no thorn. The scale shield of the Z pine is bulged, the transverse ridge is obvious, and the scale umbilical is thorny. Codenames for collected cones are explained in Table 1. (Scale = 10 mm.).

## 2.2. RNA Extraction and Sequencing

We randomly collected three to six cones for RNA isolation of each sample code to make sure that these cones and their RNA could be representative. Cones were taken from the −80°C freezer and briefly re-frozen in liquid nitrogen to further weaken tissue, cut into pieces, after which sections containing ovules (or seeds) were collected and crushed. RNA was extracted from each sample using the Bioteke Plant RNA Extraction Kit (Beijing, China). Three replications of RNA were extracted for each sample code. Purity and quality of the RNA samples was checked respectively by measuring 260 nm/280 nm UV absorption values with a Nanodrop 2000 (Thermo Fisher Scientific, Waltham, MA,

USA) and examining the RIN (RNA Integrity Number) with an Agilent 2100 Bioanalyzer (Agilent Technologies, Santa Clara, CA, USA).

After RNA extraction, magnetic oligo (dT) beads were used to purify mRNA, which was then collected using RNeasy RNA reagent. The mRNA was then cut into small fragments using the RNA Fragment Reagent (Illumina, San Diego, CA, USA) and subsequently cleaned using an RNeasy RNA Cleaning Kit (Qiagen, Germany). First-strand cDNA was then synthesized using MMLV reverse transcriptase (Takara, Japan), while second-strand cDNA synthesis was performed using DNA Polymerase I and RNase H. cDNA was finally sequenced on an Illumina Hiseq X Ten (Illumina, USA). The sequencing raw data were submitted to the NCBI Short Reads Archive (SRA) database under the BioProject accession number PRJNA482692.

### 2.3. Data Processing and Assembly

The raw RNA-seq data was screened by removing adaptor and low quality sequences using Trimmomatic [11]. Qualified reads were assembled into non-redundant transcripts by Trinity [12,13], using the following parameters: –min_contig_length 200 –min_kmer_cov 2 –min_glue 3 –seqType fq, other parameters use default settings. Contigs were assembled by Trinity, gathered and processed by TGICL [14], the parameters of TGICL are as follows: -l 40 -c 10 -v 25 -O -repeat_stringency 0.95 -minmatch 35 -minscore 35. After which unigenes were collected.

### 2.4. Functional Annotation

Unigenes were generally annotated by aligning against SwissProt (Release-2016_07) [15], NCBI non-redundant protein sequences (Nr, Release-20160314), NCBI non-redundant nucleotide sequences (Nt, Release-20140514) [16], Kyoto Encyclopedia of Genes and Genomes (KEGG, Release 59.3) [17,18], Cluster of Orthologous Groups of proteins (COG, Release-20090331) [19] and Gene Ontology (GO) [20] databases, selecting the most likely annotations. Blast2GO (v2.5.0) [21] was conducted as GO annotation tool against GO database (Release-201604) under default settings. Furthermore, analysis of GO function and KEGG pathways of differentially expressed genes was performed.

### 2.5. Differentially Expressed Genes (DEGs) and Gene Expression Pattern Analysis

Calculation of unigene read counts was performed using RNA-Seq by Expectation-Maximization (RSEM) software [22]. RSEM results were transformed into FPKM [23] values (expected number of Fragments Per Kilobase of transcript sequence per Millions base pairs sequenced), commonly used for measuring gene expression levels. DESeq was used to determine differentially expressed genes of different transcript libraries [24]. Differentially expressed genes were assigned based on a threshold value of FDR (false discovery rate) $\leq 0.001$ and $|\log_2\text{Ratio}| \geq 2$. Gene expression patterns of *P. massoniana* and the Z pine were assembled by Short Time-series Expression Miner (STEM) [25].

### 2.6. Validation by Quantitative Real-Time PCR (qRT-PCR)

Quantitative real-time PCR was applied for validating differentially expressed genes detected by our RNA-seq analysis. Primers were designed using the NCBI Primer-Blast Tool [26] and synthesized by Generay Biotech Co., Ltd. (Shanghai, China). cDNA samples for qRT-PCR were synthesized using the Vazyme HiScript II Q RT SuperMix for qPCR (Nanjing, China). qRT-PCR was carried out using an Applied Biosystems 7500 PCR cycler (Thermo Fisher Scientific Corporation, CA, USA) and Vazyme ChamQ SYBR qPCR Master Mix (Nanjing, China) as reaction reagent kit. Each sample was run in triplicate, with samples having a final volume of 20 μL: containing 10 μL of ChamQ SYBR qPCR Master Mix (2×), 0.4 μL of each primer, 2 μL of cDNA and 7.2 μL of $ddH_2O$. The reaction program was according to standard product instructions. An *Actin* gene that was discovered from RNA-seq data (Unigene69821_All) was utilized as reference gene. The qRT-PCR data was analyzed with the $2^{-\Delta\Delta Ct}$ method [27].

## 3. Results

### 3.1. Illumina Sequencing and Assembly

A total of 160 Gb raw data was obtained. It has an average of 53 million reads per library. We evaluated the quality of the original and clean sequencing data of all samples. The Q30 value has a range of 88.49~92.33 % for the original sequencing data (Table S1) and a range of 91.75~94.31% for the clean data (Table S2), indicating that this data set is ready for further assembly. All transcripts that we obtained from the staged pine cones of our two conifer species were assembled into 93,291 unigenes (see Materials and methods for details, Table 2). About 39.88% of them exceeded 2 kb in length, while 37.02% unigenes have a length from 1 kb to 2 kp and 23.1% were 100 bp to 1 kb (Figure S1). The average length of unigenes is 1987 nt, while N50 is 2494 nt.

**Table 2.** Number and length of unigenes.

| Sample | Total Number | Total Length (nt) | Mean Length (nt) | N50 |
|---|---|---|---|---|
| MA | 77,051 | 89,621,989 | 1163 | 1875 |
| MB | 75,111 | 86,567,730 | 1153 | 1838 |
| MC | 85,801 | 96,687,090 | 1127 | 1833 |
| MD | 78,553 | 88,590,120 | 1128 | 1834 |
| ME | 78,223 | 87,067,948 | 1113 | 1813 |
| MF | 76,159 | 84,583,542 | 1111 | 1800 |
| MG | 100,035 | 87,128,148 | 871 | 1400 |
| ZA | 68,015 | 80,547,061 | 1184 | 1893 |
| ZB | 76,206 | 88,922,013 | 1167 | 1882 |
| ZC | 81,570 | 90,787,807 | 1113 | 1824 |
| ZD | 74,227 | 84,044,527 | 1132 | 1823 |
| ZE | 66,146 | 76,428,873 | 1155 | 1833 |
| ZF | 73,004 | 79,225,689 | 1085 | 1789 |
| ZG | 85,392 | 86,084,303 | 1008 | 1708 |
| All | 93,291 | 185,359,493 | 1987 [1] | 2494 [2] |

[1,2] When all samples are assembled, they would express much higher abundance than single samples, therefore the data of 'All' is usually higher than others.

### 3.2. Functional Annotation of P. massoniana and Z Pine Unigenes

The assembled unigenes were annotated against the SwissProt, NCBI non-redundant protein and nucleotide sequences (Nr and Nt), Kyoto Encyclopedia of Genes and Genomes (KEGG), Cluster of Orthologous Groups of proteins (COG) and Gene Ontology (GO) databases. A total of 86,006 unigenes were annotated in *P. massoniana* and the Z pine, of which 25,150 unigenes were annotated against all six databases (Figure S2).

We were able to annotate most genes using the Nr database, using sequences from *Picea sitchensis* for the bulk of the annotation (30,943), after which *Amborella trichopoda* (7055) and Indian lotus (4504) provided most annotations, suggesting that these sequenced species are most closely related to *P. massoniana* and the Z pine (Figure S3).

We then used data from the COG, GO and KEGG databases for unigene functional prediction. Using the COG database, 30,017 unigenes could be annotated and were classified into 24 functional categories. The 'general function prediction only' was the most abundant, followed by 'transcription' and 'replication, recombination and repair,' 'function unknown' and 'signal transduction mechanisms' (Figure 3).

A total of 229,950 redundant unigenes (with 38,619 nonredundant unigenes) were annotated into 56 sub-categories under three primary GO categories: biological process, cellular component and molecular function (Figure 4). The top three sub-categories were metabolic process (23,854 unigenes), cellular process (22,439 unigenes) and catalytic activity (20,324 unigenes).

The KEGG classification placed 41,931 unigenes into 5 pathway functional categories (Figure 5): organismal systems (5474 unigenes), metabolism (37,045 unigenes), genetic information processing (9684 unigenes), environmental information processing (2871 unigenes) and cellular processes (1487 unigenes). The top three sub-categories out of a total of 18 were 'global map,' 'environmental adaptation' and 'carbohydrate metabolism,' which contains 14,748, 5350 and 4965 unigenes, respectively.

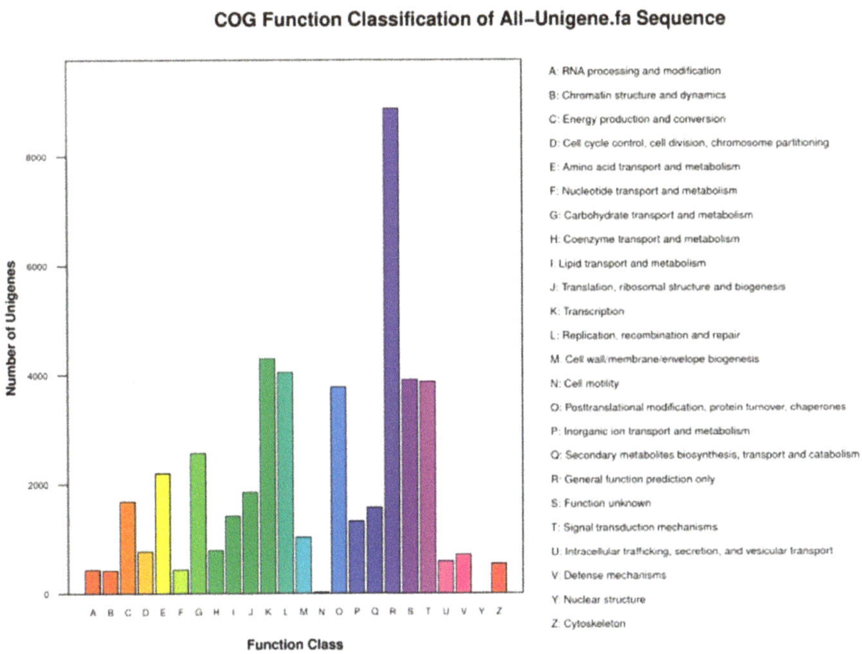


**Figure 3.** COG functional classification of *P. massoniana* and the Z pine unigenes.

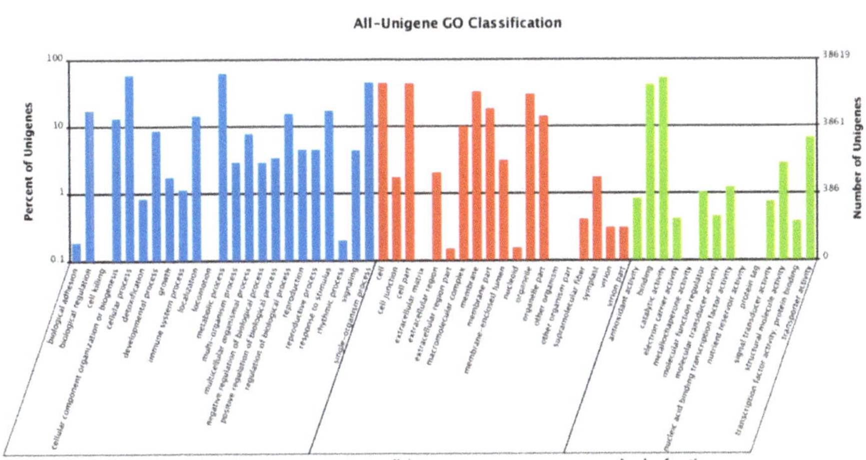


**Figure 4.** Gene ontology (GO) functional classification of *P. massoniana* and the Z pine unigenes.

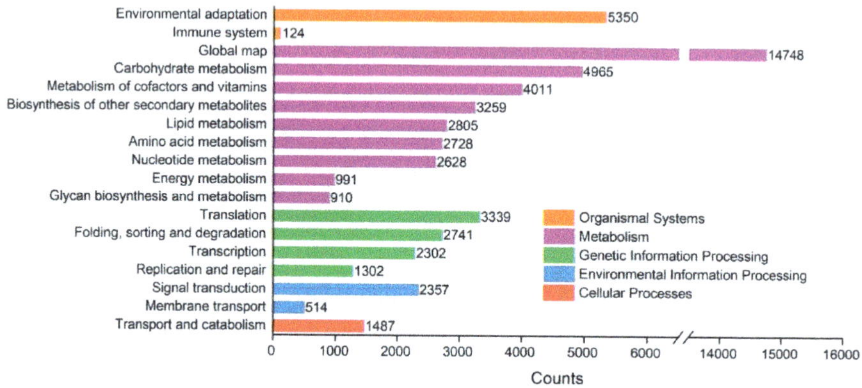


**Figure 5.** Histogram of the KEGG Pathway classification of *P. massoniana* and the Z pine unigenes.

### 3.3. *Analysis of Expected Number of Fragments per Kilobase of Transcript Sequence per Million Base Pairs Sequenced (FPKM)*

FPKM values were calculated using RSEM software. The general density distribution of expression quantity (Figure 6) was analyzed and showed that the average total of expressed mRNAs across all unigenes of *P. massoniana* and the Z pine varies between species and stages.

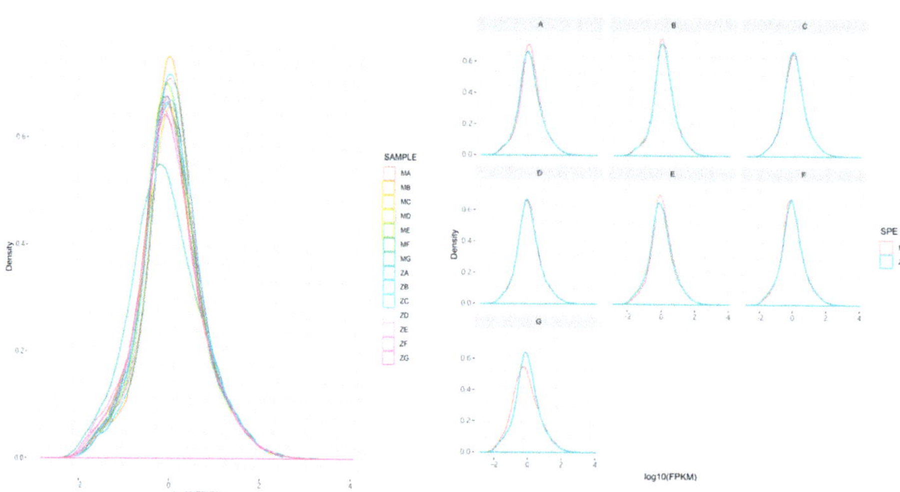


(a) (b)

**Figure 6.** (**a**) General density distribution of *P. massoniana* and Z pine unigenes by FPKM analysis; (**b**) Pairwise comparison of general density distribution of same stage *P. massoniana* and Z pine unigenes by FPKM analysis, M: *P. massoniana* (red curve), Z: Z pine (green curve), A~G: the seven successive stages of pinecone development. X-axis: logarithm to base 10 of FPKM, y-axis: density of distribution.

### 3.4. *GO Classification and KEGG Enrichment Assessment of Differentially Expressed Genes (DEGs) at Successive Pinecone Stages*

We determined and compared the number of up- and down-regulated genes between the two pine species at the seven different developmental stages of the pine cones collected (Figure 7 and Figure S4).

In comparison group A, C and G, more down-regulated genes were found than up-regulated ones, while in group B, D, E and F, there was more up-regulated genes than down-regulated ones (Figure 7).

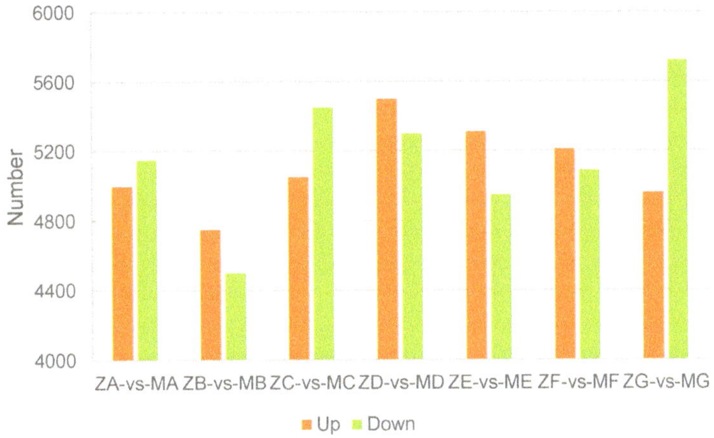


**Figure 7.** Statistics of differentially expressed genes of same stages of two pines (FDR $\leq$ 0.001 and $|\log_2\text{Ratio}| \geq 2$). Red indicates up-regulated genes while green indicates down-regulated genes. The first two bars show that compare to ZA, MA has more down-regulated genes than up-regulated ones, the rest can be read in this manner.

To gain more insight in differential regulation of genes related to the pinecone reproductive process, we performed GO classification of all DEGs at every developmental stage (Figure S5). The number of differentially expressed unigenes classified in the 'reproduction' category under 'biological process' at each stage is listed (Figure S6). Three groups displayed more up- than down-regulated genes: ZA-MA, ZD-MD and ZF-MF. While the other four groups showed more down- than up-regulated genes: ZB-MB, ZC-MC, ZE-ME and ZG-MG. The ZG-MG group contains 101 down-regulated genes, almost double the number of up-regulated genes (51 genes).

We then determined whether specific cellular processes are differentially affected at each pinecone stage by performing a KEGG pathway enrichment analysis (Figure S7).

### 3.5. Quantitative Real-Time PCR Validation

We randomly selected five unigenes for validation of the accuracy of our RNA-seq data set using qRT-PCR. The following unigenes were randomly chosen: Unigene12135_All, Unigene31229_All, Unigene5965_All, Unigene69986_All and Unigene71003_All. We tested cDNA derived from four samples: MA, ME, ZA, ZE, which were collected in different years from the two different species. Details of sequences and primers were list on Tables S3 and S4, respectively. Validation results shows a reliable correlation between RNA-seq and qRT-PCR (Figure S8).

### 3.6. Temporal Gene Expression Profiles of *P. massoniana* and the Z Pine

We analyzed gene expression dynamics of all unigenes across pinecone developmental stages for both species and clustered these into 49 unique expression profiles. The eighteen most frequently occurring profiles for each species are shown in Figure 8a. Within *P. massoniana*, the top five expression profile types are 16, 10, 31, 34 and 40, with respectively 8416, 7967, 5532, 3291 and 3092 genes showing these expression dynamics. In the Z pine the most frequently occurring profiles are profile types 40, 10, 27, 29 and 44, represented by 5610, 4336, 3441, 3371 and 3361 genes. *P. massoniana* and the Z pine shared fourteen profile types among their respective top 18 profiles (Figure 8a). In addition, we carried out a profile comparison between *P. massoniana* and the Z pine. Every single profile of *P. massoniana* is listed and similar ones of Z pine are placed on the right of it in Figure 8b.

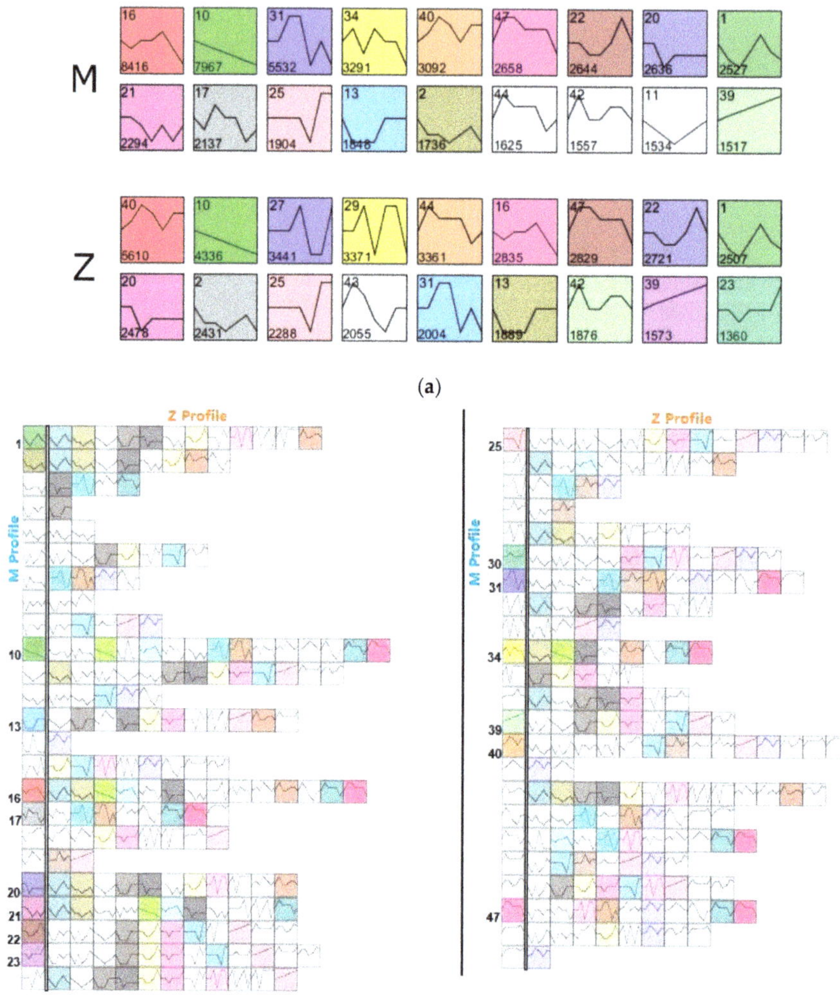


(a)

(b)

**Figure 8.** (**a**) Eighteen expression patterns of *P. massoniana* (M) and the Z pine (Z) based on the highest number of genes having these particular patterns. The number on the upper left of each square indicates profile type, the number on the lower left indicates the number of genes within each profile, the fold line shows the expression pattern, colored squares are significant profiles while white ones are insignificant profiles; (**b**) All *P. massoniana* (M) patterns compare to its similar ones in the Z pine (Z). The most left column of each part is profiles of *P. massoniana* while their similar Z pine counterparts are list on the right, significant profiles of M are marked profile numbers and colors.

### 3.7. Reproductive Genes Are Differentially Expressed between P. massoniana and the Z Pine

Next, we aimed to see whether genes related to reproduction might be differentially expressed between *P. massoniana* and the Z pine, potentially explaining the reproductive problems that the Z pine experiences. We looked for several DEGs involved in processes such as pollen development, pollen exine formation, pollen tube growth and development of the female gametophyte, endosperm, embryo and/or embryo sac according to recent reports (Table S5). Some of these genes show consistently higher expression in *P. massoniana* than in the Z pine, including: *ACA7* ($Ca^{2+}$-ATPase), *MPK4* (mitogen-activated protein kinase), *QRT2* (polygalacturonase), *TKPR1* (tetraketide alpha-pyrone reductase 1), *PI5K* (phosphatidylinositol 4-phosphate 5-kinase), *PMEs* (pectin methylesterase), *SEC6* (exocyst complex component 6), *SEC15* (exocyst complex component 15), *SWK2* (slow walker 2), *PPR* (pentatricopeptide repeat-containing protein), *EMB* (embryo), *LEA* (late embryogenesis abundant protein), *SERK* (somatic embryogenesis receptor kinase), *BLH* (*BEL1-like* homeodomain protein) (Figure 9). Some genes relate to reproduction show a similar expression level in both pines, such as: *SHT* (spermidine hydroxycinnamoyl transferase), *SEC5* (exocyst complex component 5), *SEC8* (exocyst complex component 8), *EYE* (embryo yellow), *EDD1* (embryo defective development 1) and *EDA* (embryo sac development arrest).

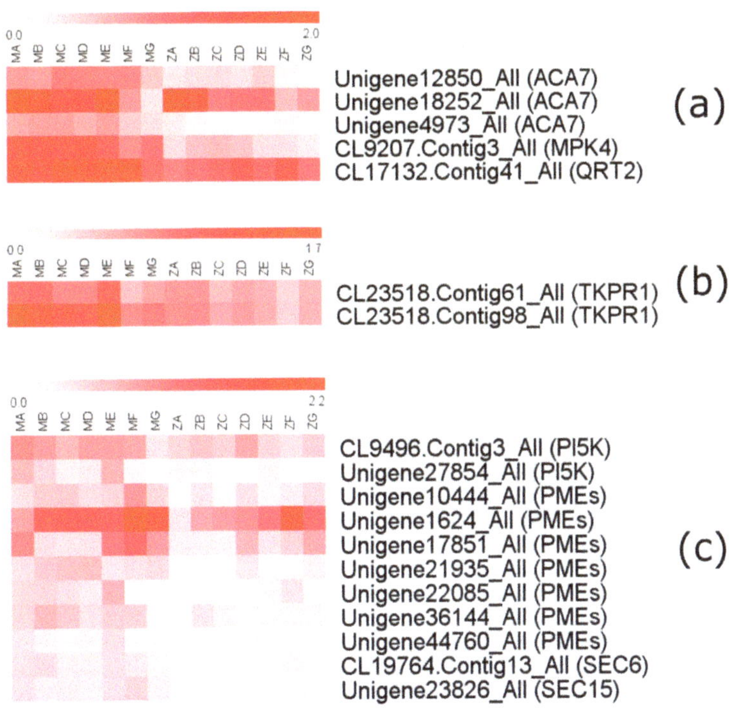


**Figure 9.** *Cont.*

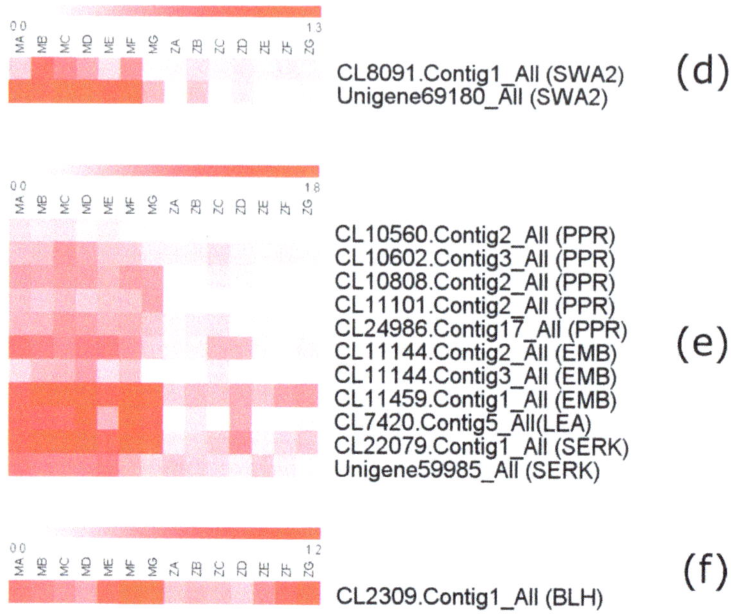


**Figure 9.** Expression levels of DEGs related to the reproductive process at successive pinecone developmental stages of *P. massoniana* and the Z pine. Expression level values in this figure have been transformed to a $\log_{10}$(FPKM+1) value. (**a**) pollen development; (**b**) pollen exine formation; (**c**) pollen tube growth; (**d**) female gametophyte; (**e**) embryo development; (**f**) embryo sac.

*ACA7* belongs to the auto-regulated $Ca^{2+}$-ATPase family, which is exclusively detected in developing flowers of *Arabidopsis* and participates in the regulation of $Ca^{2+}$ homeostasis [28]. *MPK4* plays an important role in plant growth, development and male fertility [29]. *QRT2* is necessary for pollen grain separation and is also involved in pollen development [30]. *TKPR1* takes part in a biosynthetic pathway leading to hydroxylated α-pyrone compounds [31]. *SHT* encodes an acyltransferase that conjugates spermidine to hydroxycinnamic acids, impacting the composition of the *Arabidopsis* pollen wall [32,33]. *NPG1* in *Arabidopsis* is specifically required for pollen germination [34] and not for pollen development [35]. A type B *PI5K* mediates *Arabidopsis* and *Nicotiana* pollen tube growth by regulating apical pectin secretion [36]. *PMEs* and its pro-region adjust cell wall dynamics of growing pollen tubes in *Nicotiana tabacum* [37]. The exocyst contributes to the morphogenesis of polarized cells in many eukaryotes, for example, *SEC8* facilitates the initiation and maintenance of polarized growth of pollen tubes [38]. *SWK2* has an essential role in the coordinated mitotic progression of the female gametophyte in *Arabidopsis* [39]. Absence of *CRINKLY4* could cause an inhibition of aleurone, which is in charge of differentiation normal progression over the endosperm surface development [40]. *PPR* is required for embryo and seed viability in *Arabidopsis*, its absence leading to embryo abortion [41,42]. *EYE* controls golgi-localized proteins, that have an important role in cell and organ expansion [43]. *EDD1* encodes plastid and mitochondria, functional absence mutation of *EDD1* causes embryo lethality [44]. *LEA* and *SERK* play key roles during embryogenesis and *SERK* is essential for embryogenic competence [45,46]. Misexpression of *BLH1* leads to a cell-fate switch of synergid to egg cell in the *Arabidopsis* eostre mutant embryo sac [47].

## 4. Discussion

### *4.1. Gymnosperm Gene Annotation Using Transcriptome Data*

Transcriptome analysis based on RNA sequencing is an effective way to explore the huge genomes of plants like gymnosperms. Several RNA sequencing studies related to gymnosperms have previously been reported [48–51]. Yet until now few studies have focused on the development of *Pinus* reproductive organs. In *Pinus tabuliformis*, unusual bisexual cones were found; here, the gene expression pattern of *MADS-box* transcription factors, *FT/TFL1-like* and *LFY/NDLY* genes was compared between unisexual and bisexual cones [52]. In *Pinus bungeana*, 39.62 Gb of RNA sequencing data was analyzed from two kinds of sexual cones, obtaining 85,305 unigenes, 53,944 (63.23%) of which were annotated in public databases [53].

In this study, we collected a total of 160 Gb of RNA sequence data from *P. massoniana* and its introgression hybrid at seven different stages of pinecone development. N50 is a key parameter in genome or transcriptome assembly. It is defined as the sequence length of the shortest contig at 50% of the entire genome or transcriptome length. In principle, the higher the N50 value, the better the sequencing quality. We obtained an N50 of 2494 bp in all-unigene, compared to previously obtained values of (N50 = 551 bp) for *Picea abies* [48] and (N50 = 1942 bp) for *P. bungeana* [53], which means the quality of our sequencing data improves on previously available data.

A total of 30,943 genes (47.05%, rank 1) were annotated to *Picea sitchensis* through the Nr database, with further annotations being 1174 (1.79%, rank 7) to *Pinus tabuliformis*, 1074 (1.63%, rank 8) to *Pinus taeda*, 697 (1.06%, rank 12) to *Pinus monticola*, 401 (0.61%, rank 18) to *Pinus radiata* and 376 (0.57%, rank 19) to *Picea abies*; all these species are conifers and belong to the Pinaceae family. Out of these, two (*Pinus taeda* and *Picea abies*) had their genomes sequenced [1,54]. The genome sizes of *Pinus taeda* and *Picea abies* are 21.6 Gb and 19.7 Gb, respectively. Pines have an estimated genome size ranging from 18 Gb to 40 Gb [55–57]. This indicated that a lot of novel genes in *P. massoniana* and the Z pine still completely unknown and expect to discover more in the future.

### *4.2. Impact of Introgression in Expression Levels*

Introgressive hybridization implies repeated backcrossing of hybrids with parental species [58]. Hybridization between pines exists frequently in nature [59,60]. As the pollen of pines is mainly moved by wind, it could spread to a vast area. In that case, the element consisting of individuals could be with various proportions of parental genomes. Therefore, those differences between individuals could lead to diversity on gene expression, particularly in genes relate to reproduction. In sample collection, we conducted mixture of cones for each sample code and also made a mixture apply to RNA isolation to reduce the possible expression bias through analysis process.

### *4.3. Differential Expression of Reproductive Genes Could Relate to Delayed Maturation of Z Pinecones*

Within the *Pinus* genus, some female cones take 1.5 to 3 years to mature after pollination, while for *P. massoniana* and the Z pine specifically, it takes around 1.5 years to do that. Around the middle and lower reaches of the Yangtze River, these two pines are often pollinated in April and mature cones emerge in November of the next year, a long time compared to most angiosperms. The structure of a pinecone is rather complex compared to an angiosperm flower. Therefore, more genes relate to reproduction may exist in cone than in flower and more pathways of these genes may occur in this process also.

Genes directly related to reproduction in gymnosperms have only rarely been reported. One of them is *MADS* genes, which are well studied relatively, for example in *Gnetum* spp. [61,62], *Ginkgo biloba* [63], *Picea abies* [64] and *Cryptomeria japonica* [4]. *LEAFY* is also a crucial kind of gene that involve in reproductive process in *Welwitschia mirabilis* [65] and *Pinus caribaea var. Caribaea* [66], as well as *NEEDLY* in *Pinus radiata* [67]. We collected several such genes from model plants (e.g., *Arabidopsis thaliana*) and analyzed their expression level in the two pines. We found that *ACA7*, *MPK4*, *QRT2*,

*TKPR1*, *PI5K*, *PMEs*, *SEC6*, *SEC15*, *SWK2*, *PPR*, *EMB*, *LEA*, *SERK*, *BLH1* showed a higher expression level in *P. massoniana* than in the Z pine. This result indicates that the Z pine may have a lowered expression level of genes related to pollen development, pollen exine formation, pollen tube growth and female gametophyte, embryo and/or embryo sac development, compared to *P. massoniana*. This outcome provides further understanding towards a possible molecular mechanism responsible for the altered reproduction process of the Z pine in comparison to *P. massoniana*.

## 5. Conclusions

*P. massoniana* and *P. hwangshanensis* mainly grow in southern China and produce an introgression hybrid, which we here temporarily named 'the Z pine,' on Mt. Lushan, where both species can be found. This Z pine has morphological characters derived from both parent species, yet has an ultra-low germination and ripening rate. In order to understand the molecular mechanism that might be causing this delayed reproduction, we collected cones from *P. massoniana* and the Z pine of seven successive developmental stages and determined their transcriptome. Herein we might discover differentially expressed genes underlying the observed reproductive delay. We obtained 93,291 unigenes with an average length size of 1987 bp and 2494 bp of N50. We identified significantly differentially expressed genes (DEGs) in all seven cone growth stages. We screened for DEGs related to reproduction, such as pollen tube growth, development of the female gametophyte and embryo and so forth. Several potentially vital genes were identified and the expression levels of the two pines were compared and analyzed. These results may offer insight into the molecular mechanisms of reproductive process between the two pines and several other plants that with similar differential mode.

**Supplementary Materials:** The following are available online at http://www.mdpi.com/1999-4907/10/3/230/s1. Table S1: Quality of original sequencing data of P. massoniana and the Z pine; Table S2: Quality of clean data of P. massoniana and the Z pine; Table S3: Sequences of unigenes for qRT-PCR validation; Table S4: Primers for qRT-PCR; Table S5. Genes relate to reproduction process in P. massoniana and the Z pine; Figure S1: Length distribution of all unigene sequences; Figure S2: Venn diagram of annotation against to five databases: NCBI Nr, NCBI Nt, SwissProt, COG and KEGG for all unigenes; Figure S3. The species distribution of P. massoniana and the Z pine unigenes against the NCBI Nr database; Figure S4: Differentially expressed genes analysis between P. massoniana and the Z pine of each stage; Figure S5: Gene ontology (GO) functional classification of Z pine versus P. massoniana DEGs in seven stages; Figure S6: DEGs relate to reproduction in GO classification of Z pine versus P. massoniana in seven stages; Figure S7: KEGG pathway enrichment analysis of Z pine versus P. massoniana in seven stages; Figure S8: Comparison of unigene expression results between RNA sequencing (FPKM) and qRT-PCR.

**Author Contributions:** Conceptualization, J.X., T.Y. and J.S.; Formal analysis, J.M., Y.C., L.Y., H.H., H.Z., Z.G. and J.Y.; Funding acquisition, J.X.; Investigation, J.M., Y.C., L.Y., H.H., H.Z., Z.G. and J.Y.; Project administration, J.X.; Writing—original draft, J.M.; Writing—review & editing, J.M.

**Funding:** This research was funded by National Natural Science Foundation of China (31270661) and the Priority Academic Program Development of Jiangsu Higher Education Institutions (PAPD).

**Acknowledgments:** We would like to give thanks to Youxin Du, Qiang Huang and Benzhong Zhou from Lushan Botanical Garden, Jiangxi Province and Chinese Academy of Sciences for collecting samples on the Mt. Lushan. We also thank two reviewers for insightful comments on this article. Special thanks go to editors for their help in formulating the revisions.

**Conflicts of Interest:** The authors declare no conflict of interest.

## References

1. Nystedt, B.; Street, N.R.; Wetterbom, A.; Zuccolo, A.; Lin, Y.C.; Scofield, D.G.; Vezzi, F.; Delhomme, N.; Giacomello, S.; Alexeyenko, A. The Norway spruce genome sequence and conifer genome evolution. *Nature* **2013**, *497*, 579–584. [CrossRef] [PubMed]
2. Syring, J.; Farrell, K.; Businský, R.; Cronn, R.; Liston, A. Widespread genealogical nonmonophyly in species of *Pinus subgenus* Strobus. *Syst. Biol.* **2007**, *56*, 163–181. [CrossRef] [PubMed]

3. Carlsbecker, A.; Sundström, J.F.; Englund, M.; Uddenberg, D.; Izquierdo, L.; Kvarnheden, A.; Vergara Silva, F.; Engström, P. Molecular control of normal and acrocona mutant seed cone development in Norway spruce (*Picea abies*) and the evolution of conifer ovule-bearing organs. *New Phytol.* **2013**, *200*, 261–275. [CrossRef] [PubMed]
4. Katahata, S.; Futamura, N.; Igasaki, T.; Shinohara, K. Functional analysis of *SOC1-like* and *AGL6-like MADS-box* genes of the gymnosperm *Cryptomeria japonica*. *Tree Genet. Genomes* **2014**, *10*, 317–327. [CrossRef]
5. Uddenberg, D.; Reimegård, J.; Clapham, D.; Almqvist, C.; von Arnold, S.; Emanuelsson, O.; Sundström, J.F. Early cone setting in *Picea abies* acrocona is associated with increased transcriptional activity of a *MADS* box transcription factor. *Plant Physiol.* **2013**, *161*, 813–823. [CrossRef] [PubMed]
6. Dubos, C.; Stracke, R.; Grotewold, E.; Weisshaar, B.; Martin, C.; Lepiniec, L. *MYB* transcription factors in *Arabidopsis*. *Trends Plant Sci.* **2010**, *15*, 573–581. [CrossRef] [PubMed]
7. Hennig, L.; Taranto, P.; Walser, M.; Schönrock, N.; Gruissem, W. *Arabidopsis MSI1* is required for epigenetic maintenance of reproductive development. *Development* **2003**, *130*, 2555–2565. [CrossRef] [PubMed]
8. Zhai, D.; He, Z.; Feng, J.; Zheng, Y. Study on introgression between *Pinus hwangshanensis* and *Pinus massoniana* by using inter-simple sequence repeat marker (ISSR). *For. Sci. Technol.* **2012**, *37*, 4–6. (In Chinese)
9. Luo, S.; Zou, H.; Liang, S. Study on the introgressive hybridization between *Pinus hwangshanensis* and *P. massoniana*. *Sci. Silvae Sin.* **2001**, *37*, 118–122. (In Chinese)
10. Li, S.; Chen, Y.; Gao, H.; Yin, T. Potential chromosomal introgression barriers revealed by linkage analysis in a hybrid of *Pinus massoniana* and *P. hwangshanensis*. *BMC Plant Biol.* **2010**, *10*, 37. [CrossRef] [PubMed]
11. Bolger, A.M.; Lohse, M.; Usadel, B. Trimmomatic: A flexible trimmer for Illumina sequence data. *Bioinformatics* **2014**, *30*, 2114–2120. [CrossRef] [PubMed]
12. Haas, B.J.; Papanicolaou, A.; Yassour, M.; Grabherr, M.; Blood, P.D.; Bowden, J.; Couger, M.B.; Eccles, D.; Li, B.; Lieber, M. *De novo* transcript sequence reconstruction from RNA-seq using the Trinity platform for reference generation and analysis. *Nat. Protoc.* **2013**, *8*, 1494–1512. [CrossRef] [PubMed]
13. Grabherr, M.G.; Haas, B.J.; Yassour, M.; Levin, J.Z.; Thompson, D.A.; Amit, I.; Xian, A.; Fan, L.; Raychowdhury, R.; Zeng, Q. Trinity: Reconstructing a full-length transcriptome without a genome from RNA-Seq data. *Nat. Biotechnol.* **2011**, *29*, 644–652. [CrossRef] [PubMed]
14. Pertea, G.; Huang, X.; Liang, F.; Antonescu, V.; Sultana, R.; Karamycheva, S.; Lee, Y.; White, J.; Cheung, F.; Parvizi, B. TIGR Gene Indices clustering tools (TGICL): A software system for fast clustering of large EST datasets. *Bioinformatics* **2003**, *19*, 651–652. [CrossRef] [PubMed]
15. Schneider, M.; Lane, L.; Boutet, E.; Lieberherr, D.; Tognolli, M.; Bougueleret, L.; Bairoch, A. The UniProtKB/Swiss-Prot knowledgebase and its Plant Proteome Annotation Program. *J. Proteom.* **2009**, *72*, 567–573. [CrossRef] [PubMed]
16. Pruitt, K.D.; Tatusova, T.; Maglott, D.R. NCBI Reference Sequence (RefSeq): A curated non-redundant sequence database of genomes, transcripts and proteins. *Nucleic Acids Res.* **2005**, *33*, 501–504. [CrossRef] [PubMed]
17. Kanehisa, M.; Araki, M.; Goto, S.; Hattori, M.; Hirakawa, M.; Itoh, M.; Katayama, T.; Kawashima, S.; Okuda, S.; Tokimatsu, T. KEGG for linking genomes to life and the environment. *Nucleic Acids Res.* **2008**, *36*, 480–484. [CrossRef] [PubMed]
18. Ogata, H.; Goto, S.; Sato, K.; Fujibuchi, W.; Bono, H.; Kanehisa, M. KEGG: Kyoto Encyclopedia of Genes and Genomes. *Nucleic Acids Res.* **1999**, *27*, 29–34. [CrossRef] [PubMed]
19. Tatusov, R.L.; Fedorova, N.D.; Jackson, J.D.; Jacobs, A.R.; Kiryutin, B.; Koonin, E.V.; Krylov, D.M.; Mazumder, R.; Mekhedov, S.L.; Nikolskaya, A.N.; et al. The COG database: An updated version includes eukaryotes. *BMC Bioinf.* **2003**, *4*, 41. [CrossRef] [PubMed]
20. Young, M.D.; Wakefield, M.J.; Smyth, G.K.; Oshlack, A. Gene ontology analysis for RNA-seq: Accounting for selection bias. *Genome Biol.* **2010**, *11*, R14. [CrossRef] [PubMed]
21. Conesa, A.; Terol, J.; Robles, M. Blast2GO: A universal tool for annotation, visualization and analysis in functional genomics research. *Bioinformatics* **2005**, *21*, 3674–3676. [CrossRef] [PubMed]
22. Li, B.; Dewey, C.N. RSEM: Accurate transcript quantification from RNA-Seq data with or without a reference genome. *BMC Bioinf.* **2011**, *12*, 323. [CrossRef] [PubMed]
23. Trapnell, C.; Williams, B.A.; Pertea, G.; Mortazavi, A.; Kwan, G.; Van Baren, M.J.; Salzberg, S.L.; Wold, B.J.; Pachter, L. Transcript assembly and quantification by RNA-Seq reveals unannotated transcripts and isoform switching during cell differentiation. *Nat. Biotechnol.* **2010**, *28*, 511. [CrossRef] [PubMed]

24. Anders, S. Analysing RNA-Seq data with the DESeq package. *Mol. Biol.* **2010**, *43*, 1–17.
25. Ernst, J.; Bar-Joseph, Z. STEM: A tool for the analysis of short time series gene expression data. *BMC Bioinf.* **2006**, *7*, 191. [CrossRef] [PubMed]
26. Ye, J.; Coulouris, G.; Zaretskaya, I.; Cutcutache, I.; Rozen, S.; Madden, T.L. Primer-BLAST: A tool to design target-specific primers for polymerase chain reaction. *BMC Bioinf.* **2012**, *13*, 134. [CrossRef] [PubMed]
27. Livak, K.J.; Schmittgen, T.D. Analysis of relative gene expression data using real-time quantitative PCR and the $2^{-\Delta\Delta CT}$ method. *Methods* **2001**, *25*, 402–408. [CrossRef] [PubMed]
28. Lucca, N.; León, G. *Arabidopsis ACA7*, encoding a putative auto-regulated $Ca^{2+}$-ATPase, is required for normal pollen development. *Plant Cell Rep.* **2012**, *31*, 651–659. [CrossRef] [PubMed]
29. Zeng, Q.; Chen, J.G.; Ellis, B.E. *AtMPK4* is required for male-specific meiotic cytokinesis in *Arabidopsis*. *Plant J.* **2011**, *67*, 895–906. [CrossRef] [PubMed]
30. Ogawa, M.; Kay, P.; Wilson, S.; Swain, S.M. *Arabidopsis* dehiscence zone polygalacturonase 1 (*ADPG1*), *ADPG2*, and *QUARTET2* are polygalacturonases required for cell separation during reproductive development in *Arabidopsis*. *Plant Cell* **2009**, *21*, 216–233. [CrossRef] [PubMed]
31. Grienenberger, E.; Kim, S.S.; Lallemand, B.; Geoffroy, P.; Heintz, D.; de Azevedo Souza, C.; Heitz, T.; Douglas, C.J.; Legrand, M. Analysis of TETRAKETIDE α-PYRONE REDUCTASE function in *Arabidopsis thaliana* reveals a previously unknown, but conserved, biochemical pathway in sporopollenin monomer biosynthesis. *Plant Cell* **2010**, *110*. [CrossRef]
32. Grienenberger, E.; Besseau, S.; Geoffroy, P.; Debayle, D.; Heintz, D.; Lapierre, C.; Pollet, B.; Heitz, T.; Legrand, M. A *BAHD* acyltransferase is expressed in the tapetum of *Arabidopsis* anthers and is involved in the synthesis of hydroxycinnamoyl spermidines. *Plant J.* **2009**, *58*, 246–259. [CrossRef] [PubMed]
33. Dobritsa, A.A.; Geanconteri, A.; Shrestha, J.; Carlson, A.; Kooyers, N.; Coerper, D.; Urbanczyk-Wochniak, E.; Bench, B.J.; Sumner, L.W.; Swanson, R. A large-scale genetic screen in *Arabidopsis thaliana* to identify genes involved in pollen exine production. *Plant Physiol.* **2011**, *157*, 947–970. [CrossRef] [PubMed]
34. Shin, S.; Golovkin, M.; Reddy, A.S. A pollen-specific calmodulin-binding protein, *NPG1*, interacts with putative pectate lyases. *Sci. Rep.* **2014**, *4*, 5263. [CrossRef] [PubMed]
35. Golovkin, M.; Reddy, A.S. A calmodulin-binding protein from *Arabidopsis* has an essential role in pollen germination. *Proc. Natl. Acad. Sci. USA* **2003**, *100*, 10558–10563. [CrossRef] [PubMed]
36. Ischebeck, T.; Stenzel, I.; Heilmann, I. Type B phosphatidylinositol-4-phosphate 5-kinases mediate *Arabidopsis* and *Nicotiana tabacum* pollen tube growth by regulating apical pectin secretion. *Plant Cell* **2008**, *20*, 3312–3330. [CrossRef] [PubMed]
37. Bosch, M.; Cheung, A.Y.; Hepler, P.K. Pectin methylesterase, a regulator of pollen tube growth. *Plant Physiol.* **2005**, *138*, 1334–1346. [CrossRef] [PubMed]
38. Cole, R.A.; Synek, L.; Zarsky, V.; Fowler, J.E. *SEC8*, a subunit of the putative *Arabidopsis* exocyst complex, facilitates pollen germination and competitive pollen tube growth. *Plant Physiol.* **2005**, 138. [CrossRef] [PubMed]
39. Li, N.; Yuan, L.; Liu, N.; Shi, D.; Li, X.; Tang, Z.; Liu, J.; Sundaresan, V.; Yang, W. *SLOW WALKER2*, a *NOC1/MAK21* homologue, is essential for coordinated cell cycle progression during female gametophyte development in *Arabidopsis*. *Plant Physiol.* **2009**, *151*, 1486–1497. [CrossRef] [PubMed]
40. Becraft, P.W.; Stinard, P.S.; McCarty, D.R. *CRINKLY4*: A *TNFR-like* receptor kinase involved in maize epidermal differentiation. *Science* **1996**, *273*, 1406–1409. [CrossRef] [PubMed]
41. Lurin, C.; Andrés, C.; Aubourg, S.; Bellaoui, M.; Bitton, F.; Bruyère, C.; Caboche, M.; Debast, C.; Gualberto, J.; Hoffmann, B.; et al. Genome-wide analysis of *Arabidopsis* pentatricopeptide repeat proteins reveals their essential role in organelle biogenesis. *Plant Cell* **2004**, *16*, 2089–2103. [CrossRef] [PubMed]
42. Cushing, D.A.; Forsthoefel, N.R.; Gestaut, D.R.; Vernon, D.M. *Arabidopsis emb175* and other *ppr* knockout mutants reveal essential roles for pentatricopeptide repeat (PPR) proteins in plant embryogenesis. *Planta* **2005**, *221*, 424–436. [CrossRef] [PubMed]
43. Ishikawa, T.; Machida, C.; Yoshioka, Y.; Ueda, T.; Nakano, A.; Machida, Y. *EMBRYO YELLOW* gene, encoding a subunit of the conserved oligomeric Golgi complex, is required for appropriate cell expansion and meristem organization in *Arabidopsis thaliana*. *Genes Cells* **2008**, *13*, 521–535. [CrossRef] [PubMed]
44. Berg, M.; Rogers, R.; Muralla, R.; Meinke, D. Requirement of aminoacyl-tRNA synthetases for gametogenesis and embryo development in *Arabidopsis*. *Plant J.* **2005**, *44*, 866–878. [CrossRef] [PubMed]

45. Galau, G.A.; Bijaisoradat, N.; Hughes, D.W. Accumulation kinetics of cotton late embryogenesis-abundant mRNAs and storage protein mRNAs: Coordinate regulation during embryogenesis and the role of abscisic acid. *Dev. Biol.* **1987**, *123*, 198–212. [CrossRef]
46. Hecht, V.; Vielle-Calzada, J.; Hartog, M.V.; Schmidt, E.D.; Boutilier, K.; Grossniklaus, U.; de Vries, S.C. The *Arabidopsis SOMATIC EMBRYOGENESIS RECEPTOR KINASE 1* gene is expressed in developing ovules and embryos and enhances embryogenic competence in culture. *Plant Physiol.* **2001**, *127*, 803–816. [CrossRef] [PubMed]
47. Pagnussat, G.C.; Yu, H.; Sundaresan, V. Cell-fate switch of synergid to egg cell in *Arabidopsis* eostre mutant embryo sacs arises from misexpression of the *BEL1-like* homeodomain gene *BLH1*. *Plant Cell* **2007**, *19*, 3578–3592. [CrossRef] [PubMed]
48. Chen, J.; Uebbing, S.; Gyllenstrand, N.; Lagercrantz, U.; Lascoux, M.; Källman, T. Sequencing of the needle transcriptome from Norway spruce (*Picea abies* Karst L.) reveals lower substitution rates, but similar selective constraints in gymnosperms and angiosperms. *BMC Genom.* **2012**, *13*, 589. [CrossRef] [PubMed]
49. Liu, J.J.; Sturrock, R.N.; Benton, R. Transcriptome analysis of *Pinus monticola* primary needles by RNA-seq provides novel insight into host resistance to *Cronartium ribicola*. *BMC Genom.* **2013**, *14*, 884. [CrossRef] [PubMed]
50. Cañas, R.A.; Feito, I.; Fuentemaqueda, J.F.; Ávila, C.; Majada, J.; Cánovas, F.M. Transcriptome-wide analysis supports environmental adaptations of two *Pinus pinaster* populations from contrasting habitats. *BMC Genom.* **2015**, *16*, 909. [CrossRef] [PubMed]
51. Wu, Y.; Guo, J.; Zhou, Q.; Xin, Y.; Wang, G.; Xu, L.A. *De novo* transcriptome analysis revealed genes involved in flavonoid biosynthesis, transport and regulation in *Ginkgo biloba*. *Ind. Crops Prod.* **2018**, *124*, 226–235. [CrossRef]
52. Niu, S.; Yuan, H.; Sun, X.; Porth, I.; Li, Y.; El-Kassaby, Y.A.; Li, W. A transcriptomics investigation into pine reproductive organ development. *New Phytol.* **2016**, *209*, 1278–1289. [CrossRef] [PubMed]
53. Duan, D.; Jia, Y.; Yang, J.; Li, Z.H. Comparative Transcriptome Analysis of Male and Female Conelets and Development of Microsatellite Markers in *Pinus bungeana*, an Endemic Conifer in China. *Genes* **2017**, *8*, 393. [CrossRef] [PubMed]
54. Neale, D.B.; Wegrzyn, J.L.; Stevens, K.A.; Zimin, A.V.; Puiu, D.; Crepeau, M.W.; Cardeno, C.; Koriabine, M.; Holtz-Morris, A.E.; Liechty, J.D.; et al. Decoding the massive genome of loblolly pine using haploid DNA and novel assembly strategies. *Genome Biol.* **2014**, *15*, R59. [CrossRef] [PubMed]
55. Grotkopp, E.; Rejmánek, M.; Sanderson, M.J.; Rost, T.L. Evolution of genome size in pines (*Pinus*) and its life-history correlates: Supertree analyses. *Evolution* **2004**, *58*, 1705–1729. [CrossRef] [PubMed]
56. Wakamiya, I.; Newton, R.J.; Johnston, J.S.; Price, H.J. Genome size and environmental factors in the genus *Pinus*. *Am. J. Bot.* **1993**, *80*, 1235–1241. [CrossRef]
57. Ohri, D.; Khoshoo, T.N. Genome Size in Gymnosperms. *Plant Syst. Evol.* **1986**, *153*, 119–132. [CrossRef]
58. Anderson, E.; Hubricht, L. Hybridization in *Tradescantia*. III. The evidence for introgressive hybridization. *Am. J. Bot.* **1938**, *25*, 396–402. [CrossRef]
59. Wang, B.; Mao, J.F.; Gao, J.; Zhao, W.E.; Wang, X.R. Colonization of the Tibetan Plateau by the homoploid hybrid pine *Pinus densata*. *Mol. Ecol.* **2011**, *20*, 3796–3811. [CrossRef] [PubMed]
60. Watano, Y.; Kanai, A.; Tani, N. Genetic structure of hybrid zones between *Pinus pumila* and *P. parviflora var. pentaphylla* (Pinaceae) revealed by molecular hybrid index analysis. *Am. J. Bot.* **2004**, *91*, 65–72. [CrossRef] [PubMed]
61. Shindo, S.; Ito, M.; Ueda, K.; Kato, M.; Hasebe, M. Characterization of *MADS* genes in the gymnosperm *Gnetum parvifolium* and its implication on the evolution of reproductive organs in seed plants. *Evol. Dev.* **1999**, *1*, 180–190. [CrossRef] [PubMed]
62. Becker, A.; Saedler, H.; Theissen, G. Distinct *MADS-box* gene expression patterns in the reproductive cones of the gymnosperm *Gnetum gnemon*. *Dev. Genes Evol.* **2003**, *213*, 567–572. [CrossRef] [PubMed]
63. Jager, M.; Hassanin, A.; Manuel, M.; Le Guyader, H.; Deutsch, J. *MADS-box* genes in *Ginkgo biloba* and the evolution of the *AGAMOUS* family. *Mol. Biol. Evol.* **2003**, *20*, 842–854. [CrossRef] [PubMed]
64. Carlsbecker, A.; Tandre, K.; Johanson, U.; Englund, M.; Engström, P. The *MADS-box* gene *DAL1* is a potential mediator of the juvenile-to-adult transition in Norway spruce (*Picea abies*). *Plant J.* **2004**, *40*, 546–557. [CrossRef] [PubMed]

65. Moyroud, E.; Monniaux, M.; Thévenon, E.; Dumas, R.; Scutt, C.; Frohlich, M.; Parcy, F. A link between *LEAFY* and B-gene homologues in *Welwitschia mirabilis* sheds light on ancestral mechanisms prefiguring floral development. *New Phytol.* **2017**, *216*, 469–481. [CrossRef] [PubMed]
66. Dornelas, M.; Rodriguez, A. A *FLORICAULA/LEAFY* gene homolog is preferentially expressed in developing female cones of the tropical pine *Pinus caribaea var. caribaea. Genet. Mol. Biol.* **2005**, *28*, 299–307. [CrossRef]
67. Mouradov, A.; Glassick, T.; Hamdorf, B.; Murphy, L.; Fowler, B.; Marla, S.; Teasdale, R. *NEEDLY*, a *Pinus radiata* ortholog of *FLORICAULA/LEAFY* genes, expressed in both reproductive and vegetative meristems. *Proc. Natl. Acad. Sci. USA* **1998**, *95*, 6537–6542. [CrossRef] [PubMed]

*forests*

MDPI

*Article*

# Diversity Estimation and Antimicrobial Activity of Culturable Endophytic Fungi from *Litsea cubeba* (Lour.) Pers. in China

**Fei Wu [1,†], Dingchao Yang [1,†], Linping Zhang [1,*], Yanliu Chen [1], Xiaokang Hu [2], Lei Li [3] and Junsheng Liang [3]**

1 2011 Collaborative Innovation Center of Jiangxi Typical Trees Cultivation and Utilization, Jiangxi Agricultural University, Nanchang 330045, China; wufei315@163.com (F.W.); chao973788942@163.com (D.Y.); xiaoliuchen0316@126.com (Y.C.)

2 Gannan Arboretum, Ganzhou 341200, China; kang123878@163.com

3 Hunan Academy of Forestry, Changsha 410004, China; lileijxpx@163.com (L.L.); ljscaf@163.com (J.L.)

* Correspondence: zlping619@mail.jxau.edu.cn; Tel.: +86-0791-8381-3243

† These authors contributed equally to this work.

Received: 6 December 2018; Accepted: 31 December 2018; Published: 6 January 2019

**Abstract:** Endophytes are important components of forest ecosystems, and have potential use in the development of medical drugs and the conservation of wild medicinal plants. This study aimed to examine the diversity and antimicrobial activities of endophytic fungi from a medicinal plant, *Litsea cubeba* (Lour.) Pers. The results showed that a total of 970 isolates were obtained from root, stem, leaf, and fruit segments of *L. cubeba*. All the fungal endophytes belonged to the phylum Ascomycota and could be classified into three taxonomic classes, nine orders, twelve families, and seventeen genera. SF15 (*Colletotrichum boninense*) was the dominant species in *L. cubeba*. Leaves harbored a greater number of fungal endophytes but lower diversity, while roots harbored the maximum species diversity of endophytic fungi. For the antimicrobial activities, seventeen isolates could inhibit the growth of plant pathogenic fungi, while the extracts of six endophytes showed antimicrobial activity to all the tested pathogenic fungi. Among these endophytes, SF22 (*Chaetomium globosum*) and SF14 (*Penicillium minioluteum*) were particularly effective in inhibiting seven plant pathogenic fungi growths and could be further explored for their potential use in biotechnology, medicine, and agriculture.

**Keywords:** endophytes; medicinal plants; pathogen; molecular identification; plant-microbe interaction

## 1. Introduction

The demand for new and useful compounds for disease prevention and control is ever growing [1]. Antibiotic resistance, the increasing incidence of fungal diseases, and the development of superbugs cause biodiversity loss and constantly bring challenges to the field of medicine [2,3]. Thus, there is an urgent need to find new antibiotics that are more effective, have lower toxicity, and a smaller environmental impact.

Forest ecosystems cover an area of approximately 38 million square kilometers and contain substantial resources [4,5]. Endophytes are an important component of the forest ecosystem, which inhabit the internal tissues of plants, have no detrimental effects on plants, and can sometimes improve plant growth performance [6,7]. Most of the natural compounds produced by endophytes have exhibited antimicrobial activity and, in many cases, these are related to the protection of the host from phytopathogenic microorganisms [8]. The endophyte *Beauveria bassiana* has been able to inhibit fungal pathogens by the production of bioactive metabolites [9]. The endophytic fungus *Gliocladium catenulatum* can reduce the incidence of witches' broom disease in cacao by up to 70% [10]. Furthermore, some endophytic fungi can produce the same chemical compounds as the host, such as the paclitaxel producing fungus *Taxomyces andreanae* from *Taxus brevifolia* [11,12], and the podophyllotoxin generating fungus *Fusarium oxysporum* from *Juniperus recurva* [13]. There have been over 8600 discovered bioactive metabolites of fungal origin [14]. It is estimated that there are approximately 1 million fungal species of endophytic fungi in nature [15], whereas only a small percentage of endophytes have been discovered [16]. The enormous biodiversity and abundant fungal endophytes that occur in plant tissues show the potential role of endophytes in the production of novel natural antimicrobial compounds.

*Litsea cubeba* (Lour.) Pers. (Lauraceae) is a native woody species in China, Indonesia, and other countries in Southeast Asia [17]. It is a valuable traditional Chinese medicinal plant that has been used to treat rheumatic diseases, stomach aches, and common cold for thousands of years [18,19]. The active components of *L. cubeba* were reported to be antibacterial [20], anticancer [21], and anti-inflammatory [19]. Intercropping of *L. cubeba* and *Camellia oleifera* Abel. can reduce the incidence of *C. oleifera* disease, suggesting the role of *L. cubeba* in protecting economic plants from diseases. *Colletotrichum gloeosporioides* (Penz.) Penz. & Sacc. [22], *Fusarium andiyazi* Marasas, Rheeder, Lampr., K.A. Zeller & J.F. Leslie [23], *Alternaria alternata* (Fr.) Keissl. [24], *Phomopsis* sp. [25], *Ceratosphaeria phyllostachydis* Zhang [26], *Rhizoctonia solani* Kühn [27], and *Phytophthora capsici* Leonian [28] cause diseases in main economic crops in South China, leading to a heavy decline in crop yield and quality. Currently, the associated microflora of medicinal plants is being paid increased amounts of attention for the exploitation of antimicrobial drugs [29]. However, to our knowledge, there are no reports on the biodiversity and bioactivity of endophytic fungi in *L. cubeba*. This study aimed to investigate the diversity and antimicrobial activities of endophytic fungi of *L. cubeba*, and, further, to screen them as potential biocontrol agents against seven plant pathogens.

## 2. Materials and Methods

### 2.1. Collection of Samples and Isolation of Endophytic Fungi

The leaves, branches, roots, and fruits of *Litsea cubeba* were collected from a planting base in Lichuan county of Jiangxi Province, China, in May 2016. The leaves and fruits samples were cut into small pieces of about 0.5 × 0.5 cm using a sterile knife, and the branch and root samples were cut into small segments 1 cm in length. These fragments were surface sterilized with 70% (v/v) ethanol for 3 min, 3% (v/v) NaClO for 3–5 min, and then rinsed with sterile water four times. Excess moisture was blotted by sterile filter papers [30]. Then, they were cultured on potato dextrose agar (PDA) medium supplemented with streptomycin (50 U/mL) and penicillin (30 U/mL) at 25 °C under dark conditions for 7–15 days. Pure fungal cultures were obtained by picking hyphal tips of the developing fungal colonies. The acquired isolates were preserved on PDA slants and deposited at 4 °C for identification.

### 2.2. Genomic DNA Extraction, PCR Amplification and Molecular Identification

The isolates were first identified based on the morphological characteristics of the colony culture and spores. Fungal genomic DNA was extracted from the mycelia using an Ezup Column Fungi Genomic DNA Purification Kit (Sangon Biotech, Inc., Shanghai, China) according to the manufacturer's protocol. The internal transcribed spacer (ITS) regions were amplified using the universal primers ITS1 (5′-TCCGTAGGTGAACCTGCGC-3′) and ITS4 (5′-TCCTCCGCTTATTGATATGC-3′) [31]. The reaction mixtures (50 μL) contained 25 μL 2 × Taq PCR Master mixture (Sangon Biotech, Inc., Shanghai, China), 2 μL of ITS4, 2 μL of ITS5, 2 μL of Template DNA, and 19 μL of $ddH_2O$. The reaction conditions were 94 °C for 5 min, 30 cycles at 94 °C for 50 s, 52 °C for 50 s, 72 °C for 1 min, and a final extension at 72 °C for 7 min. The PCR products were examined by electrophoresis in 1% (w/v) agarose gels and then purified using the Agarose Gel DNA Extraction Kit (Takara, Japan) and sequenced.

The resultant sequences were compared with previously deposited sequences in the GenBank, NCBI (http://www.ncbi.nlm.nih.gov) using a basic local alignment search tool (BLAST). Sequence alignment and phylogenetic analysis were conducted using MEGA version 7 [32]. Phylogenetic trees were constructed using a neighbor-joining method. The ITS gene sequences of the potential novel isolates were deposited in GenBank under the accession numbers MF962537–MF962573.

### *2.3. Estimation and Quantification of Fungal Diversity*

Fungal diversity and richness in different plant tissues were measured and quantified using various indices, including the colonization rate (*CR*), isolation rate (*IR*), and Shannon-Wiener ($H'$), Simpson's ($D_s$) diversity index and evenness index (*E*). The calculations were as follows.

$$CR = Nf/Nt \times 100, \quad (1)$$

$$IR = Ng/Nt \times 100, \quad (2)$$

$$H' = -\sum Pi \times \mathrm{Ln}(Pi), \quad (3)$$

$$D_s = 1 - \Sigma Pi^2, \quad (4)$$

$$E = H'/\mathrm{Ln}(S), \quad (5)$$

where *Nf* was the number of fragments with fungal growth, *Nt* was the total number of fragments, and *Ng* was the number of isolates of a given type isolated [33]. $Pi = n_i/N$, is the relative abundance of the endophytic fungal species, $n_i$ is the number of isolates of one species, and *N* is the total species number of isolates [34,35]. *S* was the total number of the taxa (ITS genotype) present within each sample [16].

### *2.4. Antimicrobial Activity of Endophytic Fungi*

The indicator strains include the following plant pathologens: the fungi *Colletotrichum gloeosporioides*, *F. andiyazi*, *A. alternata*, *Phomopsis* sp., *Ceratosphaeria phyllostachydis*, *R. solani*, and the Chromista *Phytophthora capsici*, provided by the Plant Pathology Laboratory, College of Forestry, Jiangxi Agricultural University, China.

A dual culture technique was applied to examine the antimicrobial activity of endophytic fungi from *L. cubeba* against fungal pathogens [36]. The mycelial discs (6 mm in diameter) of actively growing endophytes were placed at the periphery of the PDA plate. The mycelial discs of the pathogen were placed on the other side of the PDA plate, 4 cm away from the endophyte disc. The plate with only the pathogen was used as a control. Each treatment replicated 3 times. The dual culture plates were incubated for 3–8 days at 25 °C. The inhibition rate against pathogens was calculated according to the formula below.

$$\text{Inhibition rate } (\%) = (R_1 - R_2)/(R_1 - 0.6) \times 100, \quad (6)$$

where $R_1$ is the colony diameter of the control, $R_2$ is the colony diameter under experimental treatments, and 0.6 mm represents the mycelial discs.

The endophytes with high antimicrobial activity were selected and investigated for the in vitro antimicrobial activity of their extracts. Each of the endophytes were separately cultured on 200 mL PDA liquid medium at 25 °C, by shaking at 150 rpm for 8–12 days. The culture broth was collected by filtration and extracted with an equal amount of ethyl acetate three times. The organic phase was evaporated to dryness using a rotary evaporator. The dry extract was dissolved in 3 mL of methanol and formulated into 15 μg/mL of mycelia broth.

In vitro antimicrobial tests were conducted by testing the growth rate of the pathology fungi. The mycelial discs (6 mm in diameter) of the pathogen were placed in the center of the PDA plate containing 1.5 mL mycelia broth. The PDA plate without mycelia broth (containing only 1.5 mL methanol) was used as the control. The tested plates were cultured at 25 °C for 3–7 days. The formula for calculating the inhibition rate is the same as Formula (6).

### 2.5. Statistical Analyses

Statistical tests were performed using SPSS 13.0 (SPSS Inc., Chicago, IL, USA). Turkey's multiple range test was used to pairwise multiple comparisons between treatments.

## 3. Results

### 3.1. Identification and Composition of Endophyte Assemblage

A total of 970 isolates were obtained from root, stem, leaf, and fruit segments of *L. cubeba* (Table 1). The maximum number of isolates was obtained from the leaves (438 isolates), followed by stems (241 isolates), fruits (149 isolates), and roots (142 isolates). Molecular identification of the isolates was conducted based on a comparative analysis of ITS gene sequences and their similarity to reference sequences (Figure 1). The results showed that the isolated endophytic fungi could be allocated to 36 operational taxonomic units (OTUs). All of them belonged to the Ascomycota phylum and were classified into three taxonomic classes (Eurotiomycetes, Dothideomycetes, and Sordariomycetes), nine orders (Eurotiales, Botryosphaeriales, Pleosporales, Hypocreales, Chaetosphaeriales, Sordariales, Diaporthales, Xylariales, and an unassigned order), twelve families and seventeen genera. Twenty-three fungal morphotypic groups were taxonomically assigned to species, and the other 13 were classified at the genus level (Table 1). SF15 (*Colletotrichum boninense*) accounted for 39.79% of the total isolates and was the dominant species in the whole fungal endophytic community, followed by SF4 (*Botryosphaeria dothidea*) (6.60%).

### 3.2. Diversity Estimation of Endophytic Fungi

The biodiversity of endophytic fungi in *L. cubeba* was quantitatively investigated in terms of the colonization rate (*CR*), isolation rate (*IR*), Shannon-Wiener ($H'$), and Simpson's ($D_s$) diversity index and evenness index ($E$) (Table 2). The total $H'$ and $D_s$ were 2.52 and 0.82, respectively. The highest biodiversity of endophytic fungi was observed in roots ($H' = 2.74$, $D_s = 0.90$), followed by stems ($H' = 2.56$, $D_s = 0.90$), fruits ($H' = 1.99$, $D_s = 0.76$), and leaves ($H' = 1.43$, $D_s = 0.56$). The leaf samples had the highest endophytic fungi colonization rate but the lowest species evenness ($E = 0.51$) compared to the other plant parts.

### 3.3. In Vitro Antimicrobial Activity of Endophytic Fungi

The results of dual culture experiments showed that 17 isolates inhibited the growth of pathogenic fungi, which was manifested by the occurrence of the inhibition zone or mycelial atrophy of pathogens (Table 3). Among them, 10 isolates exhibited antibiotic effects on all the tested pathogenic microbes. SF22 (*Chaetomium globosum*) showed the strong activity against *Ceratosphaeria phyllostachydis*, *Phomopsis* sp., and *Alternaria alternata*, with inhibition rates of 78.43, 73.20, and 70.23%, respectively.

The results of the antimicrobial test on the fermentation products support that the fermentation products of SF14, SF22, SF23, SF27, SF29 and SF32 showed antimicrobial activity against all the tested pathogen fungi (Table 4). The antimicrobial activity of the fermentation products was stronger than the endophytic fungi. The inhibition rate of SF22 (*Chaetomium globosum*) extracts against *Ceratosphaeria phyllostachydis* was 93.24%. The inhibition rate of SF14 (*Penicillium minioluteum*) extracts against *Phomopsis* sp. was 87.87%. The inhibition rates of the fermentation products of these two isolates against the other six pathogens were over 60%.

**Table 1.** Identification, abundance, and percentage recovery of endophytic fungi isolated from different tissues of *Litsea cubeba* (Lour.) Pers.

| Endophytic Fungal Taxon | Isolate Codes | Accession Numbers | The Closet Genbank Taxa | Similarity (%) | Numbers of Isolates from Plant Tissues | | | | Total Abundance (Percentage Recovery) |
|---|---|---|---|---|---|---|---|---|---|
| | | | | | Roots | Stems | Leaves | Fruits | |
| *Aspergillus fumigatus* 1 | SF25 | MF962555 | *Aspergillus fumigatus* (KP131566.1) | 99 | 6 | 0 | 0 | 0 | 0.62 |
| *Aspergillus fumigatus* 2 | SF48 | MF962572 | *Aspergillus fumigatus* (EU833205.1) | 99 | 1 | 0 | 0 | 0 | 0.10 |
| *Botryosphaeria dothidea* | SF4 | MF962539 | *Botryosphaeria dothidea* (FJ478129.1) | 99 | 1 | 46 | 11 | 6 | 6.60 |
| *Clonostachys sp.* | SF31 | MF962561 | *Clonostachys* sp. (LC133855.1) | 99 | 5 | 2 | 3 | 0 | 1.03 |
| *Calonectria curvispora* | SF39 | MF962566 | *Calonectria curvispora* (GQ280568.1) | 99 | 25 | 2 | 0 | 0 | 2.78 |
| *Chaetomium globosum* | SF22 | MF962552 | *Chaetomium globosum* (KM268652.1) | 99 | 1 | 0 | 1 | 0 | 0.21 |
| *Colletotrichum boninense* | SF15 | MF962548 | *Colletotrichum boninense* (MF076585.1) | 99 | 0 | 1 | 3 | 0 | 0.41 |
| *Colletotrichum gloeosporioides* 1 | SF40 | MF962567 | *Colletotrichum gloeosporioides* (EU552111.1) | 99 | 2 | 32 | 284 | 68 | 39.79 |
| *Colletotrichum gloeosporioides* 2 | SF3 | MF962538 | *Colletotrichum gloeosporioides* (KU534983.1) | 99 | 4 | 0 | 41 | 4 | 5.06 |
| *Diaporthe phaseolorum* 1 | SF8 | MF962542 | *Diaporthe phaseolorum* (KX866868.1) | 99 | 7 | 37 | 18 | 4 | 6.80 |
| *Diaporthe phaseolorum* 2 | SF45 | MF962570 | *Diaporthe phaseolorum* (AF001018.2) | 99 | 0 | 0 | 0 | 9 | 0.93 |
| *Diaporthe eres* | SF30 | MF962560 | *Diaporthe eres* (KX866867.1) | 97 | 0 | 11 | 0 | 0 | 1.13 |
| *Diaporthe* sp. | SF34 | MF962563 | *Diaporthe* sp. (EF42278.1) | 97 | 0 | 7 | 0 | 0 | 0.72 |
| *Fusarium graminearum* | SF35 | MF962564 | *Fusarium graminearum* (KF624778.1) | 99 | 5 | 4 | 11 | 10 | 3.09 |
| *Nigrospora sphaerica* | SF13 | MF962546 | *Nigrospora sphaerica* (KM510416.1) | 100 | 0 | 0 | 1 | 6 | 0.72 |
| *Nemania diffusa* | SF10 | MF962543 | *Nemania diffusa* (KP133219.1) | 99 | 3 | 4 | 0 | 0 | 0.72 |
| *Phomopsis* sp.1 | SF5 | MF962540 | *Phomopsis* sp. (KP184328.1) | 99 | 0 | 8 | 29 | 0 | 3.82 |
| *Phomopsis* sp.2 | SF7 | MF962541 | *Phomopsis* sp. (JX436795.1) | 98 | 14 | 12 | 10 | 13 | 5.05 |
| *Phomopsis* sp.3 | SF21 | MF962551 | *Phomopsis* sp. (AB505410.1) | 97 | 3 | 3 | 0 | 4 | 1.03 |
| *Phomopsis* sp.4 | SF38 | MF962565 | *Phomopsis* sp. (HQ832822.1) | 99 | 1 | 22 | 0 | 0 | 2.37 |
| *Phomopsis* sp.5 | SF44 | MF962569 | *Phomopsis* sp. (HM595506.1) | 99 | 0 | 0 | 0 | 9 | 0.93 |
| *Phomopsis fukushii* | SF11 | MF962544 | *Phomopsis fukushii* (KT951302.1) | 97 | 1 | 0 | 2 | 1 | 0.41 |
| *Phyllosticta capitalensis* | SF1 | MF962537 | *Phyllosticta capitalensis* (KR056285.1) | 100 | 1 | 15 | 9 | 9 | 3.51 |
| *Pestalotiopsis* sp.1 | SF24 | MF962554 | *Pestalotiopsis* sp. (HQ607806.1) | 99 | 0 | 9 | 0 | 0 | 0.93 |
| *Pestalotiopsis* sp.2 | SF46 | MF962571 | *Pestalotiopsis* sp. (HE608797.1) | 99 | 2 | 0 | 0 | 0 | 0.21 |
| *Pestalotiopsis* sp.3 | SF49 | MF962573 | *Pestalotiopsis* sp. (EF423541.1) | 100 | 1 | 0 | 0 | 5 | 0.62 |
| *Pestalotiopsis disseminata* | SF28 | MF962558 | *Pestalotiopsis disseminate* (JQ323000.1) | 99 | 3 | 13 | 0 | 1 | 0.62 |
| *Pestalotiopsis vismiae* | SF12 | MF962545 | *Pestalotiopsis vismiae* (KM015217.1) | 99 | 2 | 1 | 3 | 0 | 1.75 |
| *Penicillium rubens* | SF18 | MF962550 | *Penicillium rubens* (LT558865.1) | 100 | 0 | 1 | 0 | 0 | 0.10 |
| *Penicillium janthinellum* | SF27 | MF962557 | *Penicillium janthinellum* (KM268648.1) | 99 | 29 | 0 | 0 | 0 | 2.99 |
| *Penicillium citrinum* | SF32 | MF962562 | *Penicillium citrinum* (LT558897.1) | 100 | 5 | 0 | 0 | 0 | 0.52 |
| *Penicillium minioluteum* | SF14 | MF962547 | *Penicillium minioluteum* (L14505.1) | 99 | 3 | 0 | 0 | 0 | 0.31 |
| *Phoma* sp. | SF26 | MF962556 | *Phoma* sp. (HQ631000.1) | 99 | 5 | 9 | 11 | 0 | 2.58 |
| *Thozetella* sp. | SF16 | MF962549 | *Thozetella* sp. (KU059840.1) | 96 | 1 | 1 | 0 | 0 | 0.21 |
| *Talaromyces* sp. | SF23 | MF962553 | *Talaromyces* sp. (KU556510.1) | 99 | 3 | 0 | 0 | 0 | 0.31 |
| *Talaromyces amestolkiae* | SF29 | MF962559 | *Talaromyces amestolkiae* (LT558956.1) | 99 | 8 | 1 | 1 | 0 | 1.03 |
| Total | | | | | 142 | 241 | 438 | 149 | 100.00 |

**Table 2.** The Index of endophytic fungi flora diversity of *Litsea cubeba* (Lour.) Pers.

| Parts | No. of Tissue | No. of Fungi | No. of Strains | No. of Genus | *CR* % | *IR* % | *H'* | $D_s$ | *E* |
|---|---|---|---|---|---|---|---|---|---|
| Root | 450 | 142 | 142 | 27 | 32.22 | 31.56 | 2.74 | 0.90 | 0.83 |
| Stem | 450 | 232 | 241 | 21 | 51.56 | 53.56 | 2.56 | 0.90 | 0.84 |
| Leaf | 450 | 413 | 438 | 16 | 91.78 | 97.33 | 1.43 | 0.56 | 0.51 |
| Fruit | 450 | 139 | 149 | 14 | 30.89 | 33.11 | 1.99 | 0.76 | 0.75 |
| Total | 1800 | 926 | 970 | 36 | 51.44 | 53.89 | 2.52 | 0.82 | 0.70 |

Abbreviations: No.: number; *CR*: colonization rate; *IR*: isolation rate; *H'*: Shannon-Wiener diversity index; $D_S$: Simpson's diversity index; *E*: evenness index.

**Table 3.** Antimicrobial activities of endophytic fungi from *Litsea cubeba* (Lour.) Pers.

| No. | Endophytic fungi | Inhibition Ratio of Pathogen Mycelium Growth (%) | | | | | | |
|---|---|---|---|---|---|---|---|---|
| | | ① | ② | ③ | ④ | ⑤ | ⑥ | ⑦ |
| SF11 | *Phomopsis fukushii* | 46.15 ± 0.33 $^{W-}$ | 55.56 ± 0.53 $^{W-}$ | 45.91 ± 0.71 $^{W+}$ | 29.52 ± 0.32 $^{W+}$ | 46.56 ± 0.52 $^{W-}$ | 55.35 ± 1.61 $^{W+}$ | — |
| SF14 | *Penicillium minioluteum* | 39.23 ± 0.27 $^{W+}$ | 58.82 ± 0.28 $^{W+}$ | 52.20 ± 0.21 $^{W+}$ | 60.95 ± 0.41 $^{W-}$ | 52.67 ± 0.47 $^{W+}$ | 53.46 ± 0.31 $^{W+}$ | 55.49 ± 0.37 $^{W+}$ |
| SF22 | *Chaetomium globosum* | 58.61 ± 0.44 $^{W+}$ | 73.43 ± 0.43 $^{W+}$ | 53.12 ± 0.14 $^{W+}$ | 70.23 ± 0.30 $^{W+}$ | 73.20 ± 0.24 $^{W+}$ | 78.43 ± 0.49 $^{W+}$ | 56.60 ± 0.35 $^{W+}$ |
| SF23 | *Talaromyces* sp. | 42.31 ± 0.27 $^{W+}$ | 64.05 ± 0.18 $^{W+}$ | 49.06 ± 0.21 $^{W+}$ | 42.86 ± 0.24 $^{W+}$ | 55.73 ± 0.27 $^{W+}$ | 52.83 ± 0.28 $^{W+}$ | 52.44 ± 0.34 $^{W-}$ |
| SF24 | *Pestalotiopsis* sp.1 | 46.92 ± 0.25 $^{W+}$ | 52.94 ± 0.31 $^{W-}$ | 56.60 ± 0.29 $^{W+}$ | 68.57 ± 0.43 $^{W-}$ | 57.25 ± 0.14 $^{W+}$ | 50.31 ± 0.28 $^{W+}$ | 39.63 ± 0.39 $^{W-}$ |
| SF27 | *Penicillium janthinellum* | 38.46 ± 0.41 $^{W+}$ | 39.22 ± 0.20 $^{W+}$ | 47.17 ± 0.41 $^{W-}$ | 31.43 ± 0.40 $^{W-}$ | 53.44 ± 0.32 $^{W+}$ | 40.88 ± 0.37 $^{W+}$ | 35.37 ± 0.21 $^{W+}$ |
| SF28 | *Pestalotiopsis disseminata* | 50.00 ± 0.23 $^{W+}$ | 58.17 ± 0.21 $^{W+}$ | 57.23 ± 0.27 $^{W+}$ | 39.05 ± 0.21 $^{W-}$ | 45.04 ± 0.21 $^{W-}$ | 48.43 ± 0.31 $^{W-}$ | 25.61 ± 0.37 $^{W-}$ |
| SF29 | *Talaromyces amestolkiae* | 60.77 ± 0.43 $^{W-}$ | 78.43 ± 0.18 $^{W-}$ | 60.38 ± 0.37 $^{W-}$ | 59.05 ± 0.19$^{W+}$ | 73.28 ± 0.27 $^{W-}$ | 64.78 ± 0.42 $^{W+}$ | 51.22 ± 0.43 $^{W-}$ |
| SF31 | *Clonostachys sp.* | 33.08 ± 0.38 $^{W+}$ | 47.06 ± 0.35 $^{W+}$ | 48.43 ± 0.41 $^{W+}$ | 28.57 ± 0.33 $^{W-}$ | 32.06 ± 0.25 $^{W-}$ | — | — |
| SF32 | *Penicillium citrinum* | 50.77 ± 0.45 $^{W+}$ | 62.75 ± 0.29 $^{W+}$ | 61.01 ± 0.32 $^{W+}$ | 39.05 ± 0.21 $^{W+}$ | — | — | 80.49 ± 0.27 $^{W-}$ |
| SF35 | *Fusarium graminearum* | 36.92 ± 0.43 $^{W+}$ | 49.67 ± 0.33 $^{W+}$ | 53.46 ± 0.31 $^{W+}$ | 37.14 ± 0.26 $^{W+}$ | 49.62 ± 0.28 $^{W+}$ | 40.88 ± 0.36 $^{W-}$ | 32.93 ± 0.17 $^{W-}$ |
| SF39 | *Calonectria curvispora* | 45.38 ± 0.25 $^{W+}$ | — | — | — | 70.13 ± 0.19 $^{W+}$ | 59.12 ± 0.18 $^{W-}$ | — |
| SF44 | *Phomopsis* sp.5 | 50.77 ± 0.40 $^{W+}$ | 71.42 ± 0.25 $^{W+}$ | 57.86 ± 0.32 $^{W-}$ | 40.95 ± 0.44 $^{W-}$ | 49.62 ± 0.36 $^{W+}$ | 55.35 ± 0.23 $^{W+}$ | — |
| SF49 | *Pestalotiopsis* sp.3 | 46.92 ± 0.24 $^{W+}$ | 70.21 ± 0.35 $^{W+}$ | 56.60 ± 0.35 $^{W+}$ | 52.38 ± 0.33 $^{W-}$ | 51.91 ± 0.39 $^{W+}$ | 57.23 ± 0.31 $^{W+}$ | — |

Note: Data presented are the means ± SD ($n$ = 3). ① *Colletotrichum gloeosporioides* (Penz.) Penz. & Sacc.; ② *Phytophthora capsici* Leonian; ③ *Fusarium andiyazi* Marasas, Rheeder, Lampr., K.A. Zeller & J.F. Leslie; ④ *Alternaria alternata* (Fr.) Keissl.; ⑤ *Phomopsis* sp.; ⑥ *Ceratosphaeria phyllostachydis* Zhang; ⑦ *Rhizoctonia solani* Kühn; $^{-}$: No inhibition zone; $^{+}$: Inhibition zone; w: Pathogen hyphae shrink; –: No inhibition.

**Table 4.** Antimicrobial activity of the metabolites of endophytic fungi from *Litsea cubeba* (Lour.) Pers.

| No. | Endophytic fungi | Inhibition Ratio of Pathogen Mycelium Growth (%) | | | | | | |
|---|---|---|---|---|---|---|---|---|
| | | ① | ② | ③ | ④ | ⑤ | ⑥ | ⑦ |
| SF14 | *Penicillium minioluteum* | $76.32 \pm 1.35^{aB}$ | $75.21 \pm 1.63^{bBC}$ | $69.00 \pm 2.40^{aC}$ | $61.08 \pm 1.85^{bD}$ | $87.87 \pm 1.97^{aA}$ | $84.73 \pm 4.03^{aA}$ | $60.01 \pm 3.03^{bD}$ |
| SF22 | *Chaetomium globosum* | $66.18 \pm 3.98^{bC}$ | $91.73 \pm 1.67^{aA}$ | $66.67 \pm 3.39^{aC}$ | $77.60 \pm 2.72^{aB}$ | $80.00 \pm 6.25^{abB}$ | $93.24 \pm 2.10^{aA}$ | $61.00 \pm 3.82^{bC}$ |
| SF23 | *Talaromyces* sp. | $79.34 \pm 3.22^{aAB}$ | $55.10 \pm 0.85^{dD}$ | $37.33 \pm 1.31^{bF}$ | $45.21 \pm 3.25^{cE}$ | $80.63 \pm 3.55^{abA}$ | $65.00 \pm 2.19^{bC}$ | $73.60 \pm 1.36^{aB}$ |
| SF27 | *Penicillium janthinellum* | $80.66 \pm 2.80^{aA}$ | $71.07 \pm 3.87^{bcAB}$ | $29.87 \pm 6.01^{bcE}$ | $42.16 \pm 3.55^{cDE}$ | $74.80 \pm 4.02^{bAB}$ | $56.08 \pm 8.85^{bcCD}$ | $64.78 \pm 3.76^{bBC}$ |
| SF29 | *Talaromyces amestolkiae* | $27.37 \pm 4.30^{dCD}$ | $64.19 \pm 6.58^{cdA}$ | $25.33 \pm 6.73^{cD}$ | $46.73 \pm 6.43^{cB}$ | $50.65 \pm 4.27^{cAB}$ | $45.00 \pm 5.53^{cdB}$ | $42.14 \pm 3.30^{cBC}$ |
| SF32 | *Penicillium citrinum* | $50.34 \pm 3.70^{cB}$ | $60.24 \pm 5.04^{dA}$ | $61.23 \pm 2.53^{aA}$ | $42.23 \pm 3.57^{cBC}$ | $32.76 \pm 1.55^{dD}$ | $33.78 \pm 2.70^{dCD}$ | $45.12 \pm 2.91^{cB}$ |

Note: Data presented are the means ± SD ($n = 3$). Means followed by the same lowercase letters within a column and by the same uppercase letters within a row do not differ significantly at $p \leq 0.05$ according to Turkey's test. ① *Colletotrichum gloeosporioides* (Penz.) Penz. & Sacc.; ② *Phytophthora capsici* Leonian; ③ *Fusarium andiyazi* Marasas, Rheeder, Lampr., K.A. Zeller & J.F. Leslie; ④ *Alternaria alternata* (Fr.) Keissl.; ⑤ *Phomopsis* sp.; ⑥ *Ceratosphaeria phyllostachydis* Zhang; ⑦ *Rhizoctonia solani* Kühn.

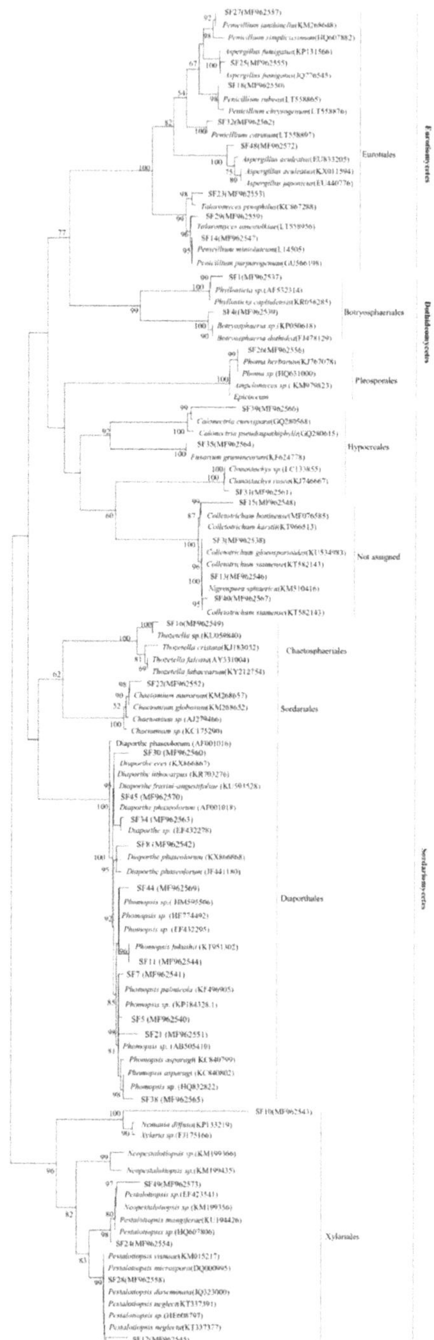


**Figure 1.** Neighbor-joining phylogenetic tree based on internal transcribed spacer (ITS)–rDNA gene sequences of endophytic fungi associated with *Litsea cubeba* (Lour.) Pers. Bootstrap percentages (>50) after 1000 replications are shown.

## 4. Discussion

Medicinal plants are legitimate targets to isolate endophytic fungi for their role in producing pharmacologically important secondary metabolites [37]. These fungal endophytes can be used to treat plant diseases. This is the first study that demonstrates the diversity, phylogeny, and bioactive potential of endophytic fungi associated with a medicinal plant, *L. cubeba*. In this study, all the fungal isolates were identified as Ascomycota, which is consistent with previous findings on *Ophiopogon japonicas* [38], *Calotropis procera* [39], and *Cannabis sativa* [35]. It is estimated that the phylum Ascomycota covers about 8% of the Earth's land and is among the most prevalent and diverse phyla of eukaryotes [37,40]. Endophytic fungi are ubiquitously distributed thoughout various classes of Ascomycota, including Eurotiomycetes, Dothideomycetes, Leotiomycetes, Pezizomycetes, and Sordariomycetes [6,41]. Katoch et al. [37] observed that the endophytic fungi in *Monarda citriodora*, a medicinal plant, were mainly distributed in the Sordariomycetes class, followed by Eurotiomycetes and Dothideomycetes. A similar presentation of classes was found in this study, indicating that endophytic fungi isolated in this study were cosmopolitan endophytes.

The fungal endophytes discovered in *L. cubeba* in this study were not identical to those reported in other studies. Ho et al. (2012) [42] isolated endophytic fungi from twigs of seven medicinal herbs belonging to the Lauraceae family (including *L. cubeba*) and found that the endophytes from *L. cubeba* belonged to six genera (*Pestalotiopsis*, *Arthrinium*, *Diaporthe*, *Xylaria*, *Hypoxylon*, and *Pyrenochaeta*). Only two genera (*Pestalotiopsis* and *Diaporthe*) were consistent with the results of the present study. This may due to the differences in sites, seasons, and climates [6].

The variation in endophytic communities was also found in spatial distribution. The endophytic community in *L. cubeba* exhibited tissue specificity. A similar phenomenon was also observed in *Dendrobium officinale* [16], which may be caused by the different external environments or by the biological differences among tissues and organs [6]. Microorganisms in the environment usually show low diversity and low abundance compared with the soil [43]. The results of the present study support this point that roots harbor the maximum species diversity of endophytic fungi. Leaves harbor a greater number of fungal endophytes but with a lower diversity than other plant samples. This may be because the large surface area and the presence of stomata in leaves exposed to the external environment provide access for the entry of fungal mycelium, so that leaves may harbor a greater number of endophytic fungi [36]. However, the substantial organic compounds in leaves were largely inaccessible to foliar microorganisms, and microorganisms may present in the leaves in the form of co-metabolism, thus limiting the diversity of endophytic fungi in leaves [4,44,45].

Some fungal endophytes have been considered as beneficial mutualisms in protecting the host from pathogens [46]. In this study, the fungal endophytes were investigated for antifungal activity using a dual culture method. The results showed that 17 isolates inhibited the growth of plant pathogenic fungi. SF22 (*Chaetomium globosum*) showed strongest anti-pathogen activity. Previous studies demonstrated that some endophytic fungi could produce metabolites with antimicrobial function [6,37]. The endophytic extracts were screened for antifungal activity, and the results indicate that there were six endophytes exhibiting strong anti-pathogen activity. The extracts of SF22 (*C. globosum*) and SF14 (*Penicillium minioluteum*) were particularly effective in inhibiting pathogen growth. The dominant fungi, SF15 (*Colletotrichum boninense*), was less efficacious, though previous studies reported that *Colletotrichum* sp. showed a broad range of antifungal activity [47]. This phenomenon showed that there was no direct relationship between antifungal activity and fungal colonization rate [36]. *Chaetomium globosum* was reported to have disease control capacity by producing chaetoviridins and chaetoglobosin [48,49]. The application of the culture filtrates of *C. globosum* to maize showed efficacy in the inhibition of northern corn leaf blight [48]. *Penicillium* sp. was also reported to be efficacious against plant pathogenic fungi [50] and, interestingly, *P. minioluteum* attracted more attention for its beneficial effects on plant stress tolerance [51]. The growth inhibitory activity against plant pathogenic fungi by these endophytes indicates that endophytic fungi have the potential to be used as biocontrol agents in the future.

## 5. Conclusions

This study is the first to investigate the diversity of endophytic fungi in *L. cubeba*. The results demonstrated that *L. cubeba* harbors a rich fungal endophytic community with antimicrobial activities. SF22 (*C. globosum*) and SF14 (*P. minioluteum*) were found to have anti-pathogenic fungi properties and, thus, could be sources of novel natural antimicrobial compounds. Meanwhile, the results highlighted the potential use of endophytes in the development of drugs and the conservation of medicinal plants.

**Author Contributions:** L.Z. designed the study; D.Y. and Y.C. carried out the experiment and analyzed the data. F.W. wrote the first draft of the manuscript; F.W., D.Y., L.Z., Y.C., X.H., L.L. and J.L. contributed with suggestions and corrections, and approved the final manuscript. F.W. and D.Y. contributed equally to this work.

**Funding:** This work was supported by the National Natural Science Foundation of China [grant numbers 31660189, 31570594], and Hunan Provincial Natural Science Foundation of China (2018JJ2217, 2018JJ3281).

**Acknowledgments:** The authors thank Key Laboratory of State Forestry Administration on Forest Ecosystem Protection and Restoration of Poyang Lake Watershed (JXAU) for providing experimental equipment support.

**Conflicts of Interest:** The authors declare no conflict of interest.

## References

1. Marr, A.K.; Gooderham, W.J.; Hancock, R.E. Antibacterial Peptides for Therapeutic Use: Obstacles and Realistic Outlook. *Curr. Opin. Pharmacol.* **2006**, *6*, 468–472. [CrossRef] [PubMed]
2. Strobel, G.; Daisy, B. Bioprospecting for Microbial Endophytes and Their Natural Products. *Microbiol. Mol. Biol. Rev.* **2003**, *67*, 491–502. [CrossRef] [PubMed]
3. Petersen, P.J.; Wang, T.; Dushin, R.G.; Bradford, P.A. Comparative in Vitro Activities of Ac98-6446, a Novel Semisynthetic Glycopeptide Derivative of the Natural Product Mannopeptimycin A, and Other Antimicrobial Agents against Gram-Positive Clinical Isolates. *Antimicrob. Agents Chemother.* **2004**, *48*, 739–746. [CrossRef] [PubMed]
4. Baldrian, P. Forest Microbiome: Diversity, Complexity and Dynamics. *Fems Microbiol. Rev.* **2017**, *41*, 109–130. [CrossRef]
5. Lewis, S.L.; Edwards, D.P.; Galbraith, D. Increasing Human Dominance of Tropical Forests. *Science* **2015**, *349*, 827–832. [CrossRef] [PubMed]
6. Porrasalfaro, A.; Bayman, P. Hidden Fungi, Emergent Properties: Endophytes and Microbiomes. *Annu. Rev. Phytopathol.* **2011**, *49*, 291–315. [CrossRef] [PubMed]
7. Suryanarayanan, T.S.; Thirunavukkarasu, N.; Rajulu, M.B.G.; Venkatesan, G.; Sukumar, R. Endophytic Fungal Communities in Woody Perennials of Three Tropical Forest Types of the Western Ghats, Southern India. *Biodivers. Conserv.* **2011**, *20*, 913–928. [CrossRef]
8. Aly, A.H.; Debbab, A.; Kjer, J.; Proksch, P. Fungal Endophytes from Higher Plants: A Prolific Source of Phytochemicals and Other Bioactive Natural Products. *Fungal Divers.* **2010**, *41*, 1–16. [CrossRef]
9. Ownley, B.H.; Gwinn, K.D.; Vega, F.E. Endophytic Fungal Entomopathogens with Activity against Plant Pathogens: Ecology and Evolution. *BioControl* **2010**, *55*, 113–128. [CrossRef]
10. Rubini, M.R.; Silva-Ribeiro, R.T.; Pomella, A.W.V.; Maki, C.S.; Araújo, W.L.; Santos, D.R.D.; Azevedo, J.L. Diversity of Endophytic Fungal Community of Cacao (*Theobroma cacao* L.) and Biological Control of *Crinipellis perniciosa*,Causal Agent of Witches' Broom Disease. *Int. J. Biol. Sci.* **2005**, *1*, 24–33. [CrossRef]
11. Stierle, A.; Strobel, G.; Stierle, D. Taxol and Taxane Production by Taxomyces Andreanae, an Endophytic Fungus of Pacific Yew. *Science* **1993**, *260*, 214–216. [CrossRef] [PubMed]
12. Stierle, A.; Strobel, G.; Stierle, D.; Grothaus, P.; Bignami, G. The Search for a Taxol-Producing Microorganism among the Endophytic Fungi of the Pacific Yew, *Taxus Brevifolia*. *J. Nat. Prod.* **1995**, *58*, 1315–1324. [CrossRef] [PubMed]
13. Kour, A.; Shawl, A.S.; Rehman, S.; Sultan, P.; Qazi, P.H.; Suden, P.; Khajuria, R.K.; Verma, V. Isolation and Identification of an Endophytic Strain of *Fusarium Oxysporum* Producing Podophyllotoxin from *Juniperus Recurva*. *World J. Microbiol. Biotechnol.* **2008**, *24*, 1115–1121. [CrossRef]
14. Bérdy, J. Bioactive Microbial Metabolites. *J. Antibiot.* **2005**, *58*, 1–26. [CrossRef] [PubMed]

15. Ganley, R.J.; Brunsfeld, S.J.; Newcombe, G. A Community of Unknown, Endophytic Fungi in Western White Pine. *Proc. Natl. Acad. Sci. USA* **2004**, *101*, 10107–10112. [CrossRef] [PubMed]
16. Jin, Z.; Li, D.; Liu, T.; Yu, F.; Zhang, Z.; Su, C.; Wang, Y.; Guo, Q.; Liu, Z. Cultural Endophytic Fungi Associated with *Dendrobium Officinale*: Identification, Diversity Estimation and Their Antimicrobial Potential. *Curr. Sci.* **2017**, *112*, 1690–1697. [CrossRef]
17. Wang, F.; Chang, Z.; Duan, P.; Yan, W.; Xu, Y.; Zhang, L.; Miao, J.; Fan, Y. Hydrothermal Liquefaction of *Litsea Cubeba* Seed to Produce Bio-Oils. *Bioresour. Technol.* **2013**, *149*, 509–515. [CrossRef] [PubMed]
18. Feng, T.; Xu, Y.; Cai, X.H.; Du, Z.Z.; Luo, S.D. Antimicrobially Active Isoquinoline Alkaloids from *Litsea Cubeba*. *Planta Med.* **2009**, *75*, 76–79. [CrossRef]
19. Lin, B.; Zhang, H.; Zhao, X.X.; Rahman, K.; Wang, Y.; Ma, X.Q.; Zheng, C.J.; Zhang, Q.Y.; Han, T.; Qin, L.P. Inhibitory Effects of the Root Extract of *Litsea cubeba* (Lour.) Pers. on Adjuvant Arthritis in Rats. *J. Ethnopharmacol.* **2013**, *147*, 327–334. [CrossRef] [PubMed]
20. Wang, H.W.; Liu, Y.Q. Chemical Composition and Antibacterial Activity of Essential Oils from Different Parts of *Litsea cubeba*. *Chem. Biodivers.* **2010**, *7*, 229–235. [CrossRef]
21. Ho, C.L.; Jie-Ping, O.; Liu, Y.C.; Hung, C.P.; Tsai, M.C.; Liao, P.C.; Wang, E.I.; Chen, Y.L.; Su, Y.C. Compositions and in Vitro Anticancer Activities of the Leaf and Fruit Oils of *Litsea cubeba* from Taiwan. *Nat. Prod. Commun.* **2010**, *5*, 617–620. [PubMed]
22. Wang, Y.; Chen, J.; Li, D.W.; Zheng, L.; Huang, J. CglCUT1 gene required for cutinase activity and pathogenicity of *Colletotrichum gloeosporioides* causing anthracnose of *Camellia oleifera*. *Eur. J. Plant Pathol.* **2017**, *147*, 103–114. [CrossRef]
23. Wulff, E.G.; Sorensen, J.L.; Lubeck, M.; Nielsen, K.F.; Thrane, U.; Torp, J. *Fusarium* spp. associated with rice Bakanae: Ecology, genetic diversity, pathogenicity and toxigenicity. *Environ. Microbiol.* **2010**, *12*, 649–657. [CrossRef] [PubMed]
24. Tanaka, A.; Shiotani, H.; Yamamoto, M.; Tsuge, T. Insertional mutagenesis and cloning of the genes required for biosynthesis of the host-specific AK-toxin in the Japanese pear pathotype of *Alternaria alternata*. *Mol. Plant Microb. Interact.* **1999**, *12*, 691–702. [CrossRef]
25. Jaegoon, L.; Donghyun, L.; Sookyoung, P.; Jaeseoun, H.; Youngjin, K. First report of *Diaporthe actinidiae*, the causal organism of stem-end rot of kiwifruit in Korea. *Plant Pathol. J.* **2001**, *17*, 110–113.
26. Boa, E.R. *Recent Research on Bamboos*; The Chinese Academy of Forestry: Hanzhou, China, 1985.
27. Nagarajkumar, M.; Bhaskaran, R.; Velazhahan, R. Involvement of secondary metabolites and extracellular lytic enzymes produced by *Pseudomonas fluorescens* in inhibition of *Rhizoctonia solani*, the rice sheath blight pathogen. *Microbiol. Res.* **2004**, *159*, 73–81. [CrossRef]
28. Foster, J.M.; Hausbeck, M.K. Managing Phytophthora Crown and Root Rot in Bell Pepper Using Fungicides and Host Resistance. *Plant Dis.* **2010**, *94*, 697–702. [CrossRef]
29. Jin, Z.; Gao, L.; Zhang, L.; Liu, T.; Yu, F.; Zhang, Z.; Guo, Q.; Wang, B. Antimicrobial Activity of Saponins Produced by Two Novel Endophytic Fungi from *Panax notoginseng*. *Nat. Prod. Res.* **2017**, *31*, 2700–2703. [CrossRef]
30. Morakotkarn, D.; Kawasaki, H.; Seki, T. Molecular Diversityof Bamboo-Associated Fungi Isolated from Japan. *FEMS Microbiol. Lett.* **2007**, *266*, 10–19. [CrossRef]
31. White, T.J.; Bruns, T.; Lee, S.; Taylor, J. Amplification and Direct Sequencing of Fungal Ribosomal Rna Genes for Phylogenetics. In *PCR Protocols: A Guide to Methods and Applications*; Academic Press: San Diego, CA, USA, 1990; Volume 18, pp. 315–322.
32. Tamura, K.; Peterson, D.; Peterson, N.; Stecher, G.; Nei, M.; Kumar, S. Mega5: Molecular Evolutionary Genetics Analysis Using Maximum Likelihood, Evolutionary Distance, and Maximum Parsimony Methods. *Mol. Biol. Evol.* **2011**, *28*, 2731–2739. [CrossRef]
33. Nascimento, T.L.; Oki, Y.; Lima, D.M.M.; Almeida-Cortez, J.S.; Fernandes, G.W.; Souza-Motta, C.M. Biodiversity of Endophytic Fungi in Different Leaf Ages of *Calotropis procera* and Their Antimicrobial Activity. *Fungal Ecol.* **2015**, *14*, 79–86. [CrossRef]
34. Yuan, Z.L.; Zhang, C.L.; Lin, F.C.; Kubicek, C.P. Identity, Diversity, and Molecular Phylogeny of the Endophytic Mycobiota in the Roots of Rare Wild Rice (*Oryza granulate*) from a Nature Reserve in Yunnan, China. *Appl. Environ. Microbiol.* **2010**, *76*, 1642–1652. [CrossRef]

35. Kusari, P.; Kusari, S.; Spiteller, M.; Kayser, O. Endophytic Fungi Harbored in *Cannabis sativa* L.: Diversity and Potential;as Biocontrol Agents against Host Plant-Specific Phytopathogens. *Fungal Divers.* **2013**, *60*, 137–151. [CrossRef]
36. Gond, S.K.; Mishra, A.; Sharma, V.K.; Verma, S.K.; Kumar, J.; Kharwar, R.N.; Kumar, A. Diversity and Antimicrobial Activity of Endophytic Fungi Isolated from *Nyctanthes arbor-tristis*, a Well-Known Medicinal Plant of India. *Mycoscience* **2012**, *53*, 113–121. [CrossRef]
37. Katoch, M.; Phull, S.; Vaid, S.; Singh, S. Diversity, Phylogeny, Anticancer and Antimicrobial Potential of Fungal Endophytes Associated with *Monarda citriodora* L. *BMC Microbiol.* **2017**, *17*, 44–57. [CrossRef] [PubMed]
38. Liang, H.; Xing, Y.; Chen, J.; Zhang, D.; Guo, S.; Wang, C. Antimicrobial Activities of Endophytic Fungi Isolated from *Ophiopogon japonicus* (Liliaceae). *BMC Complement. Altern. Med.* **2012**, *12*, 238–244. [CrossRef] [PubMed]
39. Gherbawy, Y.A.; Gashgari, R.M. Molecular Characterization of Fungal Endophytes from *Calotropis procera* Plants in Taif Region (Saudi Arabia) and Their Antifungal Activities. *Plant Biosyst.* **2014**, *148*, 1085–1092. [CrossRef]
40. Schoch, C.L.; Gi-Ho, S.; Francesc, L.G.; Townsend, J.P.; Jolanta, M.; Valérie, H.; Barbara, R.; P Brandon, M.; Frank, K.; Zheng, W. The Ascomycota Tree of Life: A Phylum-Wide Phylogeny Clarifies the Origin and Evolution of Fundamental Reproductive and Ecological Traits. *Syst. Biol.* **2009**, *58*, 224–239. [CrossRef]
41. Jumpponen, A.; Jones, K. Massively Parallel 454 Sequencing Indicates Hyperdiverse Fungal Communities in Temperate Quercus *Macrocarpa phyllosphere*. *New Phytol.* **2009**, *184*, 438–448. [CrossRef]
42. Ho, M.Y.; Chung, W.C.; Huang, H.C.; Chung, W.H.; Chung, W.H. Identification of Endophytic Fungi of Medicinal Herbs of Lauraceae and Rutaceae with Antimicrobial Property. *Taiwania* **2012**, *57*, 229–241.
43. Šamonil, P.; Valtera, M.; Bek, S.; Šebková, B.; Vrška, T.; Houška, J. Soil Variability through Spatial Scales in a Permanently Disturbed Natural Spruce-Fir-Beech Forest. *Eur. J. For. Res.* **2011**, *130*, 1075–1091. [CrossRef]
44. Lebeis, S.L. Greater Than the Sum of Their Parts: Characterizing Plant Microbiomes at the Community-Level. *Curr. Opin. Plant Biol.* **2015**, *24*, 82–86. [CrossRef] [PubMed]
45. Vorholt, J.A. Microbial Life in the Phyllosphere. *Nat. Rev. Microbiol.* **2012**, *10*, 828–840. [CrossRef] [PubMed]
46. Tan, R.X.; Zou, W.X. Endophytes: A Rich Source of Functional Metabolites. *Nat. Prod. Rep.* **2001**, *18*, 448–459. [CrossRef] [PubMed]
47. Zou, W.; Meng, J.; Lu, H.; Chen, G.; Shi, G.; Zhang, T.; Tan, R. Metabolites of *Colletotrichum gloeosporioides*, an Endophytic Fungus in Artemisia Mongolica. *J. Nat. Prod.* **2000**, *63*, 1529–1530. [CrossRef] [PubMed]
48. Zhang, G.; Wang, F.; Qin, J.; Wang, D.; Zhang, J.; Zhang, Y.; Zhang, S.; Pan, H. Efficacy Assessment of Antifungal Metabolites from *Chaetomium globosum* No. 05, a New Biocontrol Agent, against *Setosphaeria turcica*. *Biol. Control* **2013**, *64*, 90–98. [CrossRef]
49. Park, J.H.; Choi, G.J.; Jang, K.S.; Lim, H.K.; Kim, H.T.; Cho, K.Y.; Kim, J.C. Antifungal Activity against Plant Pathogenic Fungi of Chaetoviridins Isolated from *Chaetomium globosum*. *FEMS Microbiol. Lett.* **2005**, *252*, 309–313. [CrossRef]
50. Yang, L.; Xie, J.; Jiang, D.; Fu, Y.; Li, G.; Lin, F. Antifungal Substances Produced by *Penicillium oxalicum* Strain Py-1—Potential Antibiotics against Plant Pathogenic Fungi. *World J. Microbiol. Biotechnol.* **2008**, *24*, 909–915. [CrossRef]
51. Sahoo, S.; Sarangi, S.; Kerry, R.G. Bioprospecting of Endophytes for Agricultural and Environmental Sustainability. *Microb. Biotechnol.* **2017**, 429–458.

forests

MDPI

Article

# Variation in Photosynthetic Traits and Correlation with Growth in Teak (*Tectona grandis* Linn.) Clones

**Guihua Huang [1,*], Kunnan Liang [1], Zaizhi Zhou [1], Guang Yang [1] and Enarth Maviton Muralidharan [2]**

1 Research Institute of Tropical Forestry, Chinese Academy of Forestry, Guangzhou 510520, China; lkn@ritf.ac.cn (K.L.); zzzhou@ritf.ac.cn (Z.Z.); yg2018.8@ritf.ac.cn (G.Y.)
2 Kerala Forest Research Institute, Peechi, Kerala 680653, India; emmurali@kfri.org
* Correspondence: huanggh@ritf.ac.cn; Tel.: +86-20-8703-2929

Received: 24 July 2018; Accepted: 10 September 2018; Published: 10 January 2019

**Abstract:** In order to interpret the patterns of genetic variation of photosynthesis and the relationships with growth traits within gene resources of teak (*Tectona grandis* Linn.), gas exchange, and chlorophyll fluorescence parameters, growth traits of plants in nursery and field trials were measured for 20 teak clones originated from different countries. The results show that there was abundant genetic variation in gas exchange, chlorophyll fluorescence, and growth among the teak clones. The measured traits were found to have generally high heritability ($h^2$) except for intercellular concentration of carbon dioxide ($CO_2$) ($C_i$). The net photosynthetic rate ($P_n$), seedling height, and individual volume of wood were significantly correlated with each other, and seedling height was significantly correlated with plant height in field trials, suggesting that $P_n$ and seedling height can be useful in teak breeding. Teak clones 7029, 71-5, 7219, 7412, and 7122, and provenances 3070, 3074, and 3071 had higher photosynthetic rates, and can be regarded as a key resource in teak improvement programs. This work provides useful information for teak breeding and germplasm resource management.

**Keywords:** gas exchange; chlorophyll fluorescence; growth trait; genetic variation; early selection

## 1. Introduction

Teak (*Tectona grandis* Linn.) is naturally distributed in India, Thailand, Myanmar, and Laos [1,2]. Its desirable hardwood properties, fine grain, and durability have made teak the luxury timber for furniture making, carving, and building around the world [3,4]. Due to its economical importance, teak has been introduced widely in the tropical regions since the 19th century, especially in Asia, Africa, Central America, and South America [5].

As one of the most valuable wood species in international markets, teak plantations have developed rapidly in the recent decade. Developing high productivity and uniform clones that can be used for plantations in different regions has become an important objective of teak breeding. Information on variation of photosynthetic parameters and their relationship with growth traits help us understand underlying processes and responses, and will be useful in tree improvement programs. During the growth process of plants, organic compounds are generated by photosynthesis, and gradually accumulate in trunks. The photosynthetic characteristics are the main measurable indicators of plant growth rates [6]. Numerous studies on breeding for high photosynthetic ability in crops have been conducted, to improve the yield [7–9], but studies on forest trees are limited [10–12]. Chu et al. 2010 [10] studied gas exchange and chlorophyll fluorescence parameters, as well as their relationship with the growth of *Populus nigra*, and found that the species originating in Serbia, southern and east Europe can be regarded as a resource with high light-use efficiency for future breeding. Teak has broad leaves and prefers warmth and sunlight, and developing clones with high productivity and uniformity by evaluating photosynthetic characteristics can be an important goal in teak breeding. In the past, teak breeding was mainly focused on the analysis of growth indices in field experiments [13,14], and studies on photosynthetic physiology of teak are limited to those on photosynthetic responses of a single clone to simulated acid rain stress [15], photosynthetic physiological characteristics under different disturbance intensities among teak plants [16], and diurnal and seasonal photosynthetic characteristics in teak clones [17]. However, studies on teak germplasm or clones, which systematically estimate photosynthetic characteristics and correlation with growth, have not been reported.

The purpose of this study was (1) to investigate the genetic variation of photosynthetic parameters and growth traits of teak clones, (2) to reveal the correlation, if any, between photosynthetic characteristics and growth traits within the gene resources of teak, and (3) to evaluate and select superior teak resources possessing high photosynthetic efficiency for breeding.

## 2. Materials and Methods

### 2.1. Materials

A total of 20 widely cultivated teak clones propagated through tissue culture were investigated in this study. The 20 clones were selected from international provenance trial planted at Jianfeng, Hainan, China, by the Research Institute of Tropical Forestry of Chinese Academy of Forestry (RITF-CAF). A complete list of accessions with descriptions and origins is given in Table 1. Among these accessions, 10 were clones originating from India, 9 were from Myanmar sources and 1 from Nigeria.

**Table 1.** Information of 20 commercial teak clones investigated in the study.

| Clone Name | Provenance Name | Longitude Latitude | Altitude (m) | Annual Rain Fall (mm) |
|---|---|---|---|---|
| Clones from India provenances | | | | |
| 7013 | 3070 | 77°20′ E 08°00′ N | 700 | 1270 |
| 7029 | 3070 | 77°20′ E 08°00′ N | 700 | 1270 |
| 71-5 | 3071 | 76°47′ E 10°30′ N | 640 | 2030 |
| 7114 | 3071 | 76°47′ E 10°30′ N | 640 | 2030 |
| 7122 | 3071 | 76°47′ E 10°30′ N | 640 | 2030 |
| 7146 | 3071 | 76°47′ E 10°30′ N | 640 | 2030 |
| 7137 | 3071 | 76°47′ E 10°30′ N | 640 | 2030 |
| 7210 | 3072 | 76°10′ E 11°55′ N | 823 | 1270 |
| 7219 | 3072 | 76°10′ E 11°55′ N | 823 | 1270 |
| 7412 | 3074 | 74°28′ E 15°12′ N | 43 | 2032 |
| Clones from planted provenances in China (Myanmar source but no detailed records of origin) | | | | |
| Z408 | 20001 | 110°14′ E 21°07′ N | 60 | 1650 |
| 7509 | 8204 | 108°42′ E 18°51′ N | 60 | 1600 |
| 7514 | 8204 | 108°42′ E 18°51′ N | 60 | 1600 |
| 7531 | 8204 | 108°42′ E 18°51′ N | 60 | 1600 |
| 7544 | 8204 | 108°42′ E 18°51′ N | 60 | 1600 |
| 7549 | 8204 | 108°42′ E 18°51′ N | 60 | 1600 |
| 7555 | 8204 | 108°42′ E 18°51′ N | 60 | 1600 |
| 7559 | 8204 | 108°42′ E 18°51′ N | 60 | 1600 |
| 8301 | 8204 | 108°42′ E 18°51′ N | 60 | 1600 |
| Clone from Nigeria provenance | | | | |
| 3078-5 | 3078 | 03°52′ E 07°10′ N | 700 | 1900 |

### 2.2. Experimental Design and Growth Parameter Measurement

The young in vitro plantlets of teak clones were transplanted to a sterilized sand bed in the greenhouse at the Research Institute of Tropical Forestry, Chinese Academy of Forestry (RITF-CAF), in Guangzhou (113°18′ E, 23°06′ N). One month later, healthy and uniform seedlings (Ramets derived from each clone) about 6 cm in height were transplanted into plastic pots filled with a mixture of lateritic red soil, black peat, vermiculite, and perlite (2:2:1:1, v/v/v/v)—one seedling per pot. A completely randomized block design was used in this nursery experiment with 5 seedlings in one row per plot, 6 repeats in total with 40 cm × 40 cm pot space. Seedling height and collar diameter of all seedlings in the nursery were measured at the age of one year.

Field trial was carried out at Dingan in Hainan Island (110°19′ E, 19°39′ N) and a completely randomized block design was used with 6 plants in one row per plot, 6 repeats in total, with 2.5 m × 4 m space. Plant height (H) and diameter at breast height (DBH) of each plant in field trial were measured at the age of four years.

### 2.3. Physiological Parameter Measurement

Three seedlings in the nursery were randomly selected for each clone, 1 seedling per plot, 3 repeats in total, in a completely randomized block design (to make sure the test was random for all 60 selected seedlings) and 3 functional leaves per seedling exposed to sunlight were chosen for study. The gas exchange parameters including net photosynthetic rate ($P_n$), stomatal conductance ($G_s$), intercellular $CO_2$ concentration ($C_i$), and transpiration rate ($Tr$) were measured on sunny days between 8:45 and 11:30 in August 2013 with a Li-6400 portable photosynthetic apparatus (LI-COR Co. Lincoln, NE, USA) at the nursery of RITF-CAF, in Guangzhou. A leaf chamber automatic light (800 $\mu mol \cdot m^{-2} \cdot s^{-1}$) was used when testing, with $CO_2$ concentration 380 ± 15 $\mu mol \cdot mol^{-1}$, temperature of the leaf chamber 30–38 °C, and a relative humidity 58%–68% recorded by the photosynthetic apparatus under natural conditions. Three stable values were recorded for each leaf. Chlorophyll fluorescence characteristics were measured at the same time using the German PAM-2500 Walz portable fluorescence spectrometer, the saturation pulse intensity was 4500 $mol \cdot m^{-2} \cdot s^{-1}$, and actinic intensity was 1000 $mol \cdot m^{-2} \cdot s^{-1}$. The actual quantum yield PSII (Yield), non-photochemical quenching (NPQ) and maximum photochemical efficiency of PSII ($F_v/F_m$) were also measured [18,19]. The calculation formula for Yield is Yield = $(F_m' - F_t)/F_m'$, where $F_m'$ is referred to maximum fluorescence under light adaptation, and $F_t$ denotes real fluorescence at any given time. The formula of NPQ = $F_m/F_m' - 1$; $(F_v/F_m) = (F_m - F_o)/F_m$, and $F_o$, $F_m$, and $F_v$ refer to dark-adapted initial fluorescence, maximum fluorescence, and variable fluorescence, respectively. Before testing, 20 min shading treatment was carried out with a blade holder to ensure selected leaves had dark adaptation for a long enough period of time.

### 2.4. Data Analysis

Water use efficiency ($WUE$) was calculated by the formula $WUE = P_n/Tr$ [20], and the coefficient of variation was calculated by the formula C = S/X, where S is the standard deviation, and X is the overall average value of each index. Clone heritability was calculated with the formula $h^2 = 1 - 1/F$ [21], where F is test statistic of clones in variance analysis. Individual volume of wood was calculated by the formula $V = 0.4787D^2H$, where D is DBH, and H is plant height of field trial [22]. Variance and Duncan's multiple comparison analyses were conducted for each parameter, and correlation analyses (using Pearson's product-moment correlations) between photosynthetic parameters, water use efficiency, and growth index, were performed using SAS software (version 8.1).

## 3. Results

### 3.1. Gas Exchange, Chlorophyll Fluorescence, and Growth Traits of Different Teak Clones

Variance analysis of gas exchange, chlorophyll fluorescence, and growth parameters among teak clones are shown in Table 2. There is a significant difference in the photosynthetic parameters, water use efficiency, and growth index but not for intercellular $CO_2$ concentration ($C_i$). In addition, apart from $C_i$ ($h^2 = 0.145$) and NPQ ($h^2 = 0.168$), other parameters had high heritability ($h^2 = 0.670$–$0.903$), with actual quantum yield PSII (Yield) having the highest heritability ($h^2$ = 0.903), suggesting a strong genetic influence on the function, and that it is less affected by environment.

**Table 2.** Variance analysis (ANOVA) of gas exchange, chlorophyll fluorescence, water use efficiency, and growth parameters among teak clones.

| Category | Parameter | $F$ | $p$ | Heritability ($h^2$) | Variation Coefficient |
|---|---|---|---|---|---|
| Gas exchange | $P_n$ | 5.46 | <0.0001 *** | 0.817 | 0.401 |
| | $G_s$ | 4.53 | <0.0001 *** | 0.779 | 0.474 |
| | $C_i$ | 1.17 | 0.3213 $^{ns}$ | 0.145 | 0.111 |
| | $Tr$ | 3.38 | 0.0004 *** | 0.704 | 0.349 |
| Chlorophyll fluorescence | NPQ | 5.93 | <0.0001 *** | 0.168 | 0.447 |
| | Yield | 10.32 | <0.0001 *** | 0.903 | 0.294 |
| | $F_v/F_m$ | 3.03 | 0.0011 ** | 0.670 | 0.028 |
| Water use efficiency | $WUE$ | 10.32 | <0.0001 *** | 0.903 | 0.474 |
| Seedling growth at 1 year | Seedling height | 3.39 | <0.0001 *** | 0.705 | 0.116 |
| | Collar diameter | 3.29 | <0.0001 *** | 0.696 | 0.102 |
| Field growth at 4 years | H | 7.88 | <0.0001 *** | 0.873 | 0.092 |
| | DBH | 6.74 | <0.0001 *** | 0.852 | 0.167 |
| Individual volume at 4 years | V | 7.24 | <0.0001 *** | 0.863 | 0.327 |

Note: $P_n$: net photosynthetic rate, $G_s$: stomatal conductance, $C_i$: intercellular $CO_2$ concentration, $Tr$: transpiration rate, NPQ: non-photochemical quenching, Yield: the actual quantum yield PSII, $F_v/F_m$: maximum photochemical efficiency of PSII, $WUE$: water use efficiency, H: height of field growth at 4 years, DBH: diameter at breast height of field growth at 4 years, V: individual volume at 4 years. ** indicate highly significant difference at $p < 0.01$ level of probability, *** more highly significant difference at $p < 0.001$ level of probability, and $^{ns}$ no significance.

Duncan's multiple comparison analysis of photosynthetic and growth traits are listed in Tables 3–5. The ranges of the main parameters, such as $P_n$ and $F_v/F_m$, were 4.45 ± 1.62–14.47 ± 0.32 μmol·m$^{-2}$·s$^{-1}$, 0.67 ± 0.02–0.75 ± 0.01, respectively. Water use efficiency ($WUE$) was between 1.02 ± 0.36 and 6.38 ± 1.25. Apart from the maximum photochemical efficiency of PSII $F_v/F_m$ (0.028), the variation coefficients of other parameters (0.092–0.474) were great, suggesting that the teak genotypes possessed extensive variation in these traits. Results indicated that there are suitable germplasm resources for breeding of teak for high photosynthetic efficiency. Teak clones 7029, 71-5, 7219, 7412, and 7122 were selected as clones with high net photosynthetic rate based on the results.

**Table 3.** Values of gas exchange parameters among teak clones from different countries.

| Clone | $P_n$ (μmol·m$^{-2}$·s$^{-1}$) | $G_s$ (mol·m$^{-2}$·s$^{-1}$) | $C_i$ (μmol·mol$^{-1}$) | $Tr$ (mmol·m$^{-2}$·s$^{-1}$) |
|---|---|---|---|---|
| 7013 | 8.09 ± 3.76 defgh | 0.14 ± 0.09 cdefgh | 263.98 ± 7.95 ab | 3.34 ± 1.58 abcde |
| 7029 | 14.19 ± 1.17 ab | 0.23 ± 0.04 abcde | 261.92 ± 7.42 ab | 2.69 ± 0.28 cdefg |
| 71-5 | 13.50 ± 2.34 abc | 0.28 ± 0.07 ab | 290.25 ± 5.21 ab | 3.04 ± 0.70 abcdef |
| 7114 | 9.87 ± 4.09 abcdefg | 0.18 ± 0.10 bcdefg | 237.06 ± 18.08 b | 3.81 ± 1.61 abc |
| 7122 | 11.58 ± 2.02 abcd | 0.25 ± 0.07 abcde | 287.70 ± 8.14 ab | 3.91 ± 0.71 abc |
| 7146 | 9.59 ± 1.30 bcdefgh | 0.20 ± 0.03 abcdef | 280.00 ± 3.58 ab | 4.02 ± 0.59 abc |
| 7210 | 6.16 ± 0.77 efghij | 0.08 ± 0.02 fgh | 219.18 ± 28.34 b | 2.07 ± 0.56 cdefg |
| 7219 | 14.47 ± 0.32 a | 0.32 ± 0.02 a | 287.80 ± 6.16 ab | 3.97 ± 0.25 abc |
| 3078-5 | 9.66 ± 2.87 abcdefgh | 0.22 ± 0.07 abcde | 292.56 ± 4.91 ab | 3.80 ± 0.66 abc |
| 7412 | 11.01 ± 2.66 abcde | 0.29 ± 0.09 ab | 298.57 ± 18.78 ab | 4.73 ± 0.97 a |
| 7509 | 7.04 ± 2.62 efghi | 0.09 ± 0.05 fgh | 220.30 ± 45.30 b | 1.34 ± 0.80 fg |
| 7514 | 10.19 ± 0.66 abcdef | 0.25 ± 0.03 abcd | 295.67 ± 20.91 ab | 3.33 ± 0.41 abcde |
| 7531 | 5.23 ± 3.07 ghij | 0.12 ± 0.08 efgh | 281.11 ± 3.07 ab | 3.05 ± 1.80 abcdef |
| 7544 | 7.74 ± 2.27 defgh | 0.17 ± 0.09 bcdefgh | 284.34 ± 18.03 ab | 2.97 ± 1.01 abcdef |
| 7549 | 4.92 ± 3.96 ij | 0.07 ± 0.06 h | 345.30 ± 199.15 a | 0.95 ± 0.72 g |
| 7555 | 4.45 ± 1.62 j | 0.06 ± 0.04 h | 270.91 ± 41.52 ab | 2.21 ± 0.89 cdefg |
| 7559 | 5.68 ± 4.55 fghij | 0.12 ± 0.12 efgh | 272.62 ± 33.65 ab | 2.72 ± 2.32 bcdefg |
| 8301 | 7.40 ± 1.80 defgh | 0.15 ± 0.04 cdefgh | 268.42 ± 10.16 ab | 3.55 ± 0.71 abcd |
| Z408 | 8.78 ± 0.78 cdefgh | 0.25 ± 0.03 abcde | 310.87 ± 2.85 ab | 4.30 ± 0.28 ab |
| 7137 | 10.04 ± 3.10 abcdefg | 0.14 ± 0.09 cdefgh | 224.10 ± 59.08 b | 1.80 ± 0.87 defg |

Note: $P_n$: net photosynthetic rate, $G_s$: stomatal conductance, $C_i$: intercellular $CO_2$ concentration, $Tr$: transpiration rate. Values followed by the different letter of each group were significantly different at $p < 0.05$ level of probability.

**Table 4.** Values of chlorophyll fluorescence parameters and *WUE* among teak clones.

| Clone | NPQ | Yield | $F_v/F_m$ | *WUE* |
|---|---|---|---|---|
| 7013 | 0.773 ± 0.256 def | 0.42 ± 0.07 bcd | 0.69 ± 0.02 bcdef | 2.43 ± 0.06 def |
| 7029 | 1.547 ± 0.397 abcde | 0.35 ± 0.09 cdefg | 0.72 ± 0.01 abcd | 5.29 ± 0.12 ab |
| 71-5 | 1.785 ± 0.268 abc | 0.21 ± 0.05 hi | 0.70 ± 0.03 bcdef | 4.48 ± 0.30 bc |
| 7114 | 0.891 ± 0.192 cdef | 0.47 ± 0.01 ab | 0.75 ± 0.01 a | 2.60 ± 0.03 de |
| 7122 | 0.585 ± 0.104 ef | 0.44 ± 0.01 bc | 0.70 ± 0.01 bcdef | 2.96 ± 0.03 de |
| 7146 | 0.834 ± 0.172 cdef | 0.40 ± 0.03 bcde | 0.70 ± 0.02 bcdef | 2.39 ± 0.08 def |
| 7210 | 0.800 ± 0.285 def | 0.41 ± 0.05 bcde | 0.68 ± 0.01 def | 3.05 ± 0.45 de |
| 7219 | 2.165 ± 0.147 a | 0.25 ± 0.01 ghi | 0.72 ± 0.03 abcd | 3.66 ± 0.30cd |
| 3078-5 | 1.528 ± 0.060 abcde | 0.32 ± 0.01 defgh | 0.71 ± 0.02 bcde | 2.51 ± 0.31 def |
| 7412 | 0.421 ± 0.179 f | 0.56 ± 0.04 a | 0.72 ± 0.02 abcd | 2.33 ± 0.34 def |
| 7509 | 2.079 ± 0.258 a | 0.20 ± 0.05 i | 0.67 ± 0.03 f | 6.38 ± 2.82 a |
| 7514 | 1.978 ± 0.582 ab | 0.20 ± 0.07 i | 0.70 ± 0.01 bcdef | 3.09 ± 0.37 de |
| 7531 | 1.006 ± 0.547 cdef | 0.41 ± 0.10 bcde | 0.71 ± 0.01 bcde | 1.78 ± 0.30 ef |
| 7544 | 1.997 ± 0.531 ab | 0.23 ± 0.04 hi | 0.70 ± 0.01 bcdef | 2.66 ± 0.36 de |
| 7549 | 1.646 ± 0.730 abcd | 0.25 ± 0.10 ghi | 0.73 ± 0.02 ab | 5.16 ± 1.04 ab |
| 7555 | 1.578 ± 0.046 abcd | 0.32 ± 0.03 defgh | 0.69 ± 0.01 cdef | 1.02 ± 0.36 f |
| 7559 | 1.737 ± 0.446 abcd | 0.26 ± 0.06 fghi | 0.67 ± 0.02 f | 2.03 ± 0.49 ef |
| 8301 | 1.085 ± 0.166 bcdef | 0.37 ± 0.05 bcdef | 0.70 ± 0.02 bcdef | 2.07 ± 0.19 ef |
| Z408 | 1.215 ± 0.732 abcdef | 0.36 ± 0.12 cdefg | 0.71 ± 0.01 bcde | 2.04 ± 0.08 ef |
| 7137 | 1.398 ± 0.212 abcde | 0.31 ± 0.04 efghi | 0.73 ± 0.03 abc | 5.98 ± 1.25 a |

Note: NPQ: non-photochemical quenching, Yield: the actual quantum yield PSII, $F_v/F_m$: maximum photochemical efficiency of PSII, *WUE*: water use efficiency. Values followed by the different letter of each group were significantly different at $p < 0.05$ level of probability.

**Table 5.** Values of growth traits among teak clones from different countries.

| Clone | Seedling Height (cm) | Collar Diameter (mm) | H (m) | DBH (cm) | Individual Volume ($dm^3$) |
|---|---|---|---|---|---|
| 7013 | 48.67 ± 8.16 cdefg | 10.19 ± 2.18 bcdef | 4.35 ± 1.03 cdefg | 4.38 ± 1.40 bcde | 6.25 ± 4.50 efghi |
| 7029 | 61.13 ± 15.51 a | 11.12 ± 2.39 abcd | 4.87 ± 0.92 ab | 5.37 ± 1.58 ab | 11.01 ± 7.56 ab |
| 71-5 | 53.63 ± 6.85 abcd | 10.32 ± 1.79 bcde | 3.98 ± 0.66 fghi | 4.44 ± 1.59 bcde | 6.79 ± 4.80 defghi |
| 7114 | 44.33 ± 12.64 defg | 9.53 ± 2.37 def | 4.43 ± 0.68 bcdefg | 5.03 ± 1.27 abc | 9.12 ± 5.04 bcde |
| 7122 | 59.53 ± 13.65 ab | 12.63 ± 3.03a | 4.88 ± 0.60 ab | 5.32 ± 1.10 ab | 9.97 ± 4.21 bcd |
| 7146 | 48.87 ± 11.21 cdefg | 10.97 ± 2.63 abcde | 4.24 ± 1.02 defgh | 4.78 ± 1.76 bcd | 8.01 ± 6.75 bcdefg |
| 7210 | 52.07 ± 10.82 abcde | 8.65 ± 1.58 f | 4.19 ± 0.81 efgh | 3.88 ± 1.25 cdef | 6.39 ± 3.29 efghi |
| 7219 | 51.07 ± 10.03 bcdef | 9.97 ± 2.23 cdef | 3.97 ± 0.61 fghi | 4.48 ± 1.48 bcde | 8.26 ± 6.95 bcdef |
| 3078-5 | 49.80 ± 9.40 cdef | 9.95 ± 1.53 cdef | 5.07 ± 0.99 a | 6.05 ± 1.91 a | 13.59 ± 7.06 a |
| 7412 | 44.53 ± 11.45 defg | 9.95 ± 2.08 cdef | 4.28 ± 0.77 defgh | 4.68 ± 1.53 bcde | 8.63 ± 7.49 bcdef |
| 7509 | 41.07 ± 10.71 gf | 9.75 ± 2.00 def | 4.41 ± 1.13 bcdefg | 5.96 ± 5.79 a | 8.26 ± 5.85 bcdef |
| 7514 | 44.93 ± 15.40 defg | 9.28 ± 1.43 def | 4.71 ± 0.84 abcd | 4.79 ± 1.27 bcd | 10.75 ± 6.18 abc |
| 7531 | 39.40 ± 12.33 g | 9.35 ± 2.42 def | 3.65 ± 1.11 ij | 3.53 ± 1.52 ef | 2.81 ± 2.05 j |
| 7544 | 46.80 ± 9.84 cdefg | 10.76 ± 2.47 abcde | 4.36 ± 0.89 cdefg | 4.20 ± 1.43 bcdef | 5.71 ± 3.46 fghij |
| 7549 | 42.20 ± 17.77 efg | 9.40 ± 3.57 def | 4.18 ± 0.80 efgh | 4.34 ± 1.17 bcde | 5.38 ± 3.39 fghij |
| 7555 | 46.60 ± 7.40 cdefg | 11.84 ± 1.45 abc | 4.38 ± 0.63 cdefg | 4.36 ± 0.83 bcde | 6.18 ± 3.14 efghi |
| 7559 | 55.60 ± 12.16 abc | 12.33 ± 2.54 a | 4.59 ± 1.00 bcde | 5.09 ± 1.77 abc | 9.18 ± 5.61 bcde |
| 8301 | 50.47 ± 15.18 bcdef | 10.30 ± 2.81 bcdef | 4.79 ± 1.07 abc | 4.18 ± 1.11 bcdef | 8.00 ± 5.43 bcdefg |
| Z408 | 45.87 ± 10.12 defg | 12.01 ± 2.37ab | 3.97 ± 0.82 fghi | 3.62 ± 1.22 def | 4.06 ± 2.66 ij |
| 7137 | 45.00 ± 8.60 defg | 9.95 ± 2.26 cdef | 4.33 ± 0.82 cdefg | 4.23 ± 1.57 bcdef | 8.33 ± 3.83 bcdef |

Note: H: height of field growth at 4 years, DBH: diameter at breast height of field growth at 4 years. Values followed by the different letter of each group were significantly different at $p < 0.05$ level of probability.

### *3.2. Characteristics of Gas Exchange and Chlorophyll Fluorescence of Teak Resources from Different Regions*

As shown in Table 6, net photosynthetic rate ($P_n$), stomatal conductance ($G_s$), transpiration rate ($Tr$), and non-photochemical quenching (NPQ), and actual quantum yield (Yield) of PSII were significantly different among teak provenances.

**Table 6.** Variance analysis of photosynthetic parameters among teak clones from different provenances.

| Category | Parameter | Among Provenances | |
|---|---|---|---|
| | | $F$ | $p$ |
| Gas exchange | $P_n$ | 3.52 | 0.0033 ** |
| | $G_s$ | 3.19 | 0.0062 ** |
| | $C_i$ | 1.13 | 0.3667 $^{ns}$ |
| | $Tr$ | 5.40 | 0.0001 *** |
| Chlorophyll fluorescence | NPQ | 3.38 | 0.007 ** |
| | Yield | 5.59 | <0.0001 *** |
| | $F_v/F_m$ | 1.95 | 0.0758 $^{ns}$ |

Note: $P_n$: net photosynthetic rate, $G_s$: stomatal conductance, $C_i$: intercellular $CO_2$ concentration, $Tr$: transpiration rate, NPQ: non-photochemical quenching, Yield: the actual quantum yield PSII, $F_v/F_m$: maximum photochemical efficiency of PSII. ** indicate highly significant difference at $p < 0.01$ level of probability, *** more highly significant difference at $p < 0.001$ level of probability, and $^{ns}$ no significance.

Among the teak provenances (Table 7), 3070, 3074, and 3071 had higher $P_n$, 3074 had higher $G_s$ and $Tr$, while 20001 and 3074 had a higher $C_i$ value. While 8204, 3078, and 3072 showed high NPQ value (Table 8), 3074, 3071, and 3070 had higher Yield and $F_v/F_m$ values. These results suggest that different teak provenances have different photosynthetic physiological characteristics. Provenances 3070, 3074, and 3071 can be considered to have high photosynthetic rates.

**Table 7.** Values of gas exchange parameters among teak clones from different provenances.

| Provenance | $P_n$ (μmol·m$^{-2}$ s$^{-1}$) | $G_s$ (mol·m$^{-2}$ s$^{-1}$) | $C_i$ (μmol·mol$^{-1}$) | $Tr$ (mmol·m$^{-2}$ s$^{-1}$) |
|---|---|---|---|---|
| 20001 | 8.775 ± 0.776 ab | 0.248 ± 0.032 abc | 310.865 ± 2.848 a | 4.296 ± 0.280 ab |
| 3070 | 11.142 ± 4.169 a | 0.187 ± 0.075 abc | 262.948 ± 6.970 a | 3.014 ± 1.075 bc |
| 3071 | 10.914 ± 2.769 ab | 0.210 ± 0.084 abc | 263.820 ± 37.149 a | 3.317 ± 1.189 abc |
| 3072 | 10.316 ± 4.583 ab | 0.196 ± 0.133 abc | 253.485 ± 41.822 a | 3.022 ± 1.112 bc |
| 3074 | 11.011 ± 2.663 a | 0.287 ± 0.091 a | 298.570 ± 18.778 a | 4.730 ± 0.967 a |
| 8204 | 6.333 ± 3.189 bc | 0.129 ± 0.084 cd | 279.833 ± 67.715 a | 2.515 ± 1.303 cd |
| 3078 | 9.657 ± 2.869 ab | 0.223 ± 0.072 abc | 292.562 ± 4.907 a | 3.798 ± 0.657 abc |

Note: $P_n$: net photosynthetic rate, $G_s$: stomatal conductance, $C_i$: intercellular $CO_2$ concentration, $Tr$: transpiration rate. Values followed by the different letter of each group were significantly different at $p < 0.05$ level of probability.

**Table 8.** Values of chlorophyll fluorescence parameters among teak clones from different provenances.

| Provenance | NPQ | Yield | $F_v/F_m$ |
|---|---|---|---|
| 20001 | 1.215 ± 0.732 a | 0.359 ± 0.120 b | 0.706 ± 0.007 ab |
| 3070 | 1.160 ± 0.519 a | 0.385 ± 0.085 b | 0.709 ± 0.022 ab |
| 3071 | 1.098 ± 0.478 ab | 0.368 ± 0.102 b | 0.714 ± 0.028 a |
| 3072 | 1.483 ± 0.775 a | 0.329 ± 0.090 b | 0.703 ± 0.028 ab |
| 3074 | 0.421 ± 0.179 b | 0.564 ± 0.045 a | 0.718 ± 0.022 a |
| 8204 | 1.638 ± 0.550 a | 0.281 ± 0.087 b | 0.695 ± 0.025 b |
| 3078 | 1.528 ± 0.060 a | 0.318 ± 0.012 b | 0.707 ± 0.016 ab |

Note: NPQ: non-photochemical quenching, Yield: the actual quantum yield PSII, $F_v/F_m$: maximum photochemical efficiency of PSII. Values followed by the different letter of each group were significantly different at $p < 0.05$ level of probability.

### 3.3. Correlations between Photosynthetic and Growth Traits

Correlation analyses (Table 9) of teak clone parameters showed that $P_n$ had significant positive correlation with $G_s$, $Tr$, $F_v/F_m$, seedling height and individual volume, respectively. $P_n$ values can therefore be regarded as a critical parameter in teak breeding, indicating potential for faster growth.

**Table 9.** Correlation analysis among photosynthetic characteristics, growth traits, and ecological factors of teak clones.

| | $P_n$ | $G_s$ | $C_i$ | *Tr* | NPQ | Yield | $F_v/F_m$ | *WUE* | Seedling Height | Collar Diameter | H | DBH | V |
|---|---|---|---|---|---|---|---|---|---|---|---|---|---|
| $P_n$ | | | | | | | | | | | | | |
| $G_s$ | 0.891 ** | | | | | | | | | | | | |
| $C_i$ | 0.057 | 0.345 | | | | | | | | | | | |
| *Tr* | 0.568 ** | 0.802 ** | 0.286 | | | | | | | | | | |
| NPQ | 0.062 | −0.009 | 0.090 | −0.426 | | | | | | | | | |
| Yield | 0.135 | 0.228 | 0.030 | 0.580 ** | −0.944 ** | | | | | | | | |
| $F_v/F_m$ | 0.430 * | 0.405 | 0.254 | 0.318 | −0.164 | 0.391 | | | | | | | |
| *WUE* | 0.380 | 0.030 | 0.254 | 0.469 * | 0.404 | −0.396 | 0.143 | | | | | | |
| Seedling Height | 0.463 * | 0.343 | 0.029 | 0.219 | −0.061 | 0.038 | −0.184 | 0.015 | | | | | |
| Collar diameter | −0.048 | 0.039 | 0.162 | 0.112 | −0.04 | −0.114 | −0.319 | −0.262 | 0.498 * | | | | |
| H | 0.156 | 0.058 | 0.104 | 0.067 | 0.007 | −0.003 | −0.096 | 0.029 | 0.486 * | 0.258 | | | |
| DBH | 0.294 | 0.15 | 0.154 | −0.004 | 0.197 | −0.139 | −0.111 | 0.314 | 0.294 | 0.136 | 0.707 ** | | |
| V | 0.491 * | 0.369 | 0.133 | 0.208 | 0.103 | −0.036 | 0.074 | 0.183 | 0.456 * | 0.061 | 0.832 ** | 0.830 ** | |

Note: $P_n$: net photosynthetic rate, $G_s$: stomatal conductance, $C_i$: intercellular $CO_2$ concentration, *Tr*: transpiration rate, NPQ: non-photochemical quenching, Yield: the actual quantum yield PSII, $F_v/F_m$: maximum photochemical efficiency of PSII, *WUE*: water use efficiency, Seedling Height: height of seedling growth at nursery at 1 year, Collar Diameter: collar diameter of seedling growth at nursery at 1 year, H: height of field growth at 4 years, DBH: diameter at breast height of field growth at 4 years, V: individual volume at 4 years. ** indicate highly significant difference at $p < 0.01$ level of probability, * significant difference at $p < 0.05$ level of probability.

In addition, seedling height was positively correlated with collar diameter, plant height (H), and individual volume. *WUE* was significantly negatively correlated with *Tr*, suggesting that teak clones with high transpiration could show low *WUE*; $G_s$ was positively correlated with *Tr*, indicating that high stomatal conductance contributed to higher transpiration; *Tr* was positively correlated with the actual quantum yield of PSII, suggesting that the higher the transpiration rate, higher the actual quantum yield of PSII would be.

## 4. Discussion

Plant growth and yield depend largely on photosynthesis [23,24]. Plant photosynthesis is not only affected by environmental factors, but also affected by plant genetic characteristics. It is the complex process of interaction between plant genetic and environmental factors that influences photosynthetic activity [25]. To date, ecophysiological studies on photosynthesis in forest trees were those that examined the effects of stress on photosynthetic physiology [26–29], and the photosynthetic responses to light intensity [30] and $CO_2$ concentration [31]. The present study chiefly focused on systematically measuring photosynthetic gas exchange and chlorophyll fluorescence parameters, correlating the photosynthetic characteristics with growth, and providing a means of rapid evaluation of teak germplasm, for introduction, utilization, and improvement of teak resources in future breeding programs.

Our study showed that teak clones had high variation and high heritability ($h^2$) for many growth and physiological traits. The results were generally consistent with the findings reported for *Populus trichocarpa* by McKown [32]. The gas exchange, chlorophyll fluorescence, and growth parameters of teak clones were highly controlled by genetic factors, especially for the actual quantum yield (Yield) of PSII. Therefore, such a parameter has high practical significance and can be effectively used for improving the efficiency of teak breeding. However, it is worth emphasizing that intercellular $CO_2$ concentration ($C_i$) and non-photochemical quenching (NPQ) were greatly influenced by the environment.

Further analysis showed that teak clones and resources from different regions vary in their photosynthetic characteristics. In this study, teak clones 7029, 71-5, 7219, 7412, 7122, and provenances 3070, 3074, 3071, which had higher $P_n$, can be regarded as the key resource in future breeding and management programs. However, more teak clones from different provenances and countries need to be included in this kind of study in the future. Huang et al., 2016 [33] had suggested, after SSR molecular marker testing, that the Nigerian provenance 3078, investigated in this paper, may have originated from India. The present studies, that reveal their similar photosynthetic characteristics, further corroborates this conclusion.

The significantly positive correlation that the net photosynthetic rate has with seedling height, individual volume, $F_v/F_m$, $G_s$, and *Tr*, is an interesting finding of this study. In addition, seedling height was significantly and positively correlated with plant height and individual volume. Both results indicate that teak clones of high $P_n$ and high seedling height result in fast-growing clones. However, it is known that photosynthetic processes are influenced by environmental conditions such as light, temperature, water, and nutrients [25]. Photosynthetic rate is not the only limiting factor for growth [34]. These factors may affect growth differently for different clones, resulting in no significant relationship between $P_n$ and plant height or DBH of field growth at 4 years, the result being similar to previous reports [12,35].

Correlation analysis also revealed that water use efficiency was significantly but negatively correlated with *Tr*, suggesting that teak clone *WUE* may decrease when transpiration rate is high in daytime. Such results were consistent with the study by Huang et al., 2016 [17], in that diurnal variation possessed a double peaked curve, with a "midday depression" phenomenon in summer, when strong sunshine often accompanied by high temperature produces excessive transpiration, followed by decline of water use efficiency. There was no significant correlation between seedling collar diameter and other parameters, except for seedling height, consistent with the results of the study on photosynthesis and growth of *Populus nigra* [10]. At the same time, the coefficient of genetic variation of $C_i$ and $F_v/F_m$ were lower than other photosynthetic indices in the present study, similar to photosynthetic characteristics of the clones [10]. The variation coefficients of $F_v/F_m$ were small in this study (0.028) and in *Populus nigra* clones (0.024) [10]. This may be due to $CO_2$ concentration, leaf temperature, and relative humidity fluctuating significantly under natural conditions, reducing the $F_v/F_m$ compared to conditions where they remain constant [25].

Farquhar et al., 1982 [36] concluded that photosynthetic rate was controlled by stomatal factors when $P_n$, $C_i$, and $G_s$ increased or decreased at the same time. In this study, correlation analysis indicates that there was significant positive correlation between $P_n$ and $G_s$, a positive but not significant correlation between $P_n$ and $C_i$, $G_s$, and $C_i$, suggesting that the photosynthetic rate of teak was mostly controlled by stomatal factors. Plant dynamic photosynthesis was affected by many environmental factors such as light intensity, $CO_2$ concentration, leaf temperature, and relative humidity. Fluctuating environments would have a large impact on photosynthesis. Plants have a highly responsive regulatory system to make rapid photosynthetic responses to fluctuating environments, and a number of photoprotective mechanisms allow plants to maintain photosynthesis under stressful fluctuating environments [25].

For further research, the following points need to be considered in the future studies on teak. Firstly, it is desirable that more clones from different provenances be included in this kind of study in order to analyze variation among teak resources of different provenances more efficiently. Secondly, the differences in $P_n$ among teak clones in this study was greater than that seen in *Populus nigra* [10] and *Populus deltoides* clones [11]. It is to be ascertained whether such a difference was caused by inherent differences in photosynthetic characteristics between the tree species, or if is due to other reasons. Thirdly, further evaluation of differences in leaf area between teak clones is needed since tree growth is restricted not only by photosynthetic efficiency, but also by photosynthetic leaf area [37,38]. Lastly, we found that photosynthetic rates of teak plants in the field trial measured at the age of 2 years were higher than that of the potted seedlings and, therefore, correlation analysis among photosynthetic parameters, photosynthetic leaf area, and growth traits in field trials need to be executed in future teak breeding programs.

## 5. Conclusions

Our findings have at least three important implications. First, photosynthetic parameters other than intercellular $CO_2$ concentration ($C_i$) are highly controlled by genetic factors. In addition, photosynthetic parameters and growth traits in different clones revealed abundant genetic variation. Second, the net photosynthetic rate ($P_n$), seedling height, and individual volume of wood significantly correlated between each other, and seedling height was significantly correlated with plant height in field trials, suggesting $P_n$ and seedling height can help us in teak breeding. Third, teak clones 7029, 71-5, 7219, 7412, and 7122, and provenances 3070, 3074, and 3071, revealed to have higher photosynthetic rate, can be regarded as key resources for future breeding and germplasm resource management.

**Author Contributions:** G.H. designed and supervised implementation of the studies, supervised the statistical analyses, constructed the tables, wrote the manuscript, and crafted the final version. K.L. and Z.Z. carried out the statistical analyses and wrote the first draft of the manuscript. G.Y. supervised and carried out all technical aspects. E.M.M. participated in writing and editing the manuscript.

**Funding:** This work was supported by the ["Fundamental Research Funds for the Central Non-profit Research Institution of CAF" (grant number:No. CAFYBB2017ZA001-7)] and [the National Key Research and Development Program of China "Research Project on Teak Cultivation Techniques" (grant number: 2016YFD0600602)].

**Acknowledgments:** The authors are grateful to Bingshan Zeng and Zhenfei Qiu for providing the materials used in this study. We would like to thank anonymous reviewers for their valuable comments.

**Conflicts of Interest:** The authors declare no conflict of interest.

## References

1. Kaosa-ard, A. Teak (*Tectona grandis*): Its natural distribution and related factors. *Nat. Hist. Bull. Siam Soc.* **1981**, *29*, 55–74.
2. Mohanan, C.; Sharma, J.K.; Florence, E.J.M. Nursery diseases of teak in India. In Proceedings of the International Teak Symposium, Thiruvananthapuram, Kerala, India, 2–4 December 1991; Chand Basha, S., Mohanan, C., Sankar, S., Eds.; Kerala Forest Department and Kerala Forest Research Institute: Thrissur, Kerala, India, 1997; pp. 107–112.
3. Liu, P.; Yang, J.J.; Lu, H.J. *Southeast Asian Tropical Timber*; China Forestry Publishing House: Beijing, China, 1993; p. 280, ISBN 9787503809330.
4. Liang, K.N. Teak. In *Cultivation Techniques of Valuable Tree Species in South China*; Xu, D.P., Qiu, Z.W., Eds.; Guangdong Science and Technology Publishing House: Guangzhou, China, 2013; pp. 213–230, ISBN 978-7-5359-5807-5.
5. Alcantara, B.K.; Veasey, E.A. Genetic diversity of teak (*Tectona grandis* L.F.) from different provenances using microsatellite markers. *Rev. Arvore* **2013**, *37*, 747–758. [CrossRef]
6. Blake, T.J.; Yeatman, C.W. Water relations, gas exchange, and early growth rates of outcrossed and selfed Pinus *banksiana families*. *Can. J. Bot.* **1989**, *67*, 1618–1623. [CrossRef]
7. Hou, A.J.; Xu, D.C. Current advance of high photosynthetic efficiency breeding by gene engineer in plants. *China Biotechnol.* **2005**, *25*, 19–23.
8. Chen, Y.; Yuan, L.P.; Wang, X.H.; Zhang, D.Y.; Chen, J.; Deng, Q.Y.; Zhao, B.R.; Xu, D.Q. Relationship between grain yield and leaf photosynthetic rate in super hybrid rice. *J. Plant Physiol. Mol. Biol.* **2007**, *3*, 235–243.
9. Long, S.P.; Zhu, X.G.; Naidu, S.L.; Ort, D.R. Can improvement in photosynthesis increase crop yields? *Plant Cell Environ.* **2006**, *29*, 315–330. [CrossRef] [PubMed]
10. Chu, Y.G.; Su, X.H.; Huang, Q.J.; Zhang, X.H. Relationships between Photosynthetic Characteristics and Growth Traits in Gene Resources of *Populus nigra*. *Sci. Silvae Sin.* **2010**, *46*, 77–83.
11. Gao, M.; Ding, C.J.; Su, X.H.; Huang, Q.J. Comparison of photosynthetic characteristics of *Populus deltoides* and their $F_1$ hybrid clones. *For. Res.* **2014**, *27*, 721–728.
12. Sulzer, A.M.; Greenwood, M.S.; Livingston, W.H.; Adams, G. Early selection of black spruce using physiological and morphological criteria. *Can. J. For. Res.* **1993**, *23*, 657–664. [CrossRef]
13. Liang, K.N.; Lai, M.; Huang, G.H.; Ling, M.P.; Zhou, Z.Z.; Ma, H.M. Growth and adaptability of ten provenances of *Tectona grandis* at 27-year-old. *J. Cent. South Univ. For. Technol.* **2011**, *31*, 8–12.
14. Lai, M.; Liang, K.N.; Huang, G.H.; Ling, M.P.; Zhou, Z.Z.; Ma, H.M. Genetic variation and comprehensive evaluation in growth and wood relevant properties of different provenances of *Tectona grandis*. *For. Res.* **2011**, *24*, 234–238.
15. Zheng, F.X.; Wen, D.Z.; Kuang, Y.W. Effects of simulated acid rain on the growth, photosynthesis and water use efficiency in *Tectona grandis*. *J. Trop. Subtrop. Bot.* **2006**, *14*, 93–99.
16. Chen, D.X.; Ban, X.Q.; Li, Y.D.; Xiao, W.F.; Luo, T.S.; Lin, M.X.; Xu, H. Responses of gas exchange to neighborhood interference in leaves of teak (*Tectona grandis* L. f.) in a tropical plantation forest. *Acta Ecol. Sin.* **2008**, *28*, 4059–4069.
17. Huang, G.H.; Liang, K.N.; Zhou, Z.Z.; Ma, H.M. Diurnal and seasonal Photosynthetic characteristics and influencing factors in teak clones. *J. Cent. South Univ. For. Technol.* **2016**, *36*, 11–16.
18. Maxwell, K.; Johnson, G.N. Chlorophyll fluorescence—A practical guide. *J. Exp. Bot.* **2000**, *51*, 659–668. [CrossRef]
19. Schreiber, U. Pulse amplitude modulation (PAM) fluorometry and saturation pulse method. In *Chlorophyll a Fluorescence: A Signature of Photosynthesis*; Papageorgiou, G., Govind, J., Eds.; Kluwer Academic Publishers: Dordrecht, The Netherlands, 2004; pp. 279–319.
20. Wang, R.Z.; Gao, Q. Photosynthesistranspiration and water use efficiency in two divergent *Leymus chinensis* populations from northeast China. *Photosynthetica* **2001**, *39*, 123–126.
21. Duan, A.G.; Zhang, X.Q.; Zhang, J.G.; Zhong, J.D. Growth and genetic evaluation of 21-year-old Chinese fir clonal plantation. *For. Res.* **2014**, *27*, 672–676.
22. Zhou, Z.Z. Calcium Nutrition of Teak in Acidic Soils. Doctoral Thesis, Chinese Academy of Forestry, Beijing, China, 2009.

23. Yamori, W.; Kondo, E.; Sugiura, D.; Terashima, I.; Suzuki, Y.; Makino, A. Enhanced leaf photosynthesis as a target to increase grain yield: Insights from transgenic rice lines with variable Rieske FeS protein content in the cytochrome *b6/f* complex. *Plant Cell Environ.* **2016**, *39*, 80–87. [CrossRef]
24. Zhu, X.G.; Long, S.P.; Ort, D.R. Improving photosynthetic efficiency for greater yield. *Annu. Rev. Plant Biol.* **2010**, *61*, 235–261. [CrossRef]
25. Yamori, W. Photosynthetic response to fluctuating environments and photoprotective strategies under abiotic stress. *J. Plant Res.* **2016**, *129*, 379–395. [CrossRef]
26. Wu, Y.B.; Xu, J.H. Impacts of salt stress on the growth and photosynthesis of three *Fraxinus* species. *J. Nanjing For. Univ. (Nat. Sci. Ed.)* **2002**, *26*, 19–22.
27. Cao, F.L.; Cai, J.F.; Wang, G.B.; Zhang, W.X. Effects of Waterlogging Stress on the Growth and Photosynthesis of *Sapium sebiferum*. *Sci. Silvae Sin.* **2010**, *46*, 57–61.
28. Zhang, L.Y.; Wen, X.; Lin, Y.M.; Li, J.; Chen, C.; Wu, C.Z. Effect of salt stress on photosynthetic and chlorophyll fluorescent characteristics in *Alnus formosana* seedlings. *J. Fujian Coll. For.* **2013**, *33*, 193–199.
29. Feng, Y.L.; Ju, G.S.; Zhu, C.Q. Responses of photosynthesis and pv-parameters to water stress in poplar clone seedlings. *Sci. Silvae Sin.* **2003**, *39*, 30–36.
30. Zhang, W.X.; Wu, J.S.; Cao, F.L. Influence of photosynthetically active radiation on photosynthesis and photochemistry efficiency in leaves of *Ginkgo*. *J. Nanjing For. Univ. (Nat. Sci. Ed.)* **2002**, *26*, 5–9.
31. Su, P.X.; Zhang, L.X.; Du, M.W.; Bi, Y.R.; Zhao, A.F.; Liu, X.M. Photosynthetic character and water use efficiency of different leaf shapes of *Populus euphratica* and their response to $CO_2$ enrichment. *Acta Phytoecol. Sin.* **2003**, *27*, 34–40.
32. McKown, A.; Guy, R.; Klápště, J.; Geraldes, A.; Friedmann, M.; Cronk, Q.; El-Kassaby, Y.; Mansfield, S.; Douglas, C. Geographical and environmental gradients shape phenotypic trait variation and genetic structure in *Populus trichocarpa*. *New Phytol.* **2014**, *201*, 1263–1276. [CrossRef]
33. Huang, G.H.; Liang, K.N.; Zhou, Z.Z.; Ma, H.M. SSR Genotyping—Genetic diversity and fingerprinting of teak (*Tectona grandis*) clones. *J. Trop. For. Sci.* **2016**, *28*, 48–58.
34. Ceulemans, R.; Impens, I.; Steenackers, V. Variation in photosynthetic, anatomical and enzymatic leaf traits and correlations with growth in recently selected *Populus* clones. *Can. J. For. Res.* **1987**, *17*, 273–283. [CrossRef]
35. Larsen, J.B.; Wellendorf, H. Early test in *Picea abies* fullsibs by applying gas exchange, frost resistance and growth measurements. *Scand. J. For. Res.* **1990**, *5*, 369–380. [CrossRef]
36. Farquhar, G.D.; Sharkey, T.D. Stomatal conductance and photosynthesis. *Ann. Rev. Plant Physiol.* **1982**, *33*, 317–345. [CrossRef]
37. Fujimoto, R.; Taylor, J.M.; Shirasawa, S.; Peacock, J.; Dennis, E. Heterosis of *Arabidopsis* hybrids between $C_{24}$ and Col is associated with increased photosynthesis capacity. *Proc. Natl. Acad. Sci. USA* **2012**, *109*, 7109–7114. [CrossRef] [PubMed]
38. Turnbull, M.H.; Murthy, R.; Griffin, K.L. The relative impacts of day-time and night-time warming on photosynthetic capacity in Populus deltoids. *Plant Cell Environ.* **2002**, *25*, 1729–1737. [CrossRef]

*forests*

MDPI

*Article*

# Screening of Applicable SSR Molecular Markers Linked to Creeping Trait in Crape Myrtle

**Tangchun Zheng [1,2,3,4,5], Bo Qin [1,2,3,4,5,6], Suzhen Li [1,2,3,4,5,6], Ming Cai [1,2,3,4,5], Huitang Pan [1,2,3,4,5], Jia Wang [1,2,3,4,5], Tangren Cheng [1,2,3,4,5] and Qixiang Zhang [1,2,3,4,5,6,*]**

1 Beijing Key Laboratory of Ornamental Plants Germplasm Innovation & Molecular Breeding, Beijing Forestry University, Beijing 100083, China; zhengtangchun@bjfu.edu.cn (T.Z.); qinbo_bjfu@163.com (B.Q.); suzhenli@bjfu.edu.cn (S.L.); jasoncai82@163.com (M.C.); htpan2000@163.com (H.P.); wangjia8248@163.com (J.W.); chengtangren@163.com (T.C.)

2 National Engineering Research Center for Floriculture, Beijing Forestry University, Beijing 100083, China

3 Beijing Laboratory of Urban and Rural Ecological Environment, Beijing Forestry University, Beijing 100083, China

4 Engineering Research Center of Landscape Environment of Ministry of Education, Beijing Forestry University, Beijing 100083, China

5 Key Laboratory of Genetics and Breeding in Forest Trees and Ornamental Plants of Ministry of Education, Beijing Forestry University, Beijing 100083, China

6 Beijing Advanced Innovation Center for Tree Breeding by Molecular Design, Beijing Forestry University, Beijing 100083, China

* Correspondence: zqx@bjfu.edu.cn; Tel.: +86-010-6233-8005

Received: 26 April 2019; Accepted: 17 May 2019; Published: 18 May 2019

**Abstract:** Creeping plants have unique ornamental value because they have more branches and flowers and the creeping trait is rare in crape myrtle (*Lagerstroemia indica* L.). In this study, the first filial generation ($F_1$) population was derived from *Lagerstroemia fauriei* Koehne (standard) and *L. indica* "Creole" (creeping) and the backcross1 ($BC_1$) population was derived from the backcross of $F_1$ individual S82 (creeping) and *L. fauriei*. The segregation of the creeping trait was analyzed for 174 seedlings of the $BC_1$ population to examine the linkage relationship between simple sequence repeat (SSR) molecular markers and the creeping trait. Creeping genes were screened using bulked segregant analysis combined with 322 SSR primers, which were detected with good polymorphism. The results show that two SSR markers (S364 and LYS12) were detected, with genetic distances of 23.49 centimorgan (cM) and 25.86 cM from the loci controlling the plant opening angle trait and the branching angle trait, respectively. The accuracy rate for phenotypic verification using S364 and LYS12 was 76.51% and 74.14%, respectively. Our results provide basic information for the molecular marker-assisted selective breeding and cloning of the creeping gene to improve architecture diversity in the breeding of crape myrtle.

**Keywords:** *Lagerstroemia* species; simple sequence repeat markers; bulked segregant analysis; creeping trait; plant architecture

## 1. Introduction

Plant architecture is the result of long-term evolution and natural selection, which involve complex regulatory processes based on genetic and environmental interactions [1]. According to the different growth angles and branching directions, plant architecture can be roughly classified as standard, weeping, pillar, upright, brachytic, and dwarf [2]. Plant architecture traits have important application value for the improvement of crop production; for example, the cultivation of dwarf or semi-dwarf horticultural or agronomic crops improves yield and production efficiency because mechanized management is more conducive. The discovery and utilization of dwarf genes in rice, which greatly promoted grain production, led to the first "green revolution" in agriculture, demonstrating the importance of genetic resources [3,4]. The genetic regulation of plant architecture traits in woody plants is more complex than in herbaceous plants and plant architecture traits are susceptible to external environmental conditions [5]. Many studies have been performed on plant architecture for extreme plant types of trees, including pillar and weeping peach [6], dwarf and compact peach (*Prunus persica* L.) [7], weeping *Prunus mume* [8], weeping *Cercis chinensis* [9], and columnar and weeping apple (*Malus* × *domestica*) [10,11].

Molecular marker technology can reflect the differences between DNA sequences for different species and is widely used in plant genetic diversity analysis, genetic map construction, map-based cloning, and marker-assisted selective breeding [12–14]. Individual hybrids with target traits can be selected at the seedling stage using molecular marker-assisted selection, which reduces resource waste and breeding cost and speeds up the breeding process [15]. Bulked segregant analysis (BSA) is derived from near-isogenic line analysis, which was reported for the first time in 1991 [16]. SSR markers are widely used in relationship analysis, genetic diversity analysis, mapping quantitative trait locus (QTL), and so on. Four expressed sequence tag (EST)-SSR marker loci closely linked to the dwarf trait in pear were found by BSA technology. In *Actinidia chinensis* Planch, an EST-SSR marker was screened and the genetic distance between the marker and dwarf gene was 8.8 cM [17]. Three selected EST-SSR markers were used to determine genetic structure in 29 cultivars and were used for fruit color selection in *Prunus salicina* breeding [18]. In addition, DNA markers associated with the dwarf trait from *Brassica napus* [19], *Prunus persica* [20], and *Avena sativa* [21] have been studied.

*Lagerstroemia* L. (crape myrtle) belongs to the Lythraceae family, which includes at least 50 species of deciduous or evergreen shrubs or trees native to Southeast Asia [22]. China is located in the center of the worldwide *Lagerstroemia* distribution and origin [23]. Crape myrtle was first cultivated as an ornamental species in China approximately 1800 years ago [24]. Species in this genus are highly valued in landscaping for their graceful plant architecture, long-lasting flowering period, and colorful flowers during the summer [25]. When exposed to natural conditions, crape myrtle is a diploid plant (2n = 2x = 48) [26]. Crape myrtle is susceptible to powdery mildew; therefore, the initial purpose for hybrid breeding by interspecific hybridization between *L. indica* and *L. fauriei* was for disease resistance and then a series of excellent varieties with various plant architecture types and colorful flowers was selected from the resistant hybrid offspring ('Pocomoke") [27–29]. *Lagerstroemia* species are self-compatible and easily produce interspecific hybrids with related species. To further improve the ornamental value of hybrids with a larger flower diameter, *Lagerstroemia speciosa* and *L. indica* were used for interspecific hybridization and hybrid seedlings with traits of the parents were obtained; however, only plant height and plant width had the characteristics of the hybrids and no variation in flower color or flower size was observed [30]. Recently, research in America, Japan, and China has focused on breeding new cultivars with unique plant architecture, especially dwarf or potted crape myrtle. A few dwarf crape myrtle cultivars with many branches and large flowers were bred through intraspecific hybridization (*L. indica*) with the aim of generating dwarf plant architecture [31–35].

The molecular study of *Lagerstroemia* species has led to the development of a series of genomic SSR markers [26,36–40]. Two single nucleotide polymorphism (SNP) markers (M16337 and M38412) that are highly correlated with internode length and one SNP marker (M25207) that is highly correlated with primary lateral branch height were validated in the $F_1$ population of *L. indica* [41]. In addition, an SSR marker linked to the dwarf gene, with a genetic distance of 23.33 cM, was screened from the *L. fauriei* × *L. indica* "Pocomoke" $F_1$ population [35]. Currently, less is known about the plant architecture of the crape myrtle, especially the creeping trait. To analyze the linkage relation between SSR molecular markers and creeping trait, the segregation of the creeping trait was analyzed by the BSA method using the $BC_1$ population with 174 hybrids and 322 SSR primers. Our results provide an important technical and theoretical basis for plant architecture molecular marker-assisted selective breeding for the *Lagerstroemia* species.

## 2. Materials and Methods

### 2.1. Plant Materials

The $F_1$ population was derived from *L. fauriei* (♀, standard) and *L. indica* "Creole" (♂, creeping). To analyze the linkage relation between SSR markers and creeping trait, the $BC_1$ segregation population was derived from a backcross of the $F_1$ creeping individual S82 (♀) × *L. fauriei* (♂). The $F_1$ female parent was a tree (>3 m) with standard branching and the male parent was a dwarf plant (0.3–0.5 m) with creeping branching. The $F_1$ individual S82 had the same phenotype as *L. indica* "Creole". The linkage relationship between SSR markers was analyzed for 174 individuals of the $BC_1$ population. All of the materials were planted in a breeding nursery at the National Engineering Research Center for Floriculture (Beijing) (40°02′ N, 115°50′ E) (Figure 1).

**Figure 1.** Plant materials and method of measurement used in this study. (**a**) *L. fauriei*; (**b**) *L. indica* "Creole"; (**c**) S82 the creeping individual of $F_1$; (**d**,**e**) creeping and standard offsprings in $BC_1$ population (in summer); (**f**) creeping and standard offsprings in $BC_1$ population (in autumn); (**g**,**h**) methods of measuring branch angle and plant canopy angle. $\theta_1$ and $\theta_2$ are the angles of inclination of the widest position of the canopy from vertical orientation on both sides. Plant canopy angle is the sum of $\theta_1$ and $\theta_2$.

### 2.2. Phenotypic Measurement and Data Analysis

Phenotypic traits related to plant architecture, i.e., plant height (PH), plant width (PW), plant canopy angle (PCA), branching angle of the main branch (BA), the number of main branches (NMB), and branching height (BH), were measured for 174 individuals from the S82 (♀) × *L. fauriei* (♂) population at the end of the growing season. In addition, leaf length (LL) and leaf width (LW) were measured during the peak of the growing season. PH and BH were measured at the highest point, and PW was measured at horizontal direction by ruler. LL and LW were measured with an electronic vernier caliper. Each trait was measured more than three times. NMB was directly visualized. The measurements of the above phenotypic traits were described by Ye et al. [35,37].

The branching angle of the main branch corresponds to the angle between the main branch and the horizontal direction. In [42], it is reported that the angle between the horizontal direction and the line connecting the basal branch and apical branch (angle $\alpha$ in Figure 1g) is a more reliable parameter than branch growing angle (angle $\beta$ in Figure 1g). Moreover, this more reliable angle is more closely related to plant architecture and measures the branching angle of the main branch by the measurement of angle $\alpha$. PCA and BA were measured by a huge protractor suspended at the origin. The plant canopy angle ($\theta = \theta_1 + \theta_2$, Figure 1h) was measured according to Thakur et al. [43].

Genetic variation analyses were performed on phenotype traits of $BC_1$ population using SPSS 22.0 software (SPSS, Chicago, IL, USA). Statistical parameters included mean, maximum, minimum, variance, standard deviation, skewness, peakness, and coefficient of variation and plot the frequency distribution. The Pearson correlation coefficient between traits was calculated by the Correlations module in SPSS software and significant difference tests were performed at different levels.

### 2.3. DNA Extraction and Detection

Genomic DNA was extracted from fresh young leaves using the Fast DNA kit (Tiangen Biotech, Beijing, China) following the manufacturer's protocol. DNA quality and concentration were measured by 1% agarose gel electrophoresis with Gel Red [44].

### 2.4. Construction of Near-Isogenic Pools

The construction of near-isogenic pools was based on the methods reported by Michelmore et al. [16]. Five extremely standard individuals and five extremely creeping individuals (as described in Figure 1f) were randomly selected from 174 individuals in $BC_1$ population and their genomic DNA was mixed to construct standard and creeping gene pools, respectively. The standard gene pool was denoted $B_Z$, whereas the creeping gene pool was denoted $B_P$.

### 2.5. Screening SSR Markers Linked to Creeping Strains

DNA from eight $BC_1$ individuals was selected randomly to detect the polymorphism of the new SSR primers. Based on all the highly polymorphic primers in crape myrtle, two near-isogenic pools ($B_Z$, $B_P$) and three standard and three creeping individuals were then randomly selected and screened by SSRs. The 322 pairs of primers used in the experiment were developed using the transcriptome data from a previous study [45] (Table S1).

PCR products were detected by non-denaturing polyacrylamide gel electrophoresis. If the product strips with polymorphisms in two pools were similar for six plants, it was initially concluded that it may relate to plant architecture of crape myrtle. These primers were detected in parents and segregating populations and were correlated with phenotypic data and genetic distance (cM) between the SSR markers. The genes related to plant architecture were identified by Kosambi's method [46]. Furthermore, the validity of the screened polymorphic marker was verified by varieties in a germplasm resource nursery.

## 3. Results

### 3.1. Segregation Analysis of Phenotypic Variation in $BC_1$ Populations

Eight statistical parameters were calculated from eight phenotypic characters (PH, PW, PCA, BA, NMB, BH, LL, and LW), and the frequency distribution histogram was constructed using SPSS 22.0 and Excel (Figure 2). In the S82 × *L. fauriei* population, the coefficients of genetic variation were between 20.09% and 35.49%, except for the BH genetic variation coefficient. The highest degree of variation was observed for PH followed by NMB, whereas the smallest variation was observed for PCA. The variability in the measured traits among individuals was greater than 10%, indicating significant genetic variation in these traits (Table 1).

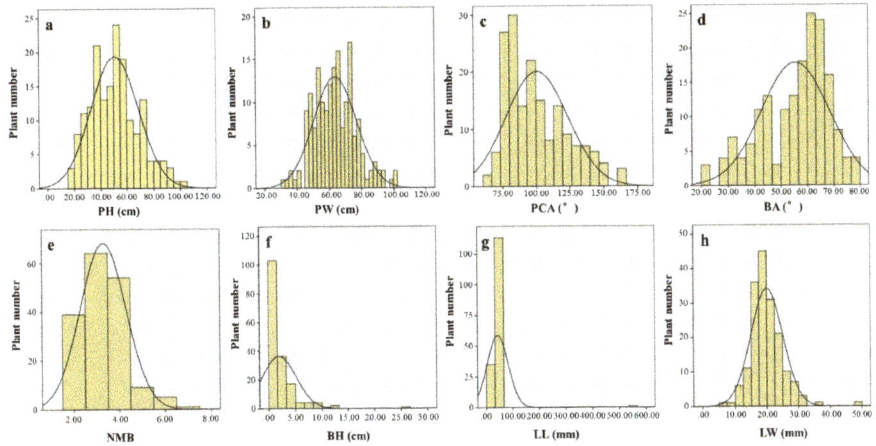

**Figure 2.** Distribution of eight phenotypic traits of 174 individuals in S82 × *L. fauriei* population. (**a**) Plant height (PH); (**b**) plant width (PW); (**c**) plant canopy angle (PCA); (**d**) branching angle of the main branch (BA); (**e**) the number of main branches (NMB); (**f**) branching height (BH); (**g**) leaf length (LL); (**h**) leaf width (LW).

**Table 1.** Descriptive statistics of phenotypic characters in S82 × *L. fauriei* population.

| Traits | Mean | Minimum | Maximum | Standard Deviation | Variance | Skewness | Peakness | Coefficient of Variation |
|---|---|---|---|---|---|---|---|---|
| PH (cm) | 50.70 | 19.00 | 101.00 | 17.99 | 322.83 | 0.44 | −0.25 | 35.49% |
| PW (cm) | 63.74 | 31.00 | 101.50 | 13.40 | 179.57 | 0.26 | 0.10 | 21.02% |
| PCA (°) | 105.64 | 60.70 | 163.45 | 21.23 | 533.00 | 0.48 | −0.18 | 20.09% |
| BA (°) | 55.19 | 21.27 | 78.90 | 12.84 | 167.95 | −0.56 | −0.35 | 23.27% |
| NMB | 3.30 | 2.00 | 7.00 | 1.01 | 1.01 | 0.71 | 0.75 | 30.57% |
| BH (cm) | 1.98 | 0.00 | 26.00 | 3.17 | 9.92 | 3.56 | 20.05 | 159.69% |
| LL (mm) | 40.29 | 19.66 | 64.19 | 8.32 | 69.30 | 0.36 | 0.32 | 20.66% |
| LW (mm) | 20.00 | 6.98 | 49.32 | 5.07 | 25.75 | 1.37 | 6.08 | 25.37% |

Note: PH: plant height; PW: plant width; PCA: plant canopy angle; BA: branching angle of the main branch; NMB: the number of main branches; BH: branching height; LL: leaf length; LW: leaf width. The phenotype units are only used for the mean, standard deviation, maximum, and minimum.

## 3.2. Correlation Analysis of Population Phenotypic Traits

The correlation between the eight phenotypic characters was analyzed by SPSS (Table 2). PCA showed a significant positive correlation with NMB and a significant negative correlation with PH and BA. BA showed a significant positive correlation with PH and a significant negative correlation with PCA and NMB.

**Table 2.** Correlation analysis of phenotypic characters in S82 × *L. fauriei* population.

| Traits | PH | PW | PCA | BA | NMB | BH | LL | LW |
|---|---|---|---|---|---|---|---|---|
| PH | 1 | | | | | | | |
| PW | 0.282** | 1 | | | | | | |
| PCA | −0.538** | 0.126 | 1 | | | | | |
| BA | 0.545** | 0.009 | −0.703** | 1 | | | | |
| NMB | −0.143 | 0.052 | 0.214* | −0.319** | 1 | | | |
| BH | 0.073 | −0.005 | −0.091 | −0.075 | −0.241** | 1 | | |
| LL | 0.383** | 0.259** | −0.12 | 0.095 | −0.008 | 0.013 | 1 | |
| LW | 0.354** | 0.210* | −0.074 | 0.058 | 0.06 | −0.042 | 0.681** | 1 |

Note: PH: plant height; PW: plant width; PCA: plant canopy angle; BA: branching angle of the main branch; NMB: the number of main branches; BH: branching height; LL: leaf length; LW: leaf width. ** means highly significant at the 0.01 level; * means significant at the 0.05 level.

### 3.3. Detection of Near-Isogenic Pools, Parents, and $BC_1$ Segregation Populations by SSR Markers

High polymorphism was detected for 322 pairs of SSR primers for eight $BC_1$ individuals (selected randomly from $BC_1$). The 322 pairs of primers with high polymorphism were screened using the standard gene pool $B_Z$ and creeping gene pool $B_P$ as templates. Finally, four pairs of SSR primers with polymorphic strips were screened in two gene pools (Figure 3).

Primers that are polymorphic in both gene pools were further amplified in the parents and other hybrids to determine strip type. Of the above four primers, only S364 and LYS12 were able to amplify strips based on the phenotype of the parents (Figure S1, Table 3) and 174 individuals of the segregating population (Figure S2 and Figure S3).

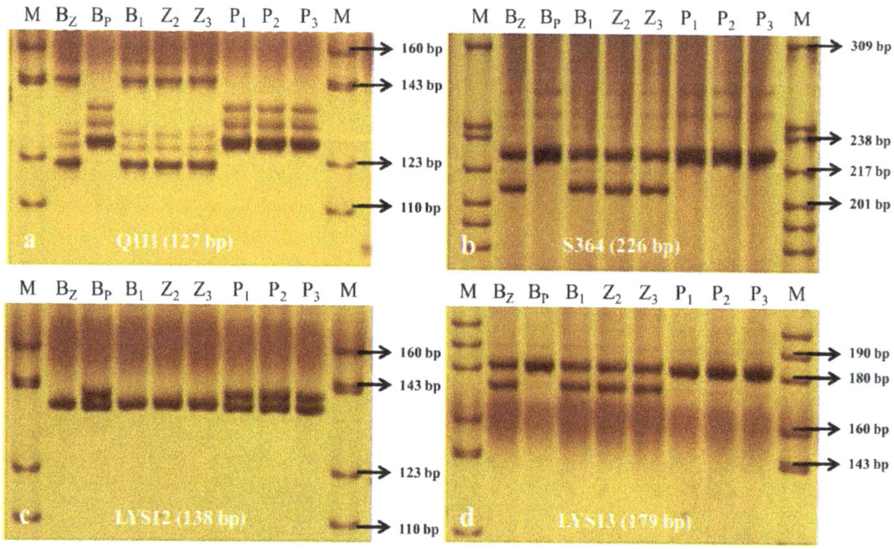


**Figure 3.** Electrophoresis of four primers' amplification products in parents and gene pools. (**a**) Electrophoretic results of Q111 primers; (**b**) electrophoretic results of S364 primers; (**c**) electrophoretic results of LYS12 primers; (**d**) electrophoretic results of LYS13 primers. $B_Z$: Standard creeping gene pool; $B_P$: Creeping gene pool; $Z_1$, $Z_2$, and $Z_3$: Three standard individuals; $P_1$, $P_2$, and $P_3$: Three creeping individuals; M: DNA marker.

**Table 3.** Phenotypes and amplified strips of parents and near isogenic pools.

| Code | Phenotype | Strip Type | Strip Type |
|---|---|---|---|
| *L. faurei* | Z | b | m |
| S82 | P | a | n |
| *L. indica* 'Creole' | P | a | n |

Note: Z: Standard individual; P: creeping individual; a: Single strip in primer S364; b: Double strip in primer S364; m: Single strip in primer LYS12; n: Double strip in primer LYS12.

### 3.4. Linkage Analysis of Markers and Phenotypes

From the continuous observation of the $BC_1$ population, we found that PH, PW, NMB, and BH changed over time, whereas the ratio between the plant width and plant height (PW/PH), PCA, and BA maintained a relatively stable state. The statistical results of the phenotypes of the $BC_1$ population and parents were listed in Table S2. A histogram of the frequency distribution for three traits (PW/PH, PCA, and BA) of the parents for the $BC_1$ population was drawn using Excel software (Figure S4). Based on the phenotypic data for parents and hybrids, the plant architecture of crape myrtle was classified into six classes using PW/PH, PCA, and BA. Artificially, PW/PH less than 1.5 were classified as G-type and all others were H-type. PCA was classified as J-type or K-type depending on whether values were less than 90°. Similarly, BA was classified as U-type or V-type depending on whether values were greater than 60° (Table 4). It can be seen that types G, J, and U correspond to plants with smaller PW/PH, smaller PCA, and larger BA, respectively; namely, plant architecture will be creeping and flat. Conversely, H-, K-, and V-types correspond to those with larger PW/PH, larger PCA, and smaller BA, which means that the plant architecture is more vertical. The segregation ratio of traits in the hybrids of the population implies that the upright type has a dominant effect on the creeping type. It is preliminarily considered that the upright/creeping type of *Lagerstroemia* species is controlled by both the major gene and the minor genes.

**Table 4.** Classification of the phenotype of 174 individuals in S82 × *L. fauriei* population.

| Traits | Grade | Type | No. of $BC_1$ |
|---|---|---|---|
| PW/PH | <1.5 | G | 122 |
| | ≥1.5 | H | 52 |
| PCA (°) | <90 | J | 98 |
| | ≥90 | K | 76 |
| BA (°) | >60 | U | 81 |
| | ≤60 | V | 93 |
| Plant architecture | Not satisfied H-, K-, V | Upright | 138 |
| | Satisfied H-, K-, V | Creeping | 36 |

Note: PH, plant height; PW, plant width; PCA, plant canopy angle; BA, branching angle of the main branch.

According to phenotypic grading standards, phenotypic statistics, and the strip type of the parents and two gene pools, the phenotypes and strips of 174 individuals were comparatively analyzed. The results showed that phenotypes G, J, and U correspond to the strip type "m" for primer S364 and strip type "b" for primer LYS12. Phenotypes H, K, and V correspond to strip type "a" for primer S364 and strip type "n" for primer LYS12 (Table S3). Statistical analysis showed that primer S364 correlated with PCA (J-type and K-type) and primer LYS12 correlated with BA (U-type and V-type).

Among the 174 tested individuals, eight individuals did not amplify the objective strips in primer S364, whereas all individuals amplified objective strips in primer LYS12. There are 127 and 129 strips that conformed to the phenotype in primers S364 and LYS12, respectively (Table 5). Based on statistical results, the genetic distance between the S364 molecular marker and the gene that controls PCA was approximately 23.49 cM. The genetic distance between the LYS12 molecular marker and the gene that controls BA was approximately 25.86 cM. Two SSR molecular markers were verified in the $BC_1$ population and the accuracy rate of phenotypic verification using S364 and LYS12 was 76.51% and 74.14%, respectively (Table 6).

**Table 5.** Statistics of two primer strip types in $BC_1$ plants.

| Marker | Strip Type | Number |
|---|---|---|
| S364 | a | 95 |
| | b | 71 |
| | – | 8 |
| LYS12 | m | 91 |
| | n | 83 |
| | – | 0 |

Note: "a" and "b" mean single strip and double strip in primer S364, respectively; "m" and "n" mean single strip and double strip in primer LYS12, respectively; "–" means no strip.

**Table 6.** Number of phenotypes and strip type of $BC_1$ plants.

| Marker | Coincidence | Inconformity | Total |
|---|---|---|---|
| S364 | 127 | 39 | 166 |
| LYS12 | 129 | 45 | 174 |

Two markers, S364 and LYS12, were further identified in 20 *Lagerstroemia* species or cultivars (16 were upright trees or shrubs and 4 were low shrubs with creeping or flat branches) and the strip types were recorded. As shown in Table 7, the comparison of plant architecture types and two marker strip types showed that 15 of 20 cultivars were identified by the S364 marker and the accuracy rate of phenotypic identification was 75%. Eighteen cultivars were identified with LYS12 marker bands and the accuracy rate of phenotypic identification was 90%. These results indicate that the two markers were able to accurately identify phenotypic traits in crape myrtle cultivars.

**Table 7.** Identification of two markers in 20 *Lagerstroemia* stocks with diverse plant architectures.

| Species and Cultivars | Description of Plant Type | Strip | Strip |
|---|---|---|---|
| *L. fauriei* | Upright, tree | b | m |
| *L. limii* | Upright, tree | b | m |
| *L. excelsa* | Upright, tree | b | m |
| *L. subcostata* | Upright, tree | b | m |
| *L. indica* "Osage" | Upright, large shrub or small tree | a | m |
| *L. indica* "Queens Lace" | Upright, large shrub or small tree | a | m |
| *L. indica* "Muskogee" | Upright, large shrub or small tree | b | m |
| *L. indica* "Comanche" | Upright, large shrub or small tree | b | m |
| *L. indica* "William Toovey" | Upright, large shrub or small tree | b | m |
| *L. indica* "Yuma" | Upright, large shrub or small tree | b | m |
| *L. indica* "Ebony Flame" | Upright, shrub | a | m |
| *L. indica* "Ebony Fire" | Upright, shrub | a | m |
| *L. indica* "Ebony Embers" | Upright, shrub | b | m |
| *L. indica* "Chisam Fire" | Upright, shrub | b | m |
| *L. indica* "Victor" | Upright, spherical | a | m |
| *L. indica* "Tonto" | Upright, semi-dwarf shrub | b | m |
| *L. indica* "Houston" | Dwarf shrubs, weeping branches | a | m |
| *L. indica* "Sacramento" | Dwarf shrubs, weeping branches | a | m |
| *L. indica* "New Orleans" | Dwarf shrubs, creeping branches | a | n |
| *L. indica* "Creole" | Dwarf shrubs, creeping branches | a | n |

## 4. Discussion

Crape myrtle is popular and widely used in gardens because of its long flowering time and rich and colorful flowers. It can be planted alone or in clusters, used for garden flower belts, or cut flowers can be displayed in vases after pruning and pinning. The lack of varieties with diverse plant architecture types, flowers with a pleasant fragrance, and early-flowering features limits the further application of crape myrtle. Plant architecture traits (e.g., dwarf, weeping, creeping, columnar, and branching angle) have attracted much attention; these have potentially important application value for the plant architecture improvement of horticultural crops. Dwarf crape myrtles are becoming increasingly popular for use as potted plants and indoor flowers [47]. However, dwarfing and the arborization of trees are complex quantitative traits, which have greater non-additive effects on genetic performance and heritability. The columnar trait is controlled by a single gene (*Co*); however, other modification genes that may also play a role have been confirmed in apple [11]. Incomplete dominant columnar traits (pillar) also exist in peach, with a small branching angle and vertical growth of branches, but no obvious genetic segregation rule was observed for dwarf and compact traits [7].

In this study, the number of upright and creeping individuals for the $F_1$ of *L. fauriei* and *L. indica* "Creole" was 140 and 52, respectively, and the ratio of the two plant types was 2.69/1. In the $BC_1$

population of S82 × *L. fauriei*, the ratio of upright (138) to creeping (36) individuals was 3.83/1. However, in the $BC_1$ population of S82 × *L. indica* "Creole", the number of upright and creeping individuals was 42 and 150, respectively, with a ratio of 1/3.57 (Table S4). In the $F_1$ and S82 × *L. fauriei* populations, the number of upright individuals was much larger than that of creeping individuals because all hybrids were derived from crosses between upright and creeping parents. Although both parents (S82 × *L. indica* "Creole") were creeping types, upright individuals still appeared in their hybrids (Table S4). The comprehensive analysis of the segregation of two plant architecture types in three genetic populations cannot determine the number of genes controlling the creeping trait. However, the analysis shows that the upright type has a dominant effect on the creeping type. The upright/creeping plant type of crape myrtle is speculated to be controlled by a major gene and also regulated by minor genes. Our recent research results show that the dwarf trait of crape myrtle is also likely controlled by a major gene and modified by minor genes based on the phenotypic data in the $F_1$ population of *L. fauriei* (♀) × *L. indica* "Pocomoke" (♂) [35]. Correlation analysis between phenotypic traits is helpful to advance the selection of plant architecture in crape myrtle seedlings, which can shorten the breeding time and enhance breeding efficiency. A significant positive correlation was observed between PH and PW. PCA is verified by individual traits in the $BC_1$ population, which is similar to the correlation obtained for the $F_1$ population generated from *L. caudate* (♀) and *L. indica* "Xiang Xueyun" (♂) [36]. Eleven quantitative characters for 192 individuals in the $F_1$ population from a cross between *L. fauriei* (♀) and *L. indica* "Creole" (♂) were measured and analyzed; the results showed that the diversity index for plant architecture was 1.05 times greater than 1, which means that there is great potential for genetic improvement of the plant architecture [48]. In addition, most quantitative characters were significantly correlated with plant architecture and branching pattern. The results of heredity for some traits in the $F_1$ population of *L. speciosa* and *L. indica* showed that the coefficient of variation in the $F_1$ generation was 14.58–40.16%, which indicates significant variation [49].

Traditional breeding methods have the disadvantages of a long cycle, heavy workload, and low effectiveness for improving tree architecture. By using molecular marker-assisted selection, healthy individual plants with target traits can be selected at the seedling stage, which reduces the waste of resources, reduces the cost of breeding, and speeds up the breeding process [15]. BSA overcomes the constraints of near-isogenic lines and saves time and effort; it is widely used in marker development and gene mapping [16]. Because BSA does not require a large population, it is very popular in the study of the plant architecture traits of woody plants, such as apple (vertical traits) and peach (columnar and vertical traits) and in the development of genes and markers for peach weeping traits [2,10,50]. In ground-cover chrysanthemum, a random amplified polymorphic DNA (RAPD) marker A-10555 linked to the creeping trait was 7.96 cM from the loci controlling creeping/standard traits [51]. In crape myrtle, an SSR marker was identified and linked to the dwarf gene with a distance of 23.33 cM between the loci and dwarf gene [35]. In this study, two SSR markers (S364 and LYS12) closely linked to creeping traits in crape myrtle were obtained according to the principle of BSA. One marker was linked to PCA, with a genetic distance of 23.49 cM, and the other was linked to BA, with a genetic distance of 25.86 cM. The two markers were verified in the $BC_1$ population, parents, species, and varieties with high accuracy, which suggests that it is feasible to use these two markers to perform molecular marker-assisted breeding of a creeping plant type for crape myrtle. Previous research indicates that the polymorphic loci assayed within 15 cm of the target locus can be identified; loci are detected with decreasing frequency as genetic distance increases [16]. Therefore, the difference between two near-isogenic pools cannot be determined with the genetic distance of 23.49 cM and 25.86 cM from creeping genes. In addition, BSA has limitations in locating quantitative traits, which can only detect large-effect QTLs and require samples with large phenotypic differences when near-isogenic pools are constructed. In the future, we can use multi-generational hybridization and backcrossing to obtain populations with more significant phenotypic segregation to construct pools and screen markers. Additionally, the number of species-specific molecular markers is a critical factor that influenced the results of this experiment. To achieve an ideal effect in the marker-assisted selection of the creeping trait, we will develop more species-specific molecular markers to enhance the coverage area of the crape myrtle genome.

## 5. Conclusions

Crape myrtle is a widely used horticultural plant with important ornamental value, but it lacks creeping varieties. This study is the first to map creeping genes in *Lagerstroemia* species. In a $BC_1$ population derived from *L. fauriei* Koehne (standard) and *L. indica* "Creole" (creeping), 174 individuals were employed to screen molecular markers linked to the creeping trait of crape myrtle among 322 SSR primers with good polymorphism using BSA and SSR technologies. Two SSR markers (S364 and LYS12), which were 23.49 cM and 25.86 cM from the loci controlling plant opening angle trait and branching angle trait, were detected and further verified in the population, parents, species, and varieties with more than 74% accuracy, respectively. Our study will lay the foundation for the QTL mapping and marker-assisted selection breeding for creeping architecture of crape myrtle.

**Supplementary Materials:** The following are available online at http://www.mdpi.com/1999-4907/10/5/429/s1. Figure S1: Electrophoretic results of S364 primers in $BC_1$ individuals; Figure S2: Electrophoretic results of LYS12 primers in $BC_1$ individuals; Figure S3: Electrophoresis of S364 and LYS12 amplification products in parents. (**a**) Electrophoretic results of S364 primers; (**b**) Electrophoretic results of LYS12 primers; Figure S4: Distribution of three traits of $BC_1$ population. P1: *L. fauriei*, P2: *L. indica* "Creole", P3: S82 individual, Table S1: Sequence of 322 pairs of primers used in the study; Table S2: Phenotypic characters of parents and $BC_1$ plants; Table S3: Phenotype and strip type of 174 individuals in $BC_1$ population; Table S4: Classification of phenotype of 174 individuals in $F_1$ population of *L. fauriei* and *L. indica* "Creole".

**Author Contributions:** Conceptualization, T.Z.; data curation, T.Z., B.Q., and H.P.; formal analysis, T.Z., B.Q., and S.L.; funding acquisition, Q.Z.; investigation, B.Q., M.C., and T.C.; methodology, B.Q., S.L., and M.C.; project administration, T.Z. and Q.Z.; resources, M.C., H.P., J.W., and T.C.; software, B.Q., S.L., H.P., and J.W.; writing—original draft, T.Z. and B.Q.; writing—review and editing, T.Z. and Q.Z.

**Funding:** This work was supported by the program for Science and Technology of Beijing (No. Z181100002418006) and Special Fund for Beijing Common Construction Project.

**Acknowledgments:** We are thankful to American Journal Experts (AJE) for suggesting professional native English speaker for our manuscript.

**Conflicts of Interest:** The authors declare no conflict of interest and this research is carried on the absence of any financial or commercial relationships that could be interpreted to a potential conflict of interest.

## Abbreviations

| | |
|---|---|
| BA | Branching angle of the main branch |
| $BC_1$ | Backcross1 |
| BH | Branching height |
| BSA | Bulked segregant analysis |
| cM | Centimorgan |
| DNA | Deoxyribonucleic acid |
| EST-SSR | Expressed sequence tag |
| $F_1$ | First filial generation |
| LL | Leaf length |
| LW | Leaf width |
| NMB | Number of main branches |
| PCA | Plant canopy angle |
| PCR | Polymerase chain reaction |
| PH | Plant height |
| PW | Plant width |
| QTL | Quantitative trait locus |
| RAPD | Random amplified polymorphic DNA |
| SNP | Single nucleotide polymorphism |
| SSR | Simple sequence repeat |

## References


1. McSteen, P.; Leyser, O. Shoot branching. *Annu. Rev. Plant Biol.* **2005**, *56*, 353–374. [CrossRef]
2. Hollender, C.A.; Dardick, C. Molecular basis of angiosperm tree architecture. *New Phytol.* **2015**, *206*, 541–556. [CrossRef]
3. Sasaki, A.; Ashikari, M.; Ueguchi-Tanaka, M.; Itoh, H.; Nishimura, A.; Swapan, D.; Ishiyama, K.; Saito, T.; Kobayashi, M.; Khush, G.S.; et al. Green revolution: A mutant gibberellin-synthesis gene in rice. *Nature* **2002**, *416*, 701–702. [CrossRef] [PubMed]
4. Peng, J.; Richards, D.E.; Hartley, N.M.; Murphy, G.P.; Devos, K.M.; Flintham, J.E.; Beales, J.; Fish, L.J.; Worland, A.J.; Pelica, F.; et al. Green revolution' genes encode mutant gibberellin response modulators. *Nature* **1999**, *400*, 256–261. [CrossRef] [PubMed]
5. Costes, E.; Gion, J. Genetics and genomics of tree architecture. *Adv. Bot. Res.* **2015**, *74*, 157–200.
6. Werner, D.J.; Chaparro, J.X. Genetic interactions of pillar and weeping peach genotypes. *Hortscience* **2005**, *40*, 18–20.
7. Scorza, R.; Bassi, D.; Liverani, A. Genetic interactions of pillar (columnar), compact, and dwarf peach tree genotypes. *J. Am. Soc. Hortic. Sci.* **2002**, *127*, 254–261. [CrossRef]
8. Zhang, J.; Zhang, Q.; Cheng, T.; Yang, W.; Pan, H.; Zhong, J.; Huang, L.; Liu, E. High-density genetic map construction and identification of a locus controlling weeping trait in an ornamental woody plant (*Prunus mume* Sieb. Et Zucc). *Dna. Res.* **2015**, *22*, 183–191. [CrossRef]
9. Roberts, D.J.; Werner, D.J.; Wadl, P.A.; Trigiano, R.N. Inheritance and allelism of morphological traits in eastern redbud (*Cercis canadensis* L.). *Hortic. Res.* **2015**, *2*, 15049. [CrossRef]
10. Dougherty, L.; Singh, R.; Brown, S.; Dardick, C.; Xu, K. Exploring DNA variant segregation types in pooled genome sequencing enables effective mapping of weeping trait in *Malus*. *J. Exp. Bot.* **2018**, *69*, 1499–1516. [CrossRef] [PubMed]
11. Baldi, P.; Wolters, P.J.; Komjanc, M.; Viola, R.; Velasco, R.; Salvi, S. Genetic and physical characterisation of the locus controlling columnar habit in apple (*Malus* × *domestica* Borkh.). *Mol. Breed.* **2013**, *31*, 429–440. [CrossRef]
12. Li, T.; Liu, J.; Xie, Y.; Wang, Q.; Meng, F. Analysis of genetic diversity in *Prunus mira* Koehne ex Sargent populations using AFLP markers. *Plant Syst. Evol.* **2014**, *300*, 475–482. [CrossRef]
13. Salazar, J.A.; Ruiz, D.; Campoy, J.A.; Sánchez-Pérez, R.; Crisosto, C.H.; Martínez-García, P.J.; Blenda, A.; Jung, S.; Main, D.; Martínez-Gómez, P. Quantitative trait loci (QTL) and Mendelian trait loci (MTL) analysis in *Prunus*: A breeding perspective and beyond. *Plant Mol. Biol. Rep.* **2014**, *32*, 1–18. [CrossRef]
14. Sun, L.; Yang, W.; Zhang, Q.; Cheng, T.; Pan, H.; Xu, Z.; Zhang, J.; Chen, C. Genome-wide characterization and linkage mapping of simple sequence repeats in mei (*Prunus mume* Sieb. Et Zucc.). *PLoS ONE* **2013**, *8*, e59562. [CrossRef]
15. Scorza, R. Theory and practice of genetically manipulating peach tree architecture. *NY Fruit Q.* **2005**, *13*, 27–31.
16. Michelmore, R.W.; Paran, I.; Kesseli, R.V. Identification of markers linked to disease-resistance genes by bulked segregant analysis: A rapid method to detect markers in specific genomic regions by using segregating populations. *Proc. Natl. Acad. Sci. USA* **1991**, *88*, 9828–9832. [CrossRef]
17. Xu, X.; Jiang, C.; Liao, J.; Gu, Q.; Liu, S.; Chen, J. Screening of EST-SSR marker linked to dwarf character in *Actinidia chinensis* Planch. *Acta Hortic. Sin.* **2010**, *37*, 553–558.
18. Gonzalez, M.; Salazar, E.; Castillo, J.; Morales, P.; Mura-Jornet, I.; Maldonado, J.; Silva, H.; Carrasco, B. Genetic structure based on EST-SSR: A putative tool for fruit color selection in Japanese plum (*Prunus salicina* L.) breeding programs. *Mol. Breed.* **2016**, *36*, 68. [CrossRef]
19. Foisset, N.; Delourme, R.; Barret, P.; Renard, M. Molecular tagging of the dwarf *BREIZH* (*Bzh*) gene in *Brassica napus*. *Theor. Appl. Genet.* **1995**, *91*, 756–761. [CrossRef]
20. Warburton, M.L.; Becerra-Velasquez, V.L.; Goffreda, J.C.; Bliss, F.A. Utility of RAPD markers in identifying genetic linkages to genes of economic interest in peach. *Theor. Appl. Genet.* **1996**, *93*, 920–925. [CrossRef]
21. Milach, S.; Rines, H.W.; Phillips, R.L. Molecular genetic mapping of dwarfing genes in oat. *Theor. Appl. Genet.* **1997**, *95*, 783–790. [CrossRef]
22. Byers, D. *Crapemyrtle: A grower's thoughts*; Owl Bay Publishers: Auburn, AL, USA, 1997.

23. Egulf, D.R.; Andrick, A. *The Lagerstroemia Handbook/Checklist: A Guide to Crape Myrtle Cultivars*; American Association of Botanical Gardens and Arboreta, Inc.: Las Cruces, NM, USA, 1978.
24. Zhang, Q. Studies on Cultivars of Crape-myrtle (*Lagerstroemia indica*) and their uses in urban greening. *J. Beijing For. Univ.* **1991**, *4*, 57–66.
25. Pooler, M.R. Molecular genetic diversity among 12 clones of *Lagerstroemia fauriei* revealed by AFLP and RAPD markers. *Hortscience* **2003**, *38*, 256–259. [CrossRef]
26. Wang, X.; Wadl, P.A.; Pounders, C.; Trigiano, R.N.; Cabrera, R.I.; Scheffler, B.E.; Pooler, M.; Rinehart, T.A. Evaluation of genetic diversity and pedigree within crape myrtle cultivars using simple sequence repeat markers. *J. Am. Soc. Hortic. Sci.* **2011**, *136*, 116–128. [CrossRef]
27. Egolf, D.R. 'Biloxi', 'Miami', and 'Wichita' Lagerstroemia. *HortScience (USA)* **1987**, *22*, 336–338.
28. Pooler, M.R.; Dix, R.L. Chickasaw', kiowa', and Pocomoke' *Lagerstroemia*. *Hortscience* **1999**, *34*, 361–363. [CrossRef]
29. Hagan, A.K.; Keever, G.J.; Gilliam, C.H.; Williams, J.D.; Creech, G. Susceptibility of crapemyrtle cultivars to powdery mildew and *Cercospora* leaf spot in *Alabama*. *J. Environ. Hortic.* **1998**, *16*, 143–147.
30. Pounders, C.; Rinehart, T.; Sakhanokho, H. Evaluation of interspecific hybrids between *Lagerstroemia indica* and *L. speciosa*. *Hortscience* **2007**, *42*, 1317–1322. [CrossRef]
31. Katsuo, K. Crape-Myrtle Named Purple Queen. U.S. Patent US PP07: 957, 1992.
32. Katsuo, K. Crape-Myrtle Named White Fairy. U.S. Patent US PP07: 968, 1992.
33. Pooler, M.R. Arapaho' and Cheyenne' *Lagerstroemia*. *Hortscience* **2006**, *41*, 855–856. [CrossRef]
34. Ye, Y.J.; Wu, J.; Feng, L.; Ju, Y.; Cai, M.; Cheng, T.; Pan, H.; Zhang, Q. Heritability and gene effects for plant architecture traits of crape myrtle using major gene plus polygene inheritance analysis. *Sci. Hortic.-Amst.* **2017**, *225*, 335–342. [CrossRef]
35. Ye, Y.J.; Liu, Y.; Cai, M.; He, D.; Shen, J.S.; Ju, Y.Q.; Bian, X.M.; Pan, H.T.; Zhang, Q.X. Screening of molecular markers linked to dwarf trait in crape myrtle by bulked segregant analysis. *Genet. Mol. Res.* **2015**, *14*, 4369–4380. [CrossRef]
36. He, D.; Liu, Y.; Cai, M.; Pan, H.T.; Zhang, Q.X.; Wang, X.Y.; Wang, X.J. Genetic diversity of *Lagerstroemia* (Lythraceae) species assessed by simple sequence repeat markers. *Genet. Mol. Res.* **2012**, *11*, 3522–3533. [CrossRef] [PubMed]
37. Cai, M.; Pan, H.; Wang, X.; He, D.; Wang, X.; Wang, X.; Zhang, Q. Development of novel microsatellites in *Lagerstroemia indica* and DNA fingerprinting in Chinese *Lagerstroemia* cultivars. *Sci. Hortic.-Amst.* **2011**, *131*, 88–94. [CrossRef]
38. Wang, J.; Dai, X.; Chen, Y.; Yang, Y.; Zhang, X.; Li, S.; Yin, T. Genomic sequencing using 454 pyrosequencing and development of an SSR primer database for '*Lagerstroemia indica*' L. *Plant Omics* **2015**, *8*, 17.
39. He, D.; Liu, Y.; Cai, M.; Pan, H.; Zhang, Q. The first genetic linkage map of crape myrtle (*Lagerstroemia*) based on amplification fragment length polymorphisms and simple sequence repeats markers. *Plant Breed.* **2014**, *133*, 138–144. [CrossRef]
40. Liu, Y.; He, D.; Cai, M.; Tang, W.; Li, X.Y.; Pan, H.T.; Zhang, Q.X. Development of microsatellite markers for *Lagerstroemia indica* (Lythraceae) and related species. *Appl. Plant Sci.* **2013**, *1*, 1200203. [CrossRef]
41. Ye, Y.; Cai, M.; Ju, Y.; Jiao, Y.; Feng, L.; Pan, H.; Cheng, T.; Zhang, Q. Identification and validation of SNP markers linked to dwarf traits using SLAF-Seq technology in *Lagerstroemia*. *PLoS ONE* **2016**, *11*, e158970. [CrossRef]
42. Bassi, D.; Rizzo, M. Peach breeding for growth habit. *Acta Hortic.* **2000**, *1*, 411–414. [CrossRef]
43. Thakur, A.K.; Uphoff, N.; Antony, E. An assessment of physiological effects of system of rice intensification (SRI) practices compared with recommended rice cultivation practices in India. *Exp. Agr.* **2010**, *46*, 77–98. [CrossRef]
44. Zhu, R.R.; Gao, Y.K.; Xu, L.J.; Zhang, Q.X. Genetic diversity of *Aquilegia* (Ranunculaceae) species and cultivars assessed by AFLPs. *Genet. Mol. Res.* **2011**, *10*, 817–827. [CrossRef]
45. Ju, Y.; Feng, L.; Wu, J.; Ye, Y.; Zheng, T.; Cai, M.; Cheng, T.; Wang, J.; Zhang, Q.; Pan, H. Transcriptome analysis of the genes regulating phytohormone and cellular patterning in *Lagerstroemia* plant architecture. *Sci. Rep.* **2018**, *8*, 15162. [CrossRef]
46. Kosambi, D.D. The estimation of map distances from recombination values. *Ann. Hum. Genet.* **2016**, *12*, 125–130.

47. Guidry, R.K.; Einert, A.E. Potted dwarf crape myrtles: A promising new floriculture crop. *Flor. Rev.* **1975**, *157*, 30.
48. Shi, J.; Chen, Z.; Qin, B.; Cai, M.; Pan, H.; Zhang, Q. Phenotypic and genetic analysis of reptant plant type, applanate branch and other related characters in *Lagerstroemia indica*. *Adv. Ornam. Hortic. China* **2016**, *1*, 100–109.
49. Jiao, Y.; Feng, L.; Ye, Y.; Ju, Y.; Pan, H.; Zhang, Q.; Wang, G.; Pan, L.; Zhu, Y. Phenotypic and genetic analysis of several traits in F1 Hybrids of *Lagerstroemia speciosa* and *L. India*. *Adv. Ornam. Hortic. China* **2017**, *1*, 292–297.
50. Hollender, C.A.; Pascal, T.; Tabb, A.; Hadiarto, T.; Srinivasan, C.; Wang, W.; Liu, Z.; Scorza, R.; Dardick, C. Loss of a highly conserved sterile alpha motif domain gene (*WEEP*) results in pendulous branch growth in peach trees. *Proc. Natl. Acad. Sci. USA* **2018**, *115*, E4690–E4699. [CrossRef]
51. Zhao, J.; Chen, F.; Teng, N.; Chen, S. Genetic analysis and RAPD marker of creeping habits in ground-cover *Chrysanthemum*. *Sci. Agric. Sin.* **2009**, *42*, 734–741.

*forests*

MDPI

*Article*

# Morphological Characteristics and Allometric Relationships of Shoot in Two Undergrowth Plants: *Polygonatum odoratum* and *Polygonatum multiflorum*

**Mirela Tulik [1,*], Jerzy Karczewski [2], Natalia Szeliga [3], Joanna Jura-Morawiec [4] and Ingeborga Jarzyna [5]**

1 Department of Forest Botany, Warsaw University of Life Sciences—SGGW, 02-776 Warsaw, Poland
2 Department of Biophysics and Plant Morphogenesis, University of Silesia, 40-032 Katowice, Poland; jerzy.karczewski@us.edu.pl
3 Lezajsk Forest District, 37-300 Lezajsk, Poland; natalia.szeliga.2017@gmail.com
4 Polish Academy of Sciences Botanical Garden—Centre for Biological Diversity Conservation in Powsin, 02-973 Warsaw, Poland; j.jura@gazeta.pl
5 Faculty of Biology, Biological and Chemical Research Centre, University of Warsaw, 02-089 Warsaw, Poland; i.jarzyna@uw.edu.pl
* Correspondence: mirela.tulik@wl.sggw.pl; Tel.: +48-22-593-8032

Received: 16 November 2018; Accepted: 17 December 2018; Published: 19 December 2018

**Abstract:** The main purpose of this investigation was to describe the spatial arrangement of shoot tissues, as seen in transverse section, and allometric relationships in two contrasting species of *Polygonatum* i.e., *Polygonatum odoratum* which commonly grows in mixed pine-oak forest with shoots rectangular in shape, and *Polygonatum multiflorum* found in oak-hornbeam forest with cylindrical shoots. The mass and length of the aerial shoots of each individual plant were measured. The shoot regions of each plant were then categorized as basal (b), central (c) or apical (a). Transverse sections of these shoot regions were subsequently cut, and the following parameters were measured: (1) Diameter of shoots, (2) thickness of the outer and inner zones of parenchyma and (3) thickness of the sclerenchyma zone. Additional allometric relationships between the various measurements were computed and determined as Pearson's correlation coefficients ($r$). Both species of *Polygonatum* differed significantly with respect to the length, diameter and thickness of the outer zone of parenchyma. Shoots of *P. multiflorum* were taller but narrower than those of *P. odoratum,* which had a significantly wider zone of outer parenchyma. Allometric relationships were stronger for *P. multiflorum,* and for both species, they were generally stronger in the basal part of the shoot. We conclude that in *P. multiflorum,* the strong correlation between the diameter and length of the shoot seems to be important to the growth in shaded environments.

**Keywords:** allometry; anatomy; *Polygonatum odoratum*; *Polygonatum multiflorum*; shape; shoot

## 1. Introduction

Spatial organization of the plant body is defined as plant architecture subject to genetic and environmental control [1]. It is widely believed that the purpose of the mechanical integrity of the shoot is to ensure the reproductive success and survival of the plant [2]. Such mechanical integrity, and the peripheral location (or peripheral concentration in the case of monocots) of vascular bundles and mechanical supporting tissues such as collenchyma and sclerenchyma also enable the plant to resist and respond to physical forces such as wind and are exacerbated by the weight of the plant itself [3]. Crook et al. [4] demonstrated that plants occupying different habitats are able to modify their structural investment so as to maintain a "constant factor of safety" against mechanical failure.

As a consequence, plants that grow in exposed sites differ from those occupying more sheltered environments in terms of their morphology and anatomy [5].

In a forest, light is one of the main factors affecting species diversity and coexistence of plants [6,7]. It influences the abundance [8] and composition [9] of species in the understory. The plants growing on the forest floor obtain only 0.5%–5% of incident light [10]; in tropical rainforests less than 1% [11] and less than 2%–5% in moderately humid deciduous forests [12].

We examined two species of *Polygonatum*, growing on the forest floor namely *Polygonatum odoratum* (Mill.) Druce and *Polygonatum multiflorum* (L.) All, both members of the monocotyledonous family Asparagaceae (APG III 2009). These differ from each other in several ways, including the number of flowers located in leaf axils, the size and shape of the shoot, and the habitats that they occupy. One of the most distinct differences between their habitats is the amount of light reaching the forest floor [13]. *P. odoratum* is native to Europe, Asia, and Northern Africa (http://e-monocot.org). It prefers semi-shade and a moderately exposed habitat, where it grows to a height of ~65 cm. Its white, tubular flowers (one, rarely two per axil) hang singly from the underside of the shoot, which is square to rectangular in transverse section. By contrast, *P. multiflorum* can grow to a height of 80 cm in deciduous European or Asian forests (beech, oak and hornbeam). Its tubular flowers are also white (usually 3–5 per leaf axils) and also hang from beneath the shoot which, unlike that of *P. odoratum*, is cylindrical in shape and round in transverse section.

Since mechanical design, whether it be at the cellular level or at the level of whole-plant architecture, appears to be fundamental to survival, the main purpose of this investigation was to describe the spatial arrangement of shoot tissues, as seen in transverse section, and allometric relationships in two contrasting species of *Polygonatum*.

## 2. Materials and Methods

Shoots of the two species used for our study (*P. multiflorum* and *P. odoratum)* were collected in the late spring of 2015 and 2017. Fifteen flowering individuals of each species were collected for each of the two years of our study. Individuals of *P. odoratum* were found in mixed pine-oak forest *Querco roboris-Pinetum* (W. Mat.1981) J. Mat. 1988) whereas *P. multiflorum* grew in sub-continental oak-hornbeam forest (*Tilio-Carpinetum* Tracz. 1962). Plants of each species were cut at ground level, protected from desiccation, and brought to the laboratory, where they were subjected to investigations.

Shoot length (L) was subsequently measured and three shoot regions determined for each, namely: Basal (b), located close to the base of the shoot; central (c), located in the middle part of the shoot; and apical (a), located near the apex (Figure 1).

Hand-cut transverse sections were then obtained from each of these shoot regions for both species. The protocol using carrot and cork, as recommended by Gärtner and Schweingruber [14], was adopted for preparing sections using a sliding microtome (Microm HM 440, GMI Inc, Ramsey, MN, USA). For the detection of lignified cell walls, sections were stained with Alcian blue and Safranin [15] and some sections were also observed under UV (ultraviolet) light. Photomicrographs were achieved with the aid of an Olympus system consisting of a BX61 motorized microscope (Olympus, Tokyo, Japan) and Cell P image analysis software (version 3.4) coupled to a Color View digital camera (Olympus Soft Imaging System GmbH, Műnster, Germeny). Based on images taken from each region of the shoot for every individual, the following parameters were measured: (1) diameter of shoot (D); (2) thickness of mechanical tissue zone (sclerenchyma zone, Ws); (3) thickness of parenchyma tissue located externally to the mechanical tissue (hereafter referred to as outer parenchyma zone (Wp)); and (4) thickness of inner parenchyma zone ($Wp_i$, calculated and expressed as the difference between the diameter and the sum of the thicknesses of outer parenchyma and sclerenchyma zones, as follows: $Wp_i = D - (Wp + Ws)$). Measurements were performed at two points on the circumference of each section and the average measurements calculated. The accuracy of these measurements was $\pm 0.01$ mm.

**Figure 1.** Location of three regions along the length of *Polygonatum* shoot categorized as basal (b), central (c) and apical (a). Region (b) was 0.5 cm above ground level.

In addition, we determined the mass (M) of the entire aerial part of individual plants collected in 2017 and measured shoot length (L). Accuracy of the measurements was $\pm 0.01$ g and $\pm 0.1$ mm, respectively. The allometric relationships between all measurements were characterized by allometric scaling laws of the form $Y = cM^b$ (c—constant, b—allometric scaling coefficient) and determined as Pearson's correlation coefficients ($r$) on data transformed to a natural logarithm. A similar method was used by Niklas [5], Weiner and Thomas [16], and Poorter et al. [17].

Significance of difference between means was assessed by Student's $t$-test. Analyses were computed on Statistica software, version 13.

## 3. Results

### 3.1. Morpho-Anatomical Analysis of Polygonatum Shoots

Shoot anatomy of both investigated species demonstrated significant similarity, despite differences in the cross-sectional shape of the shoots (rectangular in *P. odoratum* and round in *P. multiflorum*; Figure 2a, b). The shoots of both species were comprised of epidermis, ground tissue and collateral vascular bundles. The ground tissue was represented by the outer parenchyma, sclerenchyma and inner parenchyma (core). The collateral, closed vascular bundles, were irregularly scattered throughout the ground tissue (Figure 2c–e). Towards the outer parenchyma, the bundles were smaller in size, whereas those near the centre of the shoot were larger. The vascular bundles, especially those located peripherally, were enclosed within a sclerotic sheath comprised of layers of sclerenchyma fibres.

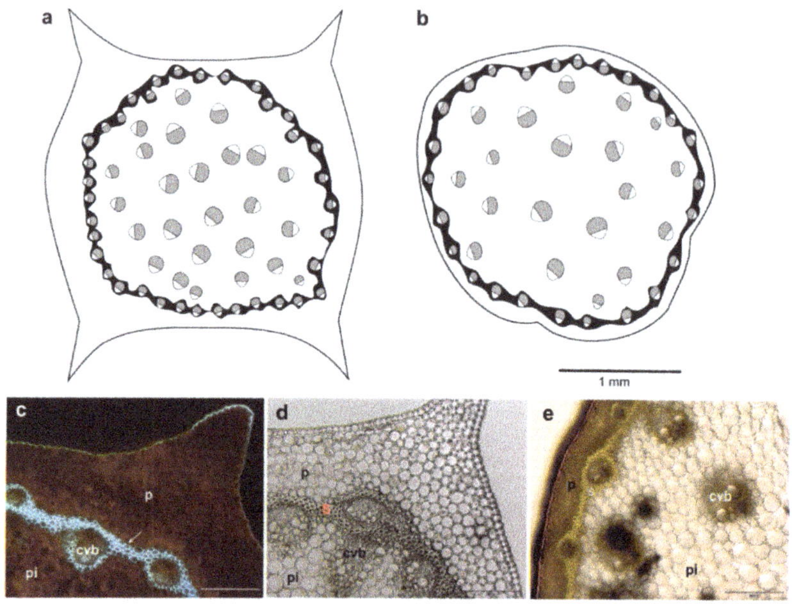


**Figure 2.** (**a**,**b**) Diagram showing shape of *Polygonatum* shoots as seen in transverse section; rectangular in *P. odoratum* (**a**) and circular in *P. multiflorum* (**b**). The sclerenchyma zone is marked black; the outer/inner parenchyma zone white. Vascular tissue is represented by scattered collateral bundles. (**c**–**e**) Photomicrographs of transverse sections of *P. odoratum* (**c**,**d**) and *P. multiflorum* (**e**) shoots. Scale bar = 200 µm. (**c**)—section viewed under UV light. Fluorescence of lignified cell walls of sclerenchyma cells is marked with arrow. p—outer parenchyma zone; s—sclerenchyma zone; pi—inner parenchyma zone; cvb—collateral vascular bundle.

No significant difference were observed in the thickness of the sclerenchyma zone between the species at basal level. Nevertheless, plants varied significantly with respect to the thickness of the outer zone of parenchyma, regardless of region (Table 1).

**Table 1.** Values of Student's *t*-test for the mean thickness of the outer parenchyma zone for the three regions (basal, central, and apical) along the shoots of *P. odoratum* (P. o) and *P. multiflorum* (P. m).

| Variable at Categorized Region | Mean | | *t*-Value | *p*-Value |
|---|---|---|---|---|
| | P. o | P. m | | |
| Basal | 0.1876 | 0.1047 | 8.6023 | 0.0000 |
| Central | 0.1589 | 0.0930 | 7.3910 | 0.0000 |
| Apical | 0.1555 | 0.0931 | 8.8375 | 0.0000 |

Means are expressed in mm, $n = 15$.

In the basal region, the mean thickness of the outer parenchyma zone of *P. odoratum* reached 0.19 mm, as compared with 0.10 mm in *P. multiflorum* (Figure 3).

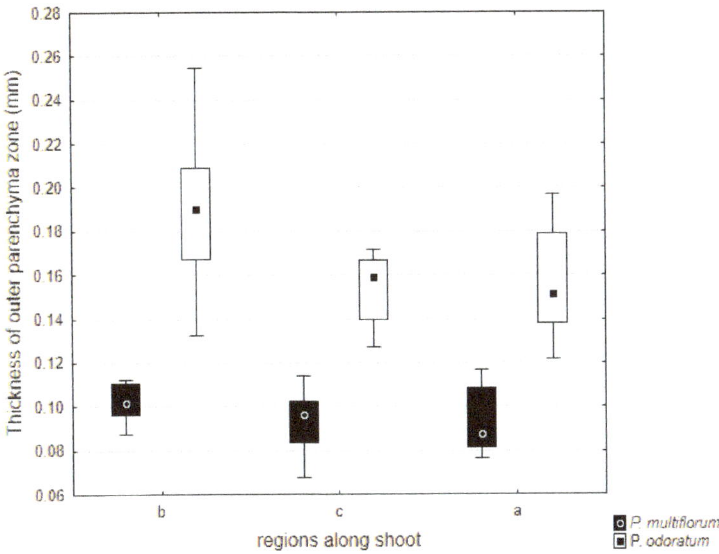


**Figure 3.** Thickness of outer parenchyma zone at basal (b), central (c) and apical (a) regions along shoot of both species.

For both species, the width of the outer parenchyma zone diminished along its length, although for *P. multiflorum*, the mean value at the central region was almost identical to that for the apical region. The mean thickness of the sclerenchyma zone in the basal region was similar: 0.055 mm for *P. odoratum* and 0.053 mm for *P. multiflorum*. In both species, the diameter of shoot and thickness of the sclerenchyma zone diminished on approaching the apical region, but we observed very significant differences between values (Figure 4).

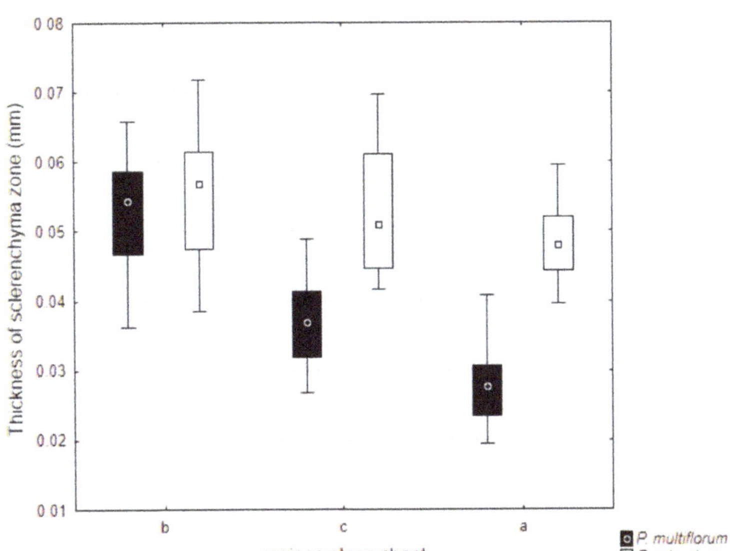


**Figure 4.** Thickness of sclerenchyma zone at basal (b), central (c) and apical (a) regions along shoot of both species. Significant differences are for central ($p < 0.0001$) and apical ($p < 0.0000001$) regions.

Moreover, the species differed significantly with respect to the radius/diameter, thickness of the outer parenchyma zone and sclerenchyma zone relative to the overall length of their shoots (Table 2).

**Table 2.** Values of Student's *t*-test for the mean radius and thickness of the outer parenchyma zone and sclerenchyma zone for the entire length of the shoot of tested plants.

| Variable | Mean | | *t*-Value | *p*-Value |
|---|---|---|---|---|
| | P. o | P. m | | |
| Radius | 1.6416 | 1.4050 | 2.0269 | 0.0456 |
| Thickness of outer parenchyma zone | 0.1674 | 0.0969 | 13.0812 | 0.0000 |
| Thickness of sclerenchyma zone | 0.05217 | 0.0395 | 5.28161 | 0.0000 |

P. o—P. odoratum; P. m—P. multiflorum. Means are expressed in mm, $n = 45$.

### 3.2. Allometric Relationships

Correlations between diameter (D) and length (L) of the shoot for each of the two species and for all three regions of the shoot (basal, central and apical) were very significant (Table 3, Figure 5). For *P. multiflorum*, values for *r* were greater than for *P. odoratum*. For both species, values for *r* were greater for the basal region than for the apical region of the shoot. Individuals of *P. multiflorum* had longer shoots, although these had smaller diameters.

**Table 3.** Values of Pearson's correlation coefficients (*r*) between diameter (D) and length (L) of the shoot for the two species *P. odoratum* and *P. multiflorum* based on three regions of the shoot (basal, central and apical).

| | *r* D–L | | |
|---|---|---|---|
| | Basal | Central | Apical |
| ***P. odoratum*** | 0.7305 ** | 0.8555 *** | 0.6224 * |
| ***P. multiflorum*** | 0.9626 *** | 0.9152 *** | 0.8511 *** |

Data transformed to natural logarithm. Asterisks denote the statistical significance of Pearson's correlation coefficient: *** $p < 0.001$, ** $p < 0.01$, * $p < 0.05$, $n = 15$.

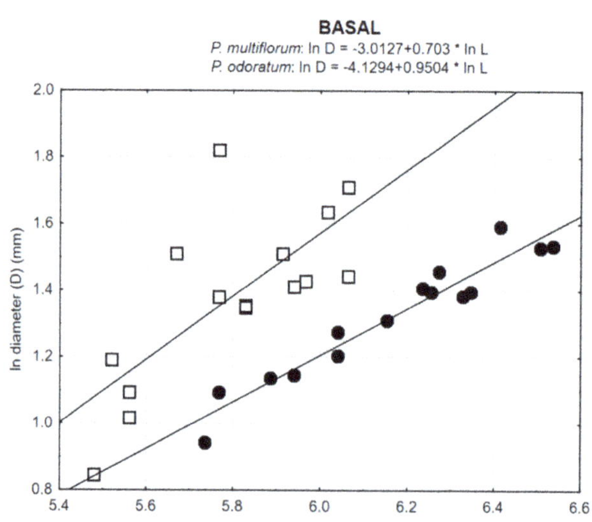


**Figure 5.** *Cont.*

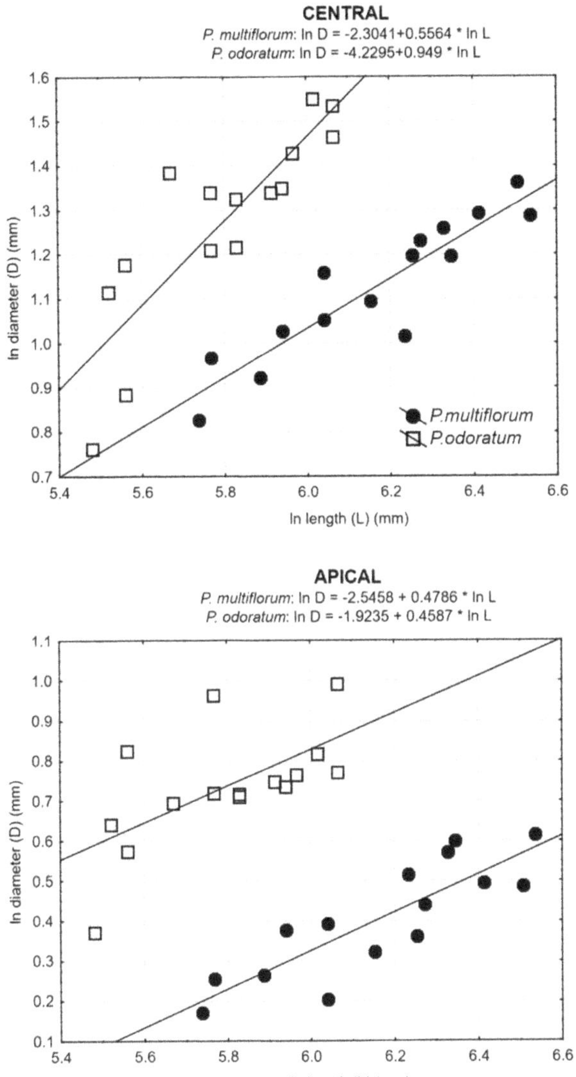


**Figure 5.** Relationships between diameter and length for all three regions of the shoot for the two species *P. odoratum* and *P. multiflorum*.

A relationship also existed between mass (M) and length (L) of the shoot (data ln-transformed). For both species, coefficients of correlation were similar: for *P. multiflorum* $r = 0.6473$ and for *P. odoratum* $r = 0.6757$, $p < 0.01$ (Figure 6). Individuals of *P. multiflorum* were generally larger, with longer shoots and greater mass.

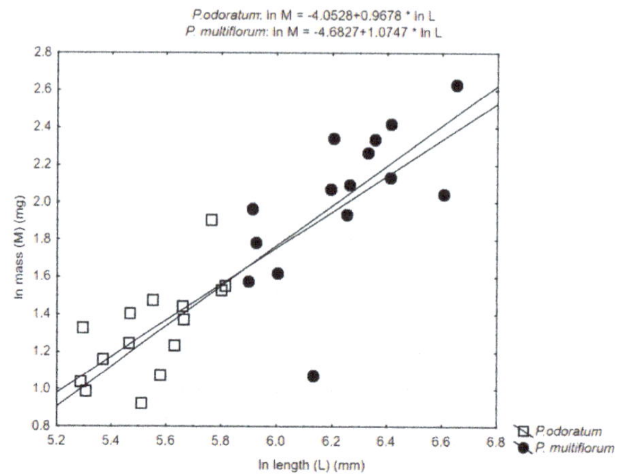


**Figure 6.** Relationships between mass and length of the shoot for the two species *P. odoratum* and *P. multiflorum*.

We also observed correlations between the thickness of the sclerenchyma zone (Ws) and that of the outer parenchyma zone (Wp) vs. shoot diameter (D), as well as shoot length (L), in both species (Table 4). Correlations were generally stronger for the basal part of the shoot than for the apical region, but this relationship was generally not very strong.

**Table 4.** Values of Pearson's correlation coefficients ($r$) between the thickness of the sclerenchyma zone (Ws) and the thickness of the outer parenchyma cylinder (Wp) vs. shoot diameter (D) and shoot length (L) for *P. odoratum* and *P. multiflorum* based on three regions of the shoot (basal, central and apical).

| | ***P. odoratum*** | | | |
|---|---|---|---|---|
| | **Ws-D** | **Ws-L** | **Wp-D** | **Wp-L** |
| basal | 0.529657 * | 0.664209 ** | NS | 0.599483 * |
| central | NS | | | |
| apical | NS | NS | NS | |
| | ***P. multiflorum*** | | | |
| | **Ws-D** | **Ws-L** | **Wp-D** | **Wp-L** |
| basal | 0.816187 *** | 0.799026 *** | NS | |
| central | NS | NS | NS | NS |
| apical | 0.522822 * | 0.741862 ** | NS | NS |

Data transformed to natural logarithm. Asterisks denote the statistical significance of Pearson's correlation coefficient: *** $p < 0.001$, ** $p < 0.01$, * $p < 0.05$, NS—non significant, $n = 15$.

## 4. Discussion

The organization and spatial distribution of tissues in the shoot of two species of *Polygonatum* may support the thesis that these plants are adapted to be resistant to mechanical failure, since they are composed of a variety of cells whose walls (made of cellulose and lignin) differ in elasticity [5,18]. Comparison of the shoot structure of *Polygonatum* with the anatomy of grasses reveals a universal solution to certain structural problems (in particular those relating to mechanical support) common to both groups of plants. Fundamentally, the most common type of aerial shoot anatomy in monocots is based on the presence of a peripherally situated cylinder of sclerenchyma fibres. Hypodermal sclerenchyma is also very common in grass. This tissue is dead at maturity, with thick, lignified walls enclosing an empty lumen [5,19]. The distribution of this tissue provides increased support and rigidity and helps the plant withstand forces, such as wind. Centralized arrangement of mechanical tissues is also observed, especially in roots, as well as the shoots of submerged, aquatic angiosperms, both of which are subject to pulling strains [20]. A sclerotic sheath (a sheath of sclerenchymatous fibres surrounding the vascular bundles) may comprise the main stiffening element in many other monocots, such as palms [21]. It is stiffer by far than the surrounding parenchymatous ground tissue in which the vascular bundles are embedded [22–24]. It is also worth noting that the shoot of grasses is usually hollow e.g., the internodes of cereals. In other words, air occurs at the 'core' of grass shoots and this is the lightest and least expensive arrangement, both in terms of material and energy expenditure. Moreover, it permits longitudinal gaseous transport and is of particular value in certain marginal plants, where like the aerenchyma of true hydrophytes, it may allow the aeration of submerged organs. The arrangement of mechanical tissue, both peripherally and around each vascular bundle is considered optimal and in the case of grasses, provides excellent protection from mechanical failure (e.g., wilting), since the weight of plant per unit of shoot volume is greatly reduced. In describing the morphological and anatomical adaptation of grass shoots to mechanical stress, Frey [25] compared it with a fundamental rule of eastern fighting, namely, "flex to win". In contrast to grasses, the shoot core of *Polygonatum* is solid and parenchymatous, consisting of thin-walled parenchyma cells. Thin-walled tissues tend to possess lower elastic moduli than thick-walled tissues [5]. Nevertheless, they become increasingly important when the shoot is thick and the amount of sclerenchyma present small. Parenchyma may provide mechanical support hydrostatically by means of cell turgidity [5].

As previously mentioned, in order to be resistant to mechanical support, the shoot must be able to bend, but not break, and therefore it is necessary that the most rigid tissue is peripherally located. As a result, some sort of compromise must be reached between the distribution of mechanical and photosynthetic tissue. The question then arises: How do both species of *Polygonatum* resolve this anatomical conflict between the requirement for support and that for photosynthesis? One possible explanation in the case of *P. multiflorum* is that the turgor pressure of the inner parenchyma cells on the inside of the cylinder of sclerenchyma reinforces the shoot. As a result, these plants may grow taller and become more slender so as to absorb more solar energy for photosynthesis and other light-dependent processes. In the case of *P. odoratum*, the outer zone of photosynthetic cortical parenchyma is wider and therefore the mechanically resistant sclerenchyma is less peripheral in its distribution. As a consequence, the shoots of this species are shorter and have a greater diameter, allowing the plant to grow at greater light intensities and owing to the wider zone of chlorophyllous cortical parenchyma (outer parenchyma), photosynthesis is enhanced. It would also appear that in the case of *Polygonatum* the compromise between mechanical support and photosynthesis is subject to the environment, especially since light conditions have a particularly strong impact both on plant size and form [26,27].

The behaviour of *P. multiflorum* may also be compared with the growth strategy of sapling trees. Young and quickly growing trees produce juvenile wood external to the pith [28]. The properties of juvenile wood are different from those of mature wood and the former is thought to be mechanically more resistant. Cells of juvenile wood tend to be shorter and have thinner walls. Therefore, juvenile wood is less dense than mature wood [29]. Young trees have slender stems that are more flexible than those of older trees, may grow rapidly towards the canopy and bend without breaking. It seems that the shoots of *P. multiflorum*, which are taller and have a smaller diameter, are better adapted to compete for light than those of *P. odoratum* (despite the fact that their mechanical tissue is represented by sclerenchyma and/or parenchyma cells, and that the pattern of biomass partitioning is intermediate to that typical of herbaceous plants and trees).

Allometric relations have frequently been reported for trees (in order to estimate their total biomass or root mass—parameters that are difficult to measure), but are less often reported for herbaceous plants, which are believed to be a result of natural selection processes and adaptive evolutionary changes [30–32]. Weiner and Thomas [16] suggested that the ability of plants to adapt to the environment modifies allometric relationships, especially in annual plants, and may depend on the prevailing growing conditions [33,34]. Jarzyna [35] investigated allometric relationships for competing and non-competing plants, discovering that those for competing plants were stronger. It is likely that in both *P. odoratum* and *P. multiflorum*, amongst others, the availability of light may be a crucial factor for explaining allometric relationships. These relationships are stronger in *P. multiflorum*, probably because this species grows in poorer light conditions (oak-hornbeam forest). In moderately humid deciduous forests, the percentage of sunlight reaching through the tree foliage to the herb and moss layer on the forest floor is less than 2%–5% [12]. Moreover, light reaches the forest floor mainly in the dormant season, while in the summer time, leaves of the canopy trees block the passage of light to the ground. Based on differences in tree crown morphology and the stand structure occupied by both investigated species of *Polygonatum*, it could be assumed that in a dense forest stand occupied by *P. mutliflorum*, less light penetrates through the canopy. Thus, with limited light availability in the understory, the taller *P. multiflorum* might invest more energy into its aboveground biomass i.e., its supporting shoot.

## 5. Conclusions

Our paper contributes towards a better understanding of the relationship that exists between plant structure and the environment, based on two contrasting species of *Polygonatum* (*P. odoratum* and *P. multiflorum*) growing on the forest floor that differ from each other in the shape of the shoot, as viewed in transverse section. We conclude that the spatial distribution of tissues in the shoots of the investigated species is typical of that found in self-supporting herbaceous plants. In *P. multiflorum*, the strong correlation between shoot diameter and length seems to be important to growth in shaded habitats.

**Author Contributions:** Conceptualization, M.T. and N.S.; Investigation and Methodology, M.T., N.S. and J.J.M.; Software, J.K. and I.J.; Writing—Review and Editing, M.T., I.J., J.J.M.

**Funding:** This research received no external funding.

**Conflicts of Interest:** The authors declare that there is no conflict of interest.

## References

1. Reinhard, D.; Kuhlemeier, C. Plant architecture. *EMBO Rep.* **2002**, *3*, 846–851. [CrossRef] [PubMed]
2. Crook, M.J.; Ennos, A.R. Mechanical differences between free-standing and supported wheat plants, *Triticum aestivum* L. *Ann. Bot.* **1996**, *77*, 197–202. [CrossRef]
3. Gartner, B.L. Root biomechanics and whole plant allocation patterns: Responses of tomato to stem flexure. *J. Exp. Bot.* **1994**, *45*, 1647–1654. [CrossRef]
4. Crook, M.J.; Ennos, A.R.; Sellers, E.K. Structural development of the shoot and root systems of two winter wheat cultivars. *J. Exp. Bot.* **1994**, *45*, 857–863. [CrossRef]
5. Niklas, K.J. *Plant allometry: The Scaling of Form and Process*; University of Chicago Press: Chicago, IL, USA, 1994; ISBN 9780226580814.
6. Grime, J.P. Evidence for the existence of three primary strategies in plants and its relevance to ecological and evolutionary theory. *Amer. Nat.* **1977**, *111*, 1169–1194. [CrossRef]
7. Schmidt, W.; Weitemeier, M.; Holzapfel, C. Vegetation dynamics in canopy gaps of beech forest on limestone—The influence of the light gradient on species richness. *Verh. Ges. Ökol.* **1996**, *25*, 253–260.
8. Elemans, M. Light, nutirents and the growth of herbaceous forest species. *Acta Oecol.* **2004**, *26*, 197–202. [CrossRef]
9. Tinya, F.; Márialigeti, S.; Király, B.; Németh, B.; Ódor, P. The effect of light conditions on herbs, bryophytes and seedlings of temperate mixed forests in Őrség, Western Hungary. *Plant Ecol.* **2009**, *204*, 69–81. [CrossRef]
10. Chazdon, R.L.; Pearcy, R.W. The importance of sunflecks for forest understory plants. *BioScience* **1991**, *41*, 760–766. [CrossRef]
11. Kitajima, K.; Mulkey, S.S.; Wright, J. Variation in crown light utilization characteristics among tropical canopy trees. *Ann. Bot.* **2005**, *95*, 535–547. [CrossRef]
12. Nũnements, Ũ. A review of light interception in plant stands from leaf to canopy in different plant functional types and in species with varying shade tolerance. *Ecol. Res.* **2010**, *25*, 693–714. [CrossRef]
13. Obmiński, Z. *Ekologia Lasu*; Państwowe Wydawnictwo Naukowe: Warszawa, Poland, 1977; pp. 260–268.

14. Gärtner, H.; Schweingruber, F.H. *Microscopic Preparation Techniques for Plant Stem Analysis*; Kessel Publishing House: Remagen, Germany, 2013; ISBN 978-3-941300-76-7.
15. Broda, B. *Metody Histochemii Roślinnej*; Państwowy Zakład Wydawnictw Lekarskich: Warszawa, Poland, 1971.
16. Weiner, J.; Thomas, S.C. Competition and allometry in three species of annual plants. *Ecology* **1992**, *73*, 648–656. [CrossRef]
17. Poorter, H.; Jagodziński, A.M.; Ruiz-Peinado, R.P.; Kuyah, S.; Luo, Y.; Oleksyn, J.; Usoltsev, V.A.; Buckley, T.N.; Reich, T.N.; Reich, P.B.; et al. Howe does biomass distribution change with size and differ among species? An analysis for 1200 plant species from five country. *New Phytol.* **2015**, *208*, 736–749. [CrossRef] [PubMed]
18. Vincent, J.F.V. The mechanical design of grass. *J. Mater. Sci.* **1982**, *17*, 856–860. [CrossRef]
19. Goodman, A.M. Mechanical adaptations of Cleavers (*Gallium aparine*). *Ann. Bot.* **2005**, *95*, 475–480. [CrossRef] [PubMed]
20. Koehl, M.A.R.; Wainwright, S.A. Mechanical adaptation of a giant kelp. *Limnol. Oceanogr.* **1997**, *22*, 1067–1071. [CrossRef]
21. Rüggeberg, M.; Bugert, I.; Speck, T. Structural and mechanical design of tissue interfaces in the giant reed *Arundo donax*. *J. R. Soc. Interface* **2010**, *7*, 499–506. [CrossRef]
22. Rich, P.M. Mechanical structure of the stem of arborescent palms. *Bot. Gaz.* **1987**, *148*, 42–50. [CrossRef]
23. Tomlinson, P.B. *The Structural Biology of Palms*; Clarendon Press: Oxford, UK, 1990; ISBN 10: 019854572X.
24. Spatz, H.C.; Beismann, H.; Brüchert, F.; Emanns, A.; Speck, T. Biomechanics of the giant reed *Arundo donax*. *Philos. Trans. R. Soc. Lond.* **1997**, *352*, 1–10. [CrossRef]
25. Frey, L. Trawy niezwyciężone. *Łąkarstwo W Pol.* **2000**, *3*, 9–20.
26. Winer, J.; Bernston, G.M.; Thomas, S.C. Competition and growth form in a woodland annual. *J. Ecol.* **1990**, *78*, 459–469. [CrossRef]
27. Watari, R.; Nagashima, H.; Hirose, T. Stem extension and mechanical stability of *Xantium canadense* grown in an open or in dense stand. *Ann. Bot.* **2014**, *114*, 179–190. [CrossRef] [PubMed]
28. Zobel, B.J.; Sprague, J.R. *Juvenille Wood in Forest Trees*; Springer-Verlag: Berlin, Germany, 1998; ISBN 978-3-642-72126-7.
29. Plomion, C.; Leprovost, G.; Stokes, A. Wood formation in trees. *Plant Physiol.* **2001**, *127*, 1513–1517. [CrossRef] [PubMed]
30. Niklas, K.J. Plant height and the properties of some herbaceous stems. *Ann. Bot.* **1995**, *75*, 133–142. [CrossRef]
31. Castelan-Estrada, M.; Vivin, P.; Gaudillére, J.P. Allometric relationships estimate seasonal above-ground vegetative and reproductive biomass of *Vitis vinifera* L. *Ann. Bot.* **2002**, *89*, 401–408. [CrossRef]
32. Enquist, B.J. Universal scaling in tree and vascular plant allometry toward a general quantitative theory linking plant form and function from cells to ecosystems. *Tree Physiol.* **2002**, *22*, 1045–1064. [CrossRef]
33. Prior, S.A.; Rogers, H.H. Soybean growth response to water supply. *J. Plant Nutr.* **1995**, *18*, 617–636. [CrossRef]
34. Wang, P.; Weiner, J.; Cahill, J.F.; Zhou, D.W.; Bian, H.F.; Song, Y.T.; Sheng, L.X. Shoot competition, root competition and reproductive allocation in *Chenopodium accuminatum*. *J. Ecol.* **2014**, *102*, 1688–1696. [CrossRef]
35. Jarzyna, I. Plant morphology and allometric relationships in competing and non-competing plants of *Tagetes patula* L. *Acta Soc. Bot. Pol.* **2002**, *71*, 149–154. [CrossRef]

forests

MDPI

Article

# Drought Impact on Leaf Phenology and Spring Frost Susceptibility in a *Quercus robur* L. Provenance Trial

**Ivica Čehulić [1], Krunoslav Sever [2], Ida Katičić Bogdan [2], Anamarija Jazbec [2], Željko Škvorc [2] and Saša Bogdan [2,*]**

1 Croatian Forest Research Institute, 10450 Jastrebarsko, Croatia; ivicac@sumins.hr
2 Faculty of Forestry; University of Zagreb, 10000 Zagreb, Croatia; ksever@sumfak.hr (K.S.); ikaticic@sumfak.hr (I.K.B.); ajazbec@sumfak.hr (A.J.); zskvorc@sumfak.hr (Ž.Š.)
* Correspondence: sbogdan@sumfak.hr

Received: 20 November 2018; Accepted: 8 January 2019; Published: 11 January 2019

**Abstract:** Research highlights: The susceptibility of oaks to late spring and early autumn frosts is directly related to their leaf phenology. Drought may alter the leaf phenology and therefore frost tolerance of oaks. However, the effects of drought on oak leaf phenology and frost resistance have not been thoroughly studied. Background and objectives: One of the consequences of climate change is an increase in the frequency of dry episodes during the vegetation period. Pedunculate oak (*Quercus robur* L.) is an economically and ecologically important forest tree species that prefers humid habitats. Therefore, knowledge of the impact of drought on this species is of great importance for the adaptation of forestry strategies and practices to altered environmental conditions. The aim of this study was to determine the impact of drought on leaf phenology and spring frost susceptibility in nine provenances. Materials and methods: One-year-old saplings originating from nine European provenances were used in the trial. The saplings were exposed to experimental drought and then re-watered in two subsequent years. Spring and autumn leaf phenology were scored. The trial was impacted by a late spring frost in the third year, and the resulting leaf frost injury was scored. The effects of drought treatment on the phenology and frost susceptibility of plants from the provenances were analysed. Results: Leaf phenology of plants from most of the studied provenances was significantly influenced by the drought treatment ($p < 0.001$). Drought induced a carry-over effect on flushing phenology, which was observed as delayed bud burst (from 0.6 to 2.4 days) in the second year and as advanced bud burst (from 0.1 to 6.3 days) in the third year. Therefore, opposite shifts in flushing phenology may be induced as a result of differences in the time span when plants sense water deficits. In contrast to flushing, autumn leaf phenology was unambiguously delayed following the drought treatments for all studied provenances (from 2.1 to 25.8 days). Differences in late frost susceptibility were predominantly caused by among-provenance differences in flushing phenology. However, the drought treatment significantly increased frost susceptibility in the plants (the rate of frost-injured plants per provenance increased from 3% to 78%). This higher susceptibility to spring frost was most likely caused by the advanced flushing phenology that resulted from the drought treatment in the previous year.

**Keywords:** pedunculate oak; drought; stress; memory; flushing; autumn leaf senescence; phenological shift; carry-over effect

## 1. Introduction

The hazards associated with climate change, including rising temperatures, decreasing precipitation, and an increasing frequency of extreme climatic events, are expected to intensify [1–3]. In general, the productivity of forest ecosystems is severely impaired by water availability, and drought may induce episodes of large-scale tree decline in temperate forests [4]. Most likely, such declines will substantially increase the necessity for artificial regeneration of temperate forests. However, the increasing severity and frequency of droughts may also impact forest regeneration, causing unacceptably high seedling mortality rates. Late spring frost is an additional abiotic factor that strongly limits forest regeneration [5]. Therefore, it is necessary to enhance our understanding of the mutual impacts of drought and frost on forest reproductive material.

Pedunculate oak (*Quercus robur* L.) is a widespread European temperate forest tree species that generally prefers fertile and moist habitats [6], although it is also tolerant to arid habitats such as the forest-steppe ecosystems in eastern Europe (e.g., see [7]). Moreover, this species is one of the most economically valuable European hardwood tree species and is a climax species in forests that harbor high biodiversity. Because the species is mainly adapted to moist habitats, its drought tolerance will become increasingly important for its survival, especially in the presence of additional stresses such as frost, competition, pests, and diseases. Theoretically, pedunculate oak may adapt to stressful environmental changes at the individual, population, and plant community levels [2] through phenotypic modifications (i.e., plasticity), natural selection, and hybridization with related xerophilous species, such as sessile oak and pubescent oak [8,9]. In general, oaks are relatively tolerant to drought [10], as they have a xeromorphic leaf structure and root structure that can cope with variability in soil water availability; additionally, oaks display the ability to rapidly resume assimilation after periods of water deficiency [11–15]. However, intraspecific variation in responses to drought occurs among oaks from different provenances. Some of this variation has been attributed to adaptability driven by natural selection in the original provenance habitats [16,17], but in other cases, the original site-level climate of provenances has not been correlated with the responses of oaks to drought (e.g., [18]). Reduced above-ground biomass production, together with a shift towards root growth, is a drought response often observed in oak species [18–20]. Other drought responses by oak include changes in the morphological (e.g., reduced leaf area and chlorophyll content) and physiological (e.g., reduced stomatal conductance and water use efficiency) properties of leaves [21]. Although pedunculate oak does not appear to possess efficient responses to leaf frost injury [22], oak populations display adaptation to frost events in the form of differences in leaf phenology [23]. Additionally, occasional frost events (together with other factors) may have led to high within-population genetic variation in pedunculate oak populations, which are often characterized by various phenological types [24]. For deciduous tree species such as *Q. robur*, the timing of bud burst in spring and leaf senescence in autumn is very important in defining the period of carbon fixation and growth during the vegetation season as well as determining survival due to the increased frequency of climatic extremes such as frosts. Among the most important adaptive traits, phenology is thought to be one of the most affected by climate change [25]. However, it is necessary to further enhance our understanding of the impacts of interactive stressful climatic events on the phenology of deciduous forest tree species. The effects of drought on oak phenology (especially that of the pedunculate oak) have not been thoroughly studied and have been mainly described as earlier growth cessation under dry conditions [17,26], which is also visible as earlier radial growth cessation [27]. Interestingly, drought may induce an after-effect (i.e., a carry-over effect) in the subsequent spring that is indicated by advanced bud burst [26,28]. The same effect of water stress on leaf phenology was also recorded in beech [29]. In one report, sessile oak from different provenances showed a drought-induced carry-over effect in the subsequent spring, which was observable as a delay in the bud burst date [30]. To the best of our knowledge, this discrepancy in observed leaf phenological shifts induced by drought in oaks has not been explained, nor reported for other forest tree species. Here, we propose that bud burst shifts that occur as a carry-over effect induced by drought have opposite directions depending on the time span over which the plants sensed the water deficit. That is, if plants sensed the water deficit early in the vegetation period (i.e., in spring), then flushing would likely be delayed in the subsequent year; in contrast, if drought were sensed late (i.e., in summer), then advanced flushing would likely occur in the subsequent year. The effects of drought on oak frost resistance have been even less thoroughly studied. In general, few papers address between-provenance genetic variation in oak frost susceptibility (e.g., pedunculate oak [17] and holm oak [31]). One study examined the effect of drought on oak frost hardiness (measured as resistance to winter freezing periods) [32], but it did not evaluate frosts that occurred in late spring or early autumn. As drought may alter phenology and therefore the susceptibility of forest tree species to late spring and early autumn frosts, it is necessary to improve our understanding of these interactive effects.

We established a short-term pedunculate oak provenance trial with seedlings originating from nine provenances along the eastern European north-south gradient, i.e., from Estonia to Italy. Plants from these provenances were experimentally treated to induce drought conditions during two growing seasons. As the trial was affected by a late spring frost in the third year, we had the opportunity to examine variation in oak frost susceptibility among provenances as well as the impact of drought on the frost susceptibility of these plants. The aims of the study were to determine the following: (1) the impact of drought on the flushing phenology of plants from various provenances; (2) the impact of drought on the autumn leaf senescence of plants from these provenances; and (3) among-provenance differences in the impact of drought on spring frost susceptibility.

## 2. Materials and Methods

### 2.1. Plant Origin and Trial Design

In autumn 2013, acorns were collected below crowns of at least ten randomly chosen mature trees in each of nine allegedly natural pedunculate oak stands. The provenances are located along a latitudinal gradient from Estonia to Italy (Table 1). Acorns were sent to the Croatian Forest Research Institute and sown in 0.5-litre polyvinyl chloride (PVC) pots filled with nursery substrate. In spring 2014, germinated seedlings were individually transplanted into 50-litre PVC pots filled with natural soil extracted from a local pedunculate oak forest (soil type: gleysol; pH = 7.6; soil texture: silty loam). The soil was homogenized and cleared of stones, leaves, sprouts, and twigs prior to filling the pots. The pots were kept outdoors during the first vegetation period; thus, they were exposed to natural local weather conditions at the nursery of the Croatian Forest Research Institute (45°40′08.27″ N; 15°38′31.96″ E; 141 m a.s.l.). In spring 2015, the pots were transferred to a greenhouse and arranged following a strip-plot experimental design. The first factor (water availability) had two levels: a control plot and a drought treatment plot (DT plot). The second factor (replication) had three levels. Each provenance was represented within each combination of the factors by various numbers of randomly placed plants (3–9 per provenance due to differences in acorn germination and seedling survival).

**Table 1.** Location and climatic characterization of the provenance sites.

| Country | Provenance Abbreviation | Latitude N | Longitude E | T (°C) | PR (mm) | bFFP |
|---|---|---|---|---|---|---|
| Estonia | ES | 58.238520 | 22.442890 | 6.1 | 568.3 | April 24th |
| Lithuania | LI | 54.537975 | 23.811175 | 6.4 | 625.3 | April 20th |
| Poland | PL | 51.190300 | 16.549010 | 8.6 | 554.7 | April 11th |
| Hungary | HU | 47.025120 | 18.263499 | 10.8 | 562.0 | April 1st |
| Croatia (Repaš) | CR (RE) | 46.139763 | 17.091610 | 10.8 | 793.2 | April 4th |
| Croatia (Koška) | CR (KO) | 45.569734 | 18.235555 | 10.9 | 719.0 | April 3th |
| Croatia (Karlovac) | CR (KA) | 45.494896 | 15.708384 | 11.4 | 994.7 | March 29th |
| Croatia (Otok) | CR (OT) | 45.094245 | 18.821524 | 11.3 | 745.3 | April 3th |
| Italy | IT | 42.755380 | 11.918041 | 14.3 | 722.8 | February 18th |

Mean annual temperature (T), mean annual precipitation (PR) and average beginning of the frost-free period in spring (bFFP) were generated with ClimateEU software (version 4.63) (http://www.ualberta.ca/~{}ahamann/data/climateeu.html) for the period of 1901 to 2009. The provenance abbreviations indicate their country of origin.

The trial included a drop-in irrigation system that allowed water availability in the main plots (i.e., control and DT plots) to be controlled. Plants in the control plot were constantly maintained in soil with a high volumetric soil moisture (VSM) content (i.e., 40%–50% VSM) during the growing period.

The VSM content was monitored by soil moisture sensors, which were set up in six representative pots (one in each combination of the factors) and connected to automatic meteorological stations.

## 2.2. Treatments

In the first growing period, the trial was established in a greenhouse that was automatically ventilated, thus preventing excessive heat and air humidity during the summer. Plants in the DT plot were deprived of water on March 30th, 2015 (Figure 1). The drought treatment was stopped, and the plants were re-watered on July 21st (Figure 1), a date when approximately 20% of the plants in the plot had visible leaf injury symptoms. All plants in the trial continued to regularly receive water until the end of the growing season. During the dormant period of the year (i.e., beginning on November 1st, 2015), the pots were transferred outside the greenhouse and exposed to normal local winter conditions. Because the plants had unexpectedly shown vigorous growth in the greenhouse and threatened to reach the ceiling, the trial was performed outside for the remainder of the study. To prevent the soil from receiving natural precipitation, each pot was covered with polystyrene panels that were shaped to fit around the plant stem and additionally sealed with polyurethane foam and duct tape.

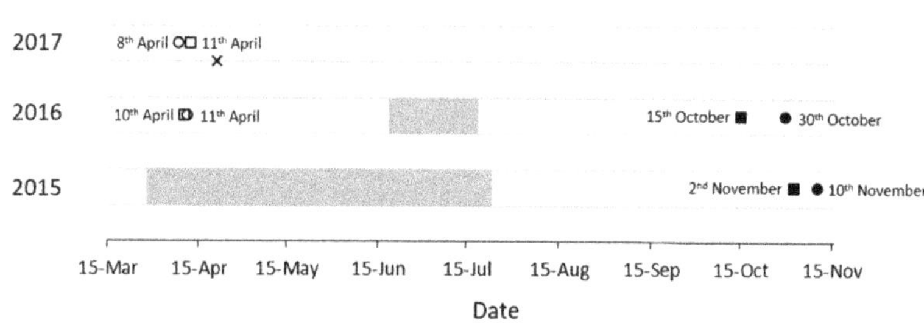


**Figure 1.** The beginning and end of growing seasons in the control and drought treatment (DT) plots in three consecutive years (2015, 2016 and 2017). Open symbols indicate the mean bud burst date in the control (square) and DT (circle) plots, whilst closed symbols indicate the mean date of autumn leaf senescence in the control (square) and drought treatment (circle). The symbol × indicates the late frost event (on 21st April 2017). Grey polygons indicate the duration of drought treatments in 2015 (from 30th March to 21st July) and 2016 (from 20th June to 16th July).

In the subsequent growing period (in 2016), the plants in the DT plot were deprived of water beginning on June 20th. The drought treatment was stopped on July 16th (Figure 1), when almost 25% of the plants showed visible leaf injury symptoms. Afterwards, the plants in the DT plot continued to receive regular watering until the end of the growing season, similar to the way plants in the control plot were treated.

## 2.3. Leaf Phenology Scoring

Leaf flushing and autumn leaf senescence were studied from 2015 to 2017 (flushing only in the last year).

Leaf flushing phenological phases were scored on all plants in the trial on a 1–8 ordinal scale:

1—bud scales completely closed; buds smaller, darker and hard;
2—buds swelling, scales slightly spaced;
3—bud burst; buds opening, scales considerably more widely spaced, green leaf tips visible;
4—folded leaf visible;
5—leaf unfolding but not yet flattened, small;
6—leaves still relatively small but with flattened blades, blade edges bent downwards, withered, lighter green or reddish;

7—leaves appear developed, larger but still more tenderly structured (thinner) than fully developed leaves and lighter green or reddish;
8—leaves fully developed, darker green, thicker.

Autumn leaf phenological phases were scored on a 0–5 ordinal scale:

0—leaves completely green with no visible yellowing;
1—up to 25% of plant leaves brown;
2—up to 50% of plant leaves brown;
3—more than 50% of plant leaves brown;
4—more than 75% of plant leaves brown;
5—leaves shed.

The day of the year when each plant reached flushing phase 3 was recorded as the bud burst date, and the day of the year when each plant reached autumn leaf phenological phase 3 was recorded as the autumn senescence date. These individual plant data were used for subsequent statistical analyses.

*2.4. Spring Frost Injury Scoring*

A late spring frost affected the trial (as well as a much wider geographic area) on April 21st, 2017. After a warm period (the mean air temperature from April 1st to April 20th was 12 °C, with a maximum temperature of 29.5 °C), the air temperature suddenly dropped from the maximum of 13.6 °C at 3 p.m. on April 20th to −0.1 °C at 8 p.m. on the same day, and the temperature continued to decrease until it reached a minimum of −5.8 °C at 5 a.m. on April 21st, after which the temperature rose to −0.3°C by 7 a.m. and continued to rise.

Several days later, there were visible leaf injuries on the plants in the trial. Leaf injuries appeared as necrotic leaf tissue visible as various shades of brown stretching from the margins of the leaf blade towards the midrib. On some leaves, these injuries were located closer to the apex, but on others, the injuries were located closer to the base. In general, there were no cases in which the whole leaf blade surface was necrotic. Additionally, in all cases, a rather low percentage of plant leaves were injured (on average, 5% of all plant leaves were partly injured). As a result, it was difficult to grade the plants based on frost injury; thus, we decided to record frost damage using a binary scale (0—plant has no visible injuries, and 1—plant has visible injuries).

*2.5. Statistical Analyses*

Individual plant data on bud burst date and autumn leaf senescence date were used for statistical analyses. All statistical analyses were performed in SAS/STAT 15.1 software, a free version of SAS University Edition by SAS Institute Inc., Cary, NC, USA. Descriptive statistics were performed using the MEANS procedure. Analyses of variance (ANOVAs) were performed using the MIXED procedure to determine the statistical significance of the factors (irrigation treatments, blocks, provenances, and the provenance by treatment interaction) according to the following linear model:

$$y_{iklj} = \mu + T_i + B_j + P_k + TP_{ik} + \varepsilon_{ijk} \text{ (ANOVA model)} \quad (1)$$

where $y_{ijkl}$—individual value of a trait; $\mu$—overall mean; $T_i$—fixed effect of treatment i, where i = 1, 2; $B_j$—random effect of block j, where j = 1, 2, 3; $P_k$—fixed effect of population k, where k = 1, 2, … , 9; $TP_{ik}$—population by treatment interaction; and $\varepsilon_{ijkl}$—random error.

Assumptions of residual normality and variance homogeneity were tested by using the Shapiro–Wilk test and Levine's test [33] with the GLM and UNIVARIATE procedures in SAS. Residuals were plotted as a function of fitted values to test for variance homogeneity, and the distribution of residuals was also tested. The significance of the irrigation treatments was tested by the Wald test and with the Satterthwaite approximation [34]. Student's *t*-test was performed to establish the significance of differences in the bud burst date between the different groups of frost-injured and uninjured plants.

## 3. Results

### 3.1. Flushing Phenology

As expected, in spring 2015 (prior to drought treatment), there were no statistically significant differences in the timing of bud burst between the control plot and the DT plot (F = 2.52; $p = 0.1131$). However, in spring 2016 (the year after the first drought treatment), the mean bud burst date of plants from almost all (except CR (OT)) provenances in the DT plot was delayed (1.5 days on average), ranging from 0.6 to 2.4 days per provenance (Figure 2). The drought treatment effect was statistically significant (F = 12.90; $p = 0.0004$). In spring 2017, the mean bud burst date of plants in the DT plot was advanced by 3 days on average (compared with the control), ranging from 0.1 to 6.3 days per provenance (Figure 2). The only exception was provenance ES, for which plants exhibited a delay in mean bud burst of 1.3 days (Figure 2). The effect of drought on the bud burst date was again highly statistically significant (F = 31.44; $p < 0.0001$).

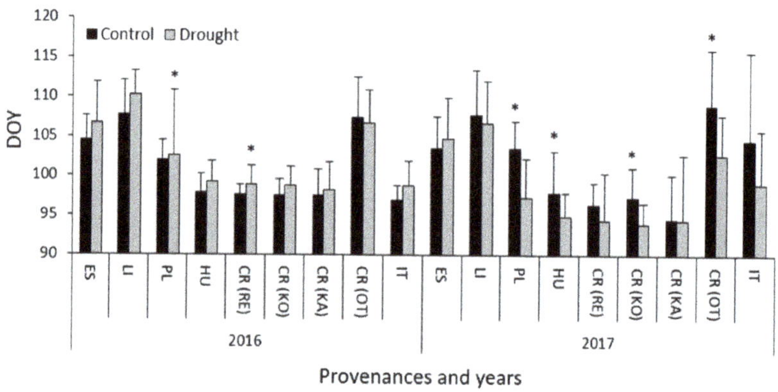


**Figure 2.** Provenance mean bud burst dates in the treatments (control and drought), expressed as day of year (DOY) in 2016 and 2017. Vertical bars indicate the standard deviation of the mean values. Asterisks above bars indicate significant differences between treatments for the same provenance. Provanance abbreviations: ES—Estonia; LI—Lithuania; PL—Poland; HU—Hungary; CR (RE)—Croatia (Repaš); CR (KO)—Croatia (Koška); CR (KA)—Croatia (Karlovac); CR (OT)—Croatia (Otok); IT—Italy.

### 3.2. Autumn Leaf Phenology

Autumn leaf phenology was scored in 2015 and 2016 after one and two years of successive drought treatments were imposed on plants in the DT plot, respectively. In 2015, the plants exhibited an average delay in senescence of 8 days (ranging from 2 (provenance ES) to 13 (provenance PL) days) (Figure 3). In 2016, leaf senescence was again delayed, in this case by an average of 15 days (ranging from 3 (LI) to 26 (CR (RE) days) (Figure 3). Based on ANOVA, the leaf senescence date of plants in the DT plot was significantly delayed compared to that in the control in both years (F = 179.00 and 318.85; $p < 0.0001$, respectively). Based on Tukey's test, the mean autumn senescence of plants from only the most northern provenances (ES and LI) did not significantly differ between the DT plot and the control plot.

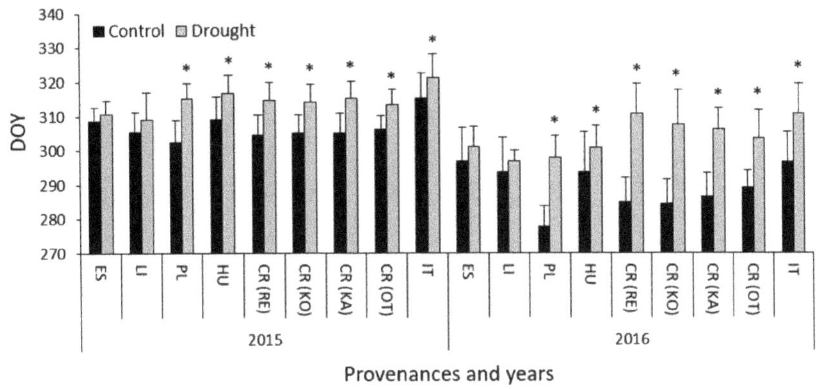


**Figure 3.** Provenance means of autumn leaf senescence in the treatments (control and drought) expressed as day of year (DOY) in 2015 and 2016. Vertical bars indicate the standard deviation of the mean values. Asterisks above bars indicate significant differences between treatments for the same provenance. Provanance abbreviations: ES—Estonia; LI—Lithuania; PL—Poland; HU—Hungary; CR (RE)—Croatia (Repaš); CR (KO)—Croatia (Koška); CR (KA)—Croatia (Karlovac); CR (OT)—Croatia (Otok); IT—Italy.

### 3.3. *Spring Frost Susceptibility*

The trial was impacted by a late spring frost occurring on April 21st, 2017. The proportion of frost-injured plants varied among provenances. More than 90% of plants from provenances CR (KO), HU, and CR (RE) were injured in the DT plot, though no plants from provenance LI were injured (Figure 4). Additionally, a larger proportion of damaged plants were observed in the DT plots for most provenances, indicating an adverse impact of drought on frost tolerance.

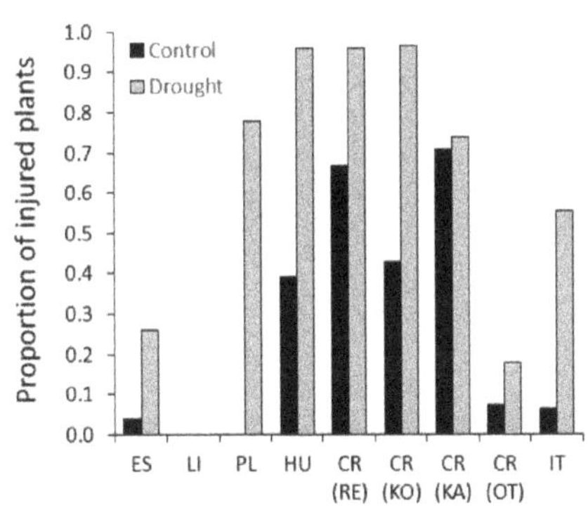


**Figure 4.** Proportions of frost-injured plants by provenance and treatment. Provanance abbreviations: ES—Estonia; LI—Lithuania; PL—Poland; HU—Hungary; CR (RE)—Croatia (Repaš); CR (KO)—Croatia (Koška); CR (KA)—Croatia (Karlovac); CR (OT)—Croatia (Otok); IT—Italy.

Based on the relationship between the state of flushing on April 21st and the incidence of frost injuries, plants that had entered flushing phases 5, 6, and 7 were injured (Figure 5). All plants that had entered flushing phase 7 were injured in both the control and DT plots (only a single plant in the DT plot had no visible symptoms of frost injury). Additionally, all plants in the DT plot that had entered flushing phase 6 were injured, whereas only 44% of plants in the same flushing phase had visible leaf injuries in the control plot. There was an obvious difference in the proportion of injured plants that had reached flushing phase 5 between the control and DT plots (Figure 5).

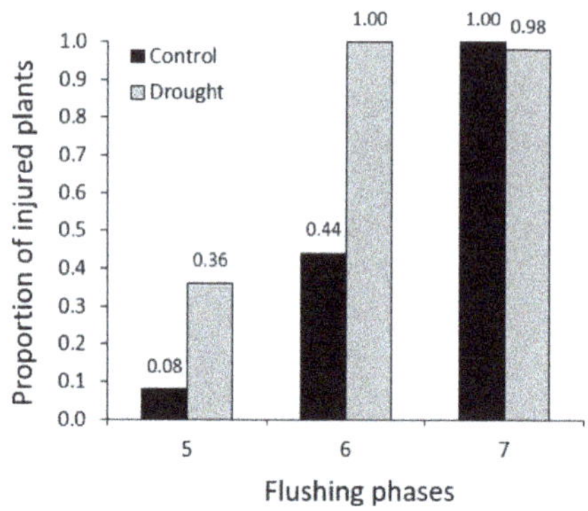


**Figure 5.** Total proportion of frost-injured plants from all provenances (in the control and drought treatments) that were in flushing phase 5 (leaf unfolding but not yet flattened, small), 6 (leaves still relatively small but with flattened blades, blade edges bent downwards, withered, lighter green or reddish), and 7 (leaves appear completely developed, larger but still more tenderly structured (thinner) than fully developed leaves and lighter green or reddish) at the time of the frost event (on 21st April 2017).

Because the frost-injured plants were those that had entered flushing phases 5, 6, and 7 on April 21st, we analysed the proportions of these plants from the studied provenances separately for the control and DT plots. These results are presented in Figure 6. The provenances with the largest proportion of frost-injured plants in the control plot (see Figure 4) were also those with the largest proportion of plants that had entered more advanced flushing phases, and vice versa (compare Figures 4 and 6).

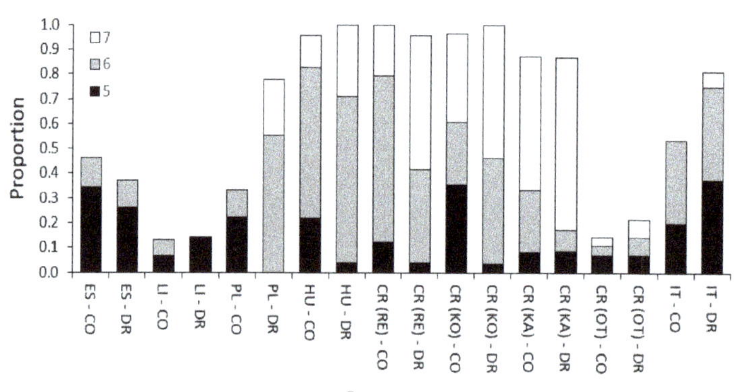


**Figure 6.** Proportions of plants in advanced flushing phases at the time of the frost event on 21st April 2017 by provenance and treatment: 5 (leaf unfolding but not yet flattened, small), 6 (leaves still relatively small but with flattened blades, blade edges bent downwards, withered, lighter green or reddish), and 7 (leaves completely developed, larger but still more tenderly structured (thinner) than fully developed leaves and lighter green or reddish). CO—control; DR—drought treatment. Provanance abbreviations: ES—Estonia; LI—Lithuania; PL—Poland; HU—Hungary; CR (RE)—Croatia (Repaš); CR (KO)—Croatia (Koška); CR (KA)—Croatia (Karlovac); CR (OT)—Croatia (Otok); IT—Italy.

Additionally, the increase in frost susceptibility in the DT plot may be explained by more advanced (i.e., earlier) flushing of plants in that plot. For example, the proportion of frost-injured plants increased by >50% in the DT plot for provenances PL (from 0% to 78%), HU (from 39% to 96%), and CR (KO) (from 43% to 96%) (Figure 4). The same provenances exhibited the greatest shifts towards more advanced flushing on April 21st (Figure 6).

These results led to the question of why all plants that had reached flushing phase 6 (FP6) suffered frost injuries in the DT plot while only 44% of the FP6 plants were damaged in the control plot. Thus, we tested for significant differences in the mean bud burst date between control plants not showing frost injury symptoms and drought-treated plants as well as between frost-injured control plants and drought-treated plants. These results are shown in Table 2.

**Table 2.** Student's *t*-test comparing the mean bud burst date between the two groups of control plants (injured or uninjured) and DT plants (all frost injured).

| Groups of Plants which were in FP 6 * on April 21st regarding the Plots and Frost Injuries | Number of Plants | Mean Day of the Bud Burst (± Standard Deviation) | Difference from the Group Number 2 *t*-Value (*p*) |
|---|---|---|---|
| 1. Uninjured control plants | 30 | 96.91 ± 1,70 | 2.898 (0.0026) |
| 2. Plants in the DT plot (they were all injured by the frost) | 55 | 95.77 ± 1.80 | |
| 3. Injured control plants | 24 | 96.26 ± 1.93 | 1.067 (0.146) |

* flushing phase 6. DT = drought treatment.

Therefore, we compared two groups of control plants with drought-treated plants. Student's *t*-test showed that uninjured control plants had started flushing significantly later, but injured control plants had not. Although all these plants were recorded as being in flushing phase 6 on April 21st, these results (Table 2) indicate that plants in the DT plot were at a more advanced flushing state similar to that of the damaged control plants.

## 4. Discussion

### 4.1. Leaf Phenology

In our study, pedunculate oak plants that were experimentally exposed to drought exhibited delayed bud burst in the subsequent spring (Figure 2). However, after the second drought treatment, the mean bud burst date of the plants was advanced (Figure 2). This discrepancy (i.e., opposite shifts in bud burst induced by water stress) was also revealed in other studies. For example, Mijnsbrugge et al. [30] reported delayed bud burst of oak exposed to drought and Kuster et al. [28] and Spieß et al. [26] reported advanced bud burst of oak exposed to drought stress in previous years. Thus, based on our results as well as those from the above-mentioned studies, drought stress has a carry-over effect on oak flushing phenology, which causes it to shift. This effect of water stress on flushing phenology was also recorded in *Fagus crenata* Blume [29]. Plant environmental responses are epigenetically regulated. Various environmental signals and stresses can induce epigenetic modifications, thereby creating a flexible "memory" system for short or prolonged periods of time [35]. Changes in epigenetic marks in trees are accompanied by morphological and physiological changes in various processes such as ageing, organ maturation, and bud set or burst [36–38]. Therefore, drought stress may have triggered an epigenetic response ("memory") resulting in the observed carry-over effect on bud burst date in the oaks from the studied provenances. Why bud burst shifts may occur in opposite directions (i.e., towards an earlier or a later date) is an open question. There may be more than one molecular mechanism behind this epigenetic "memory" phenomenon. The complexity of signalling events associated with sensing drought stress in plants is well known (e.g., see [39]). Various chemical signals (e.g., reactive oxygen species, calcium, and plant hormones) participate in the induction of stress tolerance via transduction cascades and the activation of genomic re-programming [40]. For example, abscisic acid (ABA) plays an important role in epigenetic processes such as in ABA-mediated responses to abiotic stress (see [41]). Therefore, the complexity of drought sensing and signalling pathways likely results in different epigenetic modifications and therefore phenotypic changes such as opposite shifts in bud burst date. The different time spans over which plants sense water deficits in the vegetation period presumably cause opposite carry-over effects regarding flushing phenology. Accordingly, in the first year of the experiment, we stopped watering the plants in the DT plot on March 30th, allowing them to sense the water deficit early in the vegetation period, i.e., during or shortly after flushing. In contrast, in the second year of the experiment, the plants were deprived of water in the second half of June after they had completely developed leaves. Plants from the studied pedunculate oak provenances responded with delayed bud burst following the first drought treatment (Figure 2). Combined with advanced bud burst following the second drought treatment (Figure 2), this indicated that the hypothesis in this study was supported. Of course, the molecular mechanisms underlying the epigenetic "memory" phenomenon are still poorly understood [35]; thus, it is not clear why bud burst shifts occur in opposite directions. We can only speculate that drought in different time spans induced different sensing and signalling pathways, which resulted in the observed effects on the flushing phenology of the studied oak plants. Addressing the mechanisms underlying the observed effects of drought on flushing phenology was beyond the scope of this study.

Leaf senescence was significantly delayed by drought in plants from all provenances in both the first and second year of the experiment (Figure 3). Delayed autumn leaf senescence is a known physiological response to drought and re-watering in oaks [9,30] as well as other tree species [42]. This phenomenon has been explained as the photosynthetic and growth compensation of trees during the post-drought period of recovery. Notably, leaf senescence in the most northern provenances (i.e., Estonia and Lithuania) did not significantly differ between the DT and control plots. Plants from the northern provenances may be more sensitive to environmental signalling at the end of the growing season, which may prevent them from overextending their growing season. Vitasse et al. [43] reported a similar result in a study on clinal variation in the leaf senescence of *Quercus petraea* (Matt.) Liebl. from provenances along an altitudinal gradient. Additionally, Deans and Harvey [44] reported a negative correlation between leaf yellowing and latitude in sessile oak, although this correlation was not as strong as the positive correlation between budburst date and latitude. Nevertheless, natural selection in harsher (colder) habitats appears to favour genotypes that respond with more rapid growth cessation in autumn. This relationship may be why the delayed leaf senescence of plants from the northern provenances was not as significant as that of plants from the more southern provenances.

### 4.2. Spring Frost Susceptibility

Spring frost affected the trial (as well as a much wider geographic area) on the morning of April 21st, 2017. Within a few days, symptoms of necrosis were observable on plant leaves. This incident enabled us to study variation in frost susceptibility among provenances and between the treatments in the trial. There were significant differences among the studied oak provenances in terms of spring frost susceptibility. The mean proportion of injured plants was strongly related to their state of flushing development on the particular day when frost impacted the trial. Thus, differences among provenances in the proportion of frost-injured plants were caused by differences in flushing phenology. Additionally, the difference in the proportion of frost-injured plants between the control and DT plots may also be explained by differences in flushing phenology. Plants that were in more advanced flushing phases suffered leaf injury, whereas plants that had not started flushing or were in a state of less advanced flushing did not suffer damage (Figure 5). This result confirms the report by Utkina and Rubtsov [24], who found greater resilience to spring frosts of later flushing pedunculate oaks.

The most frost-susceptible plants were clearly those in flushing phase 7 (almost fully developed leaves with completely flattened blades but still tenderly structured, light green or reddish) because all but one of these plants suffered frost injuries in both the control and the DT plots. Additionally, all plants in phenological phase 6 in the DT plot were frost injured (Figure 5). Pedunculate oak do not appear to possess effective active molecular responses to frost; instead, avoidance is their major anti-frost "strategy". This lack of active frost responses was true at least for plants from the early flushing provenances that were exposed to the frost in 2017 and suffered leaf injury. The use of escape as an adaptive response to spring frost risks has been reported in sessile oak populations along an elevation gradient [45] as well as in various pedunculate oak populations [24,46,47]. However, the presence of a few exceptions (such as a single plant in flushing phase 7 that did not show visible symptoms of leaf injury) suggests that at least some pedunculate oak genotypes possess an efficient response to freezing at the molecular level. Such a molecular response might be due to alterations in enzymatic activity [48] and the accumulation of soluble sugars, as has already been reported for various *Quercus* species [49,50]; additionally, such a response might be due to the more efficient mobilization of non-structural carbohydrates (NSCs), as reviewed by Villar-Salvador et al. [51].

There was a large increase in the proportion of frost-injured plants in the DT plot (for most of the provenances), indicating an adverse impact of drought on oak frost susceptibility (Figure 4). This result raises the question of why plants in the DT plot (drought stressed) showed increased spring frost susceptibility. Based on our results, the main reason for the increased frost susceptibility of these plants was the earlier flushing caused by the drought treatment in the previous year. Drought stress in the previous year induced a clear carry-over effect, which resulted in the earlier flushing of plants from most provenances (Figure 2; year 2017). Therefore, most of the provenances in the DT plot exhibited an increased proportion of plants at more developed flushing phases on the day when the spring frost occurred. Nonetheless, it was not as obvious why all plants at flushing phase 6 (100%) in the DT plot were harmed by the frost, while only 44% of plants in the same flushing phase in the control plot showed visible injuries. Drought stress may have impacted plants in the DT plot, resulting in an even more reduced efficiency of molecular responses to freezing temperatures. For example, the accumulation and re-mobilization of NSCs and nitrogen (N) reserves play a very important role in frost resistance (see [51] and references therein). Drought stress in the current growing season may hinder re-mobilization capacity in the next growing season by decreasing seedling N and NSC content. [51]. Oaks use NSC and N reserves from storage tissues for their early spring growth before bud burst [52], and significant reductions in these reserves in spring have been reported for oaks [48]. Therefore, the higher proportion of frost-injured saplings in the DT plot may be explained in the following way: summer drought impaired the capacity to accumulate, store, and re-mobilize NSC and N reserves (responsible for frost resistance and vessel formation in early spring), which elevated the frost susceptibility of the saplings in the DT plot.

Another possibility was that not all plants in flushing phase 6 were in the exact same state of flushing on a particular day. Specifically, injured plants may have started flushing slightly earlier (i.e., were at a slightly more advanced flushing state) than the uninjured plants, despite all of the plants being scored as in the same flushing phase on that particular day. This explanation for the difference in the proportions of damaged phase-6 plants between the DT and control plots is supported by the test results (Table 2). Uninjured phase-6 plants in the control plot started flushing significantly later (had an earlier bud burst date, on average), while injured phase-6 plants did not (Table 2). Therefore, injured phase-6 plants in the control plot and phase-6 plants in the DT plot were at a slightly more advanced state of flushing on the day of the frost, which may have rendered them more susceptible to frost injuries. Furthermore, the flushing phenology scoring method, in which a plant was graded based on the most frequent phase, may have resulted in two plants in somewhat different states of flushing being assigned the same score. For example, if most buds indicated flushing phase 6, a plant was graded as being in phase 6; however, some newly emerging leaves may have already been in phase 7. Such plants may have sustained frost injuries on those few phase-7 leaves. Although plausible, this hypothesis should be further tested.

## 5. Conclusions

The results of this study confirmed the previously reported effects of drought stress on pedunculate oak leaf phenology, i.e., delayed autumn leaf senescence and a carry-over effect on bud burst date. Although the effect of drought on autumn leaf senescence was unambiguous (i.e., delayed senescence), the mean bud burst dates were shifted in opposite directions in the two analyzed years. Our study supports the hypothesis that opposite shifts in bud burst dates occur due to the different times in the growing season at which plants sense water deficit signals. Accordingly, if oaks sense water deficits early in the growing season (i.e., during or shortly after flushing), then their bud burst date will be delayed in the subsequent spring. In contrast, if the oaks sense water deficit signals late in the growing season (i.e., summer drought), then their bud burst will shift towards an earlier date in the subsequent spring.

The results strongly indicated an adverse effect of drought stress on pedunculate oak spring frost susceptibility. The adverse effect was most likely caused by a shift in spring leaf phenology. In other words, summer drought stress induced a carry-over effect, resulting in earlier mean bud burst in the subsequent spring, which made plants from the majority of the studied oak provenances more susceptible to late-frost leaf injuries. Importantly, the impacts of drought on autumn leaf senescence, i.e., delayed leaf senescence, most likely increase the susceptibility of oaks to early autumn frosts.

However, plants from the latest-flushing provenances were the least sensitive to the above-mentioned adverse effects of drought on frost resistance. Therefore, these provenances may be considered valuable sources of forest reproductive material for the regeneration of oak forests made necessary by an increase in the frequency of drought and frost events.

**Author Contributions:** Conceptualization, K.S. and S.B.; Data curation, I.Č., A.J. and S.B.; Formal analysis, I.Č., A.J. and S.B.; Funding acquisition, S.B.; Investigation, I.Č., I.K.B. and S.B.; Methodology, K.S. and S.B.; Project administration, S.B.; Resources, I.Č. and A.J.; Supervision, S.B.; Validation, I.Č., I.K.B. and A.J.; Visualization, I.K.B. and Ž.Š.; Writing—original draft, I.Č. and S.B.; Writing—review & editing, K.S., A.J. and Ž.Š.

**Funding:** This study was fully supported by the Croatian Science Foundation under grant number IP-2014-09-4686 (Phenotypic and Epigenetic Response to Drought Stress and Adaptability of *Quercus robur* L. Populations along a Latitudinal Gradient).

**Acknowledgments:** Thank you to all those who kindly supplied acorns from their local provenances for the trial, including Dr. Moica Piazzai (Monte Rufeno Nature Reserve, Italy); Dr. Krista Takkis (Estonia); Dr. Virgilijus Baliuckas (Forest Tree Genetics and Breeding Department of the Lithuanian Research Centre for Agriculture and Forestry); Czesław Kozioł (Leśny Bank Genów Kostrzyca); Tomasz Miedzyrzecki and Bartlomiej Laski (Miękinia Forest District, Poland); Goran Perić (Forest Office Koška); staff of the Forest Office Repaš; staff of the Forest Office Karlovac and Mladen Šimunić (Forest Office Otok). The trial was maintained by employees of Nursery Production and the Division for Genetics, Forest Tree Breeding and Seed Husbandry of the Croatian Forest Research Institute. Many thanks to all of them. Thank you to the three anonymous reviewers for helpful comments and suggestions.

**Conflicts of Interest:** The authors declare no conflicts of interest.

## References

1. Parry, M.L. *Climate Change: Impacts, Adaptation and Vulnerability. Working Group II Contribution to the Fourth Assessment Report of the IPCC Intergovernmental Panel on Climate Change*; Cambridge University Press: Cambridge, UK, 2007.
2. Lindner, M.; Maroschek, M.; Netherer, S.; Kremer, A.; Barbati, A.; Garcia-Gonzalo, J.; Seidl, R.; Delzon, S.; Corona, P.; Kolström, M.; et al. Climate change impacts, adaptive capacity, and vulnerability of European forest ecosystems. *For. Ecol. Manag.* **2010**, *259*, 698–709. [CrossRef]
3. Intergovernmental Panel on Climate Change (IPCC). *Contribution of Working Groups I, II and III to the Fifth Assessment Report of the Intergovernmental Panel on Climate Change*; Climate Change 2014: Synthesis Report; IPCC: Geneva, Switzerland, 2014.
4. Bréda, N.; Huc, R.; Granier, A.; Dreyer, E. Temperate forest trees and stands under severe drought: A review of ecophysiological responses, adaptation processes and long-term consequences. *Ann. For. Sci.* **2006**, *63*, 625–644. [CrossRef]

5. Kreyling, J.; Stahlmann, R.; Beierkuhnlein, C. Spatial variation in leaf damage of forest trees after the extreme spring frost event in May 2011. *Allgemeine Forst und Jagdzeitung* **2012**, *183*, 15–22.
6. Ducousso, A.; Bordacs, S. EUFORGEN Technical Guidelines for genetic conservation and use for pedunculate and sessile oaks (*Quercus robur* and *Q petraea*). Available online: http://www.euforgen.org/publications/publication/iquercus-roburi-and-iquercus-petraeai-technical-guidelines-for-genetic-conservation-and/ (accessed on 15 November 2018).
7. Árvai, M.; Morgós, A.; Kern, Z. Growth-climate relations and the enhancement of drought signals in pedunculate oak (*Quercus robur* L.) tree-ring chronology in Eastern Hungary. *iForest* **2018**, *11*, 267–274. [CrossRef]
8. Kremer, A. Evolutionary responses of European oaks to climate change. *Irish For.* **2010**, *67*, 53–66.
9. Van der Mijnsbrugge, K.; Turcsán, A.; Maes, J.; Duchêne, N.; Meeus, S.; Van der Aa, B.; Steppe, K.; Steenackers, M. Taxon-Independent and Taxon-Dependent Responses to Drought in Seedlings from *Quercus robur* L., *Q. petraea* (Matt.) Liebl. and Their Morphological Intermediates. *Forests* **2017**, *8*, 407. [CrossRef]
10. Leuschner, C.; Backes, K.; Hertel, D.; Schipka, F.; Schmitt, U.; Terborg, O.; Runge, M. Drought responses at leaf, stem and fine root levels of competitive *Fagus sylvatica* L. and *Quercus petraea* (Matt.) Liebl. trees in dry and wet years. *For. Ecol. Manag.* **2001**, *149*, 33–46. [CrossRef]
11. Kubiske, M.E.; Abrams, M.D. Stomatal and nonstomatal limitations of photosynthesis in 19 temperate tree species on contrasting sites during wet and dry years. *Plant Cell Environ.* **1993**, *16*, 1123–1129. [CrossRef]
12. Galle, A.; Haldimann, P.; Feller, U. Photosynthetic performance and water relations in young pubescent oak (*Quercus pubescens*) trees during drought stress and recovery. *New Phytol.* **2007**, *174*, 799–810. [CrossRef]
13. Kuster, T.M.; Arend, M.; Bleuler, P.; Gunthardt-Goerg, M.S.; Schulin, R. Water regime and growth of young oak stands subjected to air-warming and drought on two different forest soils in a model ecosystem experiment. *Plant Biol.* **2013**, *15*, 138–147. [CrossRef]
14. Kuster, T.M.; Arend, M.; Gunthardt-Goerg, M.S.; Schulin, R. Root growth of different oak provenances in two soils under drought stress and air warming conditions. *Plant Soil* **2013**, *369*, 61–71. [CrossRef]
15. Turcsán, A.; Steppe, K.; Sárközi, E.; Erdélyi, É.; Missoorten, M.; Mees, G.; Mijnsbrugge, K.V. Early Summer Drought Stress during the First Growing Year Stimulates Extra Shoot Growth in Oak Seedlings (*Quercus petraea*). *Front. Plant Sci.* **2016**, *7*, 193. [CrossRef] [PubMed]
16. Bruschi, P. Geographical variation in morphology of *Quercus petraea* (Matt.) Liebl. as related to drought stress. *Plant Biosyst.* **2010**, *144*, 298–307. [CrossRef]
17. Jensen, J.; Hansen, J. Genetic variation in responses to different soil water treatments in *Quercus robur* L. *Scand. J. For. Res.* **2010**, *25*, 400–411. [CrossRef]
18. Arend, M.; Kuster, T.; Günthardt-Goerg, M.S.; Dobbertin, M. Provenance-specific growth responses to drought and air warming in three European oak species (*Quercus robur*, *Q. petraea* and *Q. pubescens*). *Tree Physiol.* **2011**, *31*, 287–297. [CrossRef] [PubMed]
19. Broadmeadow, M.S.J.; Jackson, S.B. Growth responses of *Quercus petraea*, *Fraxinus excelsior* and *Pinus sylvestris* to elevated carbon dioxide, ozone and water supply. *New Phytol.* **2000**, *146*, 437–451. [CrossRef]
20. Thomas, F.M.; Gausling, T. Morphological and physiological responses of oak seedlings (*Quercus petraea* and *Q. robur*) to moderate drought. *Ann. For. Sci.* **2000**, *57*, 325–333. [CrossRef]
21. Quero, J.L.; Villar, R.; Marañón, T.; Zamora, R. Interactions of drought and shade effects on seedlings of four *Quercus* species: Physiological and structural leaf responses. *New Phytol.* **2006**, *170*, 819–834. [CrossRef]
22. Selig, M.; Bohne, H. Late frost reactions of different populations of *Quercus robur* L. and *Tilia cordata* Mill. in Germany. *Eur. J. Hortic. Sci.* **2016**, *81*, 157–164. [CrossRef]
23. Alberto, F.; Bouffier, L.; Louvet, J.-M.; Lamy, J.-B.; Delzon, S.; Kremer, A. Adaptive responses for seed and leaf phenology in natural populations of sessile oak along an altitudinal gradient. *J. Evolut. Biol.* **2011**, *24*, 1442–1454. [CrossRef]
24. Utkina, I.A.; Rubtsov, V.V. Studies of Phenological forms of Pedunculate Oak. *Contemp. Probl. Ecol.* **2017**, *10*, 804–811. [CrossRef]
25. Bertin, R.I. Plant phenology and distribution in relation to recent climate change. *J. Torrey Bot. Soc.* **2008**, *135*, 126–146. [CrossRef]
26. Spieß, N.; Oufir, M.; Matušíková, I.; Stierschneider, M.; Kopecky, D.; Homolka, A.; Burg, K.; Fluch, S.; Hausman, J.F.; Wilhelm, E. Ecophysiological and transcriptomic responses of oak (*Quercus robur*) to long-term drought exposure and rewatering. *Environ. Exp. Bot.* **2012**, *77*, 117–126. [CrossRef]

27. Pflug, E.E.; Siegwolf, R.; Buchmann, N.; Dobbertin, M.; Kuster, T.M.; Gunthardt-Goerg, M.S.; Arend, M. Growth cessation uncouples isotopic signals in leaves and tree rings of drought-exposed oak trees. *Tree Physiol.* **2015**, *35*, 1095–1105. [CrossRef] [PubMed]
28. Kuster, T.; Dobbertin, M.; Günthardt-Goerg, M.; Schaub, M.; Arend, M. A Phenological Timetable of Oak Growth under Experimental Drought and Air Warming. *PLoS ONE* **2014**, *9*, e89724. [CrossRef]
29. Yonekura, T.; Yoshidome, M.; Watanabe, M.; Honda, Y.; Ogiwara, I.; Izuta, T. Carry-over effects of ozone and water stress on leaf phenological characteristics and bud frost hardiness of *Fagus crenata* seedlings. *Trees* **2004**, *18*, 581–588. [CrossRef]
30. Mijnsbrugge, K.; Turcsán, A.; Maes, J.; Duchêne, N.; Meeus, S.; Steppe, K.; Steenackers, M. Repeated Summer Drought and Re-watering during the First Growing Year of Oak (*Quercus petraea*) Delay Autumn Senescence and Bud Burst in the Following Spring. *Front. Plant Sci.* **2016**, *7*, 419. [CrossRef]
31. Andivia, E.; Fernández, M.; Vázquez-Piqué, J.; Alejano, R. Two provenances of *Quercus ilex* ssp. *ballota* (Desf) Samp. nursery seedlings have different response to frost tolerance and autumn fertilization. *Eur. J. For. Res.* **2012**, *131*, 1091–1101. [CrossRef]
32. Thomas, F.; Ahlers, U. Effects of excess nitrogen on frost hardiness and freezing injury of above-ground tissue in young oaks (*Quercus petraea* and *Q. robur*). *New Phytol.* **1999**, *144*, 73–83. [CrossRef]
33. Sabin, T.E.; Stafford, S.G. *Assessing the Need for Transformation of Response Variables*; Forest Research Laboratory, Oregon State University: Corvallis, OR, USA, 1990; p. 31.
34. Littell, R.C.; Milliken, G.A.; Stroup, W.W.; Wolfinger, R.D. *SAS System for Mixed Models*, 6th ed.; SAS Institute: Cary, NC, USA, 2006.
35. Bräutigam, K.; Vining, K.; Lafon-Placette, C.; Fossdal, C.; Mirouze, M.; Marcos, J.; Fluch, S.; Fraga, M.F.; Guevara, M.Á.; Abarca, D.; et al. Epigenetic regulation of adaptive responses of forest tree species to the environment. *Ecol. Evol.* **2013**, *3*, 399–415. [CrossRef]
36. Fraga, M.F.; Rodriguez, R.; Canal, M.J. Genomic DNA methylation-demethylation during aging and reinvigoration of *Pinus radiata*. *Tree Physiol.* **2002**, *22*, 813–816. [CrossRef] [PubMed]
37. Santamaria, M.E.; Hasbun, R.; Valera, M.J. Acetylated H4 histone and genomic DNA methylation patterns during bud set and bud burst in *Castanea sativa*. *J. Plant Physiol.* **2009**, *166*, 1360–1369. [CrossRef]
38. Valledor, L.; Hasbún, R.; Meijón, M. Involvement of DNA methylation in tree development and micropropagation. *Plant Cell Tissue Organ Culture* **2007**, *91*, 75–86. [CrossRef]
39. Chaves, M.; Maroco, J.; Pereira, J. Understanding plant responses to drought—From genes to the whole plant. *Funct. Plant Biol.* **2003**, *30*, 239–264. [CrossRef]
40. Farooq, M.; Wahid, A.; Kobayashi, N.; Fujita, D. Plant drought stress: Effects, mechanisms and management. *Agron. Sustain. Dev.* **2009**, *29*, 185–212. [CrossRef]
41. Chinnusamy, V.; Gong, Z.; Zhu, J. Abscisic Acid-mediated Epigenetic Processes in Plant Development and Stress Responses. *J. Integr. Plant Biol.* **2008**, *50*, 1187–1195. [CrossRef] [PubMed]
42. Arend, M.; Sever, K.; Pflug, E.; Gessler, A.; Schaub, M. Seasonal photosynthetic response of European beech to severe summer drought: Limitation, recovery and post-drought stimulation. *Agric. For. Meteorol.* **2016**, *220*, 83–89. [CrossRef]
43. Vitasse, Y.; Delzon, S.; Bresson, C.C.; Michalet, R.; Kremer, A. Altitudinal differentiation in growth and phenology among populations of temperate-zone tree species growing in a common garden. *Can. J. For. Res.* **2009**, *39*, 1259–1269. [CrossRef]
44. Deans, J.D.; Harvey, F.J. Phenologies of sixteen European provenances of sessile oak growing in Scotland. *Forestry* **1995**, *68*, 265–274. [CrossRef]
45. Dantec, C.; Ducasse, H.; Capdevielle, X.; Fabreguettes, O.; Delzon, S.; Desprez-Loustau, M. Escape of spring frost and disease through phenological variations in oak populations along elevation gradients. *J. Ecol.* **2015**, *103*, 1044–1056. [CrossRef]
46. Puchałka, R.; Koprowski, M.; Przybylak, J.; Przybylak, R.; Dąbrowski, H.P. Did the late spring frost in 2007 and 2011 affect tree-ring width and earlywood vessel size in Pedunculate oak (*Quercus robur*) in northern Poland? *Int. J. Biometeorol.* **2016**, *60*, 1143–1150. [CrossRef] [PubMed]
47. Puchałka, R.; Koprowski, M.; Gričar, J.; Przybylak, R. Does tree-ring formation follow leaf phenology in Pedunculate oak (*Quercus robur* L.)? *Eur. J. For. Res.* **2017**, *136*, 259–268. [CrossRef]
48. Ebermann, R.; Korori, S.A.A.; Lickl, E. Temperature Dependent Alterations of Peroxidase and Amylase Isoenzymes in *Quercus robur*. *Phyton* **1991**, *31*, 121–128.

49. Morin, X.; Améglio, T.; Ahas, R.; Kurz-Besson, C.; Lanta, V.; Lebourgeois, F.; Miglietta, F.; Chuine, I. Variation in cold hardiness and carbohydrate concentration from dormancy induction to bud burst among provenances of three European oak species. *Tree Physiol.* **2007**, *27*, 817–825. [CrossRef] [PubMed]
50. Holland, V.; Koller, S.; Lukas, S.; Brüggemann, W. Drought- and frost-induced accumulation of soluble carbohydrates during accelerated senescence in *Quercus pubescens*. *Trees* **2016**, *30*, 215–226. [CrossRef]
51. Villar-Salvador, P.; Uscola, M.; Jacobs, D. The role of stored carbohydrates and nitrogen in the growth and stress tolerance of planted forest trees. *New For.* **2015**, *46*, 813–839. [CrossRef]
52. Michelot, A.; Simard, S.; Rathgeber, C.; Dufrêne, E.; Damesin, C. Comparing the intra-annual wood formation of three European species (*Fagus sylvatica*, *Quercus petraea* and *Pinus sylvestris*) as related to leaf phenology and non-structural carbohydrate dynamics. *Tree Physiol.* **2012**, *32*, 1033–1045. [CrossRef] [PubMed]

*forests*

MDPI

*Article*

# Differentiation and Non-Linear Responses in Temporal Phenotypic Plasticity of Seasonal Phenophases in a Common Garden of *Crataegus monogyna* Jacq.

**Kristine Vander Mijnsbrugge [1,*] and Astrid Janssens [2]**

1 Department of Forest Ecology and Management, Research Institute for Nature and Forest, Gaverstraat 4, B-9500 Geraardsbergen, Belgium

2 Department of Bioengineering Technology TC, University of Leuven, Kleinhoefstraat 4, B-2440 Geel, Belgium; astrid-janssens@hotmail.com

* Correspondence: kristine.vandermijnsbrugge@inbo.be; Tel.: +32-478-282-137

Received: 16 February 2019; Accepted: 19 March 2019; Published: 27 March 2019

**Abstract:** Phenology in perennial plants implies the temporal occurrence of biological events throughout the year. Heritable phenotypic plasticity in the timing of the phenophases can be of importance in the adaptation of woody species to a quickly changing environment. We observed the timing of bud burst, flower opening, leaf senescence and leaf fall in two successive years in a common garden of *Crataegus monogyna* Jacq. in Belgium, consisting of six local and five non-local provenances. Data were processed with cumulative logistic mixed models. Strong auto-correlation was present among the spring phenophases as well as among the autumnal phenophases, with spring phenophases being negatively correlated with fall phenophases. The strongest between-provenance differentiation was found for the timing of bud burst in spring, followed by flower opening and finally by leaf senescence and leaf fall. Warmer spring temperatures in March 2017 advanced the timing of bud burst, and to a lesser extent of flower opening, in all provenances compared to 2016. However, the advancement was non-linear among the provenances, with the lower latitude provenances being relatively less early and the higher elevation provenances being more late than the local provenances in this year. It can be hypothesized that non-local provenances display larger temporal phenotypic plastic responses in the timing of their spring phenophases compared to local provenances when temperatures in the common garden deviate more from their home-sites.

**Keywords:** phenology; leafing out; flowering; senescence; cumulative logistic regression; hawthorn; provenance trial; non-local populations; variance analysis

---

## 1. Introduction

Woody plants are sessile and perennial organisms that are characterized by long generation times and slow migration rates [1]. Therefore, it is expected that woody species can adapt relatively quickly to changing local environmental conditions [2]. The capacity for evolutionary change depends on the standing genetic variation in tree populations [3] and climate is a major driver of evolutionary change over longer time scales [4]. Understanding the genetic basis of complex polygenic traits in woody plants that are clearly influenced by climate is therefore a current challenge in forest genetic research. Because all individuals in a common garden share the same environment, any average difference in a trait between provenances of the same species has a genetic origin. The genetic variation in fitness-related traits is typically estimated in open pollinated progeny tests in common garden experiments, including the estimation of differentiation between provenances [5,6], and the study of this variation has its place in the field of quantitative genetics [7].

The seasonal cycle in deciduous woody plant species of temperate regions is characterized by the timing of bud burst and flowering in spring and by the timing of bud formation, leaf senescence and leaf shedding in autumn. These phenophases can deviate between genetically differentiated populations and are believed to be adaptive, responding to selection induced by environmental change [5,8,9]. The timing of the phenophases marking the beginning and the end of the yearly growing season maximizes the annual growth while minimizing the risk of frost damage in spring and autumn. Frost can damage woody plants at high fitness costs [10]. Late frosts in spring can damage the soft tissues of young leaves, whereas early frosts in autumn may cause early leaf abscission, hindering the resorption of nutrients. In addition, early frosts are also known to damage the cambial zone in trees [11].

The timing of bud burst and leaf unfolding vary considerably among tree species, responding to divergent climatic conditions [3,5]. Bud burst can be the main trait that is affected by climate mediated selection [3]. Selection on timing of bud burst can be sufficiently strong to counteract the homogenizing effect of gene flow [12,13]. Rising spring temperatures advance the onset of the growing season in many woody species of the temperate zone, prolonging the growing season and thus affecting plant productivity and the global carbon balance [14–16]. On the other hand, late frosts in spring may hamper early flushing species in a global warming scenario [17]. Because of its adaptive nature and the easy assessment from an early age onward, spring flushing is frequently evaluated in common gardens [5,18]. Within the same woody species, population differentiation for bud burst, as observed in common garden experiments, typically follows clines along gradients of elevation and/or latitude of the home-sites of the populations [5,19,20] possibly due to different temperature requirements that are genetically determined [21]. The genetic variation in timing of bud burst, together with divergent selection, is believed to have allowed tree species in temperate regions to occupy large distribution ranges [22]. Within populations, individual trees display variable timing of bud burst, which is suggested to have a genetic cause [23]. Individual trees in a population can therefore be categorized as early, intermediate and late phenological forms [24]. Neutral genetic marker analysis showed that late bud burst forms of *Fagus sylvatica* L. in natural populations in Poland displayed higher within-population genetic variation in comparison to the early forms, suggesting that late spring frosts shape the neutral genetic structure of the populations [24]. Finally, no relationship was found between the timing of cambial activity and the timing of bud burst in *Quercus robur* L. [23].

Flowering and the subsequent fruit formation are part of the sexual reproductive cycle in plants. Reproductive phenology is sensitive to environmental cues such as temperature, moisture and herbivory [25]. Divergent timing of flowering can stimulate assortative mating in populations of woody plants, reducing gene flow and promoting population differentiation [25]. The timing of flowering varies strongly among woody angiosperm species and can occur before, during or after bud burst. For instance, pollen emission concurs with leaf unfolding in oaks. Assortative mating through long-distance pollen flow is therefore suggested to interact with local adaptation of bud burst [3,26]. In general, timing of bud burst and flowering are most likely auto-correlated in temperate tree species [26,27].

The emergence and growth of new spring foliage in temperate deciduous trees relies strongly on the nutrients that were resorbed during the preceding leaf senescence [28]. The timing of leaf senescence is affected by both photoperiod and temperature [29–32] while the timing of bud burst is primarily influenced by temperature [33,34]. It shows less year-to-year variability in comparison with timing of bud burst and is concomitant with less favorable conditions for photosynthesis [29]. In 59 tree species, the timing of autumnal leaf senescence displayed a pattern according to the climatic clines of the home-sites of the studied populations, which was clearer than the pattern observed in bud burst timing [5]. But, in a common garden experiment of *Quercus petraea* (Matt.) Liebl. composed of populations derived from the same geographic region but from deviating elevations, the population differentiation for autumnal leaf senescence did not correlate with the elevation of the origin [3]. There is no consensus concerning the factors controlling the leaf senescence process [29,35]. Leaves that emerged after a late spring frost in beech and oak displayed higher photosynthesis rates and a delayed leaf senescence in autumn, compensating for spring frost damage and demonstrating that long-lived trees can adapt their autumnal phenology depending on preceding productivity [36]. The spring phenophases of bud burst and flowering, and the timing of leaf senescence and leaf abscission are most probably strongly auto-correlated. Therefore, it is suggested that leaf fall can be used as a proxy for leaf senescence, assuming a certain time delay [30].

The range of phenotypes that a plant can express as a function of the environment is called phenotypic plasticity. Genetically controlled, heritable phenotypic plasticity has the potential to influence plant evolution [37]. Species that are able to adjust their phenological responses to warming spring temperatures by earlier bud burst or earlier flowering show better performance when compared with less responsive species [38]. Because of their longevity, phenotypic plasticity may play an important role in the adaptation of woody perennials to the predicted climate change [39]. Studying responses of provenances of woody species in a common garden not only allows assessment of the differentiation among the provenances that is shaped by divergent selection, e.g. [40]. Also, repeated observations in successive years allow the estimation of phenotypic plastic reactions to variable meteorological conditions in the garden over time. We planted a common garden in Belgium consisting of *Crataegus monogyna* Jacq., a common shrub species in western Europe, including local and non-local provenances, and studied the seasonal phenophases that mark the growth cycle of woody plants. Using neutral (non-adaptive) molecular markers, high levels of genetic diversity within populations, but low levels of population differentiation, were found in this species [41]. We hypothesized that: (i) the timing of bud burst, flower opening, leaf senescence and leaf fall differ between local and non-local provenances; (ii) the timing of the four phenophases display varying degrees of within-provenance and between-provenance variation; (iii) spring phenophases are strongly auto-correlated, as well as autumnal phenophases; and (iv) the non-local provenances respond in a non-linear way, in comparison with the local provenances, to the variable meteorological conditions in the common garden environment in two successive years (variable temporal phenotypic plasticity among the provenances) which can be related to the home-site conditions of the non-local provenances.

## 2. Materials and Methods

### 2.1. Common Garden

A description of the provenances, the seed collection, the growth of the plants and the planting of the common garden have already been reported in [42]. In short, the common garden consisted of six local Flemish provenances (northern part of Belgium), two Walloon provenances (southern part of Belgium), one provenance from the UK, one from Italy and one from Hungary (Table 1). The Belgian (Flemish and Walloon) provenances were collected by the authors and grown in the nursery of the research institute in Geraardsbergen, Belgium, whereas the UK, Italian and Hungarian provenances were grown in adjacent nursery beds in a Flemish forest nursery, under a sales contract. For the collection of the seeds in the Belgian populations, care was taken to collect from *C. monogyna*, excluding *Crataegus laevigata* (Poir.) DC. individuals and individuals with a putative hybridogenic background, by visually assessing the morphology of the leaves and the berries. Seeds were collected from at least 30 individuals from each population. Information on the seed collection in the commercial provenances was not available as it concerned commercial plant material without a certificate of provenance. Certificates of provenance are not compulsory for *C. monogyna* according to the Council Directive 1999/105/EC [43]. Therefore, the exact location of these commercial seed sources was also not available. In the beginning of 2008, planting stock of the 11 provenances were planted in a common garden in Londerzeel, Flanders, Belgium. The provenances were randomly mixed in a single tree plot design and planted with a spacing of 1.5 × 1.5 m. Mean monthly temperatures for 2016 and 2017 were acquired from the weather station in Groenendaal which is located at a distance of 30 km from the common garden site (Figure 1).

**Table 1.** Descriptive data of the provenances of *C. monogyna* in the common garden. No precise data on seed-stock populations were available for the commercial provenances. -: no data available.

| Region | Village | Provenance abb. | Latitude | Longitude | Altitude (m) | *n* |
|---|---|---|---|---|---|---|
| Flanders | Landegem | FL1 | 51.0546 | 3.5888 | 10 | 53 |
| Flanders | Riemst | FL2 | 50.8116 | 5.5978 | 112 | 52 |
| Flanders | Hansbeke | FL3 | 51.0747 | 3.5345 | 10 | 31 |
| Flanders | Zomergem | FL4 | 51.1195 | 3.5642 | 7.5 | 46 |
| Flanders | Melle | FL5 | 51.0030 | 3.7988 | 10 | 55 |
| Flanders | Muizen | FL6 | 51.0109 | 4.5143 | 12.5 | 52 |
| Wallonia | Smuid | WA1 | 50.0188 | 5.2662 | 370 | 53 |
| Wallonia | Viroin | WA2 | 50.0727 | 4.6072 | 305 | 49 |
| Hungary | - | HU | - | - | - | 46 |
| Italy | - | IT | - | - | - | 52 |
| United Kingdom | - | UK | - | - | - | 48 |

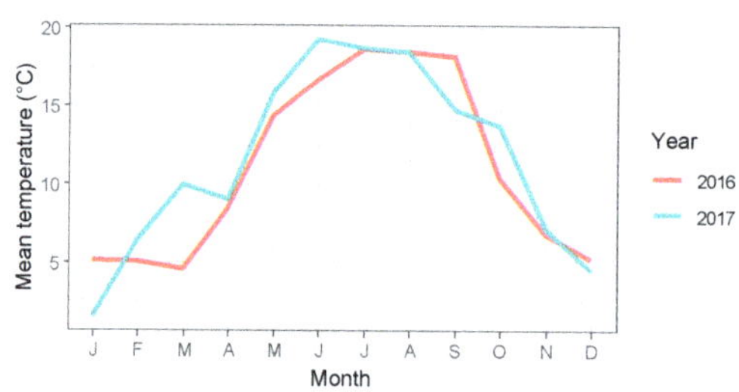


**Figure 1.** Mean monthly temperatures in 2016 and 2017 in Groenendaal, Belgium.

### 2.2. Scoring of the Four Phenophases

Observations of the phenophases were performed on the shrubs in the common garden in 2016 and 2017. Bud burst, flower opening, leaf senescence and leaf fall were scored following two 6-level and two 5-level protocols, respectively (Table 2). For all phenophases, the whole shrub was evaluated visually and a mean score level was given. Bud burst was scored in 2016 on 8, 23, 31 March and 16 April, and in 2017 on 12, 21, 26 March and 2 April. Flower opening was scored in 2016 on 28 April, 4 and 11 May, and in 2017 on 26 April, 1 and 11 May. Leaf senescence and leaf fall were scored in 2016 on 21 October and 4 November, and in 2017 on 8 and 29 October.

**Table 2.** Description of the different score levels of the four observed phenophases.

| Phenophase | Score Level | Description |
|---|---|---|
| bud burst | 1 | winter buds |
| | 2 | buds swelling |
| | 3 | buds opening and first green tips of the leaves visible |
| | 4 | leaves emerging from the bud but not yet unfolding |
| | 5 | leaves emerged and unfolding |
| | 6 | leaves fully unfolded and expanding |
| flower opening | 1 | flower buds closed and green |
| | 2 | flower buds closed and white |
| | 3 | less than half of the flowers in an inflorescence opened |
| | 4 | half to more than half of the flowers in an inflorescence opened |
| | 5 | most but not all flowers in an inflorescence opened |
| | 6 | all flowers opened |

**Table 2.** *Cont.*

| Phenophase | Score Level | Description |
|---|---|---|
| leaf senescence | 1 | (mainly) green leaves |
| | 2 | mainly light green to yellow leaves |
| | 3 | mainly yellow leaves |
| | 4 | yellow and brown leaves |
| | 5 | (mainly) brown leaves |
| leaf fall | 1 | (nearly) no leaves fallen |
| | 2 | About 1/4of the leaves are fallen |
| | 3 | About 1/2 of the leaves are fallen |
| | 4 | About 3/4 of the leaves are fallen |
| | 5 | (nearly) all leaves fallen |

### 2.3. Statistical Analysis of Phenological Data

All statistical analyses were performed in the open source software R 3.5.1. [44]. Models were fitted to examine the timing of the four phenophases in the two successive years. Each phenological trait ($T_{ph}$) was a response variable and was modeled using cumulative logistic regression in the "ordinal" package [45], as the recorded observations were in an ordinal scale. The command "clmm" in the "ordinal" package models the probability (p) of having reached maximally a given level of the ordinal response variable. The score levels for leaf senescence and fall were defined in increasing order, and the probability was modeled for having reached maximally a given score level; e.g., having reached maximally a leaf senescence score of 3 was to be interpreted as the probability of having reached scores of 1, 2 or 3. The score levels of bud burst and flower opening were defined in decreasing order, so that the probability of having reached maximally a bud burst score of 3, for example, included the probability of having reached scores of 6, 5, 4or 3. In this way, this could be interpreted as having reached a score of at least 3. Mixed models were fitted as the phenological data concerned repeated observations on the same plants. The year of observation (*Y*, categorical variable) and the provenance (*P*, categorical variable) were present in the fixed part of each model, including an interaction term between these two variables. In the four phenological models, the local Flemish provenance FL1 (Table 1) was taken as the standard provenance to which the timing of the other provenances in the common garden was compared. Day (*D*, numerical variable) was added in the fixed part to account for the different observation days. The random part (random intercept) consisted of a unique shrub identity code (*ID*). The latter accounted for the repeated observations on the same plants.

$$\log (p_{Tph}/(1 - p_{Tph})) = \alpha_i - \beta_Y.Y \text{ (fixed)} - \beta_P.P \text{ (fixed)} - \beta_{YP}.Y.P \text{ (fixed)} - \beta_D.D \text{ (fixed)} - r_{ID} \text{ (random)}$$

where $\alpha_i$ was an intercept value indicating the passing from one level of the ordinal phenological response variable to the next. $\beta_Y$, $\beta_P$ and $\beta_D$ were the estimated coefficients for the fixed covariates *Y*, *P* and *D*, and $r_{ID}$ was the random effect coefficient for all levels of the variable *ID*.

To be able to compare the timing of the four phenophases for all provenances, the days were calculated for which the different provenances in each phenophase had reached the same stage of phenological development. The DOY (day of the year) was calculated for which the probability for having reached a bud burst or flower opening score of at least 4 and having reached a maximal leaf senescence and leaf fall score of 3, attained 50% ($D_{50\%PY}$) in the observation years 2016 and 2017 and for every provenance. A $D_{50\%PY}$ for a given provenance and a given year therefore indicated the day that half of the plants of this provenance had reached minimally (bud burst and flower opening) or maximally (leaf senescence and fall) a given score level of the respective phenophase in the respective year. This calculation was based on $\log(p_{Tph}/(1 - p_{Tph}))$ being 0 for p = 50%. With 2016 as the standard level for the variable *Y*, to which 2017 is compared, the following formulas were used:

$$D_{50\%P2016} = (\alpha_i - \beta_P)/(\beta_D)$$

$$D_{50\%P2017} = (\alpha_i - \beta_{2017} - \beta_P - \beta_{2017P})/(\beta_D)$$

When the timing of a given phenophase for a given provenance differed significantly from the standard provenance FL1, the time lag was calculated between this provenance and the standard provenance. For 2016, this was inferred from the model with 2016 as the standard level of the categorical variable $Y$, whereas for 2017 this was inferred from the model with 2017 as the standard level for the variable $Y$. The time span between a provenance and the standard provenance from which it differed significantly, was calculated by subtracting the calculated $D_{50\%FL1Y}$ from $D_{50\%PY}$. Pearson correlation coefficients were calculated between the timing of the phenophases in the two observation years using the $D_{50\%P2016}$ and $D_{50\%P2017}$ values.

A significant interaction term between provenance $P$ and year $Y$ in the model statistics indicated that the time span between the timing of a phenophase for a given provenance and the standard provenance in 2016 differed significantly from the respective time span between these two provenances in 2017. The significant interaction terms were visualized in reaction norm figures, in which the timing of a phenophase for a provenance was compared between 2016 and 2017. The slope of the line connecting the timing of a phenophase between 2016 and 2017 for a provenance with a significant interaction term in the model statistics, differed significantly from the slope connecting the timings for the standard provenance FL1.

To examine the relative variance in the timing of bud burst, flower opening, leaf senescence and leaf fall among the different provenances in the common garden, in comparison with the relative variance among the shrubs within a provenance, the four phenological models were adapted by moving the provenance variable ($P$) from the fixed to the random part.

$$\log(p_{Tph}/(1 - p_{Tph})) = \alpha_i - \beta_Y.Y - \beta_D.D \text{ (fixed)} - r_P \text{ (random)} - r_{ID} \text{ (random)}$$

The relative variance between the provenances ($\sigma^2{}_P$) and the relative variance between the shrubs within a provenance ($\sigma^2{}_{ID}$) were obtained from these models.

## 3. Results

### 3.1. Timing of Bud Burst

All provenances in the common garden burst buds earlier in 2017 compared with 2016 (the covariate year was significant in the model statistics, Table 3, Figure S1). Significant differences in the timing of bud burst were observed between several provenances and the standard Flemish provenance FL1 in 2016 (Table 3) and in 2017 (Table S1). The time spans between the timing of these provenances and FL1 were calculated for both observation years (Table 4). The Italian and Hungarian provenances flushed earlier in comparison to the Flemish provenances, whereas the Walloon provenances with a higher elevation in the home-sites, flushed later (Figure 2a, Table 3).

**Table 3.** Model statistics for the timing of bud burst and flower opening. The provenance FL1 and the year 2016 are the standard levels for the categorical variables provenance and year, to which the other year, 2017, and the other provenances are compared. DOY: day of the year. Provenance abbreviations are given in Table 1.

| Variable | Bud Burst | | | | Flower Opening | | | |
|---|---|---|---|---|---|---|---|---|
| | Estimate | Std. Error | z-Value | *p*-Value | Estimate | Std. Error | z-Value | *p*-Value |
| DOY | −0.64 | 0.01 | −44.79 | <0.001 *** | −0.46 | 0.01 | −38.90 | <0.001 *** |
| 2017 | −8.29 | 0.30 | −27.26 | <0.001 *** | −2.84 | 0.24 | −12.07 | <0.001 *** |
| FL2 | −0.28 | 0.48 | −0.59 | 0.558 | −0.28 | 0.31 | −0.90 | 0.367 |
| FL3 | −0.08 | 0.55 | −0.15 | 0.878 | −0.47 | 0.35 | −1.32 | 0.188 |
| FL4 | 1.30 | 0.49 | 2.63 | 0.009 ** | 0.42 | 0.32 | 1.32 | 0.188 |
| FL5 | 0.31 | 0.47 | 0.66 | 0.513 | −0.04 | 0.30 | −0.12 | 0.905 |

**Table 3.** *Cont.*

| Variable | Bud Burst | | | | Flower Opening | | | |
|---|---|---|---|---|---|---|---|---|
| | Estimate | Std. Error | z-Value | *p*-Value | Estimate | Std. Error | z-Value | *p*-Value |
| FL6 | 0.50 | 0.48 | 1.06 | 0.290 | −0.10 | 0.31 | −0.31 | 0.753 |
| HO | −6.19 | 0.51 | −12.06 | <0.001 *** | −3.02 | 0.35 | −8.65 | <0.001 *** |
| IT | −7.14 | 0.51 | −14.14 | <0.001 *** | −3.19 | 0.32 | −10.04 | <0.001 *** |
| WA1 | 3.01 | 0.48 | 6.26 | <0.001 *** | 1.46 | 0.32 | 4.55 | <0.001 *** |
| WA2 | 3.62 | 0.49 | 7.33 | <0.001 *** | 0.69 | 0.34 | 2.04 | 0.041* |
| UK | 0.43 | 0.49 | 0.89 | 0.375 | −0.65 | 0.36 | −1.82 | 0.069 |
| 2017:FL2 | 0.56 | 0.34 | 1.64 | 0.101 | 0.39 | 0.33 | 1.19 | 0.233 |
| 2017:FL3 | 1.14 | 0.39 | 2.94 | 0.003 ** | 0.77 | 0.39 | 1.97 | 0.049 * |
| 2017:FL4 | −0.07 | 0.35 | −0.21 | 0.837 | 0.44 | 0.34 | 1.32 | 0.188 |
| 2017:FL5 | 0.10 | 0.34 | 0.31 | 0.760 | 0.50 | 0.32 | 1.56 | 0.119 |
| 2017:FL6 | 0.51 | 0.34 | 1.51 | 0.130 | 0.40 | 0.33 | 1.19 | 0.236 |
| 2017:HO | 2.44 | 0.37 | 6.54 | <0.001 *** | 1.91 | 0.36 | 5.30 | <0.001 *** |
| 2017:IT | 3.15 | 0.36 | 8.73 | <0.001 *** | 0.99 | 0.33 | 3.05 | 0.002 ** |
| 2017:WA1 | 1.01 | 0.35 | 2.91 | 0.004 ** | 0.90 | 0.38 | 2.40 | 0.016 * |
| 2017:WA2 | 0.62 | 0.36 | 1.75 | 0.080 | 1.59 | 0.39 | 4.04 | <0.001 *** |
| 2017:UK | 0.94 | 0.35 | 2.73 | 0.006 ** | 1.21 | 0.39 | 3.11 | 0.002 * |

Significant results: *** $p < 0.001$; ** $p < 0.01$; * $p < 0.05$.

**Table 4.** Differences in timing between the standard local provenance FL1 and the other provenances in the common garden in 2016 and 2017 for the four phenophases. Time spans are shown only for the provenances that differed significantly from the standard provenance FL1 in the models (Table 3, Table S1). Negative values indicate earlier timing of the phenophase, positive values indicate later timing. Provenance abbreviations are given in Table 1.

| Phenophase | Provenance | Difference in Timing with FL1 (days) | |
|---|---|---|---|
| | | 2016 | 2017 |
| bud burst | FL3 | | 1.7 |
| | FL4 | 2 | 1.9 |
| | FL6 | | 1.6 |
| | HO | −9.7 | −5.9 |
| | IT | −11.2 | −6.3 |
| | WA1 | 4.7 | 6.3 |
| | WA2 | 5.7 | 6.7 |
| | UK | | 2.2 |
| flower opening | FL4 | | 1.9 |
| | HO | −6.6 | −2.4 |
| | IT | −7 | −4.8 |
| | WA1 | 3.2 | 5.2 |
| | WA2 | 1.5 | 5 |
| | UK | | 1.2 |
| leaf senescence | HO | 15.7 | 11.9 |
| | IT | 9 | 12.3 |
| | UK | | −7.5 |
| leaf fall | FL5 | 4.4 | |
| | HO | 6.8 | 8 |
| | IT | 8.7 | 6.3 |

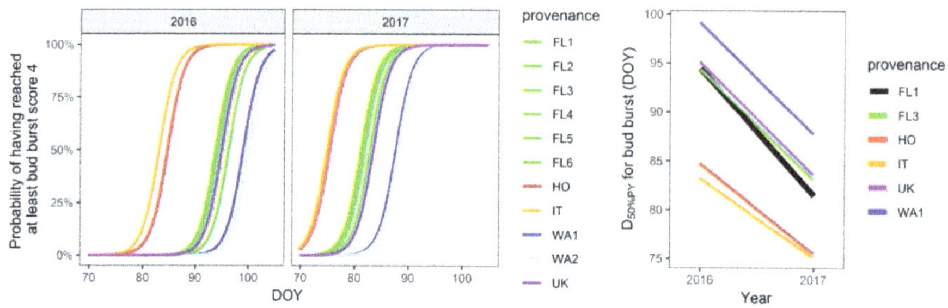


(a) (b)

**Figure 2.** Modeled timing of bud burst for the different provenances in 2016 and 2017: (**a**) Modeled probability of having reached at least bud burst score 4; (**b**) $D_{50\%PY}$ values for the provenances (*P*) with a significant interaction term in the model statistics, for the years (*Y*) 2016 and 2017. $D_{50\%PY}$ values indicate the modeled day of the year (DOY) when half of the plants of a provenance attain a bud burst score of at least four.

A significant interaction term between provenance and year in the model indicated a significant relative change in time span (between the timing of bud burst for the respective provenance and the timing for the standard provenance FL1) between the two observation years (Table 3). A significant interaction term for a provenance was visualized as the slope of a line connecting the timings (expressed as $D_{50\%PY}$ values) for this provenance in the two observation years differing from the slope of the standard provenance FL1 (Figure 2b). In 2016, the southern European provenances burst their buds about 10 days earlier than the Flemish provenance FL1, whereas this time lag between the southern European provenances and FL1 was reduced to about 6 days difference in 2017, resulting in less steep slopes for the provenances HO and IT compared to FL1 in Figure 2b (provenance abbreviations in Table 1). Whereas in 2017 the difference between the timing of bud burst in the southern European provenances and FL1 was reduced in comparison with 2016, the time lag between the Walloon provenance WA1 and FL1 increased in 2017 in comparison with 2016, resulting in a less steep slope for WA1 compared with FL1 in Figure 2b (provenance abbreviations in Table 1). The UK provenance and the Flemish provenance FL3 also displayed significant interaction terms (Tables 1 and 3, Figure 2b).

### 3.2. Timing of Flower Opening

In general, all provenances in the common garden opened their flowers earlier in 2017 compared with 2016 (the covariate year was significant in the model statistics, Table 3, Figure S2). The southern European provenances and the Walloon provenances differed significantly in the timing of this phenophase when compared with the standard local provenance FL1 in 2016 and in 2017 (Table 3, Table S1 and Figure 3a). In addition, the Flemish FL4 and the UK provenance differed significantly from FL1 in 2017 (provenance abbreviations in Table 1). The time spans between these provenances and FL1 were calculated for both observation years (Table 4). Similar to bud burst, the Italian and Hungarian provenances opened their flowers earlier in comparison with the Flemish provenances, whereas the Walloon provenances tended to flower later (Figure 3a, Table 3). All time spans for flower opening between FL1 and the provenances that differed significantly from FL1, were smaller than for bud burst in both observation years.

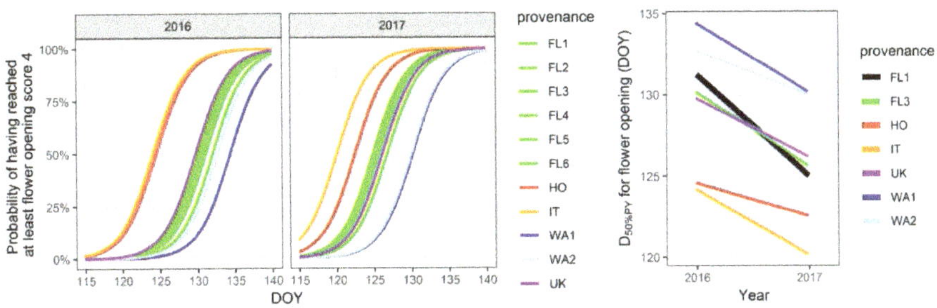


(**a**) (**b**)

**Figure 3.** Modeled timing of flower opening for the different provenances in 2016 and 2017: (**a**) Modeled probability of having reached at least a flower opening score of four; (**b**) $D_{50\%PY}$ values for the provenances (*P*) with a significant interaction term in the model statistics, for the years (*Y*) 2016 and 2017. $D_{50\%PY}$ values indicate the modeled DOY when half of the plants of a provenance attain a flower opening score of at least four.

Significant interaction terms between provenance and year in the model (Table 3) are visualized in Figure 3b. The pattern was comparable to bud burst. In 2016 the southern European provenances opened their flowers about 6 to 7 days earlier than the Flemish provenance FL1. In 2017, this time lag between the southern European provenances and FL1 was reduced to about 2 to 5 days, resulting in less steep slopes than FL1 in Figure 3b. Similar to bud burst, the time lag between the Walloon provenances and FL1 increased in 2017 compared with 2016 (from 2 to 3 days in 2016 to about 5 days in 2017), also resulting in less steep slopes than FL1 in Figure 3b. One extra local provenance, FL4, and the UK provenance differed significantly in timing of flower opening from FL1 in 2017 (Table S1). The UK provenance displayed a significant interaction term (Table 3, Figure 3b), whereas among the Flemish provenances only FL3 had a significant interaction term, visualized as a less steep slope than FL1 in Figure 3b.

### 3.3. Timing of Leaf Senescence and Leaf Fall

For the timing of leaf senescence and leaf fall, fewer provenances differed significantly from the standard provenance FL1 in comparison with the timing of bud burst and flower opening (Table 5, Table S2, Figure 4, Figure 5a, Figure S3 and Figure S4). For the southern European provenances, leaf senescence in 2016 occurred 9 to 16 days later than the local provenance FL1, and around 12 days later in 2017 (Table 4). Timing of leaf fall was modeled for these provenances 6 to 9 days later than the local provenance FL1 in both years (Table 4). In 2017, the timing of leaf senescence in the UK provenance was 7.5 days earlier than FL1, whereas in 2016, the timing of leaf fall in the Flemish provenance FL5 was 4 days later (Table 4).

**Table 5.** Model statistics for the timing of leaf senescence and leaf fall. The provenance FL1 and the year 2016 are the standard levels for the categorical variables provenance and year, to which the other year, 2017, and the other provenances are compared. Provenance abbreviations are given in Table 1.

| Variable | Leaf Senescence | | | | Leaf Fall | | | |
|---|---|---|---|---|---|---|---|---|
| | Estimate | Std. Error | z-Value | *p*-Value | Estimate | Std. Error | z-Value | *p*-Value |
| DOY | 0.10 | 0.01 | 14.97 | <0.001 *** | 0.27 | 0.01 | 31.11 | <0.001 *** |
| 2017 | −1.32 | 0.33 | −4.06 | <0.001 *** | 1.89 | 0.29 | 6.45 | <0.001 *** |
| FL2 | −0.07 | 0.35 | −0.19 | 0.851 | −0.13 | 0.46 | −0.28 | 0.782 |
| FL3 | 0.13 | 0.41 | 0.32 | 0.747 | 0.49 | 0.54 | 0.91 | 0.366 |
| FL4 | −0.10 | 0.35 | −0.29 | 0.771 | −0.81 | 0.48 | −1.70 | 0.090 |

**Table 5.** *Cont.*

| Variable | Leaf Senescence | | | | Leaf Fall | | | |
|---|---|---|---|---|---|---|---|---|
| | Estimate | Std. Error | z-Value | *p*-Value | Estimate | Std. Error | z-Value | *p*-Value |
| FL5 | −0.16 | 0.34 | −0.48 | 0.635 | −1.17 | 0.46 | −2.56 | 0.010* |
| FL6 | −0.06 | 0.35 | −0.16 | 0.874 | 0.17 | 0.47 | 0.36 | 0.721 |
| HO | −1.60 | 0.35 | −4.61 | <0.001 *** | −1.82 | 0.47 | −3.83 | <0.001 *** |
| IT | −0.92 | 0.33 | −2.76 | 0.006 ** | −2.34 | 0.47 | −5.03 | <0.001 *** |
| WA1 | 0.00 | 0.34 | −0.01 | 0.996 | −0.84 | 0.46 | −1.83 | 0.068 |
| WA2 | 0.10 | 0.35 | 0.29 | 0.768 | −0.23 | 0.47 | −0.49 | 0.624 |
| UK | −0.05 | 0.35 | −0.15 | 0.884 | −0.47 | 0.47 | −1.00 | 0.317 |
| 2017:FL2 | 0.73 | 0.45 | 1.62 | 0.106 | 0.79 | 0.41 | 1.94 | 0.052 |
| 2017:FL3 | 0.16 | 0.53 | 0.31 | 0.757 | 0.01 | 0.47 | 0.02 | 0.984 |
| 2017:FL4 | 0.20 | 0.45 | 0.44 | 0.658 | 0.43 | 0.41 | 1.04 | 0.299 |
| 2017:FL5 | 0.09 | 0.44 | 0.21 | 0.834 | 0.40 | 0.39 | 1.02 | 0.308 |
| 2017:FL6 | −0.06 | 0.46 | −0.13 | 0.898 | 0.22 | 0.41 | 0.53 | 0.598 |
| 2017:HO | 0.39 | 0.46 | 0.86 | 0.391 | −0.34 | 0.41 | −0.84 | 0.402 |
| 2017:IT | −0.33 | 0.44 | −0.76 | 0.450 | 0.66 | 0.40 | 1.65 | 0.100 |
| 2017:WA1 | 0.47 | 0.45 | 1.05 | 0.294 | 0.72 | 0.40 | 1.80 | 0.072 |
| 2017:WA2 | 0.19 | 0.46 | 0.41 | 0.683 | 0.99 | 0.41 | 2.42 | 0.016 * |
| 2017:UK | 0.82 | 0.45 | 1.81 | 0.070 | 1.13 | 0.41 | 2.76 | 0.006 ** |

Significant results: *** $p < 0.001$; ** $p < 0.01$; * $p < 0.05$.

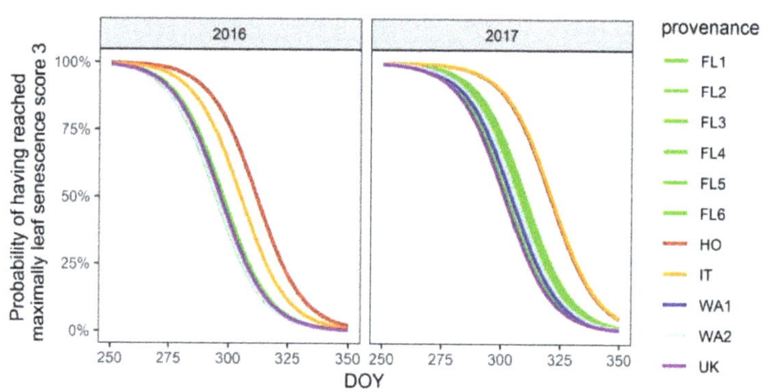


**Figure 4.** Modeled timing of leaf senescence for the different provenances in 2016 and 2017. Modeled probability of having reached a maximal leaf senescence score of three.

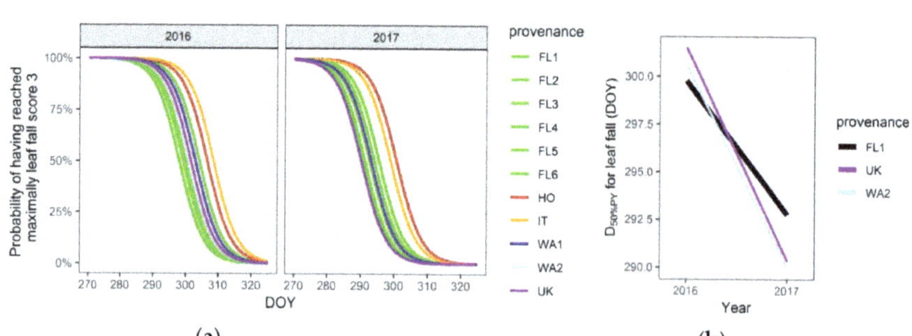


**Figure 5.** Modeled timing of leaf fall for the different provenances in 2016 and 2017: (**a**) Modeled probability of having reached a maximal leaf fall score of three; (**b**) $D_{50\%PY}$ values for the provenances (*P*) with a significant interaction term in de model statistics, for the years (*Y*) 2016 and 2017. $D_{50\%PY}$ values indicate the modelled DOY when half of the plants of a provenance attain a maximal leaf fall score of three.

### 3.4. Correlation and Variance Analysis

Pearson correlation coefficients between the timing of the different phenophases in the two observation years revealed, in general, high and significant correlations. The spring phenophases bud burst and flower opening displayed the highest correlation coefficients (0.98 *** and 0.97 *** in 2016 and 2017, respectively, Table 6). In comparison, the autumnal phenophases, leaf senescence and leaf fall, displayed lower correlation coefficients (0.79 ** in 2016 and 0.9 *** in 2017, Table 6). Both spring phenophases were correlated with leaf senescence (−0.87 *** and −0.86 *** for bud burst and flower opening in 2016, and −0.82 ** and −0.75 ** in 2017, respectively, Table 6), with later spring flushing and flowering occurring with earlier leaf senescence. In comparison, the spring phenophases were less correlated with leaf fall (−0.7 * and −0.65 * for bud burst and flower opening in 2016, and −0.78 ** and −0.65 * in 2017, respectively, Table 6). Intra-phenophase correlation coefficients, between 2016 and 2017, were obviously high, with bud burst displaying the highest correlation coefficient (0.97 ***), followed by flower opening (0.92 ***), leaf fall (0.89 ***) and finally leaf senescence (0.86 ***).

**Table 6.** Correlations between the timing of the four phenophases bud burst (Bb), flower opening (Fo), leaf senescence (Se) and leaf fall (Fa) in 2016 and 2017. Pearson correlation coefficients and corresponding *p*-values are indicated above and below the diagonal respectively. A correlation coefficient with a corresponding p-value below 0.001 is indicated in bold and is underlined, between 0.001 and 0.01 is in bold and between 0.01 and 0.05 is underlined.

| | Bb2016 | Fo2016 | Se2016 | Fa216 | Bb2017 | Fo2017 | Se2017 | Fa2017 |
|---|---|---|---|---|---|---|---|---|
| Bb2016 | | **0.98** | **−0.87** | −0.7 | **0.97** | **0.95** | **−0.85** | **−0.8** |
| Fo2016 | < 0.001 | | **−0.86** | −0.65 | **0.94** | **0.92** | **−0.81** | −0.73 |
| Se2016 | < 0.001 | < 0.001 | | **0.79** | **−0.83** | −0.7 | **0.86** | **0.9** |
| Fa2016 | 0.017 | 0.030 | 0.003 | | −0.64 | −0.53 | **0.75** | **0.89** |
| Bb2017 | < 0.001 | < 0.001 | 0.002 | 0.036 | | **0.97** | **−0.82** | **−0.78** |
| Fo2017 | < 0.001 | < 0.001 | 0.019 | 0.097 | < 0.001 | | **−0.75** | −0.65 |
| Se2017 | < 0.001 | 0.002 | < 0.001 | 0.008 | 0.002 | 0.008 | | **0.9** |
| Fa2017 | 0.003 | 0.010 | < 0.001 | < 0.001 | 0.004 | 0.030 | < 0.001 | |

The relative variability in timing of the phenophases between the provenances was compared with the relative variability between the different shrubs within a provenance (Figure 6). In 2016, the between-provenance variance was relatively highest in the phenophase bud burst, and lower in decreasing order in flower opening, leaf senescence and leaf fall. The relative within-provenance variance increased accordingly. In 2017, the between-provenance variance for bud burst was lower compared with 2016 but was still higher than the relative within-provenance variance in this year. For flower opening, the between-provenance variance in 2017 was lower compared with 2016 and was as high as the relative within-provenance variance in this year. For leaf senescence and leaf fall in 2017, the relative variances attributable to the differentiation between the provenances were lower than in 2016, and thus displayed correspondingly higher relative within-provenance variances. In general, the spring phenophases displayed the largest relative between-provenance variances, in comparison with the autumnal phenophases.

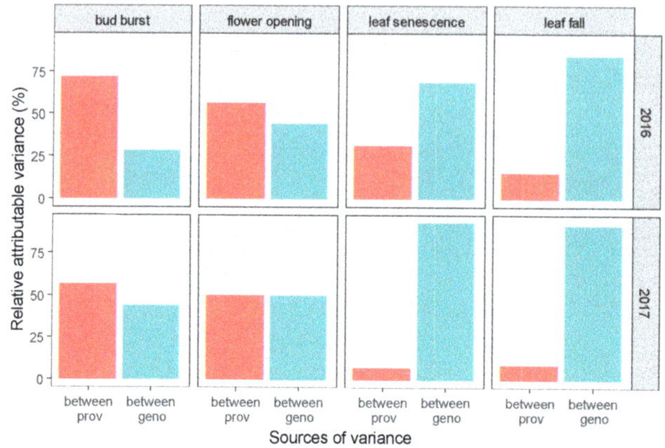


**Figure 6.** Relative between-provenance (between prov) and within-provenance (between geno) variance for the timing of bud burst, flower opening, leaf senescence and leaf fall in 2016 and 2017.

## 4. Discussion

### *4.1. Timing of the Phenophases*

Our results showed that differentiation between local and non-local provenances of *C. monogyna* is present in the phenological traits marking the seasonality in woody plants, as observed in a common garden. Differentiation in phenological traits in common gardens has been found in many tree species [46,47]. As shown before [42], the southern European provenances burst their buds earlier (lower latitude), the Walloon provenances are later (higher elevation), and the UK provenance is similar to the local Flemish provenances (lower longitude). We also detected, although in a lower order of magnitude, differentiation for bud burst between certain local provenances in both observation years. This finding may be related to results from Danish common gardens consisting of local populations of insect-pollinated shrub species (*Cornus sanguinea* L. *Malus sylvestris* (L.) Mill. and *Rosa dumalis* Bechst.), where differentiation in bud burst on a local scale, with very little spring temperature deviation between the home-sites of the populations, was suggested to be driven not only by natural selection but also by neutral processes [48]. Despite the fact that timing of bud burst and flower opening are strongly correlated, the time spans between the timing of bud burst from, on the one hand the standard local Flemish provenance and on the other hand the non-local provenances from the Walloon region (higher elevation), southern Europe (lower latitude) and the UK (lower longitude), were larger in comparison to the respective time spans between the timing of flower opening. This phenomenon has already been detected in the studied common garden [42] and proved to be consistent for two additional observation years. As hypothesized before [42], the timing of flower opening may be less sensitive to natural selection and local adaptation due to two reasons. Shrubs start flowering only after several crucial years of establishment and seedling development, and a year of reduced reproduction due to an improper timing of flowering may be less detrimental to a woody plant than reduced growth (accompanied by unfavorable competition with neighboring plants) because of an improper timing of bud burst. Therefore, the timing of flowering may be more responsive to the local micro-climate.

Leaf senescence and leaf fall in the southern European provenances were delayed compared with the local provenances. Together with an earlier bud burst, this implies a longer growing season for these provenances (spring phenophases were negatively correlated with autumnal phenophases). It is questionable whether the longer growing season is advantageous or disadvantageous. In a reciprocal common garden experiment of *Populus fremontii* S. Watson, southern populations planted in colder climates set buds relatively later in comparison with the same genotypes planted in a common garden at their home-site [39]. This later bud set is described as an inability to avoid early autumnal frosts and is interpreted as non-adaptive phenotypic plasticity, possibly caused by a lesser sensitivity to photoperiod as a cue to initiate bud set in more southern populations [39]. Bud flush in the southern population was found to be later in the colder common garden compared to the home-site common garden (but still earlier than the local populations from the cold environment) and was therefore interpreted as adaptive plasticity [39]. In our experiment, the higher elevation provenances of the Walloon region displayed a later bud burst in comparison to the local provenances, but no earlier nor later leaf senescence and leaf fall. This finding is likely in line with [34] who found no correlation between temperatures of the source sites along an altitudinal gradient and timing of leaf senescence in a common garden for *Acer pseudoplatanus* L. and *Fraxinus excelsior* L.. This may be due to a higher sensitivity of leaf senescence to photoperiod, compared to that of bud burst [29,30]. Higher elevations in the Walloon part of Belgium imply a generally colder climate compared to the local climate of the common garden site, but a negligible difference in photoperiod. When compared with the timing of bud burst and flower opening, a relatively smaller contribution of the between-provenance variation in the variance analysis indicates a weaker population differentiation for the autumnal phenophases. These results are in line with findings of [39] and [40] who both found a stronger influence of genotypic effects for bud set in poplar compared with bud burst, whereas bud burst showed stronger population-level effects relative to bud set. Although occurring in the same vegetative organ of the plant, the correlation between the timing of leaf senescence and leaf fall was smaller compared with the correlation between the timing of bud burst and flower opening, the latter implying a correlation between a vegetative (leaf bud) and a generative (flower) organ, which may suggest less tight genetic control for the autumnal phenophases.

### 4.2. Non-Linear Temporal Responses in Timing of the Phenophases

Inter-annual variation over long time periods in the timing of bud burst in temperate tree species has been widely modeled and discussed, e.g., [49]. Still, it remains difficult to accurately predict bud burst on smaller time scales. We studied the temporal phenotypic plasticity in the phenological responses on a small time scale, i.e., the responses to local meteorological conditions in the common garden, by observing the variability of the phenological responses on the same shrubs in two successive years. The non-linear inter-annual response of timing of bud burst was expressed in five significant interaction terms between the variables provenance and year in the modeling analysis, including four non-local provenances. For flower opening we found a comparable number of significant interaction terms (one extra Walloon provenance) with the same provenances involved as for bud burst, most probably giving expression to the auto-correlation of both phenophases. For leaf fall, only two provenances displayed significant interaction terms in the models, both being non-local, and for leaf senescence there were none. The lesser sensitivity of the autumnal phenophases to plasticity compared with the spring phenophases can be due to a higher sensitivity to photoperiod as a stable cue to initiate these processes [40]. The relatively high presence of non-local provenances among the significant interaction terms in the phenological models may be indicative of their non-local origin. The non-linear temporal responses in non-local provenances can be interpreted as a reaction to the prevailing growth conditions that deviate from the home-site conditions that they are adapted to. Considering the timing of bud burst, the time span between the southern European provenances and the local standard provenance in 2016 was larger than in 2017. For the Walloon provenances we observed an opposite response. As temperature is a well-known determinant for the timing of bud burst, the warmer temperatures in March 2017 have likely advanced the timing of bud burst in all provenances compared to 2016. It is by now well-known that woody plants shift to earlier bud burst dates upon increased warming [50]. However, the advancement that we observed in the non-local provenances was not linear in comparison with the local Flemish provenances, with the southern European provenances being relatively less early and the Walloon provenances being relatively later than the local provenances in 2017. In transplant experiments of *Populus fremontii*, the magnitude of phenotypic plasticity in bud flush and bud set was found to be correlated with the home-site climates [39]. Planting the southern European provenances of *C. monogyna* in the common garden in Belgium implied a transfer to a cooler environment, with the cooler spring of 2016 inducing a longer time span in the timing of bud burst with the local provenances than the warmer spring of 2017. The warmer spring in 2017 drove the higher elevation provenances to a longer time span in the timing of bud burst with the local provenances, compared with the cooler spring in 2016. Together, these results suggest a larger plastic reaction in the timing of bud burst in the common garden environment may imply that prevailing meteorological conditions deviate more from the home-site conditions.

## 5. Conclusions

In Europe, populations of woody species at the southern limits of their natural range have maintained relatively large levels of genetic diversity and are therefore considered as appropriate gene pools for assisted migration towards the north as a climate adaptation strategy [51]. On the other hand, results from Danish (high latitude in Europe) common garden trials with local populations for several shrub species indicate substantial genetic variation and evolutionary potential, questioning the need for assisted migration specifically for widespread and generally occurring woody species [48]. In addition, *C. monogyna*, as a key component of old hedgerows, displayed large levels of genetic diversity in nuclear and chloroplast markers in the UK (medium latitude in Europe) [41]. Although limited in time, our two-year study showed that all provenances of the common shrub species *C. monogyna* adjust their phenological responses to the prevailing temperatures and that non-local provenances tend to react non-linearly relative to the local provenances, with larger temporal spring plasticity coinciding with a larger difference in climatic conditions between home-sites and the common garden environment. Our results can be interpreted as an extra argument in the debate over assisted migration, suggesting that for widespread species planted stock of non-local origin may tend to "over-react" to variable environmental conditions at the site to which they are transplanted. This over-reaction in plastic response may be caused by the environmental conditions that deviate more from the optima these provenances are adapted to at their home sites. Our results therefore stress the importance of carefully reflecting on assisted migration projects, and considering at least the addition of local planting stock when deciding to transport populations of trees and shrubs over longer distances in anticipation of the predicted climate change.

**Supplementary Materials:** The following are available online at http://www.mdpi.com/1999-4907/10/4/293/s1, Figure S1: Scores of bud burst in the common garden on four observation days in the years 2016 (a) and 2017 (b). Provenance abbreviations are in Table 1. Descriptions of score levels are in Table 2; Figure S2: Scores of flower opening in the common garden on three observation days in the years 2016 (a) and 2017 (b). Provenance abbreviations are in Table 1. Descriptions of score levels are in Table 2; Figure S3: Scores of leaf senescence in the common garden on two observation days in the years 2016 (a) and 2017 (b). Provenance abbreviations are in Table 1. Descriptions of score levels are in Table 2; Figure S4: Scores of leaf fall in the common garden on two observation days in the years 2016 (a) and 2017 (b). Provenance abbreviations are in Table 1. Descriptions of score levels are in Table 2; Table S1: Model statistics for bud burst and flower opening. The provenance FL1 and the year 2017 are the standard levels for the categorical variables provenance and year, to which the other year, 2016, and the other provenances are compared. Provenance abbreviations are in Table 1; Table S2: Model statistics for leaf senescence and leaf fall. The provenance FL1 and the year 2017 are the standard levels for the categorical variables provenance and year, to which the other year, 2016, and the other provenances are compared. Provenance abbreviations are in Table 1.

**Author Contributions:** K.V.M. conceptualized the study; K.V.M. and A.J. performed the field work and the statistical analysis and wrote the manuscript.

**Funding:** This research received no external funding.

**Acknowledgments:** We specifically express our gratitude to Stefaan Moreels for general technical assistance throughout the whole project, Wim Stevens for maintenance of the common garden and Joris Vernaillen for help with the scoring.

**Conflicts of Interest:** The authors declare no conflict of interest.

## References


1. Aitken, S.N.; Yeaman, S.; Holliday, J.A.; Wang, T.L.; Curtis-McLane, S. Adaptation, migration or extirpation: Climate change outcomes for tree populations. *Evol. Appl.* **2008**, *1*, 95–111. [CrossRef]
2. Kremer, A.; Potts, B.M.; Delzon, S. Genetic divergence in forest trees: Understanding the consequences of climate change. *Funct. Ecol.* **2014**, *28*, 22–36. [CrossRef]
3. Firmat, C.; Delzon, S.; Louvet, J.M.; Parmentier, J.; Kremer, A. Evolutionary dynamics of the leaf phenological cycle in an oak metapopulation along an elevation gradient. *J. Evol. Biol.* **2017**, *30*, 2116–2131. [CrossRef]
4. Erwin, D.H. Climate as a driver of evolutionary change. *Curr. Biol.* **2009**, *19*, R575–R583. [CrossRef]
5. Alberto, F.J.; Aitken, S.N.; Alia, R.; Gonzalez-Martinez, S.C.; Hanninen, H.; Kremer, A.; Lefevre, F.; Lenormand, T.; Yeaman, S.; Whetten, R.; et al. Potential for evolutionary responses to climate change evidence from tree populations. *Glob. Chang. Biol.* **2013**, *19*, 1645–1661. [CrossRef]
6. Klisz, M.; Ukalski, K.; Ukalska, J.; Jastrzebowski, S.; Puchalka, R.; Przybylski, P.; Mionskowski, M.; Matras, J. What can we learn from an early test on the adaptation of silver fir populations to marginal environments? *Forests* **2018**, *9*, 441. [CrossRef]
7. De Villemereuil, P.; Gaggiotti, O.E.; Mouterde, M.; Till-Bottraud, I. Common garden experiments in the genomic era: New perspectives and opportunities. *Heredity* **2016**, *116*, 249–254. [CrossRef]
8. Gienapp, P.; Reed, T.E.; Visser, M.E. Why climate change will invariably alter selection pressures on phenology. *Proc. R. Soc. B-Biol. Sci.* **2014**, *281*, 20141611. [CrossRef]
9. De Kort, H.; Vandepitte, K.; Honnay, O. A meta-analysis of the effects of plant traits and geographical scale on the magnitude of adaptive differentiation as measured by the difference between Qst and Fst. *Evol. Ecol.* **2013**, *27*, 1081–1097. [CrossRef]
10. Vitasse, Y.; Lenz, A.; Korner, C. The interaction between freezing tolerance and phenology in temperate deciduous trees. *Front. Plant Sci.* **2014**, *5*, 541. [CrossRef]
11. Begum, S.; Nakaba, S.; Yamagishi, Y.; Oribe, Y.; Funada, R. Regulation of cambial activity in relation to environmental conditions: Understanding the role of temperature in wood formation of trees. *Physiol. Plant.* **2013**, *147*, 46–54. [CrossRef]
12. Savolainen, O.; Pyhäjärvi, T.; Knürr, T. Gene flow and local adaptation in trees. *Annu. Rev. Ecol. Evol. Syst.* **2007**, *38*, 595–619. [CrossRef]
13. Whittet, R.; Cavers, S.; Cottrell, J.; Rosique-Esplugas, C.; Ennos, R. Substantial variation in the timing of pollen production reduces reproductive synchrony between distant populations of *Pinus sylvestris* L. In scotland. *Ecol. Evol.* **2017**, *7*, 5754–5765. [CrossRef]
14. Dragoni, D.; Schmid, H.P.; Wayson, C.A.; Potter, H.; Grimmond, C.S.B.; Randolph, J.C. Evidence of increased net ecosystem productivity associated with a longer vegetated season in a deciduous forest in south-central Indiana, USA. *Glob. Chang. Biol.* **2011**, *17*, 886–897. [CrossRef]
15. Menzel, A.; Sparks, T.H.; Estrella, N.; Koch, E.; Aasa, A.; Ahas, R.; Alm-Kubler, K.; Bissolli, P.; Braslavska, O.; Briede, A.; et al. European phenological response to climate change matches the warming pattern. *Glob. Chang. Biol.* **2006**, *12*, 1969–1976. [CrossRef]
16. Zohner, C.M.; Renner, S.S. Common garden comparison of the leaf-out phenology of woody species from different native climates, combined with herbarium records, forecasts long-term change. *Ecol. Lett.* **2014**, *17*, 1016–1025. [CrossRef] [PubMed]
17. Duputie, A.; Rutschmann, A.; Ronce, O.; Chuine, I. Phenological plasticity will not help all species adapt to climate change. *Glob. Chang. Biol.* **2015**, *21*, 3062–3073. [CrossRef] [PubMed]
18. Aitken, S.N.; Bemmels, J.B. Time to get moving: Assisted gene flow of forest trees. *Evol. Appl.* **2016**, *9*, 271–290. [CrossRef]
19. Le Corre, V.; Kremer, A. The genetic differentiation at quantitative trait loci under local adaptation. *Mol. Ecol.* **2012**, *21*, 1548–1566. [CrossRef] [PubMed]
20. Gomory, D.; Ditmarova, L.; Hrivnak, M.; Jamnicka, G.; Kmet', J.; Krajmerova, D.; Kurjak, D. Differentiation in phenological and physiological traits in European beech (*Fagus sylvatica* L.). *Eur. J. For. Res.* **2015**, *134*, 1075–1085. [CrossRef]
21. Vitasse, Y.; Bresson, C.C.; Kremer, A.; Michalet, R.; Delzon, S. Quantifying phenological plasticity to temperature in two temperate tree species. *Funct. Ecol.* **2010**, *24*, 1211–1218. [CrossRef]

22. Pluess, A.R.; Frank, A.; Heiri, C.; Lalaguee, H.; Vendramin, G.G.; Oddou-Muratorio, S. Genome-environment association study suggests local adaptation to climate at the regional scale in *Fagus sylvatica*. *New Phytol.* **2016**, *210*, 589–601. [CrossRef]
23. Puchalka, R.; Koprowski, M.; Gricar, J.; Przybylak, R. Does tree-ring formation follow leaf phenology in pedunculate oak (*Quercus robur* L.)? *Eur. J. For. Res.* **2017**, *136*, 259–268. [CrossRef]
24. Kraj, W.; Sztorc, A. Genetic structure and variability of phenological forms in the european beech (*Fagus sylvatica* L.). *Ann. For. Sci.* **2009**, *66*, 1–7. [CrossRef]
25. Jordan, C.Y.; Ally, D.; Hodgins, K.A. When can stress facilitate divergence by altering time to flowering? *Ecol. Evol.* **2015**, *5*, S962–S973. [CrossRef]
26. Soularue, J.P.; Kremer, A. Assortative mating and gene flow generate clinal phenological variation in trees. *BMC Evol. Biol.* **2012**, *12*, 79. [CrossRef] [PubMed]
27. Soularue, J.P.; Kremer, A. Evolutionary responses of tree phenology to the combined effects of assortative mating, gene flow and divergent selection. *Heredity* **2014**, *113*, 485–494. [CrossRef] [PubMed]
28. Fu, Y.S.; Campioli, M.; Vitasse, Y.; De Boeck, H.J.; Van den Berge, J.; AbdElgawad, H.; Asard, H.; Piao, S.; Deckmyn, G.; Janssens, I.A. Variation in leaf flushing date influences autumnal senescence and next year's flushing date in two temperate tree species. *Proc. Natl. Acad. Sci. USA* **2014**, *111*, 7355–7360. [CrossRef] [PubMed]
29. Delpierre, N.; Dufrene, E.; Soudani, K.; Ulrich, E.; Cecchini, S.; Boe, J.; Francois, C. Modelling interannual and spatial variability of leaf senescence for three deciduous tree species in France. *Agric. For. Meteorol.* **2009**, *149*, 938–948. [CrossRef]
30. Estiarte, M.; Penuelas, J. Alteration of the phenology of leaf senescence and fall in winter deciduous species by climate change: Effects on nutrient proficiency. *Glob. Chang. Biol.* **2015**, *21*, 1005–1017. [CrossRef]
31. Rohde, A.; Bastien, C.; Boerjan, W. Temperature signals contribute to the timing of photoperiodic growth cessation and bud set in poplar. *Tree Physiol.* **2011**, *31*, 472–482. [CrossRef]
32. Rohde, A.; Storme, V.; Jorge, V.; Gaudet, M.; Vitacolonna, N.; Fabbrini, F.; Ruttink, T.; Zaina, G.; Marron, N.; Dillen, S.; et al. Bud set in poplar—Genetic dissection of a complex trait in natural and hybrid populations. *New Phytol.* **2011**, *189*, 106–121. [CrossRef] [PubMed]
33. Polgar, C.A.; Primack, R.B. Leaf-out phenology of temperate woody plants: From trees to ecosystems. *New Phytol.* **2011**, *191*, 926–941. [CrossRef] [PubMed]
34. Vitasse, Y.; Delzon, S.; Bresson, C.C.; Michalet, R.; Kremer, A. Altitudinal differentiation in growth and phenology among populations of temperate-zone tree species growing in a common garden. *Can. J. For. Res.* **2009**, *39*, 1259–1269. [CrossRef]
35. Panchen, Z.A.; Primack, R.B.; Gallinat, A.S.; Nordt, B.; Stevens, A.D.; Du, Y.J.; Fahey, R. Substantial variation in leaf senescence times among 1360 temperate woody plant species: Implications for phenology and ecosystem processes. *Ann. Bot.-Lond.* **2015**, *116*, 865–873. [CrossRef] [PubMed]
36. Zohner, C.M.; Rockinger, A.; Renner, S.S. Increased autumn productivity permits temperate trees to compensate for spring frost damage. *New Phytol.* **2019**, *221*, 789–795. [CrossRef]
37. Nicotra, a.B.; Atkin, O.K.; Bonser, S.P.; Davidson, A.M.; Finnegan, E.J.; Mathesius, U.; Poot, P.; Purugganan, M.D.; Richards, C.L.; Valladares, F.; et al. Plant phenotypic plasticity in a changing climate. *Trends Plant Sci.* **2010**, *15*, 684–692. [CrossRef] [PubMed]
38. Cleland, E.E.; Allen, J.M.; Crimmins, T.M.; Dunne, J.A.; Pau, S.; Travers, S.E.; Zavaleta, E.S.; Wolkovich, E.M. Phenological tracking enables positive species responses to climate change. *Ecology* **2012**, *93*, 1765–1771. [CrossRef] [PubMed]
39. Cooper, H.F.; Grady, K.C.; Cowan, J.A.; Best, R.J.; Allan, G.J.; Whitham, T.G. Genotypic variation in phenological plasticity: Reciprocal common gardens reveal adaptive responses to warmer springs but not to fall frost. *Glob. Chang. Biol.* **2019**, *25*, 187–200. [CrossRef] [PubMed]
40. Evans, L.M.; Kaluthota, S.; Pearce, D.W.; Allan, G.J.; Floate, K.; Rood, S.B.; Whitham, T.G. Bud phenology and growth are subject to divergent selection across a latitudinal gradient in *Populus angustifolia* and impact adaptation across the distributional range and associated arthropods. *Ecol. Evol.* **2016**, *6*, 4565–4581. [CrossRef] [PubMed]
41. Brown, J.A.; Beatty, G.E.; Finlay, C.M.V.; Montgomery, W.I.; Tosh, D.G.; Provan, J. Genetic analyses reveal high levels of seed and pollen flow in hawthorn (*Crataegus monogyna* Jacq.), a key component of hedgerows. *Tree Genet. Genomes* **2016**, *12*, 58. [CrossRef]

42. Vander Mijnsbrugge, K.; Onkelinx, T.; De Cuyper, B. Variation in bud burst and flower opening responses of local versus non-local provenances of hawthorn (*Crataegus monogyna* Jacq.) in Belgium. *Plant Syst. Evol.* **2015**, *301*, 1171–1179. [CrossRef]
43. Council of the European Union. Council directive 1999/105/ec of 22 december 1999 on the marketing of forest reproductive material. *Off. J. Eur. Communion* **2000**, *L11*, 17–40.
44. R Core Team. *R: A Language and Environment for Statistical Computing*; R Foundation for Statistical Computing: Vienna, Austria, 2018; Available online: http://www.R-project.org/ (accessed on 5 January 2018).
45. Christensen, R.H.B. Ordinal: Regression Models for Ordinal Data. R Package Version 2015.6-28. 2015. Available online: http://www.Cran.R-project.Org/package=ordinal/ (accessed on 5 January 2015).
46. Alberto, F.J.; Derory, J.; Boury, C.; Frigerio, J.M.; Zimmermann, N.E.; Kremer, A. Imprints of natural selection along environmental gradients in phenology-related genes of *Quercus petraea*. *Genetics* **2013**, *195*, 495–512. [CrossRef] [PubMed]
47. Salmela, M.J.; Cavers, S.; Cottrell, J.E.; Iason, G.R.; Ennos, R.A. Spring phenology shows genetic variation among and within populations in seedlings of scots pine (*Pinus sylvestris* L.) in the Scottish highlands. *Plant Ecol. Divers* **2013**, *6*, 523–536. [CrossRef]
48. Lobo, A.; Hansen, J.K.; Hansen, L.N.; Kjaer, E.D. Differences among six woody perennials native to northern Europe in their level of genetic differentiation and adaptive potential at fine local scale. *Ecol. Evol.* **2018**, *8*, 2231–2239. [CrossRef]
49. Chmielewski, F.M.; Rotzer, T. Response of tree phenology to climate change across Europe. *Agric. For. Meteorol.* **2001**, *108*, 101–112. [CrossRef]
50. Franks, S.J.; Weber, J.J.; Aitken, S.N. Evolutionary and plastic responses to climate change in terrestrial plant populations. *Evol. Appl.* **2014**, *7*, 123–139. [CrossRef] [PubMed]
51. Hampe, A.; Petit, R.J. Conserving biodiversity under climate change: The rear edge matters. *Ecol. Lett.* **2005**, *8*, 461–467. [CrossRef] [PubMed]

*forests*

Article

# Effects of Lime Application and Understory Removal on Soil Microbial Communities in Subtropical *Eucalyptus* L'Hér. Plantations

**Songze Wan [1], Zhanfeng Liu [2], Yuanqi Chen [3], Jie Zhao [4], Qin Ying [1] and Juan Liu [1,*]**

1 Collaborative Innovation Center of Jiangxi Typical Trees Cultivation and Utilization, College of forestry, Jiangxi Agriculture University, Nanchang 330045, China; swan0722@jxau.edu.cn (S.W.); yingqinww@163.com (Q.Y.)
2 Key Laboratory of vegetation Restoration and Management of Degradation Ecosystems, South China Botanical Garden, Chinese Academy of Sciences, Guangzhou 510650, China; liuzf@scbg.ac.cn
3 Hunan Province Key Laboratory of Coal Resources Clean-utilization and Mine Environment Protection, Hunan University of Science and Technology, Xiangtan 411201, China; chenyq2016@163.com
4 Key Laboratory of Agro-ecological Processes in Subtropical Region, Institute of Subtropical Agriculture, Chinese Academy of Sciences, Changsha 410125, China; jzhao@isa.ac.cn
* Correspondence: liujuan@jxau.edu.cn

Received: 7 March 2019; Accepted: 13 April 2019; Published: 16 April 2019

**Abstract:** Soil microorganisms play key roles in ecosystems and respond quickly to environmental changes. Liming and/or understory removal are important forest management practices and have been widely applied to planted forests in humid subtropical and tropical regions of the world. However, few studies have explored the impacts of lime application, understory removal, and their interactive effects on soil microbial communities. We conducted a lime application experiment combined with understory removal in a subtropical *Eucalyptus* L'Hér. plantation. Responses of soil microbial communities (indicated by phospholipid fatty acids, PLFAs), soil physico-chemical properties, and litter decomposition rate to lime and/or understory removal were measured. Lime application significantly decreased both fungal and bacterial PLFAs, causing declines in total PLFAs. Understory removal reduced the fungal PLFAs but had no effect on the bacterial PLFAs, leading to decreases in the total PLFAs and in the ratio of fungal to bacterial PLFAs. No interaction between lime application and understory removal on soil microbial community compositions was observed. Changes in soil microbial communities caused by lime application were mainly attributed to increases in soil pH and $NO_3^-$-N contents, while changes caused by understory removal were mainly due to the indirect effects on soil microclimate and the decreased soil dissolved carbon contents. Furthermore, both lime application and understory removal significantly reduced the litter decomposition rates, which indicates the lime application and understory removal may impact the microbe-mediated soil ecological process. Our results suggest that lime application may not be suitable for the management of subtropical *Eucalyptus* plantations. Likewise, understory vegetation helps to maintain soil microbial communities and litter decomposition rate; it should not be removed from *Eucalyptus* plantations.

**Keywords:** lime application; understory removal; microbial community; forest management; *Eucalyptus*

## 1. Introduction

Soil microorganisms play essential roles in regulating ecosystem processes and maintaining ecosystem functions and services, such as litter decomposition [1], nutrient cycling [2], primary production, and climate regulation [3]. Both biotic and environmental factors drive the activity, structure, and diversity of soil microbial communities, which are controlled by many factors including edaphic conditions [4]. Soil microbes generally respond quickly to environmental changes. Although numerous previous studies have explored the soil microbial responses to environmental changes in various forest types, more quantitative research estimating the effects of forest management practices on soil microbial community compositions is needed [5,6]. Information from such studies is valuable in both theoretical and practical aspects, which will improve our understanding of the mechanisms underlying soil carbon and nutrient cycling in forest ecosystems and may support sustainable forest management.

Acrisol is a common type of soil, distributed mainly in humid tropical/subtropical and even warm temperate regions, such as Southern Asia, South China, and Southeast US [7]. Lime application is one of the most widely applied management practices for amending Acrisol [8]. In addition to lowering soil acidity, lime application could improve soil structure, and thus typically promote agriculture and forest productivities [4,9]. Lime application also affects soil biological and biochemical properties, including microbial activities and carbon and nitrogen mineralization [7]. Previous studies investigating the effects of liming on soil microbial community composition reported mixed results. For example, both Pennanen et al. (1998) and Pawlett et al. (2009) found that lime application significantly reduced bacterial phospholipid fatty acids (PLFAs) and did not affect fungal PLFAs [10,11]. While Bruneau et al. (2010) reported that the density of bacteria was increased by liming [12]. Kamal et al. (2010) found that lime application led to increases in the relative abundance of prokaryotes, bacilli, and actinomycetes but decreases in micromycetes [13]. On the contrary, Wang et al. (2007) reported an inhibitory effect of lime on actinomycetes that could facilitate soil carbon mineralization [14]. Although the acid tolerance of individual microbial species varies widely, it is well recognized that low pH inhibits the growth of soil bacteria but favors fungi that are more resistant to acidity. Therefore, the soil microbial community generally shifts from fungi to bacteria as the pH increases after lime application [15]. However, most of these studies focused on the soil microbial communities in agricultural soil [8,16]. We still have a poor understanding of how fungal and bacterial communities respond to lime application in forests, especially for *Eucalyptus* plantations in South China where soils are strongly acidic.

*Eucalyptus urophylla* S.T.Blake is a fast-growing species and is widely planted in South China. Due to the high demand for wood and fiber, the planting area of *E. urophylla* is expanding [17]. Within an *E. urophylla* plantation, the understory vegetation is typically dominated by a native fern *Dicranopteris dichotoma* (Thunb.) Bernh., which usually forms a mat-like dense layer beneath the open-canopy areas [6,17]. Several studies have revealed that understory vegetation plays important roles in maintaining both the aboveground and belowground biodiversity, the microclimate, and the nutrient cycling [1,6,18]. Furthermore, understory fern biomass can occupy up to 20% of the total biomass in *Eucalyptus* plantations [19], indicating that understory plants could make a substantial contribution to the regulation of soil processes. Although the number of such studies is growing recently, the importance of understory vegetation in driving ecosystem structure and functions remains unfocused in most ecological studies [20,21]. In practice, understory vegetation is usually removed to reduce competition for nutrients against plantation trees, which could facilitate the growth of tree seedlings [22] and cause significant changes in soil microclimates [20] or nutrient availabilities [23]. To date, however, only a few studies quantitatively determined the effects of understory removal on soil microbial community composition, especially in combination with other forest management practices, and the effects tend to be conflicting. Some studies reported significant alterations in the composition of soil microbial communities [6,24], while others did not [25]. Therefore, the effects of understory removal on soil microbial communities are highly context dependent and need to be further studied.

In the present study, a lime application experiment combined with understory removal was conducted in a subtropical *E. urophylla* plantation to examine the main and interactive effects of lime application and understory removal on soil microbial community composition. We hypothesized that: (1) Lime application would increase soil pH, leading to negative effects on fungal biomass but positive effects on bacterial biomass, hence, changing the microbial composition, because most fungi are moderately acidophilic but bacterial species could grow within wider pH ranges; and (2) understory removal would result in negative effects on soil microbial biomass and composition, due to reductions of belowground resource inputs and changes in soil microclimates.

## 2. Materials and Methods

### 2.1. Site Description

This study was conducted in three six-year old *E. urophylla* monoculture plantations at Heshan National Field Research Station of Forest Ecosystem (112°50′ E, 22°34′ N), Chinese Academy of Science (CAS), Guangdong Province, China. Each of the three plantations occupied about 1 ha. The distance between any two plantations exceeded 1 km. The climate in this region is typically subtropical monsoon, with a humid hot season from April to September and a dry cold season from October to March. The mean annual temperature and rainfall are 21.7 °C and 1700 mm, respectively. The soil is classified as an Ultisol developed from sandstone [26]. *Eucalyptus* plantations were reestablished in 2005 after all previous trees (*Pinus elliottii* Engelm.) had been cut, with saplings planted at a spacing of 3 m × 2 m (1650 trees per ha). The understory vegetation within the plantation was dominated (almost 100%) by *D. dichotoma*, accompanied by some rarely occurring species such as *Rhodomyrtus tomentosa* (Aiton) Hassk. and *Miscanthus sinensis* Andersson.

### 2.2. Experiment Design

A split-plot design was used and four treatments were replicated three times, with the lime application as the main-plot factor and understory removal as the sub-plot factor. In December 2011, two paired main-plots were established in each plantation, and within each main plot, a sub-plot was distinguished. The main-plot size was 10 m × 10 m, while the sub-plot size was 5 m × 5 m. Lime application and understory removal were arranged for the main plot and the sub-plot, respectively. Therefore, the four treatments were: Control without lime application and with intact understory (CK), understory removal (UR), lime application (LA), and lime application combined with understory removal (LUR). Around each plot, polyvinyl chloride (PVC) boards were embedded at a depth of 80 cm to exclude disturbance from roots outside the plots and to eliminate the movement of water and nutrients. For plots receiving lime application, lime was added to the soil surface on 15 December 2011 and 15 August 2012 in the form of pulverized lime (60 kg per 100 $m^2$ as CaO in total per year). Understory vegetation (including the above- and belowground components) was physically removed from plots for the understory removal treatments (i.e., UR and LUR) prior to the start of the experiment. Any understory plant germinated during the experimental period was removed on a monthly basis.

### 2.3. Soil Sampling and Analysis

Soils were sampled on 0, 10, 60, 147, 275, and 377 days after the treatments were applied, with the day 0 sample as the baseline. The surface litter was carefully removed before sampling. In each plot, five soil cores (2.5 cm in diameter) at 0–10 cm depth from random sampling points were combined as one composite sample. After carefully removing the surface organic materials and visible roots, each composite sample was passed through a 2 mm sieve and divided into two subsamples; one subsample was stored at 4 °C for soil physical–chemical analysis, and the other was stored at −20 °C for PLFA analysis.

Soil pH was measured every month during the experimental period. Three soil cores at 0–10 cm depth were sampled in each plot and combined as a composite sample. Soil pH was determined in a 1:2.5 soil-water slurry using a combination glass electrode. Soil temperature was measured with the DS19221 temperature logger iButton® (Dallas Semiconductor Corp., Dallas, TX, USA) every 2 h at 5 cm soil depth in situ from December 2011 to December 2012. Soil moisture contents (SMC% g of water per 100 g of dry soil) was determined gravimetrically after the soil was dried for 48 h at 105 °C.

The soil microbial community was characterized by the phospholipid fatty acid (PLFA) analysis [27,28]. Briefly, field-moist, sieved soil samples equivalent to 8 g dry mass were extracted in a chloroform–methanol–phosphate buffer (1:2:0.8 *v/v/v*), and the extracted lipids were fractionated into neutral lipids, glycolipids, and polar lipids on silica acid columns by successive elution with chloroform, acetone, and methanol. The methanol fraction (containing phospholipids) was subjected to mild alkaline methanolysis to transform the fatty acids into free methyl esters and was analyzed on a gas chromatograph (Hewlett-Packard 6890, Agilent, Santa Clara, CA, USA). Peaks were identified using bacterial fatty acid standards and MIDI peak identification software (MIDI, Inc., Newark, DE, USA). Concentrations of each PLFA were standardized relative to 19:0 internal reference concentrations. Fungal biomass was considered to be indicated by 18:2ω6,9 [28]; bacterial biomass was considered to be indicated by the following 10 PLFAs: i15:0, a15:0, 15:0, i16:0, 16:1w7, i17:0, a17:0, 17:0, cy17:0, and cy19:0; total PLFAs and the ratio of fungal to bacterial PLFAs were also calculated. All PLFAs were calculated as ng $g^{-1}$ dry soil.

### 2.4. Litter Decomposition Experiment

Litter bags (26 cm × 15 cm with a mesh size of 1 mm × 1 mm) were used to measure *Eucalyptus* litter decomposition rates. Fresh *Eucalyptus* litter was collected from litter traps (1 m × 1 m nylon net) in the experimental plantation. Each litter bag contained 10 g oven-dried litter, and six litter bags were placed on the soil surface in each subplot of the four treatments in December 2011. Afterward, one litter bag in each subplot was harvested bimonthly until all the six bags were harvested. The retrieved litter decomposition residuals were collected and oven-dried at 65 °C to a constant mass, and then weighed for the mass. The litter decomposition rate was calculated as the bimonthly mass loss rate: Litter decomposition rate (%) = (mass loss bimonthly/initial weight of litter) × 100%, where the initial weight of litter in each litter bag was 10 g.

### 2.5. Statistical Analysis

Repeated-measure ANOVAs by the general linear model were employed to examine the effects of time and treatments throughout the experimental period. Soil physico-chemical properties and microbial biomass measured during the experiment were analyzed using linear mixed-effect models, with lime application and understory removal as categorical fixed effects, and replicated ($n = 3$) as a random effect. Redundancy analysis (RDA) was performed to determine the relationships between microbial communities (PLFA profile) and soil physico-chemical properties. The most discriminating soil property variables were selected by the "forward selection" procedure. Data were transformed (natural log, square root, or rank) when necessary to meet assumptions of normality and homogeneity of variance. Statistical significance was determined at $P \leq 0.05$. All univariate analyses were performed in SPSS 18.0 (SPSS Inc., Chicago, IL, USA). RDA was performed with the CANOCO 4.5 software (Ithaca, NY, USA). The forward selection was implemented based on the Monte Carlo permutation ($n = 499$).

## 3. Results

### 3.1. Soil Physico-Chemical Properties

Contents of soil organic carbon (SOC), total nitrogen (TN), dissolved organic carbon (DOC), as well as soil pH, did not differ among plots before the treatments were applied (Table 1), indicating no spatial heterogeneity between the experimental plots. Soil $NH_4{}^+$-N contents varied with time but did not differ among the four treatments during the experimental period (Table 2). Soil pH, however, was significantly higher in the LA treatment (4.19–4.56 units) than in the control (3.78–3.83 units) or the UR treatment (3.57–3.87 units) across the sampling events (Figure 1). Soil $NO_3{}^-$-N contents were increased by LA, but decreased by UR. Soil moisture contents (SMC) were significantly lower in both LA and UR treatments than in the control. For DOC, there was a slight decline in the UR treatment, but the trend was minimal ($P = 0.06$). In addition, no interaction between LA and UR on any soil physico-chemical property measured in this study was observed (Table 2).

**Table 1.** Soil characteristics before treatments were applied.

| Treatments | SOC (g kg$^{-1}$) | TN (g kg$^{-1}$) | DOC (μg C g$^{-1}$) | pH |
|---|---|---|---|---|
| Control | 14.43 (0.66) | 1.86 (0.32) | 288.64 (74.26) | 3.97 (0.04) |
| UR | 15.78 (1.82) | 1.72 (0.53) | 268.11 (88.39) | 4.05 (0.04) |
| LA | 17.06 (1.95) | 1.68 (0.37) | 249.22 (62.16) | 4.14 (0.02) |
| LUR | 15.64 (1.21) | 1.71 (0.47) | 246.54 (46.55) | 4.16 (0.01) |
| ANOVA | ns | ns | ns | ns |

SOC, soil organic carbon; TN, total nitrogen; DOC, dissolved organic carbon; Control, no lime application and no understory removal; UR, understory removal; LA, lime application; LUR, lime application with understory removal. Values are means (with SE in parentheses; $n$ = 3). Statistical significance was determined at $P < 0.05$. ns, no significant difference between treatments using ANOVA.

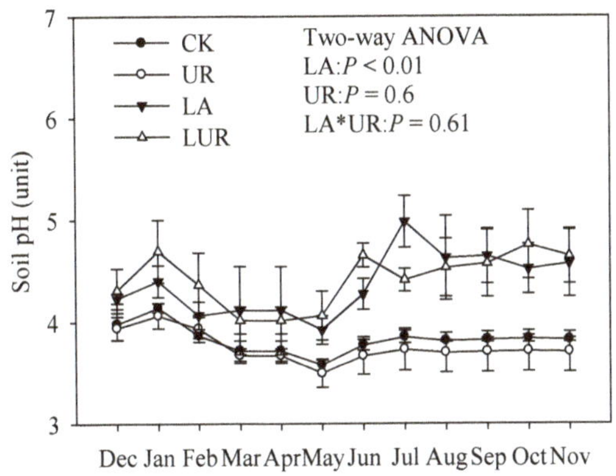


**Figure 1.** Soil pH under different treatments during the experimental period. CK, control without lime application and with intact understory; UR, understory removal; LA, lime application; LUR, lime application with understory removal. Values are means ± SE, $n$ = 3.

**Table 2.** Effects of time, lime application, understory removal, and the interaction of lime application and understory removal on soil moisture content, soil temperature at 0–5cm depth, dissolved organic carbon, and inorganic nitrogen content, $n$ = 3.

| Factors | Time | | LA | | UR | | LA × UR | |
|---|---|---|---|---|---|---|---|---|
| | $F$ | $P$ | $F$ | $P$ | $F$ | $P$ | $F$ | $P$ |
| SMC | 15.14 | <0.01 | 20.45 | <0.01 | 9.4 | <0.01 | 2.56 | 0.112 |
| ST | 12.15 | <0.01 | 3.57 | 0.09 | 36.4 | <0.01 | 1.05 | 0.34 |
| DOC | 1.45 | 0.2 | 0.95 | 0.33 | 0.18 | 0.06 | 0.23 | 0.63 |
| $NH_4^+$–N | 7.15 | <0.01 | 0.27 | 0.6 | 2.31 | 0.14 | 2.2 | 0.99 |
| $NO_3^-$–N | 26.01 | <0.01 | 27.13 | <0.01 | 13.31 | <0.01 | 0.99 | 0.32 |

Time, sampling time; LA, lime application; UR, understory removal; LA × UR, interactions between lime application and understory removal; SMC, soil moisture content; ST, soil temperature; DOC, dissolved organic carbon. Results are from a three-way factorial ANOVA, with factors of time (levels: Dec 2011, Feb 2012, May 2012, Sep 2012, and Dec 2012), lime application (levels: limed, not limed), and understory removal (levels: understory removed, understory not removed).

### 3.2. Soil Microbial Community

Soil microbial biomass (as indicated by the total PLFAs, i.e., the sum of fungal and bacterial PLFAs) ranged from 453 to 2701 ng g$^{-1}$ under different treatment during the experimental period. Sampling time had a significant effect on total PLFAs, bacterial PLFAs, fungal PLFAs, and the F:B ratio (ratio of fungal to bacterial PLFAs) (Figure 2, Table 3). Across sampling time, the LA treatment significantly decreased both fungal and bacterial PLFAs and hence the total PLFAs, but had no significant effect on the F:B ratio (Figure 2, Table 3). The UR treatment, however, significantly decreased the fungal and total PLFAs without affecting bacterial PLFAs, and thus reduced the F:B ratio (Figure 2, Table 3). There were no significant interaction effects of LA and UR on the composition of soil microbial communities during the entire experimental period. In addition, no interaction effect between sampling time and LA was observed, while the interaction between sampling time and UR on the total PLFAs was remarkable (Table 3).

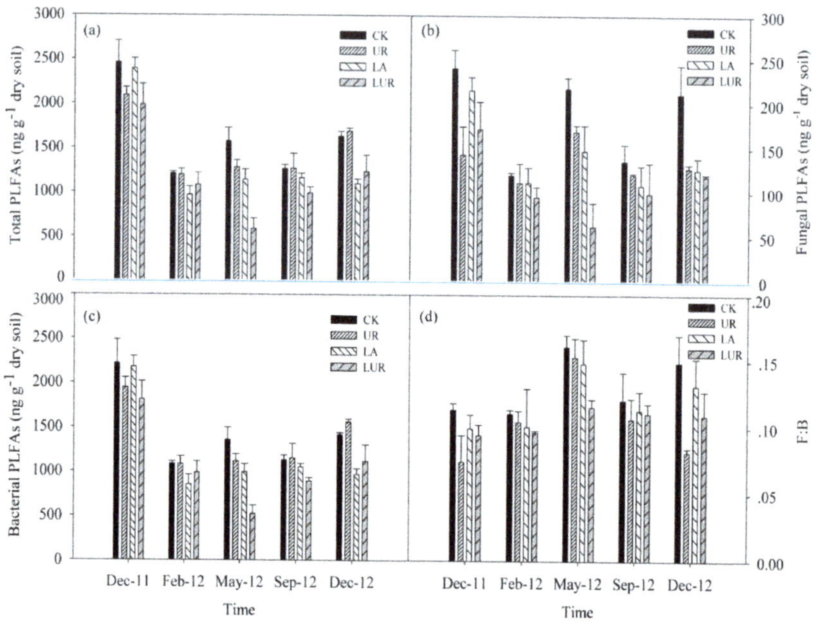


**Figure 2.** Soil microbial phospholipid fatty acids (PLFAs) under different treatments. (**a**) Soil Total PLFAs; (**b**) Soil Fungal PLFAs; (**c**) Soil Bacterial PLFAs; (**d**) the Ratio of Fungal PLFAs to Bacterial PLFAs. UR, understory removal; LA, lime application; LUR, lime application with understory removal. F:B indicates the ratio of fungal to bacterial PLFAs. Values are means ± SE, $n = 3$.

**Table 3.** Effects of time, lime application, understory removal, and interactions between time, lime application and understory removal on soil microbial PLFAs, $n = 3$.

| Variable | Time | LA | UR | LA × UR | Time × LA | Time × UR | Time × LA × UR |
|---|---|---|---|---|---|---|---|
| | *P* | *P* | *P* | *P* | *P* | *P* | *P* |
| Total PLFAs | <0.01 | <0.01 | 0.02 | 0.65 | 0.64 | 0.02 | 0.81 |
| Bacteria PLFAs | <0.01 | <0.01 | 0.14 | 0.79 | 0.24 | 0.07 | 0.94 |
| Fungal PLFAs | <0.01 | <0.01 | <0.01 | 0.46 | 0.06 | 0.08 | 0.21 |
| F:B | <0.01 | 0.76 | 0.02 | 0.74 | 0.65 | 0.26 | 0.38 |

Time, sampling time; LA, lime application; UR, understory removal; LA × UR, interactions between lime application and understory removal; F:B indicates the ratio of fungal to bacterial PLFAs. Results are from a three-way factorial ANOVA, with factors of time (levels: Dec 2011, Feb 2012, May 2012, Sep 2012, and Dec 2012), lime application (levels: limed, not limed), and understory removal (levels: understory removed, understory not removed).

### *3.3. Relationships between Soil Physico-Chemical Properties and Microbial Community*

Redundancy analysis showed that soil microclimate ($P < 0.01$), soil DOC ($P < 0.01$), soil pH ($P = 0.02$), and soil $NO_3^-$-N content ($P = 0.02$) were significantly correlated with whole soil microbial communities (Figure 3). Specifically, soil pH and SMC were correlated negatively to 18:2ω6,9, i16:0, 15:0, and a17:0 (Figure 3). Soil temperature was correlated positively to i17:0 and negatively to a15:0, i15:0, and 16:1ω7 (Figure 3). Soil $NO_3^-$-N content and soil DOC, however, were correlated negatively to i17:0, but correlated positively to a15:0, i15:0, and 16:1ω7 (Figure 3). All the environmental data explained 92.4% of the total variance, with the first axis (Axis 1) explaining 66.3% of the variance and the second axis (Axis 2) explaining another 26.1% (Figure 3).

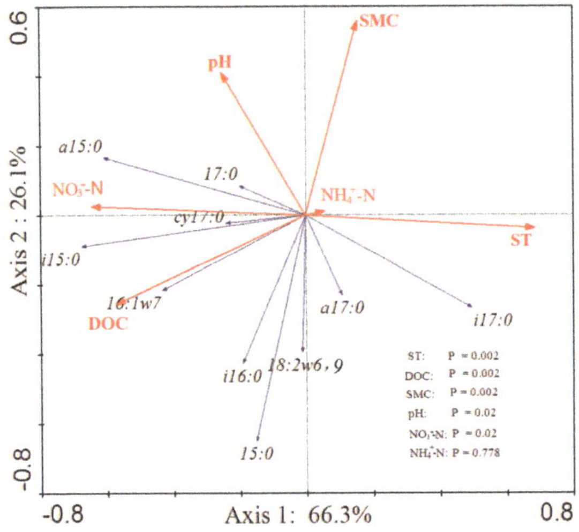


**Figure 3.** Redundancy analysis of soil microbial PLFA biomarkers. Ordination diagrams presenting species scores and environmental factor scores (vectors). ST, soil temperature at 0–5 cm depth; DOC, dissolved organic carbon; SMC, soil moisture content.

### 3.4. Litter Decomposition

One year after the treatments were applied, both LA and UR treatments significantly reduced litter mass loss (both $P < 0.01$). There was no significant interaction of LA and UR on the litter mass loss during the experimental period (Figure 4).

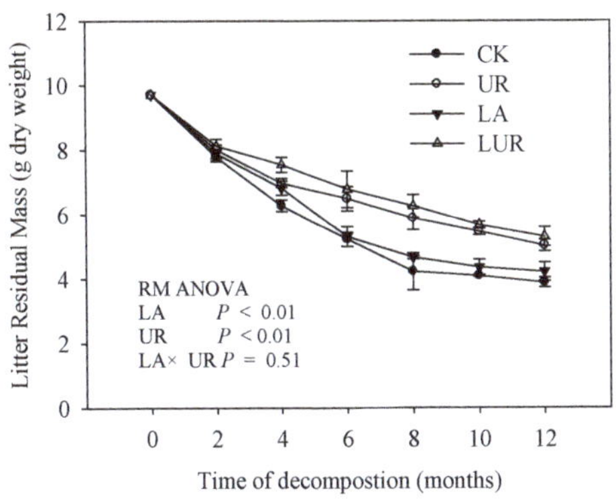


**Figure 4.** Mass of *Eucalyptus* litter remaining per litter bag under different treatments. Vertical bars represent SEs. LA, lime application; UR, understory removal. The inserted $P$ values were from repeated-measure ANOVA.

## 4. Discussion

### 4.1. The Impacts of Lime Application on Soil Microbial Community

Liming is commonly used to improve environmental conditions for the development of acid-intolerant microbes, resulting in increased microbial biomass and activities [4,29]. In this study, however, we observed a reduction of soil fungal and bacterial PLFAs after lime application. Consistent with our results, some other studies have also found that lime application inhibited the soil microbial biomass in both field- and lab-based conditions. For instance, Pawlett et al. (2009) reported that liming decreased soil microbial PLFAs and bacterial PLFAs [11]. Lin et al. (2018) also found that some soil bacterial groups were consistently decreased by a 27-year lime application in South China, and lime application may potentially affect the N-fixation ability by altering the soil community structure of soil diazotrophs [30]. Lime application did not affect soil the F:B ratio in this study, due to the decreases in both fungal and bacterial PLFAs (Figure 2, Table 3). Likewise, Xue et al. (2010) observed that liming had minimal effects on the F:B ratio, though soil microbial community structure had been significantly affected [31].

The most likely reason for lime application inducing alterations in soil microbial biomass may be associated directly with the increased soil pH, as shown by the fungal biomarker 18:2ω6,9, and the bacterial biomarkers *a*17:0, *i*16:0, and 15:0 were correlating negatively with soil pH (Figure 3). It is well recognized that most fungi are moderately acidophilic [7], and therefore, the fungal activities would be inhibited and the microbial communities would typically shift from fungal to bacterial as soil pH increased after lime application. Furthermore, the signature of *a*17:0, *i*16:0, and 15:0 had dominated the decreased bacterial biomass with higher pH in liming soils in the present study. In addition, our results showed that lime application reduced the soil water content, which might in turn suppress the $NO_3^-$-N leaching from soil. The increased inorganic N might have "toxicity effects or salt effects" on soil microbes and consequently suppress the microbial biomass [32]. In line with our results, studies from Wu et al. (2011) and Cox et al. (2010) also demonstrated that the microbial diversity and biomass were reduced by the increased N availability [19,33].

### 4.2. The Impacts of Understory Removal on Soil Microbial Community

Soil microorganisms are greatly affected by both the quantity and quality of resources, as well as the soil habitat. Soil sampling time had significant effects on soil microbial biomass and composition in this study (Figure 2, Table 3), probably due to the seasonal variations in soil temperature and/or moisture. The present study region is within a typical subtropical monsoon climate, with distinct wet–hot and dry–cold seasons. Over 83% and 66% of annual precipitation and radiation inputs to this region occur during the wet–hot season (from April to September). The temporal changes in soil microbial biomass and composition could be ascribed to the changes in moisture, soil temperature, and the substrate availability, as indicated by other studies [24].

In addition, soil sampling time had a significant interactive effect with UR on soil total PLFAs, which should be considered when conducting "before-after-control-impact" (BACI) experiments in the monsoon climate regime region. Our findings showed that understory removal significantly reduced total PLFAs, fungal PLFA, and the F: B ratio. The effects of understory plant removal on soil microbial communities have also been observed in other forest ecosystems. For instance, Winsome et al. (2017) reported that understory removal significantly reduced soil microbial biomass in a ponderosa pine plantation [34]. Tong et al. (2015) found that understory removal apparently decreased soil fungal PLFAs and the F:B ratio in a Chinese fir plantation [35]. Yin et al. (2016) found that the biomass of all microbial community groups (i.e., fungi, bacteria, arbuscular mycorrhizal fungi, and actinomycetes) decreased with decreasing understory herbs biomass in subtropical plantations, which suggested that understory vegetation would exert strong controls on soil microbial communities [36]. Zhao et al. (2012) and Giuggiola et al. (2018) also suggested that the indirect effects of understory removal on soil microclimate wound be the most likely factor affecting soil microbial community composition [20,37].

Furthermore, RDA showed that the increased soil temperature and decreased soil moisture derived from the understory removal treatment resulted in changes in soil microbial composition. This may be attributed to the fact that the soil microbial community was significantly correlated with soil temperature and moisture, and all of the environmental variables explained up to 92.4% of the total variance in soil microbial communities in this study (Figure 3).

It is generally accepted that soil microbial community is controlled by the bottom-up force [38,39]. Normally, understory vegetation is considered as a minor portion of biomass in a forest ecosystem and contributes less to the bottom-up force on soil decomposers than overstory plants. Wu et al. (2011) reported that the DOC remained unaffected by understory removal in both two-year and 24-year *Eucalyptus* plantations [19], which indicated that the bottom-up control induced by understory vegetation was minor. However, there was an apparent trend that understory removal decreased DOC ($P = 0.06$) (Table 2), and thus resulted in the decrease in fungal PLFAs in the six-year *Eucalyptus* plantation in this study, which was proven by the positive correlation between the fungal PLFA biomarker (i.e., 18:2ω6,9) and the DOC (Figure 3). Additionally, understory removal decreased soil moisture ($P < 0.01$), and subsequently increased the fungal PLFAs, as indicated by the negative relationship between the fungal PLFA biomarker 18:2ω6,9 and soil moisture (Figure 3). Consequently, the suppressing effect of understory removal on soil fungal PLFAs could be ascribed to the decrease in DOC. Furthermore, our previous study in the same experiment site found that understory removal significantly reduced the soil $CO_2$ flux, and the biomass of understory fern occupied up to 20% of the *Eucalyptus* plantation [17], which indicated that the understory dominated by *D. dichotoma* might have a bottom-up control on soil microbial communities. Since the ecological functions of understory vegetation in a forest ecosystem largely depend on their species [21] and biomass [6,40], the differentiated results from different ecosystems indicate that the effects of understory vegetation on soil microbial communities are context dependent and require further studies.

## 5. Conclusions

In conclusion, soil microbial communities in this study were affected by both lime application and understory removal, but without any interaction. Lime application increased soil pH and soil $NO_3^-$-N content, which decreased the soil fungal and bacterial PLFAs. Understory presence is favorable for sustaining soil microclimates and providing resources for soil microorganisms. Understory removal inhibited soil total PLFAs and fungal PLFAs, and hence altered the soil microbial communities. In addition, both lime application and understory removal significantly reduced the litter decomposition rates (Figure 4), which indicates the lime application and understory removal may impact the microbe-mediated soil ecological process (i.e., soil nutrient cycling). We propose that lime application may not be suitable for the management of subtropical *Eucalyptus* plantations, while the understory fern *D. dichotoma* functions as a facilitator in *Eucalyptus* plantations and should not be removed. In general, our findings in this study would be helpful for the development of management practices that optimize the yield and sustainability of subtropical plantations.

**Author Contributions:** S.W. and Z.L. designed the experiment. S.W., Y.C., and J.Z. performed the study data collection. J.L., S.W, and Q.Y. analyzed the results. J.L. and Z.L. reviewed the article. S.W. wrote the final article. All authors approved the final version of the manuscript.

**Funding:** This study was funded by the National Science Foundation of China (41867007 and 31500341), Guangdong Provincial Key Laboratory of Applied Botany, South China Botanical Garden, Chinese Academy of Science (AB2016003), the Open Foundation of the State Key Laboratory of Urban and Regional Ecology of China (SKLURE2016-2-5), and the Youth Innovation Promotion Association CAS.

**Acknowledgments:** We thank Shenglei Fu and Weixing Zhang from Henan University for their comments and discussions on the early version of the manuscript. We also thank Xiaoli Wang and Xiaolin Zhu for their assistance in samples analysis and data collection.

**Conflicts of Interest:** The authors declare no conflict of interest.

## References

1. Wan, S.; Zhang, C.; Chen, Y.; Zhao, J.; Wang, X.; Wu, J.; Zhou, L.; Lin, Y.; Liu, Z.; Fu, S. The understory fern *Dicranopteris dichotoma* facilitates the overstory Eucalyptus trees in subtropical plantations. *Ecosphere* **2014**, *5*, 1–12. [CrossRef]
2. Leff, J.W.; Jones, S.E.; Prober, S.M.; Barberán, A.; Borer, E.T.; Firn, J.L.; Harpole, W.S.; Hobbie, S.E.; Hofmockel, K.S.; Knops, J.M.H.; et al. Consistent responses of soil microbial communities to elevated nutrient inputs in grasslands across the globe. *Proc. Natl. Acad. Sci. USA* **2015**, *112*, 10967–10972. [CrossRef]
3. Delgado-Baquerizo, M.; Maestre, F.T.; Reich, P.B.; Jeffries, T.C.; Gaitan, J.J.; Encinar, D.; Berdugo, M.; Campbell, C.D.; Singh, B.K. Microbial diversity drives multifunctionality in terrestrial ecosystems. *Nat. Commun.* **2016**, *7*, 10541. [CrossRef]
4. Narendrula-Kotha, R.; Nkongolo, K.K. Microbial response to soil liming of damaged ecosystems revealed by pyrosequencing and phospholipid fatty acid analyses. *PLoS ONE* **2017**, *12*, e0168497. [CrossRef]
5. Li, J.; Li, Z.; Wang, F.; Zou, B.; Chen, Y.; Zhao, J.; Mo, Q.; Li, Y.; Li, X.; Xia, H. Effects of nitrogen and phosphorus addition on soil microbial community in a secondary tropical forest of China. *Biol. Fertil. Soils* **2015**, *51*, 207–215. [CrossRef]
6. Zhao, J.; Wan, S.; Fu, S.; Wang, X.; Wang, M.; Liang, C.; Chen, Y.; Zhu, X. Effects of understory removal and nitrogen fertilization on soil microbial communities in Eucalyptus plantations. *For. Ecol. Manag.* **2013**, *310*, 80–86. [CrossRef]
7. Zhao, J.; Zhao, C.; Wan, S.; Wang, X.; Zhou, L.; Fu, S. Soil nematode assemblages in an acid soil as affected by lime application. *Nematology* **2015**, *17*, 179. [CrossRef]
8. Bramble, D.S.E.; Gouveia, G.A.; Ramnarine, R. Organic residues and ammonium effects on $CO_2$ emissions and soil quality indicators in limed acid tropical soils. *Soil Syst.* **2019**, *3*, 16. [CrossRef]
9. Bakina, L.G.; Chugunova, M.V.; Zaitseva, T.B.; Nebol'sina, Z.P. The effect of liming on the complex of soil microorganisms and the humus status of a soddy-podzolic soil in a long-term experiment. *Eurasian Soil Sci.* **2014**, *47*, 110–118. [CrossRef]
10. Pennanen, T.H.; Vanhala, P.; Kiikkila, O.; Neuvonen, S.; Baath, E. Structure of a microbial community in soil after prolonged addition of low levels of simulated acid rain. *Appl. Environ. Microb.* **1998**, *64*, 2173–2180.
11. Pawlett, M.; Hopkins, D.W.; Moffett, B.F.; Harris, J.A. The effect of earthworms and liming on soil microbial communities. *Biol. Fertil. Soils* **2009**, *45*, 361–369. [CrossRef]
12. Bruneau, P.M.C.; Davidson, D.A.; Grieve, I.C.; Young, I.M.; Naoise, N. The effects of soil horizons and faunal excrement on bacterial distribution in an upland grassland soil. *Fems Microb. Ecol.* **2010**, *52*, 139–144. [CrossRef]
13. Kamal, S.; Prasad, R.; Varma, A. Soil microbial diversity in relation to heavy metals. In *Soil Heavy Metals;* Springer: Berlin, Heidelberg, 2010; volume 19, pp. 31–63.
14. Wang, Y.; Shi, J.; Wang, H.; Lin, Q.; Chen, X.; Chen, Y. The influence of soil heavy metals pollution on soil microbial biomass, enzyme activity, and community composition near a copper smelter. *Ecotoxicol. Environ. Saf.* **2007**, *67*, 75–81. [CrossRef]
15. Haynes, R.J.; Naidu, R. Influence of lime, fertilizer and manure applications on soil organic matter content and soil physical conditions: A review. *Nutr. Cycl. Agroecosyst.* **1998**, *51*, 123–137. [CrossRef]
16. Shi, L.; Guo, Z.; Liang, F.; Xiao, X.; Peng, C.; Zeng, P.; Feng, W.; Ran, H. Effect of liming with various water regimes on both immobilization of Cadmium and improvement of bacterial communities in Contaminated Paddy: A field experiment. *Int. J. Environ. Res. Public Health* **2019**, *16*, 498. [CrossRef]
17. Wan, S.; Zhang, C.; Chen, Y.; Zhao, J.; Zhu, X.; Wu, J.; Zhou, L.; Lin, Y.; Liu, Z.; Fu, S. Interactive effects of understory removal and fertilization on soil respiration in subtropical Eucalyptus plantations. *J. Plant Ecol.* **2015**, *3*, 284–290. [CrossRef]
18. Wang, F.; Chen, F.; Wang, G.G.; Mao, R.; Fang, X.; Wang, H.; Bu, W. Effects of experimental nitrogen addition on nutrients and nonstructural carbohydrates of dominant understory plants in a Chinese fir plantation. *Forests* **2019**, *10*, 155. [CrossRef]

19. Wu, J.; Liu, Z.; Chen, D.; Huang, G.; Zhou, L.; Fu, S. Understory plants can make substantial contributions to soil respiration: Evidence from two subtropical plantations. *Soil Biol. Biochem.* **2011**, *43*, 2355–2357. [CrossRef]
20. Giuggiola, A.; Zweifel, R.; Feichtinger, L.M.; Vollenweider, P.; Bugmann, H.; Haeni, M.; Rigling, A. Competition for water in a xeric forest ecosystem—Effects of understory removal on soil micro-climate, growth and physiology of dominant Scots pine trees. *For. Ecol. Manag.* **2018**, *409*, 241–249. [CrossRef]
21. Nilsson, M.-C.; Wardle, D.A. Understory vegetation as a forest ecosystem driver: Evidence from the northern Swedish boreal forest. *Front. Ecol. Environ.* **2005**, *3*, 421–428. [CrossRef]
22. Wang, J.; Hai, R.; Long, Y.; Duan, W. Establishment and early growth of introduced indigenous tree species in typical plantations and shrubland in South China. *For. Ecol. Manag.* **2009**, *7*, 1293–1300. [CrossRef]
23. Li, Y.L.; Wang, S.L.; Yan, S.K. Short-term effects of understory vegetation removal on nutrient cycling in litter layer of Chinese fir plantation. *Chin. J. Appl. Ecol.* **2011**, *22*, 2560.
24. Wu, J.; Liu, Z.; Wang, X.; Sun, Y.; Zhou, L.; Lin, Y.; Fu, S. Effects of understory removal and tree girdling on soil microbial community composition and litter decomposition in two Eucalyptus plantations in South China. *Funct. Ecol.* **2011**, *25*, 921–931. [CrossRef]
25. Urcelay, C.; Díaz, S.; Gurvich, D.E.; Chapin, F.S., III; Cuevas, E.; Domínguez, L.S. Mycorrhizal community resilience in response to experimental plant functional type removals in a woody ecosystem. *J. Ecol.* **2010**, *97*, 1291–1301. [CrossRef]
26. FAO. *World Referernce Base for Soil Resources 2006*; World Soil Resources Report 103; FAO: Rome, Italy, 2006.
27. Bossio, D.; Scow, K. Impacts of carbon and flooding on soil microbial communities: Phospholipid fatty acid profiles and substrate utilization patterns. *Microb. Ecol.* **1998**, *35*, 265–278. [CrossRef]
28. Frostegård, Å.; Tunlid, A.; Bååth, E. Use and misuse of PLFA measurements in soils. *Soil Biol. Biochem.* **2011**, *43*, 1621–1625. [CrossRef]
29. Frostegård, Å.; Bååth, E.; Tunlio, A. Shifts in the structure of soil microbial communities in limed forests as revealed by phospholipid fatty acid analysis. *Soil Biol. Biochem.* **1993**, *25*, 723–730. [CrossRef]
30. Lin, Y.; Ye, G.; Liu, D.; Ledgard, S.; Luo, J.; Fan, J.; Yuan, J.; Chen, Z.; Ding, W. Long-term application of lime or pig manure rather than plant residues suppressed diazotroph abundance and diversity and altered community structure in an acidic Ultisol. *Soil Biol. Biochem.* **2018**, *123*, 218–228. [CrossRef]
31. Xue, D.; Huang, X.; Yao, H.; Huang, C. Effect of lime application on microbial community in acidic tea orchard soils in comparison with those in wasteland and forest soils. *J. Environ. Sci.* **2010**, *22*, 1253–1260. [CrossRef]
32. Thirukkumaran, C.M.; Parkinson, D. Microbial respiration, biomass, metabolic quotient and litter decomposition in a lodgepole pine forest floor amended with nitrogen and phosphorous fertilizers. *Soil Biol. Biochem.* **2000**, *32*, 59–66. [CrossRef]
33. Cox, F.; Barsoum, N.; Lilleskov, E.A.; Bidartondo, M.I. Nitrogen availability is a primary determinant of conifer mycorrhizas across complex environmental gradients. *Ecol. Lett.* **2010**, *13*, 1103–1113. [CrossRef]
34. Winsome, T.; Silva, L.C.R.; Scow, K.M.; Doane, T.A.; Powers, R.F.; Horwath, W.R. Plant-microbe interactions regulate carbon and nitrogen accumulation in forest soils. *For. Ecol. Manag.* **2017**, *384*, 415–423. [CrossRef]
35. He, T.X.; Li, Y.P.; Zhang, F.Y.; Wang, Q.K. Effects of understory removal on soil respiration and microbial community composition structure in a Chinese fir plantation. *Chin. J. Plant Ecol.* **2015**, *39*, 797–806.
36. Yin, K.; Zhang, L.; Chen, D.; Tian, Y.; Zhang, F.; Wen, M.; Yuan, C. Understory herb layer exerts strong controls on soil microbial communities in subtropical plantations. *Sci. Rep.* **2016**, *6*, 27066. [CrossRef]
37. Zhao, J.; Wan, S.; Li, Z.A.; Shao, Y.; Xu, G.; Liu, Z.; Zhou, L.; Fu, S. Dicranopteris-dominated understory as major driver of intensive forest ecosystem in humid subtropical and tropical region. *Soil Biol. Biochem.* **2012**, *49*, 78–87. [CrossRef]
38. Schuldt, A.; Bruelheide, H.; Buscot, F.; Assmann, T.; Erfmeier, A.; Klein, A.-M.; Ma, K.; Scholten, T.; Staab, M.; Wirth, C.; et al. Belowground top-down and aboveground bottom-up effects structure multitrophic community relationships in a biodiverse forest. *Sci. Rep.* **2017**, *7*, 4222. [CrossRef]

39. Benítez, E.; Paredes, D.; Rodríguez, E.; Aldana, D.; González, M.; Nogales, R.; Campos, M.; Moreno, B. Bottom-up effects on herbivore-induced plant defences: A case study based on compositional patterns of rhizosphere microbial communities. *Sci. Rep.* **2017**, *7*, 6251. [CrossRef]
40. Brown, S.; Lenart, M.; Mo, J.; Kong, G. Structure and Organic Matter Dynamics of a Human-Impacted Pine Forest in a MAB Reserve of Subtropical China. *Biotropica* **1995**, *27*, 276–289. [CrossRef]

*forests*

MDPI

*Article*

# Effects of Phosphate Solubilizing Bacteria on the Growth, Photosynthesis, and Nutrient Uptake of *Camellia oleifera* Abel.

**Fei Wu [1,2], Jianrong Li [3], Yanliu Chen [1], Linping Zhang [1,2,*], Yang Zhang [1], Shu Wang [1], Xin Shi [4], Lei Li [5] and Junsheng Liang [5]**

1 Key Laboratory of State Forestry Administration on Forest Ecosystem Protection and Restoration of Poyang Lake Watershed, Jiangxi Agricultural University, Nanchang 330045, China; wufei315@163.com (F.W.); xiaoliuchen0316@126.com (Y.C.); zhangyang0558@163.com (Y.Z.); wangshujx2012@163.com (S.W.)

2 2011 Collaborative Innovation Center of Jiangxi Typical Trees Cultivation and Utilization, Jiangxi Agricultural University, Nanchang 330045, China

3 South China Botanical Garden, Chinese Academy of Sciences, Guangzhou 510650, China; lijianrong@scbg.ac.cn

4 Research Institute of Tropical Forestry, Chinese Academy of Forestry, Guangzhou 510520, China; hb-shixin@126.com

5 Hunan Academy of Forestry, Changsha 410004, China; lileijxpx@163.com (L.L.); ljscaf@163.com (J.L.)

* Correspondence: zlping619@mail.jxau.edu.cn; Tel.: +86-0791-8381-3243

Received: 14 March 2019; Accepted: 15 April 2019; Published: 20 April 2019

**Abstract:** Phosphorus (P) is a necessary nutrient for plant growth and plays an important role in plant metabolisms; however, the majority of P in soil is in insoluble forms. Phosphate solubilizing bacteria (PSB) can convert the insoluble phosphates into plant-available forms and may have the potential for use in sustainable agricultural practices. This study examined the effects of two native PSB, namely *Bacillus aryabhattai* (JX285) and *Pseudomonas auricularis* (HN038), and a mixture of both strains (1:1) on the growth of *Camellia oleifera* Abel. seedlings. The results showed a significant promotion of the growth of *C. oleifera* plants by three inoculation treatments. All the PSB inoculation treatments could improve the leaf nitrogen (N) and P content and had positive effects on the available N, P, and potassium (K) content of rhizosphere soil. A co-inoculation of the two native PSB strains caused a synergistic effect and achieved the best benefit. In conclusion, *B. aryabhattai* and *P. auricularis* could be used as biological agents instead of chemical fertilizers for agricultural production to reduce environmental pollution and increase the yield of tea oil.

**Keywords:** phosphate solubilizing bacteria; nutrition; oil tea

## 1. Introduction

Phosphorus (P) is an essential nutrient for plant growth and development. Although a large amount of P is present in soil, the majority of it is unavailable to plants [1]. In agro-forestry practices, it is a fact that many problems arise in the application of phosphate fertilizer [2]. On one hand, phosphate rock is a nonrenewable resource and may run out in 50–100 years due to anthropogenic exploitation [3]. On the other hand, about 70% of the phosphate fertilizer in soluble forms that is applied to the soil is rapidly combined with cations such as $Ca^{2+}$, $Mg^{2+}$, $Al^{3+}$, and $Fe^{3+}$ and converted to insoluble forms [4]. Therefore, improving the absorption and use of P by crops is of great significance from both the ecological and economical perspective [5].

Phosphate-dissolving microorganisms are microbial resources closely associated with plant nutrition and account for 10% of all soil microorganisms [6,7]. Phosphate solubilizing bacteria (PSB) are a group of these microorganisms that can transform insoluble P compounds into available forms by secreting organic acids, and they may be used as inoculants to enhance P availability for plants [8]. In addition, they can also promote plant growth via producing hormones, such as cytokinin and indole acetic acid [9,10]. Despite some PSB present in plant rhizospheres and in soil, the amount of P released by these microorganisms is usually insufficient to meet the demand of growing plants [11]. High-efficiency PSB have the potential for making a great contribution to the decrease of environmental pollution and promoting ecological balance by replacing chemical fertilizers [12]. Consequently, there is an urgent need to investigate the effects of selected high-efficiency PSB on plant nutrition and growth.

*Camellia oleifera* Abel. (Theaceae), a unique edible oil tree species to China, is one of the world's four famous woody oil plants [13]. Tea oil obtained from seeds has an unsaturated fatty acid content of up to 90%, much higher than vegetable oil, peanut oil, and soybean oil, and contains special physiologically active substances, such as camellia, which have extremely high nutritional values [14]. *Camellia oleifera* generally grows in the mountains and hills of subtropical regions in southern China and is also of great value in soil and water conservation and maintaining ecological balance [15]. However, the acid soil in southern China has a low available P content and limits the growth and productivity of *C. oleifera* [16]. In our previous study, we screened two high-efficiency PSB strains (JX285 and HN038) from the rhizosphere soil of *C. oleifera* [17,18]. The present study's aim was to determine the impacts of JX285 and HN038 on the growth, photosynthesis, and nutrient uptake of *C. oleifera*. The results may provide a theoretical basis for the development of microbial fertilizers for use in agroforestry.

## 2. Materials and Methods

### 2.1. Strains, Plant Material, and Growth Medium

The two PSB strains, JX285 (*Bacillus aryabhattai*) and HN38 (*Pseudomonas auricularis*), were isolated from the rhizosphere of *C. oleifera* in our previous experiments [17,18]. The PSB inoculum was prepared as follows: the two PSB strains were individually grown in liquid Luria-Bertani (LB) medium (containing tryptone 10 g/L, yeast extract 5 g/L, and NaCl 10 g/L) with shaking at 180 rpm for 24 h at 28 °C, and then the broth was centrifuged at 10,000 r/min for 10 min. The supernatant was discarded, and the pellet was re-suspended and washed with sterilized water three times. The final concentration of PSB inoculum was adjusted to $10^8$ colony-forming units (CFU) $mL^{-1}$.

The seeds of *C. oleifera* were provided by the Jiangxi Academy of Forestry Research, China. The seeds were first surface sterilized with 0.3% potassium permanganate solution for 1 h and rinsed with sterilized water three times. They were then germinated on wet gauze at 30 °C. The germinated seeds were transplanted into 1 kg pots with sterilized sand. Finally, uniform seedlings were selected and transferred to a new pot containing 3 kg of growth medium. Each pot was planted with one seedling.

The growth medium was a mixture of krasnozem soil, sand, and peat soil (6:3:1, *v*/*v*). The soil used in the experiment was collected from the campus of Jiangxi Agricultural University (China) and was sieved after being air dried (1 mm). The growth medium was autoclaved at 0.11 MPa and 121 °C for 2 h. The basic physicochemical properties of the growth medium were as follows: pH 5.25 (soil:water—1:5), 43.71 mg/kg organic matter, 94.68 mg/kg available nitrogen (N), 2.80 mg/kg available P, and 15.80 mg/kg available potassium (K).

### 2.2. Experimental Design and Growth Condition

A pot experiment was conducted using a two-factor completely randomized block design containing five inoculation treatments and three P treatments [19]. The inoculation treatments were as follows: inoculated with *P. auricularis* HN038 (T1), inoculated with *B. aryabhattai* JX285 (T2), inoculated with the mixture of HN038 and JX285 (1:1) (T3), inoculated with LB medium (CK1), and inoculated with sterile water (CK2). The P treatments were as follows: no phosphate fertilizer (0 g/kg), (P1), added 3 g calcium superphosphate (1 g/kg) (P2), added 6 g calcium superphosphate (2 g/kg) (P3). Each treatment contained 5 replicates. Furthermore, in order to prevent the plant grow from stopping due to a lack of nutrient requirements, urea and potassium nitrate were added to the soil to reach

300 mg/kg N and 200 mg/kg K, respectively. The pot was placed in a greenhouse with 12 h of light per day. After five months of growth, the growth and photosynthetic parameters were measured and the seedlings were harvested.

### 2.3. Plant Growth Measurement

Three healthy seedlings with consistent growth were selected for each treatment to measure their growth parameters. A measuring tape (Swordfish, China) was used to measure the plant height. The leaves, stems, and roots were harvested and dried at 75 °C to a constant weight and weighed to calculate the total biomass of each plant.

### 2.4. Measurement of Gas Exchange Parameters and Relative Chlorophyll Content

A LI-6400 portable photosynthesis system (Li-Cor, Lincoln, NE, USA) was used to determine the photosynthetic parameters. When the seedlings were growing vigorously (mid-September) and the temperature was relatively stable (09:00–11:00 am), healthy leaves in the upper part of the seedlings were chosen to determine the net photosynthetic rate (Pn) and the transpiration rate (Tr). The water use efficiency (WUE) was calculated as WUE = Pn/Tr. A chlorophyll meter (SPAD-502) was used to measure the relative chlorophyll content of the plants. A random measurement method was used between each treatment to eliminate the inevitable errors caused by different measurement times.

### 2.5. Measurement of Plant Nutrient and Soil Nutrient

The dry leaves were finely ground and homogenized to determine the N and P concentrations. The P content was assayed by the dry ash digestion method [20]. The N content was determined by the Kjeldahl digestion method [21].

Samples of rhizosphere soil were collected to test the soil's nutrient status. The N content was measured using the alkaline hydrolysis method [22]. The P content was determined by the molybdenum ruthenium anti-colorimetric method [23]. The K content was measured by atomic absorption spectrometry [24].

### 2.6. Data Analysis

Statistical tests were performed using SPSS 13.0 (SPSS Inc., Chicago, IL, USA). A two-way ANOVA was performed to analysis the significance of P application, the PSB inoculation treatment and the interaction of P × PSB treatment. A one-way ANOVA was employed to examine the differences among the different inoculation treatments under each P level. Duncan's multiple range test was performed at $p = 0.05$ in case of significant impact by factor.

## 3. Results

### 3.1. Plant Height and Biomass

The two-way ANOVA results showed that plant height and biomass were significantly ($p \leq 0.01$) influenced by the interaction between P application and PSB inoculation (Table 1). The plant height and biomass were significantly ($p \leq 0.05$) improved by the single and mixed inoculation of two PSB strains compared to the non-inoculation control under each P level (Table 1). Co-inoculated plants showed a greater plant height and dry weight than those of plants inoculated with single PSB strains and non-inoculated plants. There was no significant difference in plant height and biomass between the CK1 and CK2 treatments. Moreover, the degree of the increase by PSB inoculation differed to some extent under different P levels. The maximum positive effects of the PSB inoculation on plant height and biomass were observed at the P3 level.

**Table 1.** Effect of PSB on plant height and biomass of *C. oleifera* seedlings.

| P Application | PSB Inoculation | Plant Height (cm) | Biomass (g) |
|---|---|---|---|
| P1 | T1 | 22.5 ± 0.9 b | 10.13 ± 0.90 c |
| | T2 | 22.7 ± 1.9 b | 12.00 ± 0.60 b |
| | T3 | 26.2 ± 1.2 a | 14.27 ± 1.40 a |
| | CK1 | 15.7 ± 1.2 c | 6.67 ± 0.61 d |
| | CK2 | 17.8 ± 0.3 c | 7.00 ± 0.72 d |
| P2 | T1 | 34.5 ± 1.0 c | 23.40 ± 1.59 c |
| | T2 | 36.7 ± 1.8 b | 24.40 ± 0.72 b |
| | T3 | 37.5 ± 1.0 a | 26.97 ± 1.66 a |
| | CK1 | 16.5 ± 2.1 d | 7.67 ± 1.01 d |
| | CK2 | 18.2 ± 0.9 d | 7.97 ± 1.53 d |
| P3 | T1 | 32.3 ± 1.5 a | 15.73 ± 0.90 b |
| | T2 | 33.5 ± 0.9 a | 16.47 ± 0.99 b |
| | T3 | 34.5 ± 0.8 a | 19.13 ± 0.92 a |
| | CK1 | 16.9 ± 1.3 b | 7.24 ± 0.47 c |
| | CK2 | 17.9 ± 2.1 b | 7.50 ± 0.95 c |
| Two-way ANOVA results | | | |
| P | | 0.00 ** | 0.00 ** |
| PSB | | 0.00 ** | 0.00 ** |
| P × PSB | | 0.00 ** | 0.00 ** |

Data are means ± SD ($n = 3$). P: phosphorus; PSB: phosphate solubilizing bacteria; P1: no phosphate fertilizer (0 g/kg); P2: added 3 g calcium superphosphate (1 g/kg); P3: added 6 g calcium superphosphate (2 g/kg); T1: inoculated with *P. auricularis* HN038; T2: inoculated with *B. aryabhattai* JX285; T3: inoculated with the mixture of HN038 and JX285 (1:1); CK1: inoculated with LB medium; CK2: inoculated with sterile water. Different lowercase letters within the same column indicate significant differences ($p \leq 0.05$). **, significant effect at $p \leq 0.01$.

### 3.2. Gas Exchange Parameters

The Pn was significantly ($p \leq 0.05$) influenced by P treatment and PSB treatment, while the Tr and WUE were significantly ($p \leq 0.05$) influenced by the interaction between P application and PSB inoculation (Table 2). At the same P application level, the Pn, Tr, and WUE were significantly ($p \leq 0.05$) enhanced by the single and mixed inoculation of PSB strains (Table 2). The Pn, Tr, and WUE of plants co-inoculated with *B. aryabhattai* and *P. auricularis* were higher than that of other treatments. The inoculation of single or mixed PSB strains promoted Pn and Tr more effectively at a high P level (P3 treatment), while the PSB's best improvement effect on WUE appeared at the intermediate level (P2 treatment).

**Table 2.** Effect of PSB on the photosynthetic characteristics of *C. oleifera* leaves.

| P Application Levels | PSB Inoculation | Pn [μmol/(m$^2$·s)] | Tr [μmol/(m$^2$·s)] | WUE (μmol/mmol) | SPAD |
|---|---|---|---|---|---|
| P1 | T1 | 3.41 ± 0.07 c | 2.47 ± 0.07 a | 1.38 ± 0.03 b | 48.40 ± 1.64c |
| | T2 | 3.64 ± 0.05 b | 2.59 ± 0.06 a | 1.40 ± 0.03 b | 51.37 ± 0.64 b |
| | T3 | 4.48 ± 0.20 a | 2.69 ± 0.09 a | 1.66 ± 0.07 a | 64.83 ± 2.70 a |
| | CK1 | 2.46 ± 0.25 d | 1.58 ± 0.09 b | 1.74 ± 0.10 a | 43.63 ± 2.21 d |
| | CK2 | 2.79 ± 0.43 d | 1.73 ± 0.03 b | 1.62 ± 0.23 a | 44.83 ± 1.72 d |
| P2 | T1 | 4.14 ± 0.35 c | 2.20 ± 0.09 ab | 1.88 ± 0.08 a | 62.93 ± 2.40 b |
| | T2 | 4.73 ± 0.40 b | 2.66 ± 0.07 a | 1.78 ± 0.14 a | 68.87 ± 1.66 a |
| | T3 | 5.69 ± 0.36 a | 2.80 ± 0.04 a | 2.03 ± 0.21 a | 71.30 ± 1.64 a |
| | CK1 | 2.93 ± 0.45 d | 1.87 ± 0.04 b | 1.56 ± 0.21 b | 47.83 ± 1.50 c |
| | CK2 | 3.07 ± 0.36 d | 1.83 ± 0.04 b | 1.68 ± 0.16 b | 48.31 ± 1.14 c |
| P3 | T1 | 3.76 ± 0.10 c | 2.37 ± 0.03 ab | 1.58 ± 0.02 a | 52.73 ± 2.01 c |
| | T2 | 3.93 ± 0.29 b | 2.42 ± 0.05 a | 1.62 ± 0.10 a | 57.87 ± 0.40 b |
| | T3 | 4.82 ± 0.20 a | 2.87 ± 0.06 a | 1.68 ± 0.04 a | 66.17 ± 2.15 a |
| | CK1 | 2.89 ± 0.23 d | 1.79 ± 0.07 b | 1.61 ± 0.10 a | 46.23 ± 1.00 d |
| | CK2 | 3.12 ± 0.16 d | 1.90 ± 0.10 b | 1.59 ± 0.02 a | 47.36 ± 0.92 d |
| Two-way ANOVA results | | | | | |
| P | | 0.00 ** | 0.03 * | 0.00 ** | 0.00 ** |
| PSB | | 0.01 * | 0.00 ** | 0.01 ** | 0.00 ** |
| P × PSB | | 0.17 NS | 0.00 ** | 0.17 NS | 0.00 ** |

Data are means ± SD ($n = 3$). Pn: net photosynthetic rate, Tr: transpiration rate, WUE: water use efficiency, SPAD: relative chlorophyll content. Different lowercase letters within the same column indicate significant differences ($p \leq 0.05$). *, significant effect at $0.01 < p \leq 0.05$; **, significant effect at $p \leq 0.01$; NS, no significant effect.

### 3.3. Chlorophyll Content

The chlorophyll content was significantly ($p \leq 0.01$) influenced by the interaction between P application and PSB inoculation (Table 2). The single and mixed inoculations of PSB strains could significantly ($p \leq 0.05$) increase the chlorophyll content of leaves regardless of the amount of P applied in the medium (Table 2). The co-inoculated plants had a higher chlorophyll content than that of either plants inoculated with single PSB strains or non-inoculated plants. The best promoted effect of the PSB strains on chlorophyll content was found at the intermediate P level. No significant difference in chlorophyll content was observed between the CK1 and CK2 treatments.

### 3.4. Nitrogen and Phosphorus Content in Leaves

The N content of the leaves was significantly ($p \leq 0.05$) influenced by P treatment and PSB treatment, while the P content of the leaves was significantly ($p \leq 0.01$) influenced by the interaction between P treatment and PSB treatment (Table 3). The trend of the changes in the N content of leaves was similar to that of the P content of leaves. The N and P content of the leaves increased to varying degrees by the single and mixed inoculations of PSB strains under different P application levels (Table 3). At the low (P1) and intermediate (P2) P levels, the N and P content of leaves were significantly increased by the PSB inoculation. The mixed inoculation of PSB strains increased the leaf N and P content more effectively than the single inoculation. However, there were no significant differences in the N and P content with the inoculation of PSB strains at a high P level (P3).

**Table 3.** Effect of PSB on the nitrogen content of *C. oleifera* leaves.

| P Application Levels | PSB Inoculation | Nitrogen Content (mg/kg) | Phosphorus Content (g/kg) |
|---|---|---|---|
| P1 | T1 | 1771.36 ± 86.22 b | 0.50 ± 0.03 b |
| | T2 | 1924.55 ± 128.97 a | 0.56 ± 0.02 a |
| | T3 | 1989.24 ± 100.17 a | 0.55 ± 0.01 a |
| | CK1 | 1761.09 ± 102.14 b | 0.44 ± 0.02 c |
| | CK2 | 1788.55 ± 85.44 b | 0.46 ± 0.02 c |
| P2 | T1 | 1849.29 ± 73.71 b | 0.59 ± 0.03 b |
| | T2 | 2045.05 ± 50.33 a | 0.61 ± 0.02 b |
| | T3 | 2122.31 ± 101.49 a | 0.66 ± 0.01 a |
| | CK1 | 1791.32 ± 195.53 c | 0.47 ± 0.03 c |
| | CK2 | 1886.69 ± 52.92 b | 0.48 ± 0.01 c |
| P3 | T1 | 1750.58 ± 105.83 b | 0.50 ± 0.04 a |
| | T2 | 1766.99 ± 141.89 b | 0.51 ± 0.05 a |
| | T3 | 1973.33 ± 95.04 a | 0.59 ± 0.04 a |
| | CK1 | 1752.88 ± 92.92 b | 0.49 ± 0.06 a |
| | CK2 | 1831.46 ± 83.27 b | 0.51 ± 0.05 a |
| Two-way ANOVA results | | | |
| | P | 0.01 * | 0.00 ** |
| | PSB | 0.00 ** | 0.00 ** |
| | P × PSB | 0.65 NS | 0.01 ** |

Data are means ± SD ($n = 3$). Different lowercase letters within the same column indicate significant differences ($p \leq 0.05$). *, significant effect at $0.01 < p \leq 0.05$; **, significant effect at $p \leq 0.01$; NS, no significant effect.

### 3.5. Nitrogen, Phosphorus, and Potassium Content of Rhizosphere Soil

The soil available N content was significantly ($p \leq 0.01$) influenced by P treatment and PSB treatment (Table 4). A significant increase in the available N content was found in the treatment of the inoculated PSB strains compared to the non-inoculated control ($p \leq 0.05$) (Figure 1). The maximum available N in the soil was found in the co-inoculated treatment under the intermediate P level. Moreover, no significant difference was observed in the available N content between the CK1 and CK2 treatments.

**Table 4.** Results of a two-way ANOVA of the effects of phosphate solubilizing bacteria (PSB) inoculation, phosphorus (P) application and their interactions on the soil parameters.

| Index | P | PSB | P × PSB |
|---|---|---|---|
| Soil N content | 0.00 ** | 0.00 ** | 0.06 $^{NS}$ |
| Soil P content | 0.00 ** | 0.00 ** | 0.00 ** |
| Soil K content | 0.08 $^{NS}$ | 0.00 ** | 0.00 ** |

**, significant effect at $p \leq 0.01$; NS, no significant effect.

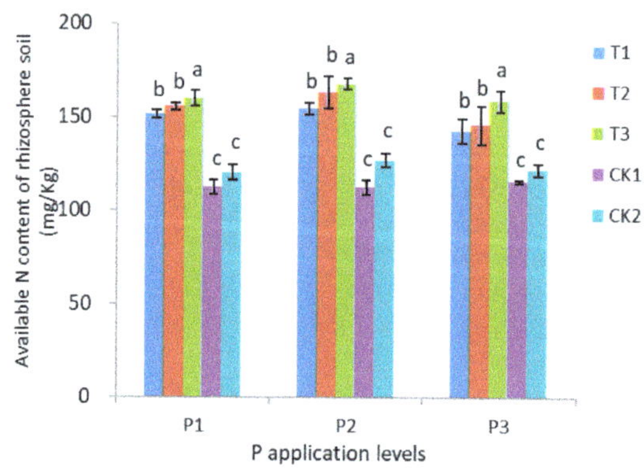


**Figure 1.** Effects of PSB on the available nitrogen content of *C. oleifera*. Note: Data are means ± SD ($n = 3$). P1: no phosphate fertilizer (0 g/kg); P2: added 3 g calcium superphosphate (1 g/kg); P3: added 6 g calcium superphosphate (2 g/kg); T1: inoculated with *P. auricularis* HN038; T2: inoculated with *B. aryabhattai* JX285; T3: inoculated with the mixture of HN038 and JX285 (1:1); CK1: inoculated with LB medium; CK2: inoculated with sterile water. Different lowercase letters within the same column indicate significant differences ($p \leq 0.05$).

The results showed that the soil available P content was significantly ($p \leq 0.01$) influenced by the interaction between P treatment and PSB treatment (Table 4). Different inoculation treatments caused different degrees of increase in the available P content in the soil (Figure 2). Under the intermediate and high P levels, the inoculation of single and mixed PSB strains significantly increased the content of available P in the soil ($p \leq 0.05$). Conversely, there were no significant differences in the available P content among the inoculation treatments under a low P level ($p > 0.05$).

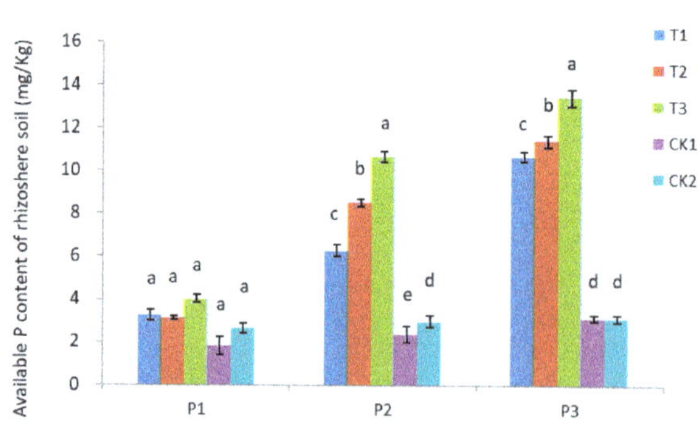


**Figure 2.** Effects of PSB on available phosphorus content of *C. oleifera*. Note: Data are means ± SD ($n = 3$). Different lowercase letters within the same column indicate significant differences ($p \leq 0.05$).

The soil available K content was significantly ($p \leq 0.01$) influenced by the interaction between P treatment and PSB treatment and (Table 4). An inoculation of PSB strains increased the available K content in the soil (Figure 3). The co-inoculated treatment had a higher available K content than single and non-inoculation treatments under each P level.

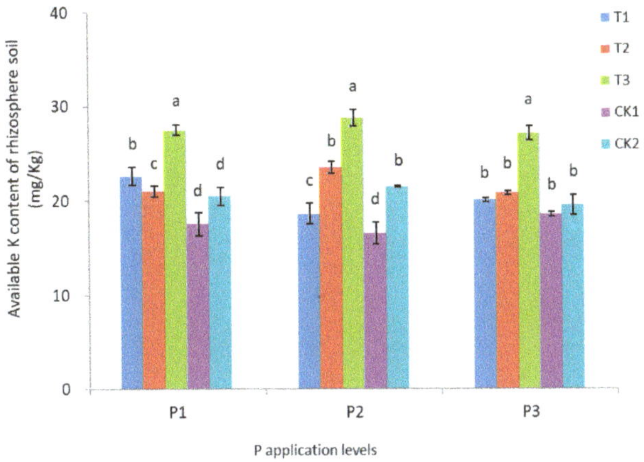


**Figure 3.** Effects of PSB on the available potassium content of *C. oleifera*. Note: Data are means ± SD ($n = 3$). Different lowercase letters within the same column indicate significant differences ($p \leq 0.05$).

## 4. Discussion

Phosphorus deficiency in soil is an important limiting factor in global agro-forestry production [25,26]. Some microorganisms can increase the concentration of available P by secreting organic acids and various degrading enzymes (phytase, nuclease, phosphatase, etc.) to decompose insoluble phosphate in the soil [27,28]. However, these microorganisms release little P in their natural state [11]. Therefore, it is necessary to artificially inoculate some high-efficiency P solubilizing microorganisms to increase the amount of available P. In our previous experiments, we isolated native PSB strains from the rhizosphere soil of *C. oleifera* and screened two strains (JX285 and HN38) with a high phosphorus solubilizing ability [17,18]. The results of the present study showed that the single and mixed inoculations of JX285 and HN38 had positive effects on the growth, photosynthetic capacity, N and P content in the leaves of *C. oleifera*, and available N, P, and K content in the soil.

Plant growth is the most obvious characteristic for evaluating the effects of PSB [29]. In this study, the single and mixed inoculations of two native PSB strains increased the plant height and biomass of *C. oleifera*, supporting the findings reported by previous studies [10,30]. This might be due to the PSB strains of JX285 and HN38 dissolving the insoluble phosphate in the soil and enhancing the available P content by producing organic acid and extracellular phosphatases [27,28]. Another possibility might be related to the metabolism of PSB, producing a variety of plant hormones, acids, and vitamins [30].

Plants use the light energy absorbed by chlorophyll molecules to drive photosynthesis [31]. This study showed the beneficial effect of PSB inoculation on the chlorophyll content and photosynthetic capacity of *C. oleifera* that is consistent with a previous observation in rice [32]. The higher Tr and WUE found in plants inoculated with PSB strains indicated that PSB improved the water status of *C. oleifera* [33]. An enhancement of Tr suggested an increased water uptake capacity, providing additional water for transpiration and improving soluble nutrient uptake [34].

Phosphorus is an essential nutrient, being a component of vital molecules in plants, and is involved in many metabolic processes [28]. PSB may convert insoluble P compounds into soluble forms by the processes of chelation, acidification, and exchange reactions [30]. In the current study, the increased available P content in the rhizosphere soil by inoculating PSB strains was only found in the intermediate and high insoluble P levels, while no significant difference was observed between the inoculated and non-inoculated plants under a low P level. This may result from the uptake of P by *C. oleifera* for growth under a low P level, as supported by the improved leaf P content and the growth performance of PSB under a low P level. Under a high P level, the available P released by PSB was sufficient for the plants' growth, which led to the PSB with no effect on the leaf P concentration. PSB not only solubilize and mineralize P from insoluble compounds but also release other nutrients [1,35]. In this study, the available N and K content of rhizosphere soil and the leaf N content were promoted by the PSB inoculation, demonstrating that PSB elevated the amounts of available N, P, and K in the soil and subsequently provided better nutrition for plant growth [36,37].

This study also found that the inoculation effect of mixed PSB strains was better than that of single strains. According to previous studies, there may be a synergistic effect between different microorganisms [38,39]. On one hand, the combination of different PSB strains may have a higher and more stable cell activity than a single strain [40]. On the other hand, the combination could better exploit the limited P sources in soil [30]. However, the mechanisms of a synergistic interaction remain to be explored.

## 5. Conclusions

In this study, we probed the effects of single and mixed inoculations of two native PSB strains, JX285 and HN38, on the growth of *C. oleifera* seedlings. The results showed the positive influence of JX285 and HN38 on plant growth, photosynthetic capacity, the N and P content of the leaves, and the available N, P, and K content of rhizosphere soil. The two PSB strains in the co-inoculation acted synergistically with each other and strengthened the beneficial effects on plant growth performance. The use of PSB as inoculants may provide an alternative to chemical fertilizers and promote sustainable agroforestry.

**Author Contributions:** L.Z. conceived the idea and established the direction; F.W. wrote the first draft of the manuscript; J.L. (Jianrong Li), Y.C., S.W., Y.Z., L.L. and X.S. carried out the experiment. S.W., J.L. (Junsheng Liang), and Y.C. analyzed the data. All authors contributed with suggestions and corrections, and approved the final manuscript.

**Funding:** This research was funded by the National Natural Science Foundation of China (31660189, 31570594), and Hunan Provincial Natural Science Foundation of China (2018JJ2217, 2018JJ3281).

**Conflicts of Interest:** The authors declare no conflict of interest.

## References

1. Oteino, N.; Lally, R.D.; Kiwanuka, S.; Lloyd, A.; Ryan, D.; Germaine, K.J.; Dowling, D.N. Plant growth promotion induced by phosphate solubilizing endophytic Pseudomonas isolates. *Front. Microbiol.* **2015**, *6*, 745. [CrossRef]
2. Ågren, G.I. Stoichiometry and nutrition of plant growth in natural communities. *Annu. Rev. Ecol. Evol. Syst.* **2008**, *39*, 153–170. [CrossRef]
3. Cordell, D.; Drangert, J.O.; White, S. The story of phosphorus: Global food security and food for thought. *Glob. Environ. Chang.* **2009**, *19*, 292–305. [CrossRef]
4. Skinner, M.F.; Attiwill, P.M. The productivity of pine plantations in relation to previous land use. *Plant Soil* **1981**, *61*, 329–339. [CrossRef]
5. Abbas, M.; Shah, J.A.; Irfan, M.; Memon, M.Y. Remobilization phosphorus in wheat cultivars under induced phosphorus deficiency. *J. Plant Nutr.* **2018**, *41*, 1–12. [CrossRef]
6. Fujita, Y.; Venterink, H.O.; Van Bodegom, P.M.; Douma, J.C.; Heil, G.W.; Hölzel, N.; Jabłońska, E.; Kotowski, W.; Okruszko, T.; Pawlikowski, P.; et al. Low investment in sexual reproduction threatens plants adapted to phosphorus limitation. *Nature* **2013**, *505*, 82–86. [CrossRef]
7. Gyaneshwar, P.; Kumar, G.N.; Parekh, L.J.; Poole, P.S. Role of soil microorganisms in improving p nutrition of plants. *Plant Soil* **2002**, *245*, 83–93. [CrossRef]
8. Khan, M.S.; Zaidi, A.; Wani, P.A. Role of phosphate-solubilizing microorganisms in sustainable agriculture—A review. *Agron. Sustain. Dev.* **2007**, *27*, 29–43. [CrossRef]
9. Coutinho, F.P.; Felix, W.P.; Yano-Melo, A.M. Solubilization of phosphates in vitro by *Aspergillus* spp. and *Penicillium* spp. *Ecol. Eng.* **2012**, *42*, 85–89. [CrossRef]

10. Wang, T.; Liu, M.Q.; Li, H.X. Inoculation of phosphate-solubilizing bacteria *Bacillus thuringiensis* b1 increases available phosphorus and growth of peanut in acidic soil. *Acta Agric. Scand.* **2014**, *64*, 252–259.
11. Collavino, M.M.; Sansberro, P.A.; Mroginski, L.A.; Aguilar, O.M. Comparison of in vitro solubilization activity of diverse phosphate-solubilizing bacteria native to acid soil and their ability to promote *Phaseolus vulgaris* growth. *Biol. Fertil. Soils* **2010**, *46*, 727–738. [CrossRef]
12. Zak, D.; Goldhammer, T.; Cabezas, A.; Gelbrecht, J.; Gurke, R.; Wagner, C.; Reuter, H.; Augustin, J.; Klimkowska, A.; Mclnnes, R. Top soil removal reduces water pollution from phosphorus and dissolved organic matter and lowers methane emissions from rewetted peatlands. *J. Appl. Ecol.* **2018**, *55*, 311–320. [CrossRef]
13. Ping, L.; Wang, K.; Zhou, C.; Xie, Y.; Yao, X.; Yin, H. Seed transcriptomics analysis in *Camellia oleifera* uncovers genes associated with oil content and fatty acid composition. *Int. J. Mol. Sci.* **2018**, *19*, 118.
14. Lin, C.Y.; Fan, C.L. Fuel properties of biodiesel produced from *Camellia oleifera* Abel oil through supercritical-methanol transesterification. *Fuel* **2011**, *90*, 2240–2244. [CrossRef]
15. Li, J. Soil, water and nutrient Loss in young plantation of the inter-cropped tea oil (*Camellia oleifera*) with different crops. *Scientia Silvae Sinicae* **2008**, *44*, 167–172, (In Chinese with English Abstract).
16. Huang, W.; Liu, J.; Wang, Y.P.; Zhou, G.; Han, T.; Li, Y. Increasing phosphorus limitation along three successional forests in southern China. *Plant Soil* **2013**, *364*, 181–191. [CrossRef]
17. Wang, S.; Zhang, L.P.; Zhang, Y.; Hao, F.F.; Hu, D.N. Screening, identification and phosphate solubilizing capability of phosphate solubilizing bacteria in rhizosphere of *Camellia oleifera* Abel at red soil region. *For. Res.* **2015**, *28*, 409–416, (In Chinese with English Abstract).
18. Huang, F.L.; Zhang, Y.; Zhang, L.P.; Wang, S.; Feng, Y.; Rong, N.H. Complete genome sequence of Bacillus megaterium JX285 isolated from *Camellia oleifera* rhizosphere. *Comput. Biol. Chem.* **2019**, *79*, 1–5. [CrossRef]
19. Kalavrouziotis, I.K.; Koukoulakis, P.H. Soil pollution under the effect of treated municipal wastewater. *Environ. Monit. Assess.* **2011**, *184*, 6297–6305. [CrossRef]
20. Bowman, R.A. A sequential extraction procedure with concentrated sulfuric acid and dilute base for soil organic phosphorus. *Soil Sci. Soc. Am. J.* **1989**, *53*, 362–366. [CrossRef]
21. Wu, F.; Zhang, H.; Fang, F.; Liu, H.; Tang, M. Arbuscular mycorrhizal fungi alter nitrogen allocation in the leaves of *Populus* × *canadensis* 'Neva'. *Plant Soil* **2017**, *421*, 477–491. [CrossRef]
22. Mcgrath, R. Protein measurement by ninhydrin determination of amino acid released by alkaline hydrolysis. *Anal. Biochem.* **1972**, *49*, 95–102. [CrossRef]
23. Pu, P.; Zhang, M.; Zhang, L.N. A study on temperature and time conditions of colorimetric method in measuring soil available phosphorus. *Adv. Mater. Res.* **2014**, *838–841*, 2047–2051. [CrossRef]
24. Billings, G.K. The determination of potassium in oil field brines by atomic absorption spectrometry: Geological note. *Clin. Infect. Dis.* **1965**, *13*, 532–534.
25. Turner, B.L.; Brenes-Arguedas, T.; Condit, R. Pervasive phosphorus limitation of tree species but not communities in tropical forests. *Nature* **2018**, *555*, 367–370. [CrossRef] [PubMed]
26. Nikolic, N.; Kostic, L.; Djordjevic, A.; Nikolic, M. Phosphorus deficiency is the major limiting factor for wheat on alluvium polluted by the copper mine pyrite tailings: A black box approach. *Plant Soil* **2011**, *339*, 485–498. [CrossRef]
27. Parekh, L.J.; Gyaneshwar, P.; Kumar, G.N.; Archana, G.; Poole, P.S.; Collins, M.D.; Hutson, R.A. Involvement of a phosphate starvation inducible glucose dehydrogenase in soil phosphate solubilization by *Enterobacter asburiae*. *FEMS Microbiol. Lett.* **1999**, *171*, 223–229.
28. Bashan, Y.; Kamnev, A.A.; de-Bashan, L.E. A proposal for isolating and testing phosphate-solubilizing bacteria that enhance plant growth. *Biol. Fertil. Soils* **2013**, *49*, 1–2. [CrossRef]
29. Ashmore, M.R. Assessing the future global impacts of ozone on vegetation. *Plant Cell Environ.* **2005**, *28*, 949–964. [CrossRef]
30. Yu, X.; Liu, X.; Zhu, T.H.; Liu, G.H.; Mao, C. Isolation and characterization of phosphate-solubilizing bacteria from walnut and their effect on growth and phosphorus mobilization. *Biol. Fertil. Soils* **2011**, *47*, 437–446. [CrossRef]
31. Wu, F.; Zhang, H.; Fang, F.; Tang, M. Nutrient allocation and photochemical responses of *Populus* × *canadensis* 'Neva' to nitrogen fertilization and exogenous *Rhizophagus irregularis* inoculation. *Acta Physiol. Plant.* **2018**, *40*, 152. [CrossRef]

32. Panhwar, Q.A.; Radziah, O.; Zaharah, A.R.; Sariah, M.; Razi, I.M. Role of phosphate solubilizing bacteria on rock phosphate solubility and growth of aerobic rice. *J. Environ. Biol.* **2011**, *32*, 607–612. [PubMed]
33. Wu, F.; Zhang, H.; Fang, F.; Wu, N.; Zhang, Y.; Tang, M. Effects of nitrogen and exogenous *Rhizophagus irregularis* on the nutrient status, photosynthesis and leaf anatomy of *Populus* × *canadensis* 'Neva'. *J. Plant Growth Regul.* **2017**, *36*, 824–835. [CrossRef]
34. Brodribb, T.J.; Field, T.S.; Jordan, G.J. Leaf maximum photosynthetic rate and venation are linked by hydraulics. *Plant Physiol.* **2007**, *144*, 1890–1898. [CrossRef] [PubMed]
35. Prodhan, M.A.; Finnegan, P.M.; Lambers, H. How does evolution in phosphorus-impoverished landscapes impact plant nitrogen and sulfur assimilation? *Trends Plant Sci.* **2019**, *24*, 69–82. [CrossRef] [PubMed]
36. Kerkhoff, A.J.; Fagan, W.F.; Elser, J.J.; Enquist, B.J. Phylogenetic and Growth Form Variation in the Scaling of Nitrogen and Phosphorus in the Seed Plants. *Am. Nat.* **2006**, *168*, E103–E122. [CrossRef]
37. Diao, C.P.; Wang, Z.H.; Li, S.S.; Liu, L.; Wang, S.; Huang, N. Differences in grain nitrogen contents of high-yielding wheat cultivars and relation to NPK uptake and utilization in drylands. *J. Plant Nutr. Fertil.* **2018**, *24*, 285–295.
38. Bayer, A.S.; Chow, A.W.; Morrison, J.O.; Guze, L.B. Bactericidal synergy between Penicillin or Ampicillin and aminoglycosides against antibiotic-tolerant lactobacilli. *Antimicrob. Agents Chemother.* **1980**, *17*, 359–363. [CrossRef]
39. Kiratisin, P.; Apisarnthanarak, A.; Kaewdaeng, S. Synergistic activities between carbapenems and other antimicrobial agents against *Acinetobacter baumannii* including multidrug-resistant and extensively drug-resistant isolates. *Int. J. Antimicrob. Agents* **2010**, *36*, 243–246. [CrossRef]
40. Rudresh, D.L.; Shivaprakash, M.K.; Prasad, R.D. Effect of combined application of rhizobium, phosphate solubilizing bacterium and *Trichoderma* spp. on growth, nutrient uptake and yield of chickpea (*Cicer aritenium* L.). *Appl. Soil Ecol.* **2005**, *28*, 139–146. [CrossRef]

MDPI
St. Alban-Anlage 66
4052 Basel
Switzerland
Tel. +41 61 683 77 34
Fax +41 61 302 89 18
www.mdpi.com

*Forests* Editorial Office
E-mail: forests@mdpi.com
www.mdpi.com/journal/forests

www.ingramcontent.com/pod-product-compliance
Lightning Source LLC
Chambersburg PA
CBHW061812210926
51850CB00006B/678
*9783039367566*